Organosilicon Chemistry

Organosilicon Chemistry

Edited by Lane Powell

Larsen & Keller
www.larsen-keller.com

Organosilicon Chemistry
Edited by Lane Powell
ISBN: 979-8-88836-064-4 (Hardback)

Larsen & Keller
Published by Larsen and Keller Education,
5 Penn Plaza,
19th Floor,
New York, NY 10001, USA

Cataloging-in-Publication Data

Organosilicon chemistry / edited by Lane Powell.
p. cm.
Includes bibliographical references and index.
ISBN 979-8-88836-064-4
1. Organosilicon compounds. 2. Chemistry, Organic. 3. Organic compounds.
4. Silicon compounds. I. Powell, Lane.
QD305.S54 O74 2023
547.08--dc23

For more information regarding Larsen and Keller Education and its products, please visit the publisher's website www.larsen-keller.com

Table of Contents

Preface

Silicon (Si) is a metalloid belonging to the carbon family (Group 14). It is a hard tetravalent metalloid and has a blue-grey metallic sheen. Silicon is a brittle crystalline solid having properties of a semiconductor. Organometallic compounds that contain carbon–silicon bonds are called organosilicon compounds. The area of chemistry under which the preparation and properties of organometallic compounds are studied is called organosilicon chemistry. Some of the prominent properties of organosilicon compounds include being colourless, flammable and hydrophobic. They have various applications in agriculture and commercial production. Within agriculture, they are used to increase the effectiveness of various herbicides and fungicides. They are also used to make antifoamers, caulks, adhesives and coatings. This book includes some of the vital pieces of works being conducted across the world, on various topics related to organosilicon chemistry. It will also provide interesting topics for research, which interested readers can take up. This book is a resource guide for experts as well as students.

This book has been the outcome of endless efforts put in by authors and researchers on various issues and topics within the field. The book is a comprehensive collection of significant researches that are addressed in a variety of chapters. It will surely enhance the knowledge of the field among readers across the globe.

It gives us an immense pleasure to thank our researchers and authors for their efforts to submit their piece of writing before the deadlines. Finally in the end, I would like to thank my family and colleagues who have been a great source of inspiration and support.

Editor

1

Generation of Bis(ferrocenyl)silylenes from Siliranes

Yang Pan [1], Shogo Morisako [2], Shinobu Aoyagi [1] and Takahiro Sasamori [1,2,*]

[1] Graduate School of Science, Nagoya City University, Nagoya, Aichi 467-8501, Japan; pan@nsc.nagoya-cu.ac.jp (Y.P.); aoyagi@nsc.nagoya-cu.ac.jp (S.A.)

[2] Division of Chemistry, Faculty of Pure and Applied Sciences, University of Tsukuba, Tsukuba, Ibaraki 305-8571, Japan; morisako.shogo.gf@u.tsukuba.ac.jp

* Correspondence: sasamori@chem.tsukuba.ac.jp

† Dedicated to Professor Yosuke Yamamoto on the occasion of his 65th birthday.

Abstract: Divalent silicon species, the so-called silylenes, represent attractive organosilicon building blocks. Isolable stable silylenes remain scarce, and in most hitherto reported examples, the silicon center is stabilized by electron-donating substituents (e.g., heteroatoms such as nitrogen), which results in electronic perturbation. In order to avoid such electronic perturbation, we have been interested in the chemistry of reactive silylenes with carbon-based substituents such as ferrocenyl groups. Due to the presence of a divalent silicon center and the redox-active transition metal iron, ferrocenylsilylenes can be expected to exhibit interesting redox behavior. Herein, we report the design and synthesis of a bis(ferrocenyl)silirane as a precursor for a bis(ferrocenyl)silylene, which could potentially be used as a building block for redox-active organosilicon compounds. It was found that the isolated bis(ferrocenyl)siliranes could be a bottleable precursor for the bis(ferrocenyl)silylene under mild conditions.

Keywords: silirane; silylene; ferrocenyl group

1. Introduction

Silylenes, i.e., divalent silicon analogues of carbenes, represent highly attractive reactive intermediates due to their considerable potential as building blocks for organosilicon compounds [1–7]. There are several examples of stable silylenes, which are usually stabilized thermodynamically by introduction of heteroatom-based substituents such as amino groups. For example, silicon analogues of *N*-heterocyclic carbenes (NHCs), *N*-heterocyclic silylenes (NHSis) [8–17], have been isolated as stable compounds (e.g., **I** [8,11] and **II** [9,13]) as well as the *N*-substituted cyclic silylene **III** [14] (Chart 1). However, such thermodynamically stabilized silylenes are too stable, i.e., they usually do not exhibit considerable reactivity as, e.g., building blocks for organosilicon compounds. Conversely, carbon-substituted silylenes, which do not contain a stabilizing electron-donating ligand, are highly reactive due to their considerable electrophilicity on account of their low-lying LUMOs [18–24]. Such highly reactive silylenes can be stabilized kinetically by sterically demanding substituents and subsequently being isolated under an inert atmosphere [25–28]. For example, dicoordinated, carbon-substituted silylenes [29–32] such as **IV** [29] and **V** [32] that bear sterically demanding substituents have been isolated and well characterized (Chart 1). However, it is very difficult to synthesize and isolate such silylenes as they usually require laborious and technically exigent synthetic procedures. In this context, easily accessible and bottleable precursors for carbon-substituted silylenes would be an attractive research target. In most cases, silylenes can be generated by reduction of a dihalosilane or photolysis of a trisilane [4,7,33]. When the substituents on the central silicon atom offer insufficient steric protection, the products from the aforementioned reactions usually afford complex mixtures of oligosilanes. Conversely, most cases of the reduction of sterically hindered dihalosilanes or the photolysis of trisilanes that bear bulky substituents result in the facile formation of a disilene, i.e., the formal dimer of a silylene [33–39]. To prepare a stable silylene precursor, it should thus be required to introduce very bulky substituents on the silicon atom to avoid facile dimerization of the generated silylenes [33]. On the other hand, too bulky substituents can be expected to negatively affect the inherent reactivity of the resulting silylene. In order to find the right balance in this trade-off relationship, transient silylenes with a combination of very bulky and less hindered groups have been designed. Our group has already reported the design and synthetic utility of the sterically demanding 2,5-bis(3,5-di-*t*-butylphenyl)-1-ferrocenyl (Fc*) group [40–42], and this group should be an appropriate substituent for bis(ferrocenyl)silylene **1** (Chart 1), which bears Fc* and ferrocenyl (Fc) groups on the silicon center [43]. Silylene **1** can be expected to (i) be very reactive due to its low-lying LUMO level, and (ii) show intriguing redox activity on account of its two ferrocenyl moieties.

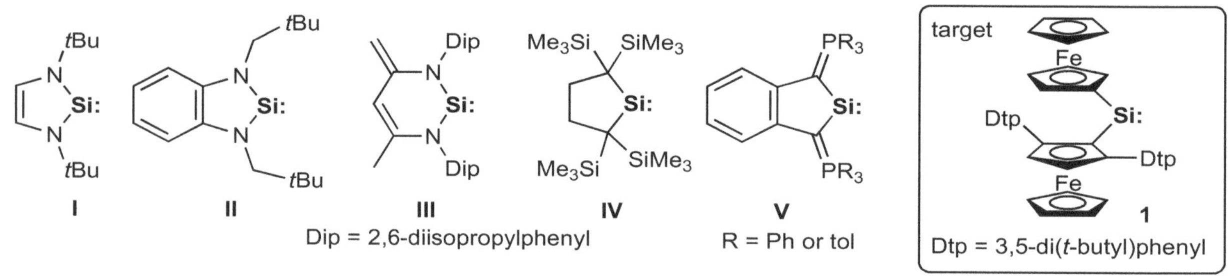

Chart 1. Examples of isolable dicoordinated silylenes.

2. Results and Discussion

Reaction of ferrocenyl (Fc) lithium with $SiCl_4$ in toluene resulted in the formation of $FcSiCl_3$ (**2**) [44], which was used in the following reactions without further purification due to its lability toward moisture. Subsequently, **2** was treated in Et_2O at r.t. with Fc*Li (**3**) [40–42], which was prepared by the reaction of Fc*Br (Fc* = 2,5-bis(3,5-di(*t*-buthyl)phenyl)-1-ferrocenyl) with *n*-BuLi in Et_2O, to give the corresponding bis(ferrocenyl)dichlorosilane (**4**) as a stable orange solid in 44% yield (Scheme 1). Although single crystals suitable for an X-ray diffraction (XRD) analysis could not be obtained due to partial hydrolysis that furnished Fc*FcSi(OH)$_2$ [45], **4** was characterized by spectroscopic techniques.

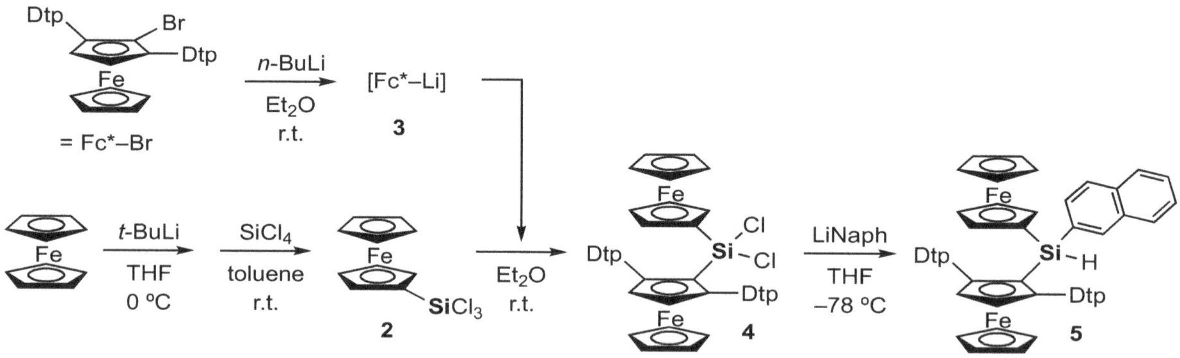

Scheme 1. Preparation of bis(ferrocenyl)dichlorosilane **4** and its reduction with lithium naphthalenide.

As the reduction of diorganodichlorosilanes represents an efficient method for the generation of silylenes [4,33], we speculated that the reduction of bis(ferrocenyl)dichlorosilane **4** might furnish silylene **1**. However, the generation of **1** by the reduction of **4** was not observed when common metal reductants such as KC_8, Na, or Li were employed, and only complex mixtures including Fc*H were obtained [41]. Conversely, when **4** was treated with lithium naphthalenide in THF at −78 °C, 2-naphthylsilane **5** was isolated in 30% yield, and its structure was unambiguously determined by spectroscopic and XRD analyses [46]. The formation of **5** should most likely be interpreted in terms of a C-H insertion of the reactive intermediate of silylene **1** with naphthalene. It should be noted here that it has previously been reported that isolable or transient silylenes react with aromatic compounds such as naphthalene and benzene to give the corresponding silepine or silirane derivatives via C-C insertion and [1+2]cycloaddition reactions, respectively [33,47–50]. Thus, the formation of **5** suggests the unique reaction manner of a silylene with an aromatic compound. Moreover, an isolable zwitterionic silylene has been reported to cleave a C-H bond of benzophenone via a 1,3-hydrogen shift [51,52].

At this point, we suspected that **5** was formed by the reaction of silylene **1** with naphthalene via the corresponding [1+2]cycloadduct, silirane **6**, which would be a conceivable intermediate in the reaction of a silylene with naphthalene. In order to clarify the reaction mechanism, the potential energy surfaces (PESs) of the reactions of silylene **1** with naphthalene were calculated at the B3PW91/6-311G(2d,p) level of theory (E_{ZPE}: relative energies including zero-point energy corrections) [53]. The [1+2]cycloaddition of **1** with naphthalene was found to be slightly exothermic (ΔE_{ZPE} = 6.12 kcal/mol) to give silirane **6** with a barrier of $\Delta E_{ZPE}{}^{\ddagger}$ = 21.6 kcal/mol. Unexpectedly, silirane **6** was not connected to **5** on the PES in these calculations. Instead, the highly exothermic (ΔE_{ZPE} = 37.4 kcal/mol) direct C-H insertion pathway of the reaction of **1** with naphthalene to furnish **5** was found to have a higher barrier of $\Delta E_{ZPE}{}^{\ddagger}$ = 29.5 kcal/mol. The PES also showed that silirane **6** could give **1** with concomitant elimination of naphthalene via the reverse reaction of the formation of **6** ($\Delta E_{ZPE}{}^{\ddagger}$ = 22.3 kcal/mol). Thus, it can be concluded **5** and **6** should be the thermodynamic and kinetic products, respectively, in the reaction of silylene **1** with naphthalene. At this stage, however, other plausible pathways for the formation of **5**, e.g., by the reaction of dichlorosilane **4** with lithium naphthalenide, which would not include the generation of silylene **1**, cannot be excluded from consideration (Scheme 2). However, based on the results of the theoretical calculations, silirane derivatives with a Si-C-C three-membered ring skeleton could potentially be considered appropriate precursors for silylenes such as **1**.

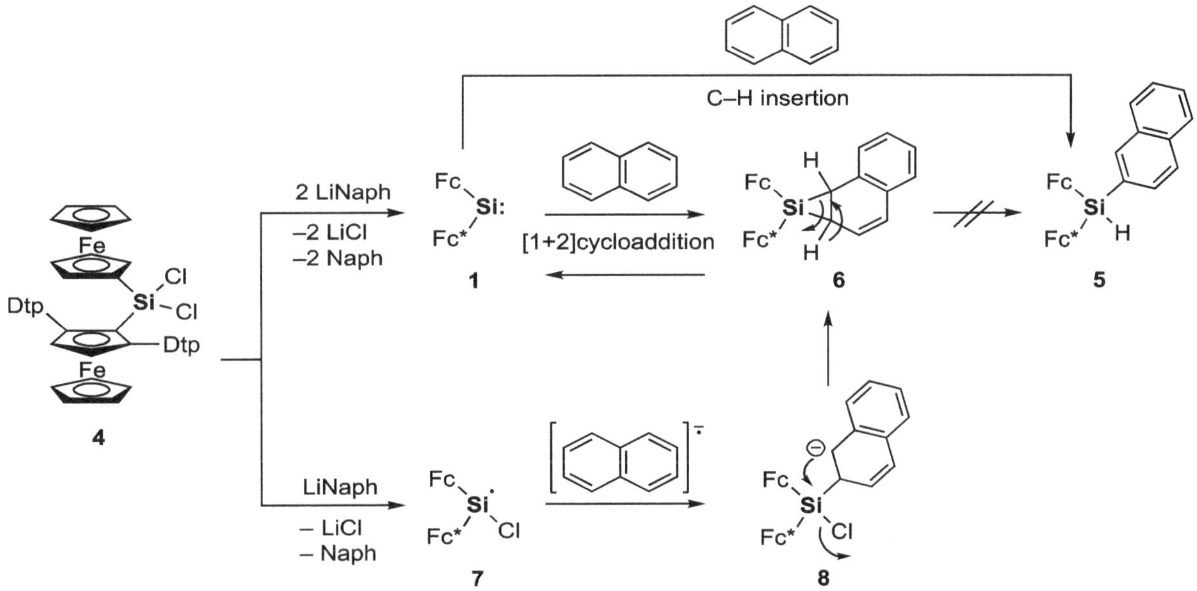

Scheme 2. Plausible mechanisms for the formation of bis(ferrocenyl)(2-naphthyl)silane **5**.

When a THF/Et_2O solution of dichlorosilane **4** was reduced with sodium in the presence of an excess of cyclohexene and a small amount of naphthalene, a mixture of siliranes **9a** and **9b** was obtained (86% yield, **9a**:**9b** = 66:34). Even though it was difficult to purify each product by common separation techniques such as gel permeation chromatography (GPC), column chromatography, or recrystallization, a few crystals of **9a** and **9b** were obtained by recrystallization from hexane and pentane, respectively. Thus, siliranes **9a** and **9b** were fully characterized by spectroscopic analyses, while **9a** was structurally characterized by XRD analysis [46]. The similar up-field shifted ^{29}Si NMR chemical shifts of **9a** (δ_{Si} = −67.2) and **9b** (δ_{Si} = −69.8) also indicated a silirane structure for **9b**, suggesting **9b** could be a stereoisomer of **9a** (Scheme 3). At this stage, **9a** and **9b** could not be separated from each other due the close similarity of their chemical properties. Heating a cyclohexene solution of a mixture of **9a** and **9b** (94:6 ratio) shifted the compositional ratio in favor of **9b** (**9a**:**9b** = 45:55), suggesting that **9a** and **9b** represent the kinetic and thermodynamic products in the reaction of **1** with cyclohexene, respectively. Both **9a** and **9b** are stable under ambient conditions in the solid state and in C_6D_6 solution. Interestingly, the conversion of **9a** to **9b** occurred in the solid state. After heating a mixture of solid **9a** and **9b** (ca. 74:26) at 120 °C for 30 min under reduced pressure, the ^1H NMR spectrum of the solid dissolved in C_6D_6 showed a **9a**:**9b** ratio of ca. 45:55, suggesting a transformation of **9a** to **9b** even in the solid state.

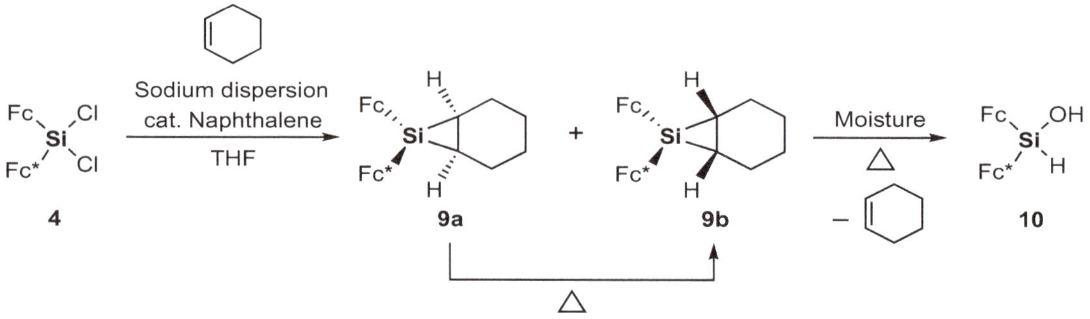

Scheme 3. Synthesis of bis(ferrocenyl)siliranes **9a** and **9b**.

The XRD structure of **9a** (Figure 1) shows that the cyclohexane moiety of **9a** is oriented toward the crowded space close to the Fc* group; **9b** could thus potentially exhibit a more stable geometry wherein the cyclohexyl group is oriented toward the less bulky ferrocenyl moiety. Theoretical calculations at B3PW91-D3(BJ)/6-311G(3d) level of theory suggest that **9b** is by 0.78 kcal/mol more stable than **9a**.

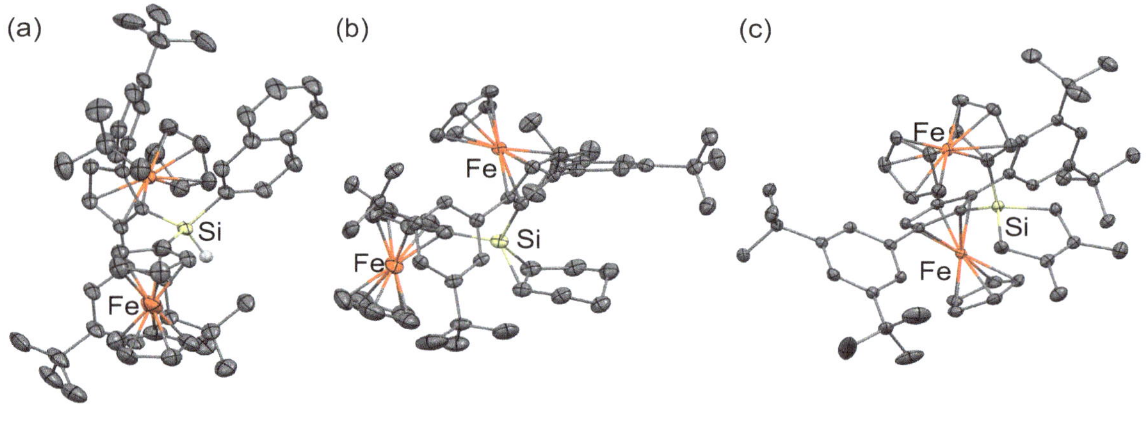

Figure 1. Molecular structures of (**a**) **5**, (**b**) **9a**, and (**c**) **14** with thermal ellipsoids at 50% probability; hydrogen atoms other than that of the Si-H moiety are omitted for clarity.

As described above, heating of the cyclohexene solution of siliranes **9a** and **9b** at 75 °C resulted in the conversion of **9a** to **9b** with concomitant formation of minor amounts of Fc*FcSiH(OH) (**10**) [46], suggesting the generation of silylene **1** in the equilibrium state. The formation of small amounts of **10** should most likely be interpreted in terms of a hydrolysis of silylene **1** generated by thermolysis of **9a** and/or **9b** due to the inevitable contamination with a small amount of moisture. Dissolving a mixture of **9a** and **9b** (66:34) in an excess amount of methanol at r.t. afforded **11** (28%), **12** (10%), and **13** (15%) as shown in Scheme 4 [6,46]. The formation of **11** suggests the generation of silylene **1** at r.t. However, siliranes **9a** and/or **9b** would probably undergo alcoholysis with MeOH to give **12** and **13** under these conditions. Thus, siliranes **9a** and **9b** could be appropriate thermal precursors for silylene **1** on heating, although they are sensitive toward protic solvents. Conversely, heating of a mixture of **9a** and **9b** (25:86) at 60 °C for 41 h in the presence of an excess of 2,3-dimethyl-1,3-butadiene afforded silolene **14** as the corresponding [1+4]cycloadduct of silylene **1** with 2,3-dimethyl-1,3-butadiene in 44% isolated yield, suggesting that the thermolysis of both **9a** and **9b** affords silylene **1** at this temperature. Thus, it can be concluded that siliranes **9a** and **9b** could work as synthetic precursors for bis(ferrocenyl)silylene **1** in such pericyclic reactions.

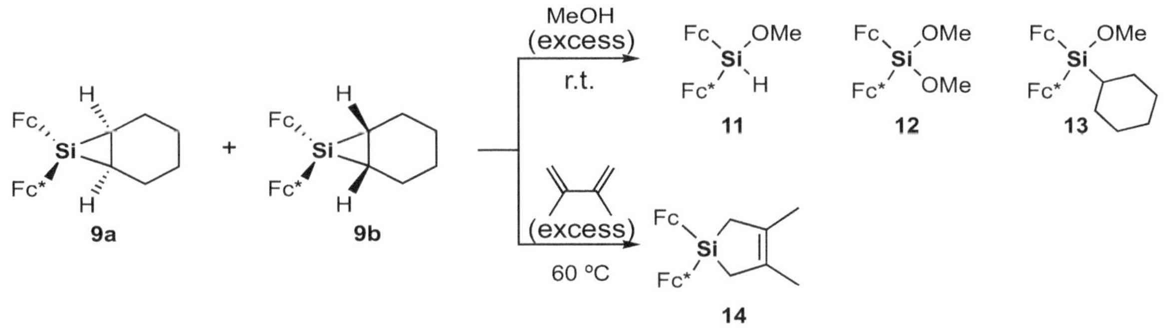

Scheme 4. Thermolysis of bis(ferrocenyl)siliranes **9a** and **9b** as well as trapping reactions of the transient silylene **1**.

Subsequently, we performed theoretical calculations on the dissociation energies of siliranes bearing several organic substituents at B3PW91-D3(BJ)/6-311G(3d) level of theory (Scheme 5). The dissociation energies of **9a** and **9b** are ca. 17 kcal/mol, which are smaller than those of phenyl- and methyl-substituted siliranes **17** (26.0 kcal/mol) and **19** (27.6 kcal/mol). Considering the small dissociation energy of less bulky bis(ferrocenyl)silirane **15** (14.0 kcal/mol), the facile generation of silylene **1** from **9a** and **9b** under mild conditions could be explained, not by the steric hindrance due to the bulky Fc* group, but the electronic effect of the ferrocenyl groups. Even though the electronic perturbation from the ferrocenyl groups toward the silirane moiety of **9a** and **9b** remains unclear at present, the electron-donating properties of the ferrocenyl groups can be expected to lower the dissociation energy of the silirane skeleton.

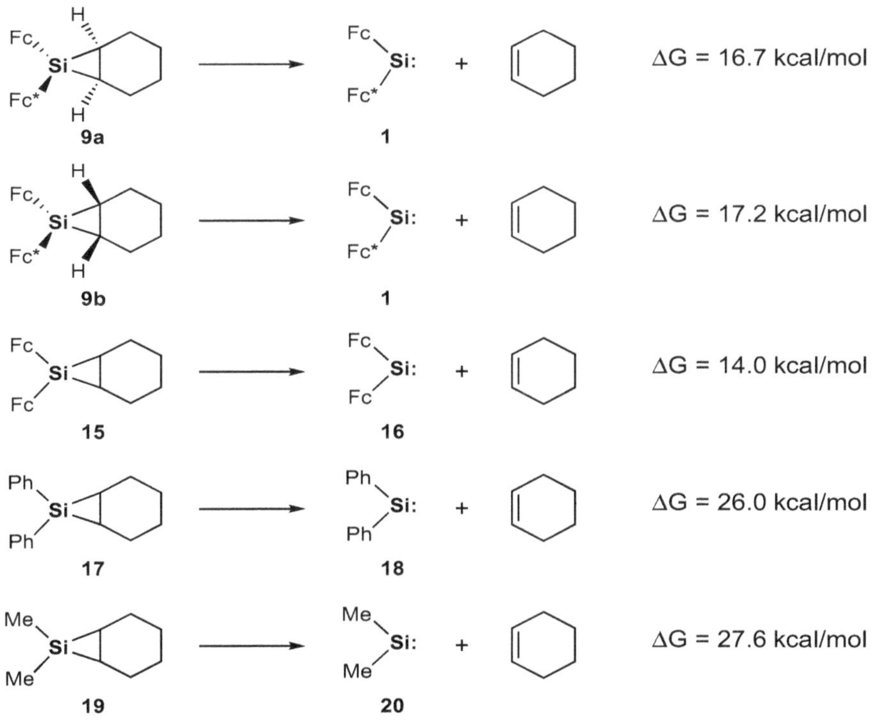

Scheme 5. Calculated dissociation energies of siliranes **9a** and **9b**, as well as **15**, **17**, and **19** to give the corresponding silylene and cyclohexene.

3. Materials and Methods

3.1. General Information

All manipulations were carried out under an argon atmosphere using either Schlenk line techniques or glove boxes. All solvents were purified by standard methods. Trace amounts of water and oxygen remaining in the solvents were thoroughly removed by bulb-to-bulb distillation from potassium mirror prior to use. ^{1}H, ^{13}C, and ^{29}Si-NMR spectra were measured on a JEOL ECZ-500R (^{1}H: 500 MHz, ^{13}C: 126 MHz, ^{29}Si: 99 MHz) or on a Bruker AVANCE-400 spectrometer (^{1}H: 400 MHz, ^{13}C: 101 MHz, ^{29}Si: 79 MHz). Signals arising from residual C_6D_5H (7.15 ppm) in C_6D_6 or $CHCl_3$ (7.26 ppm) in $CDCl_3$ were used as the internal standards for the ^{1}H NMR spectra, while those of C_6D_6 (128.0 ppm) and $CDCl_3$ (77.0 ppm) were used for the ^{13}C NMR spectra, and external $SiMe_4$ (0.0 ppm) for the ^{29}Si NMR spectra. High-resolution mass spectra (HRMS) were obtained from a JEOL JMS-T100LP (DART) or JMS-T100CS (APCI) mass spectrometer (DART). Gel permeation chromatography (GPC) was performed on an LC-6AD (Shimadzu Corp., Kyoto, Japan) equipped with JAIGEL-1H and 2H (Japan Analytical Industry Co., Ltd., Tokyo, Japan) columns using toluene as the eluent. All melting points were determined on a Büchi Melting Point Apparatus M-565 and are uncorrected. Fc*Br (Fc* = 2,5-[3,5-(tBu)$_2$C$_6$H$_3$]$_2$-1-ferrocenyl) was prepared according to literature procedures [40–42]. 1-Bromo-3,5-di-*tert*-butylbenzene was gifted by MANAC Inc., Tokyo, Japan.

3.2. Syntheses and Reactions

3.2.1. Bis(ferrocenyl)dichlorosilane **4**

Ferrocene (1.00 g, 5.38 mmol) was dissolved in THF (4.0 mL) and cooled to 0 °C. During 30 min, a pentane solution of *t*-BuLi (3.6 mL, 1.6 M in pentane, 5.76 mmol) was added dropwise. Then, hexane (10 mL) was added to the reaction mixture, and the solution was kept at −78 °C for 15 min. The resulting orange precipitate including ferrocenyl lithium was filtered off, and then washed with small portions of hexane. The orange precipitate was then dissolved in toluene (4 mL), and the solution was added to $SiCl_4$ (884 mg, 5.27 mmol). After stirring the mixture for 3 h at room temperature, the resulting inorganic salts were removed and the solvent of the filtrate was evaporated under reduced pressure to give ferrocenyltrichlorosilane **2** as an orange solid that was used for the subsequent reactions without further purification. An ether solution (2 mL) of Fc*Br (635 mg, 994 μmol) was treated with *n*-BuLi (0.5 mL, 2.60 M in hexane, 1.3 mmol) at 0 °C. After stirring for 6 h at room temperature, the solution of the resulting Fc*-Li (**3**) was added to **2** at 0 °C. After stirring for 12 h at room temperature, the solution was filtered and the filtrate was evaporated under reduced pressure. The obtained orange solid was subjected to GPC (toluene) to give bis(ferrocenyl)dichlorosilane **4** (366 mg, 434 μmol, 44%). **4**: orange solid. Mp. 107–118 °C. ^{1}H-NMR (500 MHz, benzene-d_6) δ 7.70 (d, J = 1.7 Hz, 4H), 7.49 (t, J = 1.7 Hz, 2H), 4.44 (s, 2H), 4.39 (s, 5H), 4.12 (s, 5H), 3.87 (s, 4H), 1.42 (s, 36H); ^{13}C-NMR (126 MHz, benzene-d_6) δ 150.0, 137.2, 128.3, 128.1, 127.9, 126.1, 121.3, 99.7, 75.1, 74.6, 72.0, 71.2, 69.8, 69.2, 64.5, 35.1, 31.8; ^{29}Si-NMR (79 MHz, benzene-d_6) δ 11.9. HRMS (APCI), *m/z*: Found: 844.23901 ([M$^+$]), Calcd. for $C_{48}H_{58}Cl_2Fe_2Si$ ([M$^+$]): 844.23837.

3.2.2. Bis(ferrocenyl)naphthylsilane **5**

A THF solution of LiNaph (0.2 mL, 1.1 M THF solution, 0.22 mmol) was added dropwise to a THF solution (0.2 mL) of bis(ferrocenyl)dichlorosilane **4** (47.7 mg, 56.5 μmol) at −78 °C. After stirring for 24 h at room temperature, the solvent was removed under reduced pressure, and the residue was dissolved in benzene. The resulting inorganic salts were removed by filtration and the filtrate was evaporated under reduced pressure to give **5** as an orange solid (quant. as judged by ^{1}H-NMR spectrum). The orange solid of **5** was purified by GPC (toluene) to give **5** in pure form (15.1 mg, 16.7 μmol, 30%). **5**: orange solid. Mp. 195–200 °C. ^{1}H-NMR (500 MHz, benzene-d_6) δ 8.20 (s, 1H), 7.90 (d, J = 8.0 Hz, 1H), 7.74 (d, J = 1.7 Hz, 2H), 7.66 (d, J = 8.0 Hz, 1H), 7.63–7.64 (m, 1H), 7.61 (d, J = 1.7 Hz,

2H), 7.59 (q, J = 2.1 Hz, 1H), 7.42 (t, J = 1.7 Hz, 2H), 7.24–7.27 (m, 2H), 6.21 (s, 1H), 4.65 (d, J = 1.1 Hz, 2H), 4.34 (s, 5H), 4.16 (d, J = 1.1 Hz, 1H), 4.05 (s, 1H), 3.99 (s, 1H), 3.88 (s, 5H), 3.82 (d, J = 1.1 Hz, 1H), 1.37 (s, 18H), 1.20 (s, 18H);^{29}Si-NMR (99 MHz, benzene-d_6) δ—19.5. HRMS (APCI), *m/z*: Found: 902.36587 ([M$^+$]), Calcd. for $C_{58}H_{66}Fe_2Si$ ([M$^+$]): 902.36326.

3.2.3. Siliranes **9a** and **9b**

A mixture of sodium (dispersion in mineral oil, 7.2 mg, 313 μmol), naphthalene (25.5 mg, 199 μmol), and cyclohexene (0.2 mL, 1.85 mmol) was added to a THF/Et_2O (1:1, 0.4 mL) solution of **4** (92.3 mg, 109 μmol). After stirring for 12 h at room temperature, the solvent and residual cyclohexene were removed under reduced pressure, before the residue was dissolved in cyclohexane. After filtration, the solvent of the filtrate was evaporated under reduced pressure. The obtained orange solid was subjected to GPC (toluene) to give a mixture of **9a** and **9b** (66:34 ratio as judged by the ^{1}H-NMR spectrum, 80.5 mg, 94.0 μmol, total yield: 86%). Continuous recrystallization of the mixture from hexane afforded a few crystals of **9a**, which could be isolated by the filtration. After evaporating the solvent of the filtrate, the residue was heated at 120 °C for a few hours under the reduced pressure. The following recrystallization of the resulting solids from pentane afforded a few crystals of **9b** in pure form. **9a**: orange solid. Mp. 174.1 °C (decomp). ^{1}H-NMR (400 MHz, benzene-d_6); δ 7.93 (d, J = 1.9 Hz, 4H), 7.53 (t, J = 1.9 Hz, 2H), 4.75 (s, 2H), 4.24 (br s, 2H), 4.11 (t, J = 1.9 Hz, 2H), 4.03 (s, 5H), 3.86 (br s, 5H), 1.99 (br s, 2H), 1.73 (br s, 2H), 1.49 (s, 36H), 1.26-1.27 (br, 6H); ^{13}C-NMR (101 MHz, benzene-d_6) δ 150.16, 139.86, 125.32, 120.40, 96.84, 75.04, 73.10, 72.66, 70.79, 69.16, 68.96, 66.73, 35.21, 31.91, 24.95, 22.66, 11.96; ^{29}Si-NMR (79 MHz, benzene-d_6) δ—67.6. HRMS (DART), *m/z*: Found: 856.38901 ([M$^+$]), Calcd. for $C_{54}H_{68}Fe_2Si$ ([M$^+$]): 856.38826. **9b**: orange solid. Mp. 151.2 °C (decomp). δ 7.94 (d, J = 1.8 Hz, 4H), 7.52 (t, J = 1.8 Hz, 2H), 4.73 (s, 2H), 4.16 (s, 5H), 4.10 (t, J = 1.8 Hz, 2H), 3.99 (t, J = 1.8 Hz, 2H), 3.82 (s, 5H), 2.06 (br s, 2H), 1.73–1.79 (br m, 4H), 1.52 (br s, 4H), 1.43 (s, 36H); ^{13}C-NMR (101 MHz, benzene-d_6) δ 150.25, 139.50, 125.27, 121.23, 97.72, 75.99, 73.13, 71.95, 71.12, 69.16, 68.23, 64.17, 35.13, 31.93, 23.92, 21.63, 14.47; ^{29}Si-NMR (79 MHz, benzene-d_6) δ—69.8. Continuous recrystallizations of a mixture of **9a** and **9b** from hexanes afforded a very small amount of single crystals of **9a** in pure form. Spectroscopic and XRD analyses of the single crystals thus obtained enabled us to identify **9a**. Silirane **9b** was identified based on the spectral data including ^{29}Si NMR data, which was similar to those of **9a**.

3.2.4. Thermolysis of Siliranes **9a** and **9b** in Cyclohexene

Cyclohexene (0.7 mL) was added to a mixture of **9a** and **9b** (94:6 ratio as judged by the ^{1}H NMR spectrum). After heating at 75 °C for 30 min, the residual cyclohexene was removed under reduced pressure. The ^{1}H NMR spectrum of the residue in C_6D_6 showed the signals for **9a** and **9b** (45:55 ratio).

3.2.5. Thermolysis of Siliranes **9a** and **9b** in the Solid State

A mixture of the orange solids of **9a** and **9b** (74:26 ratio as judged by the ^{1}H-NMR spectrum) was heated at 120 °C for 30 min under evacuation in an NMR tube (5 mm diameter) equipped with a J-Young© tap. The ^{1}H-NMR spectrum of the residue in C_6D_6 showed the signals for **9a** and **9b** (45:55 ratio).

3.2.6. Reaction of Siliranes **9a** and **9b** with Methanol

MeOH (0.40 mL) was added to a mixture of siliranes **9a** and **9b** (66:34, 24.0 mg, 28.0 μmol) at room temperature. After stirring for 72 h at room temperature, the solvent was removed under reduced pressure. The obtained residue was purified by column chromatography on silica gel (hexane/benzene = 4:1) to give **11** (6.3 mg, 7.8 μmol, 28%), **12** (2.3 mg, 2.8 μmol, 10%) and **13** (3.8 mg, 4.3 μmol, 15%) **11**: orange solid. Mp. 161–164 °C. ^{1}H-NMR (400 MHz, benzene-d_6) δ 7.98 (d, J = 1.8 Hz, 2H), 7.82 (d, J = 1.8 Hz, 2H), 7.51 (q, J = 1.8 Hz, 2H), 5.59 (s, 1H), 4.80 (d, J = 2.3 Hz, 1H), 4.74 (d, J = 2.3 Hz, 1H), 4.28 (s, 5H), 4.00 (td, J = 2.3, 1.1 Hz, 1H), 3.95 (td, J = 2.3, 1.1 Hz, 1H), 3.90–3.91 (m, 1H), 3.85 (s, 5H), 3.62 (s, 3H), 3.23 (t, J = 1.1 Hz, 1H), 1.44 (d, J = 1.3 Hz, 36H). HRMS (DART), *m/z*: Found: 806.32643 ([M$^+$]), Calcd. for $C_{49}H_{62}Fe_2OSi$ ([M$^+$]): 806.32688. **12**: orange solid. Mp. 67–71 °C. ^{1}H-NMR

(400 MHz, benzene-d_6) δ 7.84 (d, J = 1.8 Hz, 4H), 7.51 (t, J = 1.8 Hz, 2H), 4.59 (s, 2H), 4.34 (s, 5H), 4.03 (t, J = 1.8 Hz, 2H), 4.01 (s, 5H), 3.82 (t, J = 1.8 Hz, 2H), 3.38 (s, 6H), 1.46 (s, 36H). HRMS (DART), *m/z*: Found: 836.33713 ([M$^+$]), Calcd. for $C_{50}H_{64}Fe_2O_2Si$ ([M$^+$]): 836.33744. **13**: orange solid. Mp. 98–103 °C. ^{1}H-NMR (400 MHz, benzene-d_6) δ 7.73 (d, J = 1.8 Hz, 2H), 7.65 (d, J = 1.8 Hz, 2H), 7.47–7.48 (m, 2H), 4.54 (d, J = 2.3 Hz, 1H), 4.52 (d, J = 2.3 Hz, 1H), 4.27 (s, 5H), 4.08–4.10 (m, 3H), 4.01 (s, 5H), 3.94 (s, 1H), 3.29 (s, 3H), 2.27 (br d, J = 13.8 Hz, 1H), 2.16 (br d, J = 13.8 Hz, 1H), 1.82 (br d, J = 13.8 Hz, 1H), 1.75 (br d, J = 13.8 Hz, 2H), 1.55–1.60 (br, 1H), 1.45 (d, J = 1.8 Hz, 36H), 1.40–1.41 (br, 1H), 1.39–1.39 (br, 1H), 1.08–1.30 (br, 3H). HRMS (DART), *m/z*: Found: 888.40586 ([M$^+$]), Calcd. for $C_{55}H_{72}Fe_2OSi$ ([M$^+$]): 888.40513.

3.2.7. Reaction of Siliranes **9a** and **9b** with 2,3-Dimethyl-1,3-butadiene

2,3-Dimethyl-1,3-butadiene (0.50 mL, 4.40 mmol) was added to a mixture of **9a** and **9b** (25:86 ratio, 5.7 mg, 6.7 μmol). After stirring for 41 h at 60 °C, the residual 2,3-dimethyl-1,3-butadiene was removed under reduced pressure to give **14** (quant. as judged by the ^{1}H-NMR spectrum). The obtained orange solid of **14** was purified by column chromatography on silica gel (hexane/benzene = 4:1) to give **14** in pure form (2.5 mg, 2.9 μmol, 44%). **14**: orange solid. Mp. 242 °C (decomp). ^{1}H-NMR (400 MHz, benzene-d_6) δ 7.70 (d, J = 1.8 Hz, 4H), 7.50 (t, J = 1.8 Hz, 2H), 4.53 (s, 2H), 4.36 (s, 5H), 4.17 (t, J = 1.8 Hz, 2H), 4.06 (t, J = 1.8 Hz, 2H), 3.86 (s, 5H), 2.02 (d, J = 17.6 Hz, 2H), 1.77 (m, 8H), 1.40 (s, 36H) ^{29}Si-NMR (79 MHz, benzene-d_6) δ 2.15. HRMS (DART), *m/z*: Found: 856.38994 ([M$^+$]), Calcd. for $C_{54}H_{68}Fe_2Si$: 856.38826.

3.2.8. Characterization of Fc*(Fc)Si(OH)$_2$

Orange solid. Mp. 144-151 °C. ^{1}H-NMR (400 MHz, benzene-d_6) δ 7.84 (d, J = 1.9 Hz, 4H), 7.50 (t, J = 1.9 Hz, 2H), 4.57 (s, 2H), 4.34 (s, 5H), 3.99 (q, J = 1.9 Hz, 7H), 3.85 (t, J = 1.9 Hz, 2H), 2.68 (s, 2H), 1.40 (s, 36H). HRMS (APCI), *m/z*: Found: 808.31153 ([M$^+$]), Calcd. for $C_{48}H_{60}Fe_2O_2Si$ ([M$^+$]): 808.30614.

3.3. Computational Methods

The level of theory and the basis sets used for the structural optimization are given in the main text. Frequency calculations confirmed minimum energies for all optimized structures. All calculations were carried out on the *Gaussian 16* (Revision C.01) program package [53]. Computational time was generously provided by the Supercomputer Laboratory in the Institute for Chemical Research of Kyoto University.

3.4. X-ray Crystallographic Analysis

Single crystals of **5**, **9a**, **10**, **11**, **13**, **14**, and Fc*FcSi(OH)$_2$ were obtained by recrystallization from Et_2O (**5**, **9a**,**10**, **14**, Fc*FcSi(OH)$_2$) at room temperature or from Et_2O/MeOH (**11**, **13**) at −30 °C. Intensity data for **9a** were collected on a RIGAKU Saturn70 CCD(system) with VariMax Mo Optics using Mo-Kα radiation (λ = 0.71073 Å), while those for others were collected at the BL02B1 beamline of SPring-8 (2018A1167, 2018B1668, 2018B1179, 2019A1057, 2019A1677, 2019B1129, 2019B1784, 2020A1056, 2020A1644, 2020A1650, 2020A0834) on a PILATUS3 X CdTe 1M camera using synchrotron radiation (λ = 0.4148 Å). The structures were solved using SHELXT-2014 and refined by a full-matrix least-squares method on F^2 for all reflections using the programs of SHELXL-2016 [54,55]. All non-hydrogen atoms were refined anisotropically, and the positions of all hydrogen atoms were calculated geometrically and refined as riding models. Supplementary crystallographic data were deposited at the Cambridge Crystallographic Data Centre (CCDC) under deposition numbers CCDC-2044338-2044344 and can be obtained free of charge from via www.ccdc.cam.ac.uk/data_request.cif.

Crystallographic data for **5** (CCDC-2044338): $C_{58}H_{66}Fe_2Si$, FW 902.89, crystal size 0.01 × 0.01 × 0.01 mm^3, temperature −180 °C, λ = 0.4148 Å, triclinic, space group P–1 (#2), a = 12.072(5) Å, b = 12.297(5) Å, c = 16.770(7) Å, α = 84.253(5)°, β = 80.861(5)°, γ = 89.400(5)°, V = 2445.4(18) Å^3, Z = 2, D_{calcd} = 1.226 g cm^{-3}, μ = 0.158 mm^{-1}, θ_{max} = 15.729°, 57,864 reflections measured, 11,360 independent reflections (R_{int} = 0.0717), 692 parameters refined, GOF = 1.029, completeness = 99.4%, R_1 [$I > 2\sigma(I)$] = 0.0651, wR_2 (all data) = 0.1685, largest diff. peak and hole 1.262 and –1.498 e Å^{-3}. **9a** (CCDC-2044339): $C_{54}H_{68}Fe_2Si$, FW 856.87, crystal size 0.15 × 0.10 × 0.08 mm^3, temperature –170 °C, λ = 0.71073 Å, triclinic, space group P–1 (#2), a = 10.1310(3) Å, b = 15.7957(4) Å, c = 16.3047(4) Å, α = 67.199(2)°, β = 76.969(2)°, γ = 76.374(2)°, V = 2311.22(11) Å^3, Z = 2, D_{calcd} = 1.231 g cm^{-3}, μ = 0.688 mm^{-1}, θ_{max} = 30.835°, 41513 reflections measured, 9056 independent reflections (R_{int} = 0.0429), 551 parameters refined, GOF = 1.045, completeness = 99.6%, R_1 [$I > 2\sigma(I)$] = 0.0540, wR_2 (all data) = 0.1443, largest diff. peak and hole 1.694 and –1.089 e Å^{-3}. **10** (CCDC-2044340): $C_{48}H_{60}Fe_2OSi$, FW 792.75, crystal size 0.02 × 0.01 × 0.01 mm^3, temperature –180 °C, λ = 0.4148 Å, triclinic, space group P–1 (#2), a = 11.551(5) Å, b = 13.356(6) Å, c = 15.732(7) Å, α = 104.761(6)°, β = 108.360(6)°, γ = 100.704(6)°, V = 2131.9(17) Å^3, Z = 2, D_{calcd} = 1.235 g cm^{-3}, μ = 0.177 mm^{-1}, θ_{max} = 15.694°, 50,376 reflections measured, 9876 independent reflections (R_{int} = 0.0869), 491 parameters refined, GOF = 1.016, completeness = 99.6%, R_1 [$I > 2\sigma(I)$] = 0.0449, wR_2 (all data) = 0.1211, largest diff. peak and hole 0.549 and –0.861 e Å^{-3}. **11** (CCDC-2044341): $C_{49}H_{62}Fe_2OSi$, FW 806.78, crystal size 0.01 × 0.01× 0.01 mm^3, temperature –180 °C, λ = 0.4148 Å, triclinic, space group P–1 (#2), a = 9.748(8) Å, b = 12.421(10) Å, c = 18.807(15) Å, α = 78.213(8)°, β = 78.790(11)°, γ = 85.260(9)°, V = 2185(3) Å^3, Z = 2, D_{calcd} = 1.227 g cm^{-3}, μ = 0.173 mm^{-1}, θ_{max} = 15.765°, 52,665 reflections measured, 9471 independent reflections (R_{int} = 0.0912), 553 parameters refined, GOF = 1.040, completeness = 92.1%, R_1 [$I > 2\sigma(I)$] = 0.0879, wR_2 (all data) = 0.2863, largest diff. peak and hole 1.659 and –1.131 e Å^{-3}. **13** (CCDC-2044342): $C_{59}H_{82}Fe_2O_2Si$, FW 963.04, crystal size 0.01 × 0.01 × 0.01 mm^3, temperature –180 °C, λ = 0.4148 Å, triclinic, space group P–1 (#2), a = 10.29(5) Å, b = 15.09(7) Å, c = 16.60(8) Å, α = 104.05(4)°, β = 96.46(5)°, γ = 92.32(5)°, V = 2478(21) Å^3, Z = 2, D_{calcd} = 1.291 g cm^{-3}, μ = 0.158 mm^{-1}, θ_{max} = 9.863°, 14,683 reflections measured, 2889 independent reflections (R_{int} = 0.2584), 577 parameters refined, GOF = 1.049, completeness = 98.3%, R_1 [$I > 2\sigma(I)$] = 0.1088, wR_2 (all data) = 0.3010, largest diff. peak and hole 0.320 and –0.426 e Å^{-3}. Only preliminary data have been obtained, since the single crystals of this compound have been obtained only in very poor quality despite several attempts of careful recrystallization probably due to the partial hydrolysis of the Si-OMe moiety. **14** (CCDC-2044343): $C_{54}H_{68}Fe_2Si$, FW 856.87, crystal size 0.01 × 0.01 × 0.01 mm^3, temperature –180 °C, λ = 0.4148 Å, triclinic, space group P–1 (#2), a = 10.546(10) Å, b = 12.346(12) Å, c = 18.329(19) Å, α = 92.52(2)°, β = 97.949(12)°, γ = 99.949(16)°, V = 2322(4) Å^3, Z = 2, D_{calcd} = 1.225 g cm^{-3}, μ = 0.164 mm^{-1}, θ_{max} = 13.916°, 37,917 reflections measured, 7520 independent reflections (R_{int} = 0.0674), 529 parameters refined, GOF = 1.031, completeness = 99.3%, R_1 [$I > 2\sigma(I)$] = 0.0437, wR_2 (all data) = 0.1260, largest diff. peak and hole 0.771 and –0.648 e Å^{-3}. Fc*FcSi(OH)$_2$ (CCDC-2044344): $C_{48}H_{60}Fe_2O_2Si$, FW 808.75, crystal size 0.25 × 0.20 × 0.10 mm^3, temperature –170 °C, λ = 0.71075 Å, triclinic, space group P–1 (#2), a = 11.4879(2) Å, b = 13.2748(1) Å, c = 15.6366(2) Å, α = 104.3460(1)°, β = 108.2530(1)°, γ = 101.1560(1)°, V = 2096.58(5) Å^3, Z = 2, D_{calcd} = 1.281 g cm^{-3}, μ = 0.758 mm^{-1}, θ_{max} = 26.999°, 39,810 reflections measured, 9082 independent reflections (R_{int} = 0.0365), 718 parameters refined, GOF = 1.127, completeness = 99.2%, R_1 [$I > 2\sigma(I)$] = 0.0365, wR_2 (all data) = 0.0826, largest diff. peak and hole 0.395 and –0.437 e Å^{-3}.

4. Conclusions

Bis(ferrocenyl)siliranes **9a** and **9b** were prepared by the reduction of the corresponding dichlorosilane with lithium naphthalenide in the presence of an excess of cyclohexene. Siliranes **9a** and **9b** are appropriate precursors for bis(ferrocenyl)silylenes upon heating under mild conditions, i.e., they can be considered as bottleable synthetic precursors for silylenes. Further investigations into the creation of redox-active organosilicon compounds that bear ferrocenyl moieties by using siliranes **9a** and **b** as silylene precursors are currently in progress in our laboratory.

Author Contributions: Conceptualization, Y.P. and T.S.; resources and funding acquisition, T.S. and S.A.; XRD analyses, T.S. and S.M.; theoretical calculations (*Gaussian 16*). T.S.; experiments, Y.P.; writing—original draft preparation, Y.P.; review and editing, T.S.; project administration, T.S. All authors have read and agreed to the published version of the manuscript.

Acknowledgments: We acknowledge the generous assistance of the Research Equipment Sharing Center at Nagoya City University, Spring-8 (BL02B1: 2018A1167, 2018B1668, 2018B1179, 2019A1057, 2019A1677, 2019B1129, 2019B1784, 2020A1056, 2020A1644, 2020A1650, 2020A0834), and the Supercomputer Laboratory in the Institute for Chemical Research of Kyoto University for their technical support as well as the resources used. We would like to thank Toshiaki Noda and Hideko Natsume (Nagoya University) for the expert manufacturing of custom-tailored glassware. Moreover, we would like to express our appreciation to MANAC Inc. for a kind gift of 1-bromo-3,5-di-*tert*-butylbenzene.

References and Notes

1. Raoufmoghaddam, S.; Zhou, Y.-P.; Wang, Y.; Driess, M. N-heterocyclic silylenes as powerful steering ligands in catalysis. *J. Organomet. Chem.* **2017**, *829*, 2–10. [CrossRef]
2. Tacke, R.; Ribbeck, T. Bis(amidinato)- and bis(guanidinato)silylenes and silylenes with one sterically demanding amidinato or guanidinato ligand: Synthesis and reactivity. *Dalton Trans.* **2017**, *46*, 13628–13659. [CrossRef]
3. Zhou, Y.-P.; Driess, M. Isolable Silylene Ligands Can Boost Efficiencies and Selectivities in Metal-Mediated Catalysis. *Angew. Chem. Int. Ed.* **2018**, *58*, 3715–3728. [CrossRef] [PubMed]
4. Fujimori, S.; Inoue, S. Small Molecule Activation by Two-Coordinate Acyclic Silylenes. *Eur. J. Inorg. Chem.* **2020**, 3131–3142. [CrossRef]
5. Shan, C.; Yao, S.; Driess, M. Where silylene–silicon centres matter in the activation of small molecules. *Chem. Soc. Rev.* **2020**, *49*, 6733–6754. [CrossRef] [PubMed]
6. Zark, P.; Schäfer, A.; Mitra, A.; Haase, D.; Saak, W.; West, R.; Müller, T. Synthesis and reactivity of *N*-aryl substituted *N*-heterocyclic silylenes. *J. Organomet. Chem.* **2010**, *695*, 398–408. [CrossRef]
7. Reiter, D.; Holzner, R.; Porzelt, A.; Altmann, P.J.; Frisch, P.; Inoue, S. Disilene−Silylene Interconversion: A Synthetically Accessible Acyclic Bis(silyl)silylene. *J. Am. Chem. Soc.* **2019**, *141*, 13536–13546. [CrossRef]
8. Denk, M.; Lennon, R.; Hayashi, R.; West, R.; Belyakov, A.V.; Verne, H.P.; Haaland, A.; Wagner, M.; Metzler, N. Synthesis and Structure of a Stable Silylene. *J. Am. Chem. Soc.* **1994**, *116*, 2691–2692. [CrossRef]
9. Gehrhus, B.; Lappert, M.F.; Heinicke, J.; Boese, R.; Bläser, D. Synthesis, structures and reactions of new thermally stable silylenes. *J. Chem. Soc. Chem. Commun.* **1995**, 1931–1932. [CrossRef]
10. Schäfer, A.; Saak, W.; Weidenbruch, M.; Marsmann, H.; Henkel, G. Compounds of Germanium and Tin, 22—Reactions of a Silyene with a Germylene and a Stannylene: Formation of a Digermene with an Unusual Arrangement of the Substituents and of a Stannane. *Chem. Ber.* **1997**, *130*, 1733–1737. [CrossRef]
11. Moser, D.F.; Guzei, I.A. Robert West Crystal Structure of the Stable Silylene, N, N'-di-tert-butyl-1,3-diaza-2-silacyclopent-4-en-2-ylidene. *Main Group Met. Chem.* **2001**, *24*, 811–812. [CrossRef]
12. Becker, J.S.; Staples, R.J.; Gordon, R.G. Analysis of the crystal structures of 1,3-di-tert-butyl-2,3-dihydro-1*H*-1,3,2-diazasilol-2-ylidene and 1,3-di-*tert*-butyl-2,2-dichloro-1,3-diaza-2-sila-4-cyclopentene. *Cryst. Res. Technol.* **2004**, *39*, 85–88. [CrossRef]
13. Hill, N.J.; Moser, D.F.; Guzei, I.A.; West, R. Reactions of Stable Silylenes with Organic Azides. *Organometallics* **2005**, *24*, 3346–3349. [CrossRef]
14. Driess, M.; Yao, S.; Brym, M.; van Wüllen, C.; Lentz, D. A New Type of N-Heterocyclic Silylene with Ambivalent Reactivity. *J. Am. Chem. Soc.* **2006**, *128*, 9628–9629. [CrossRef] [PubMed]
15. Kong, L.; Zhang, J.; Song, H.; Cui, C. N-Aryl substituted heterocyclic silylenes. *Dalton Trans.* **2009**, 5444–5446. [CrossRef] [PubMed]
16. Mitra, A.; Brodovitch, J.-C.; Krempner, C.; Percival, P.W.; Vyas, P.; West, R. A Silyl Radical formed by Muonium Addition to a Silylene. *Angew. Chem. Int. Ed.* **2010**, *49*, 2893–2895. [CrossRef]
17. Kosai, T.; Ishida, S.; Iwamoto, T. A Two-Coordinate Cyclic (Alkyl)(amino)silylene: Balancing Thermal Stability and Reactivity. *Angew. Chem. Int. Ed.* **2016**, *55*, 15554–15558. [CrossRef]
18. Takeda, N.; Suzuki, H.; Tokitoh, N.; Okazaki, R.; Nagase, S. Reaction of a Sterically Hindered Silylene with Isocyanides: The First Stable Silylene−Lewis Base Complexes. *J. Am. Chem. Soc.* **1997**, *119*, 1456–1457. [CrossRef]

19. Gau, D.; Kato, T.; Saffon-Merceron, N.; Cossío, F.P.; Baceiredo, A. Stable Phosphonium Sila-ylide with Reactivity as a Sila-Wittig Reagent. *J. Am. Chem. Soc.* **2009**, *131*, 8762–8763. [CrossRef]
20. Filippou, A.C.; Chernov, O.; Blom, B.; Stumpf, K.W.; Schnakenburg, G. Stable N-Heterocyclic Carbene Adducts of Arylchlorosilylenes and Their Germanium Homologues. *Chem. Eur. J.* **2010**, *16*, 2866–2872. [CrossRef]
21. Rodriguez, R.; Gau, D.; Contie, Y.; Kato, T.; Saffon-Merceron, N.; Baceiredo, A. Synthesis of a Phosphine-Stabilized Silicon(II) Hydride and Its Addition to Olefins: A Catalyst-Free Hydrosilylation Reaction. *Angew. Chem. Int. Ed.* **2011**, *50*, 11492–11495. [CrossRef] [PubMed]
22. Suzuki, K.; Matsuo, T.; Hashizume, D.; Tamao, K. Room-Temperature Dissociation of 1,2-Dibromodisilenes to Bromosilylenes. *J. Am. Chem. Soc.* **2011**, *133*, 19710–19713. [CrossRef] [PubMed]
23. Li, W.; Köhler, C.; Yang, Z.; Stalke, D.; Herbst-Irmer, R.; Roesky, H.W. Synthesis of Cyclic Alkyl(amino) Carbene Stabilized Silylenes with Small N-Donating Substituents. *Chem. Eur. J.* **2019**, *25*, 1193–1197. [CrossRef] [PubMed]
24. Nagase, S. Theory and Calculations of Molecules Containing Heavier Main Group Elements and Fullerenes Encaging Transition Metals: Interplay with Experiment. *Bull. Chem. Soc. Jpn.* **2014**, *87*, 167–195. [CrossRef]
25. Fischer, R.C.; Power, P.P. π-Bonding and the Lone Pair Effect in Multiple Bonds Involving Heavier Main Group Elements: Developments in the New Millennium. *Chem. Rev.* **2010**, *110*, 3877–3923. [CrossRef] [PubMed]
26. Rivard, E.; Power, P.P. Multiple Bonding in Heavier Element Compounds Stabilized by Bulky Terphenyl Ligands. *Inorg. Chem.* **2007**, *46*, 10047–10064. [CrossRef] [PubMed]
27. Power, P.P. Stable Two-Coordinate, Open-Shell (d1–d9) Transition Metal Complexes. *Chem. Rev.* **2012**, *112*, 3482–3507. [CrossRef]
28. Yoshifuji, M. Sterically protected organophosphorus compounds of unusual structures. *Pure Appl. Chem.* **2017**, *89*, 281–286. [CrossRef]
29. Kira, M.; Ishida, S.; Iwamoto, T.; Kabuto, C. The First Isolable Dialkylsilylene. *J. Am. Chem. Soc.* **1999**, *121*, 9722–9723. [CrossRef]
30. Abe, T.; Tanaka, R.; Ishida, S.; Kira, M.; Iwamoto, T. New Isolable Dialkylsilylene and Its Isolable Dimer That Equilibrate in Solution. *J. Am. Chem. Soc.* **2012**, *134*, 20029–20032. [CrossRef]
31. Kobayashi, R.; Ishida, S.; Iwamoto, T. An Isolable Silicon Analogue of a Ketone that Contains an Unperturbed Si= O Double Bond. *Angew. Chem., Int. Ed.* **2019**, *58*, 9425–9428. [CrossRef] [PubMed]
32. Asay, M.; Inoue, S.; Driess, M. Aromatic Ylide-Stabilized Carbocyclic Silylene. *Angew. Chem. Int. Ed.* **2011**, *50*, 9589–9592. [CrossRef] [PubMed]
33. Mizuhata, Y.; Sasamori, T.; Tokitoh, N. Stable Heavier Carbene Analogues. *Chem. Rev.* **2009**, *109*, 3479–3511. [CrossRef] [PubMed]
34. Sasamori, T.; Tokitoh, N. A New Family of Multiple-Bond Compounds between Heavier Group 14 Elements. *Bull. Chem. Soc. Jpn.* **2013**, *86*, 1005–1021. [CrossRef]
35. Iwamoto, T.; Ishida, S. Multiple Bonds with Silicon: Recent Advances in Synthesis, Structure, and Functions of Stable Disilenes. In *Functional Molecular Silicon Compounds II: Low Oxidation States*; Scheschkewitz, D., Ed.; Structure and Bonding; Springer International Publishing: Cham, Switzerland, 2014; pp. 125–202. ISBN 978-3-319-03734-9.
36. Ishida, S.; Iwamoto, T. Recent advances in η2-disilene and η2-disilyne mononuclear transition metal complexes and related compounds. *Coord. Chem. Rev.* **2016**, *314*, 34–63. [CrossRef]
37. Matsuo, T.; Hayakawa, N. π-Electron systems containing Si= Si double bonds. *Sci. Technol. Adv. Mater.* **2018**, *19*, 108–129. [CrossRef]
38. Rammo, A.; Scheschkewitz, D. Functional Disilenes in Synthesis. *Chem. Eur. J.* **2018**, *24*, 6866–6885. [CrossRef]
39. Sasamori, T. Ferrocenyl-substituted low-coordinated heavier group 14 elements. *Dalton Trans.* **2020**, *49*, 8029–8035. [CrossRef]
40. Sasamori, T.; Suzuki, Y.; Sakagami, M.; Miyake, H.; Tokitoh, N. Structure of stable telluradiphosphirane bearing bulky ferrocenyl ligands. *Chem. Lett.* **2014**, *43*, 1464–1466. [CrossRef]
41. Sakagami, M.; Sasamori, T.; Sakai, H.; Furukawa, Y.; Tokitoh, N. 1,2-Bis(ferrocenyl)dipnictenes: Bimetallic Systems with a Pn= Pn Heavy π-Spacer (Pn: P, Sb, and Bi). *Bull. Chem. Soc. Jpn.* **2013**, *86*, 1132–1143. [CrossRef]
42. Sasamori, T.; Suzuki, Y.; Tokitoh, N. Isolation and Structural Characterization of a Lewis Base-Free Monolithioferrocene. *Organometallics* **2014**, *33*, 6696–6699. [CrossRef]

43. Suzuki, Y.; Sasamori, T.; Guo, J.-D.; Tokitoh, N. A Redox-Active Bis(ferrocenyl)germylene and Its Reactivity. *Chem. Eur. J.* **2018**, *24*, 364–368. [CrossRef] [PubMed]
44. Herberhold, M.; Ayazi, A.; Milius, W.; Wrackmeyer, B. Silyl derivatives of ferrocene with pending indenyl or fluorenyl substituents at silicon. *J. Organomet. Chem.* **2002**, *656*, 71–80. [CrossRef]
45. Fc*(Fc)Si$(OH)_2$ was characterized based on the spectroscopic and XRD analyses. See the Material and Methods section.
46. Crystallographic data for **5**, **9a**, **10**, **11**, **13**, **14**, and Fc*FcSi$(OH)_2$ were deposited at the Cambridge Crystallographic Data Centre (CCDC) under the numbers CCDC-2044338-2044344. These data can be obtained free of charge via www.ccdc.cam.ac.uk/structures.
47. Suzuki, H.; Tokitoh, N.; Okazaki, R. A Novel Reactivity of a Silylene: The First Examples of [1 + 2] Cycloaddition with Aromatic Compounds. *J. Am. Chem. Soc.* **1994**, *116*, 11572–11573. [CrossRef]
48. Kira, M.; Ishida, S.; Iwamoto, T.; Kabuto, C. Excited-State Reactions of an Isolable Silylene with Aromatic Compounds. *J. Am. Chem. Soc.* **2002**, *124*, 3830–3831. [CrossRef] [PubMed]
49. Fukuoka, T.; Uchida, K.; Sung, Y.M.; Shin, J.-Y.; Ishida, S.; Lim, J.M.; Hiroto, S.; Furukawa, K.; Kim, D.; Iwamoto, T.; et al. Near-IR Absorbing Nickel(II) Porphyrinoids Prepared by Regioselective Insertion of Silylenes into Antiaromatic Nickel(II) Norcorrole. *Angew. Chem. Int. Ed.* **2014**, *53*, 1506–1509. [CrossRef] [PubMed]
50. Kosai, T.; Ishida, S.; Iwamoto, T. Transformation of azulenes to bicyclic [4]dendralene and heptafulvene derivatives via photochemical cycloaddition of dialkylsilylene. *Chem. Commun.* **2015**, *51*, 10707–10709. [CrossRef]
51. Xiong, Y.; Yao, S.; Driess, M. Versatile Reactivity of a Zwitterionic Isolable Silylene toward Ketones: Silicon-Mediated, Regio- and Stereoselective C–H Activation. *Chem. Eur. J.* **2009**, *15*, 5545–5551. [CrossRef]
52. Azhakar, R.; Ghadwal, R.S.; Roesky, H.W.; Mata, R.A.; Wolf, H.; Herbst-Irmer, R.; Stalke, D. Reaction of N-Heterocyclic Silylenes with Thioketone: Formation of Silicon–Sulfur Three (Si–C–S)- and Five (Si–C–C–C–S)-Membered Ring Systems. *Chem. Eur. J.* **2013**, *19*, 3715–3720. [CrossRef]
53. Frisch, M.J.; Trucks, G.W.; Schlegel, H.B.; Scuseria, G.E.; Robb, M.A.; Cheeseman, J.R.; Scalmani, G.; Barone, V.; Petersson, G.A.; Nakatsuji, H.; et al. *Gaussian 16, Revision C.01*; Gaussian, Inc.: Wallingford, CT, USA, 2016.
54. Sheldrick, G.M. A short history of *SHELX*. *Acta Cryst.* **2008**, *64*, 112–122. [CrossRef]
55. Sheldrick, G.M. SHELXT—Integrated space-group and crystal-structure determination. *Acta Cryst.* **2015**, *71*, 3–8. [CrossRef] [PubMed]

2

Role of the Plasma Activation Degree on Densification of Organosilicon Films

Rita C. C. Rangel, Nilson C. Cruz and Elidiane C. Rangel *

Laboratory of Technological Plasmas (LaPTec), São Paulo State University (UNESP), Science and Technology Institute of Sorocaba (ICTS), Av. Três de Março, 511, 18087-180 Sorocaba, Brazil; rita.rangel@gmail.com (R.C.C.R.); nilson.cruz@unesp.br (N.C.C.)

* Correspondence: elidiane@sorocaba.unesp.br or elidiane.rangel@unesp.br

Abstract: The possibility of controlling the density of organosilicon films was investigated by tuning the plasma activation degree without providing extra energy to the structure, as usually reported in the literature. For this purpose, thin films were deposited in plasmas fed with hexamethyldisiloxane/Ar mixtures at a total pressure of 9.5 Pa. The power of the radiofrequency excitation signal, *P*, ranged from 50 to 300 W to alter the average energy of the plasma species while the electrical configuration was chosen to avoid direct ion bombardment of the growing films. In this way, it was possible to evaluate the effect of P on the film properties. Thickness and deposition rate were derived from profilometry data. X-ray energy dispersive and infrared spectroscopies were, respectively, applied to analyze the chemical composition and molecular structure of the layers. Surface topography and roughness were determined by atomic force microscopy while nanoindentation was used to evaluate the mechanical properties of the films. From electrochemical impedance spectroscopy the total resistance to the flow of electrolyte species was derived. The main alteration observed in the structure with changing P is related to the proportion of the methyl functional which remains connected to the Si backbone. Chain crosslinking and film density are affected by this structural modification induced by homogeneous and heterogeneous plasma reactions. The density increase resulted in a film with hardness comparable to that of the silica and more resistant to the permeation of oxidative species, but preserving the organosilicon nature of the structure.

Keywords: $SiO_xC_yH_z$; HMDSO; PECVD; densification; organosilicon; corrosion barrier

---

1. Introduction

Organosilicons are amorphous materials, usually referred to as $SiO_xC_yH_z$, considered as a specific class of (inorganic) polymers. This designation derives from their structures composed of a polysiloxane $(–Si–O–)_n$ inorganic backbone surrounded by methyl functionals, forming methylsilyl $(Si(CH_3)_x)$ groups. Weaker intermolecular forces differentiate them from other covalently bonded polymers, resulting in a structure with larger flexibility. In addition, methyl groups are free to rotate around the main backbone creating unoccupied volumes in the network, which have relevant implications for the material properties.

Thin films produced from organosilicons are normally characterized by high electrical resistance and optical transparency, in the visible region of the electromagnetic spectrum, enabling them for applications in microelectronics and integrated optics [1,2]. Furthermore, the combination of transparency, flexibility and moderate permeability to gas/vapor of such films favors a number of applications including encapsulation of organic electronic devices [3,4], improvement of packages performance (food, electrical devices, drugs) [5], reduction of water absorption and degradation [6,7], and protection of metal surfaces against corrosion [8,9]. The rotation of methyl groups may result in low surface tension materials, which are suitable for coating of fabrics [10,11], dental devices [12] and biocompatible [13] and anti-bio fouling surfaces [1,14].

Amongst the broad range of methodologies [1,15–17] used to grow $SiO_xC_yH_z$ films, plasma-based deposition is very attractive for allowing the deposition of films with tailored properties by controlling the energetic conditions of the plasma and the reactions at the film-substrate interface [18–21]. In this context, several reports have demonstrated the suitability of plasmas containing hexamethyldisiloxane, HMDSO, ($[CH_3]_3SiOSi[CH_3]_3$), an inexpensive, easy to handle and nontoxic compound, to prepare transparent Si-based films. By varying the plasma conditions, which include deposition time, pressure and chemical composition of the gas mixture and power of the excitation signal, it is possible to deposit films with structures varying from predominantly (inorganic) oxide, SiO_2 [22–24], to organosilicon [23–26], being the inorganic form denser, harder, scratch resistant and with better barrier properties. However, factors such as the adhesion of the silica to metal or polymer substrates, internal stress and especially defects generated in the deposition process or by substrate finishing affect the efficiency and physical stability of the oxide layer and, consequently, its practical application.

As stability problems are less common in organosilicon coatings [22–26], the improvement of its properties, mainly by the enhancement of density, becomes an issue of practical relevance. The densification of organosilicon films can be achieved by post-deposition treatments as reported, for instance, by Supiot et al. [27]. They have shown that organic groups are removed from organosilicon films exposed to oxygen plasmas, allowing the crosslinking of the structure. However, most of the proposed treatments caused the almost complete transformation of the organosilicon into an SiO_2-like structure, which resulted on the same stability problems mentioned above.

The densification of organosilicon films deposited in remote thermal plasma was influenced by the substrate temperature. A reduction of the deposition rate with increasing process temperature is reported [28]. Under low deposition rates and higher temperatures, the residence time of the particles adsorbed on the surface increases, allowing the filling of the structure valleys. The result is the creation of denser films with fewer voids. In addition, ionic bombardment during film growth is also an important factor that provides extra energy and may lead to dissociation of agglomerates, contributing to structural densification [29]. In the work of Massier et al. [30] it is demonstrated that the temperature required for obtaining a film with high packing density can be reduced by associating ion bombardment to the deposition process. The enhancement of density of $SiO_xC_yH_z$ films has also been proposed in the work of Blanchard et al. [21]. By changing the plasma excitation power and, consequently, the self-bias of the sample holder, the role of the power on plasma homogeneous (gas phase energetics) and heterogeneous (surface ion bombardment) reactions is discussed. The authors have proposed the densification is not entirely related to ion bombardment during film deposition but also to the plasma activation state.

Considering that the role of the plasma and plasma-surface reactions on the structure densification is not clearly understood and is still a subject under discussion [21,31], the main objective of this work is to investigate the densification of $SiO_xC_yH_z$ films by changing the degree of plasma activation without providing extra energy to the structure, as normally reported in the literature. In this sense, an electrical configuration is proposed in which radiofrequency power is used to excite the plasma without enhancing direct ion bombardment of the growing layer. The excitation power has been chosen as the parameter to be varied since it impacts directly the energy of plasma species, and thus, may represent an easier way to densify the structure as compared to other procedures commonly used. Therefore, the specific hypothesis to be tested here is whether it is possible to control the crosslinking degree of organosilicon layers by adjusting the plasma power without inducing, deliberately, ion bombardment or heating of the substrates.

2. Materials and Methods

2.1. Experimental Apparatus and Procedures

Figure 1 shows the homemade system used for film deposition. It consists of a 6 L-stainless steel vacuum chamber fitted with two parallel circular electrodes, also of stainless steel. The bottom electrode is composed of two concentric discs of 16 cm (external) and 5 cm (internal) in diameter. The inner electrode, which is electrically isolated from external electrode, is used as sample holder.

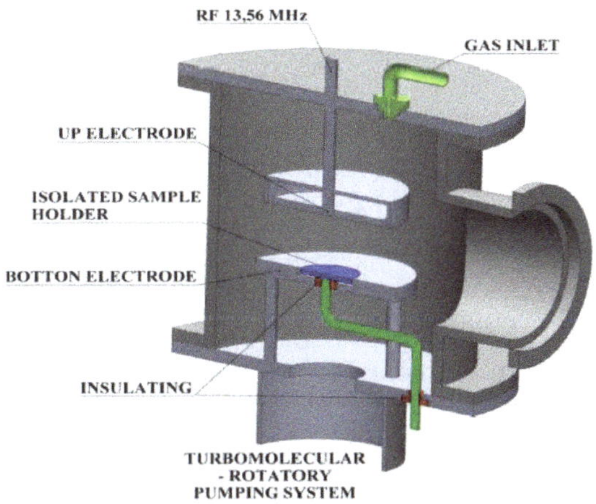

Figure 1. Experimental apparatus employed for film deposition.

Polished silicon and carbon steel specimens (99.630% Fe; 0.040% C; 0.008% Si; 0.240% Mn; 0.012% P; 0.006% S; 0.043% Al; 0.001% Cu; 0.005% Ni; 0.015% Cr) were the substrate materials. Before the experiments, all substrates were cleaned by sonication in ethanol and subsequently dried with N_2.

The clean substrates were placed on the sample holder and system pressure was reduced down to 1.9 Pa with a rotary vane vacuum pump. The deposition atmosphere was created by introducing HMDSO (4 sccm) and Ar (45 sccm), producing a final pressure of 15.7 Pa with the aid of a throttle valve. The plasma was generated by applying a radiofrequency (RF) signal (13.56 MHz) to the upper electrode (10.5 cm in diameter), while the walls of the reactor and the external bottom electrode were grounded. Samples were maintained at floating potential during depositions. Deposition time was 900 s in all the experiments and the exciting power (P) ranged from 50 to 300 W.

2.2. Characterization Techniques

2.2.1. Profilometer

Film thickness (h) was determined from the height of a step built by masking part of Si substrates with a glass slide during the depositions. The step height was measured in, at least, seven different regions with the aid of an Alpha-Step 500 KLA Tencor profilometer (KLA, Milpitas, CA, USA). Deposition rate (R) was evaluated as the ratio between thickness and deposition time.

2.2.2. Infrared and X-ray Energy Dispersive Spectroscopies

The chemical structure of films deposited on polished silicon plates was analyzed by infrared reflectance-absorbance spectroscopy (IRRAS) in a spectrometer Bruker Vertex 70v (Bruker, Billerica, MA, USA). Spectra were recorded with resolution of 4 cm^{-1} in the range from 4000 to 375 cm^{-1}. Each presented spectrum is the average of 32 scans acquired with each sample. The chemical composition was determined by X-ray Energy Dispersive Spectroscopy (EDS) using an X-ray Dry SD Hyper (EX-94410T1L11, JEOL's DrySDTM, Peabody, MA, USA detector coupled to a scanning electron microscope JEOL JSM 6010 LA (Tokyo, Japan) using 4 keV electron beams.

2.2.3. Contact Angle

Surface wettability was evaluated by contact angle measurements using deionized water as probe liquid. The presented results are the average (and standard deviation) of seven measurements in each sample using a goniometer Ramé-Hart 100-00 (Ramè-hart instrument co., Succasunna, NJ, USA) under controlled temperature and humidity conditions.

2.2.4. Nanoindentation and Atomic Force Microscopy

Mechanical properties of films deposited on polished Si were measured by nanoindentation in a Hysitron TI 750 Ubi^{TM} system (Bruker, Minneapolis, MN, USA) fitted with a Performech^{TM} control scanning unit (Bruker, Minneapolis, MN, USA). Experiments were performed in 15 representative positions of each sample using a Berkovich tip and a maximum load of 5000 μN. Hardness and elastic modulus have been calculated from the unload portion of load-displacement curves using the model proposed by Oliver and Pharr theory [32]. Surface topography was inspected by atomic force microscopy (Bruker's Hysitron TriboScope, Bruker, Minneapolis, MN, USA) using the nanoindenter and the same tip employed in the nanoindentation experiments. Scans were carried out over areas of 5 μm × 5 μm with a 2 μN normal force. The average quadratic roughness (RMS) was calculated from line profiles from three representative positions of each sample.

2.2.5. Electrochemical Impedance Spectroscopy

Corrosion experiments were performed by electrochemical impedance spectroscopy (EIS) using a PARSTAT 4000 (Ametek Scientific Instruments, Oak Ridge, TN, USA) potentiostat/galvanostat with a three electrodes electrochemical cell, a frequency response analyzer and VersaStudio software (2.41.2). Samples were immersed in 3.5%, 0.6 M NaCl solution. Platinum/rhodium and silver plates were used, respectively, as counter and reference electrodes whereas the working electrodes were carbon steel substrates coated with the films deposited in this work. Data were acquired at open circuit potential, varying the frequency of the sinusoidal signal (10 mV) from 10^4 to 10^{-2} Hz.

3. Results and Discussion

3.1. Thickness and Deposition Rate

Figure 2a shows thickness and deposition rate of the films as a function of the plasma excitation power. Since deposition time was kept constant (900 s) in all the experiments, *h* and *R* presented the same tendencies.

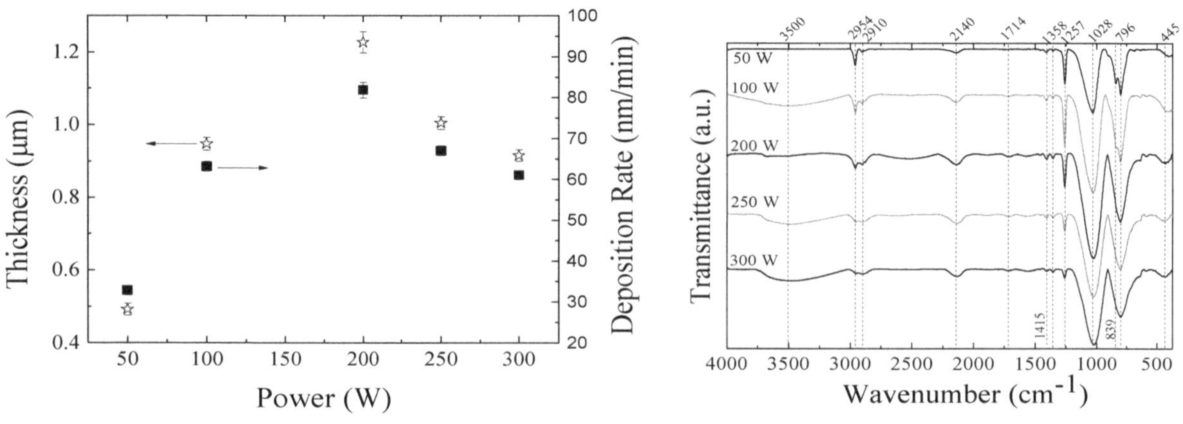

(**a**) (**b**)

Figure 2. (**a**) Thickness and deposition rate of the films as a function of the plasma excitation power. (**b**) Infrared spectra of films deposited in plasmas excited with diverse powers.

Both deposition rate and thickness increase with increasing power up to 200 W and then decreasing with the application of higher P. It is interesting to point out that, whenever the applied power was higher than 100 W, the deposition rate exceeded 60 nm/min. As a comparison, Vendemiatti et al. [6] reported a deposition rate of 21 nm/min for films deposited from atmospheres of 70% HMDSO, 15% O_2 and 15% Ar applying 150 W of radiofrequency to the substrate holder. Similarly, Wavhal et al. [33] reported an increment in deposition rate, which ranged from 25 to 300 nm/min, of organosilicon films in plasmas from different mixtures of HMDSO and O_2 as the power applied to the sample holder was increased. Despite the differences in electrical configuration and plasma composition, deposition rates obtained in the present work are within the range of values previously reported.

The trend reversal observed in the curves of Figure 2a can be explained in terms of the dependency of the plasma parameters on P. The density of electrons with enough energy to cause activation of species increases with P when other plasma parameters are kept constant [34]. Thus, the overall plasma activity [35] is expected to increase with the increase of power intensifying all the processes, including the degree of monomer fragmentation. Under severe fragmentation, the deposition rate is expected to decrease [33] mostly due to gas evolution, resulting from the breakage of chemical bonds, and the consequent reduction of the size of molecular fragments available to be deposited on the substrate. Furthermore, as the density of charged species in the plasma is increased, another relevant factor, the ion bombardment, turns out to influence the deposition process. That is because, even in a floating potential, the impact of electrons on the insulating layer, creates a negative potential, the so-called self-bias potential, which accelerates the positive ions towards the sample holder [36] as schematically illustrated in Figure 3d. This mechanism contributes to charges annihilation, energy transfer and surface heating. Thus, film deposition takes place under ion bombardment. However, in such a case, the bombardment is ruled by the plasma activation degree as it depends on the negative potential developed on the insulating layer, which in turn, is a function of the density of plasma species. Therefore, it is expected a general increase of the bombardment intensity with increasing plasma activation and hence an increase of the sample holder temperature.

In polymeric structures, the deposition of energy by ionic collisions can induce ionization, atomic displacement, bond fragmentation and heating. The emission of chain termination and lateral groups, for instance, is a process that affects deposition rate not only because it causes species loss, but also chain crosslinking. Thus, the fall trends, observed in the curves of Figure 2a for applied power values higher than 200 W, can also be ascribed to a balance between homogeneous and heterogeneous plasma reactions and indicate a possible densification of the structure induced by crosslinking of chains.

3.2. Molecular Structure and Chemical Composition

Figure 2b shows the IRRAS spectra of the films. The wavenumbers of the main contributions and their correspondent assignments are summarized in Table 1. The emergence of the band at 3500 cm^{-1} indicates O–H bonds connected to Si [37]. The presence of Si–OH groups has important implications for the structure configuration and stability. Firstly, if OH binds to Si radical during the deposition (homogeneous reactions), it inhibits the establishment of connections between neighboring chains, resulting in a material with higher degree of porosity. Furthermore, Si–OH terminations represent a reactive point where degradation may initiate under the presence of water or moisture [38]. Both are detrimental points when barrier properties and chemical stability of the material are considered.

The presence of organic groups is indicated by the absorption bands at 2954, 2910, 1415 and 1358 cm^{-1}, associated to stretching of C–H bonds in CH_3 or CH_2 groups. Interestingly, the intensity of these bands decreases even with the increase of film thickness as the power ranged from 50 to 200 W, pointing to a reduction in the concentration of organics with increasing P. All spectra revealed a contribution at 2140 cm^{-1} that is ascribed to Si–H stretching. Since this functional is not present in the precursor molecule, its presence indicates multiple-step reactions in the plasma phase.

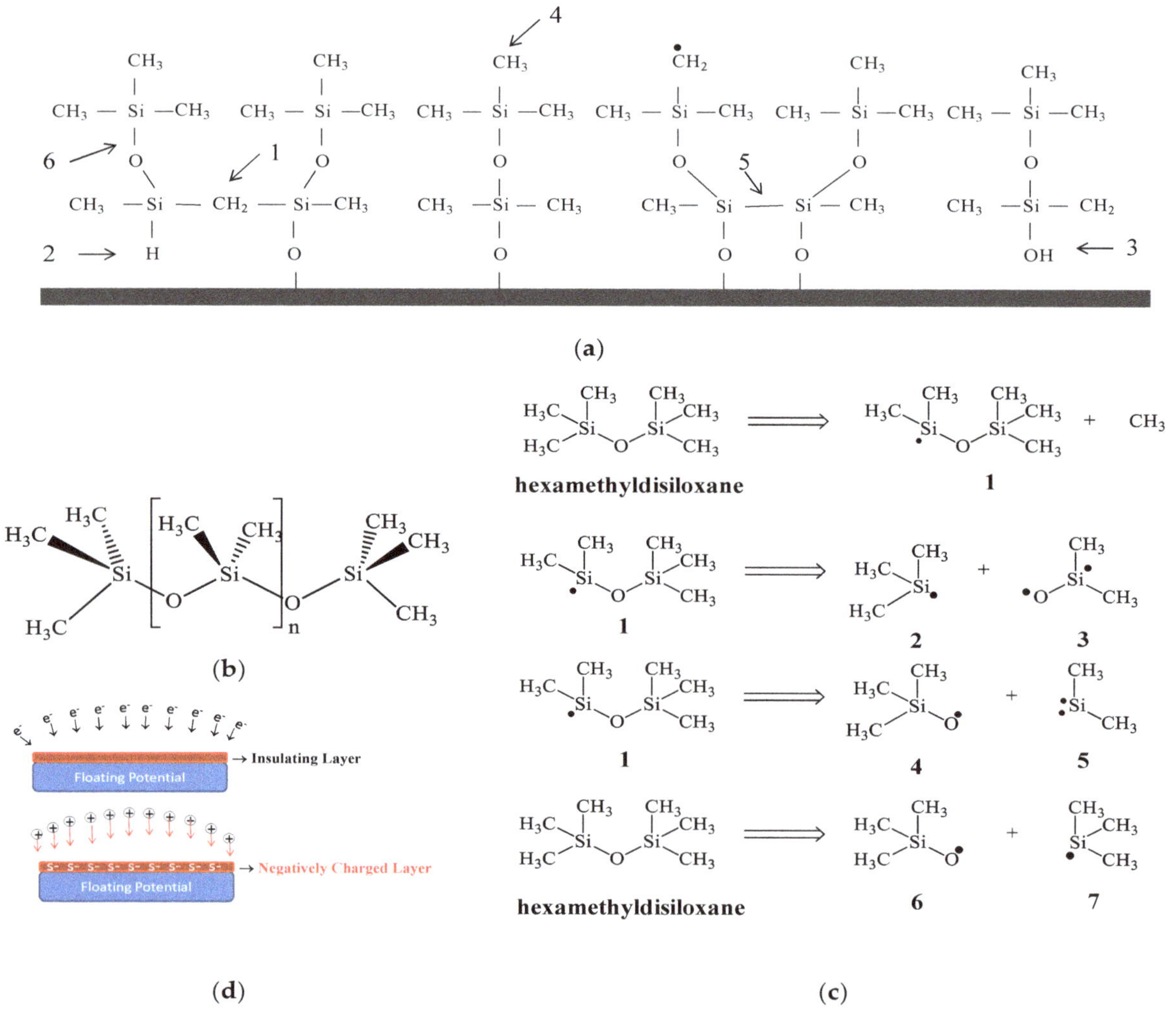

Figure 3. (**a**) Schematic representation of the structure of the films showing (1) chain crosslinking and the incorporation of (2) Si–H, (3) O–H and (4) C–H groups in an organosilicon structure. Adapted from Blanchard et al. [21]. (**b**) Chemical structure of the conventional polydimethylsiloxane, PDMS. (**c**) Chemical structure of hexamethyldisiloxane (HMDSO) and its possible fragmentation routes under the action of the plasma. (**d**) Charge annihilation mechanism in a dielectric surface induced by collisions of electron and subsequently of ions, as proposed by Yasuda [36].

Table 1. Wavenumber and assignments of the bands detected in the infrared absorption spectra of the films.

Wavenumber (cm^{-1})	Mode
3500	OH in Si–OH
2954	ν C–H in CH_3 and CH_2
2910	ν C–H in CH_2
2140	Si–H
1714	ν C=O in CH_2O
1415	δ C–H in Si–$(CH_3)_x$
1358	δ C–H_2 in Si–CH_2–Si
1257	δ C–H_3 in Si–$(CH_3)_x$
1028	ν Si–O in Si–O–Si and Si–O–C
839	ν Si–C in Si–$(CH_3)_3$
796	ν Si–C in Si–$(CH_3)_2$ and Si–O in ν Si–O–Si
445	Si–O–Si (bending)

Nevertheless, the main bands related to Si containing functionals appear in the low wavenumber region of the spectra. The band at 1028 cm^{-1}, attributed to vibration of Si–O in Si–O–Si groups, is characteristic of HMDSO. A slight shift of this band to lower wavenumbers is observed with increasing *P*, indicating structure densification [39–41]. Since the shift is observed only for $P > 200$ W it correlates well with the reversal of the trends of thickness and deposition rate (Figure 2a).

Another important band appears at 1257 cm^{-1}. This contribution, related to methylsilyl groups, Si–$(CH_3)_x$, indicates the formation of a structure composed by a main chain of Si–O–Si surrounded by methyl groups, independently of the deposition condition employed here. The resulting structure, schematically represented in Figure 3a, is very similar to that of conventional polydimethylsiloxane (Figure 3b). Furthermore, the comparison of the methylsilyl bands (1257 cm^{-1}) in the spectra of the films prepared with 100 and 250 W, which presented roughly the same thickness, demonstrates an intensity reduction corroborating the idea of organics removal with increasing *P*.

The loss of methyl groups may arise from the homogenous plasma reactions by direct fragmentation of HMDSO as illustrated in Figure 3c. Under the impact of electrons [42], there is creation of a diversity of reactive fragments which can further react in the plasma phase generating groups with molecular weight higher than that of HMDSO [43].

The fragments can also connect directly to the surface acting as a precursor of film growth. The resulting structure will depend on the fragmentation-recombination degree of the species in the plasma phase (homogenous reactions) as well as the plasma-surface reactions (heterogeneous reactions). It means that ion bombardment induced by charges accumulation can also remove lateral groups and chain terminations such as H or even CH_3. According to Rao [44], the reduction in the intensity of the methylsilyl band is also associated with the increase of the structure crosslinking. Thus, it is postulated here an increment in the density of the films with increasing *P* due to an enhancement of the crosslinking degree. This result is in accordance with the shift of the band at 1028 cm^{-1} as mentioned above and explains, together with the loss of molecular –$(CH_3)_x$ fragments, the fall in the deposition rate and in the film thickness for the depositions with applied power higher than 200 W.

Groups $Si(CH_3)_x$ are also identified by the bands at 839 ($Si(CH_3)_3$) and 796 ($Si(CH_3)_2$) cm^{-1}. The enlargement and loss of resolution of these bands suggest changes (disordering) in the vicinity of the chemical bonds due to greater diversity of fragments incorporated in the structure [37]. The band at 796 cm^{-1}, which may also have the contribution of stretching vibration of Si–O in Si–O–Si [45], indicates the presence of the $Si(CH_3)_2$ chain propagation unit, characteristic of the PDMS structure. High proportions of this group suggest formation of a structure with long chains. As illustrated in the schematic representation of the structure in Figure 3a, chain interconnection may occur via Si–CH_2–Si (indicated by 1 in Figure 3a), Si–Si (indicated by 5) and Si–O (indicated by 6 in Figure 3a) groups. It shows that, the loss of H itself may favor the connection of neighboring chains. Nevertheless, all of the illustrated possibilities involve a reduction in the proportion of methyl groups of the film. The absorption of O–H group by pendant bonds (point 3 in Figure 3a) thus inhibits chain reticulation.

Figure 4a shows the relative density of $Si(CH_3)_x$ groups as a function of *P*, calculated by the method proposed by Lanford and Rand [46] considering the absorption band at 1257 cm^{-1}. The relative density of $Si(CH_3)_x$ groups progressively fall with *P*, confirming the loss of CH_x and H groups. Chain crosslinking becomes probable, depending on the concentration and proximity of the radicals. When dangling bonds are not consumed by crosslinking, they remain active and react with atmospheric groups when the sample is exposed to air. This mechanism partially explains the presence of Si–OH functional in the organosilicon structure. The residual atmosphere in the deposition chamber (1.9 Pa) and the presence of O in the precursor compound may also account for O–H incorporation during the film deposition (homogeneous reactions).

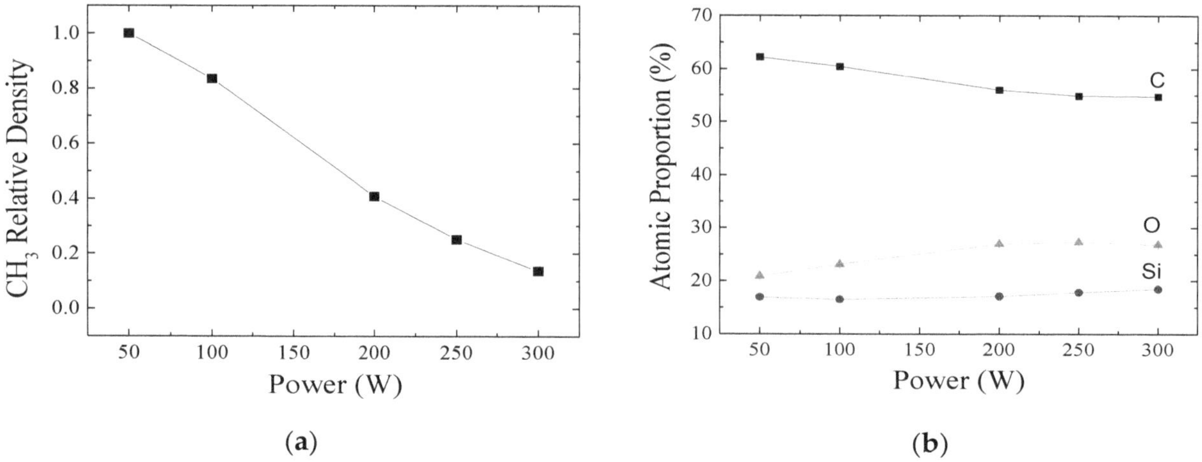

Figure 4. (**a**) Relative density of $Si(CH_3)_x$ groups in the films as a function of P. Data obtained using the model proposed by Lanford and Rand [46] and the intensity of the absorption band at 1257 cm^{-1} in Figure 2b. (**b**) Atomic proportion of C, O and Si determined by Energy Dispersive Spectroscopy (EDS) in the films as a function of P.

Figure 4b shows the proportions of C, O and Si in the films as a function of P, derived from EDS data. Together with the information on chemical structure, the detection of these species corroborates the proposal of an organosilicon material. Proportions very similar to these, derived from XPS (X-ray photoelectron spectroscopy) results, were reported for organosilicon films in the work of Blanchard et al. [21] and Montarsolo et al. [47]. Furthermore, considering the chemical structure of the conventional PDMS, which is schematically depicted in Figure 3b, and disregarding H that is not detected by EDS, structures containing 50% C, 25% O and 25% Si are expected while the HMDSO molecule, which is depicted in Figure 3c, contains 67% C, 11% O and 22% Si. Therefore, the precursor molecule should undertake reactions for carbon abstraction, resulting in the film stoichiometry observed here. The small fragments generated by homogeneous or heterogeneous reactions may undertake multiple step reactions in the plasma, resulting in stable neutral species, such as, C_2H_2, CH_4, C_2H_6 and C_3H_8, or may be further oxidized by residual atmosphere in the reactor, generating, for instance, CO, CO_2, OH or H_2O groups. All of them are very prone to be removed from the chamber by the vacuum system.

As a necessary condition for developing the study proposed here, the organosilicon nature of the material is preserved with the enhancement of P. Only slight variations are detected indicating reduction in the proportion of C and elevation in the contents of Si and O. Interestingly, the variations of the proportions of C and O have roughly the same intensity. However, the contrary trends show that the variations are caused by the same phenomenon. In other words, the loss of methyl groups decreases the proportion of C in the film, which, in turn, causes the O content to proportionally increase. Therefore, it can be inferred from results of Figure 4b that it is possible to slightly change the chemical composition of organosilicon films by controlling the plasma activation degree. It is further observed that the system is operating in a low energy regime once the increment in the energy supplied to the plasma enhances the deposition rate (in the range 50–200 W), while preserving the same nature of the precursor fragments.

Therefore, the overall analysis of the molecular structure and chemical composition revealed a material composed by C, H, O and Si organized as an organosilicon structure. Structural densification induced by crosslinking is suggested by the shift of the Si–O band (1028 cm^{-1}) towards smaller wavenumbers, by a reduction of the intensity of the bands related to the methylsilyl functionals and of the proportion of C with the elevation of P. Alterations in the porosity, mechanical and barrier properties are expected as a consequence.

3.3. *Wettability and Surface Topography*

Figure 5a shows the contact angle, θ, of the films as a function of P. The dotted lines present the results for the bare carbon steel. After film deposition all the surfaces became hydrophobic with contact angles ranging from 90 to 110°. The comparison of such results with those obtained for polytetrafluorethylene, Teflon®, (θ ~100–115°) [48] indicates the formation of non-polar surfaces upon film deposition. In general, films deposited using the lowest powers (50–100 W) are less receptive to water than those deposited using higher P.

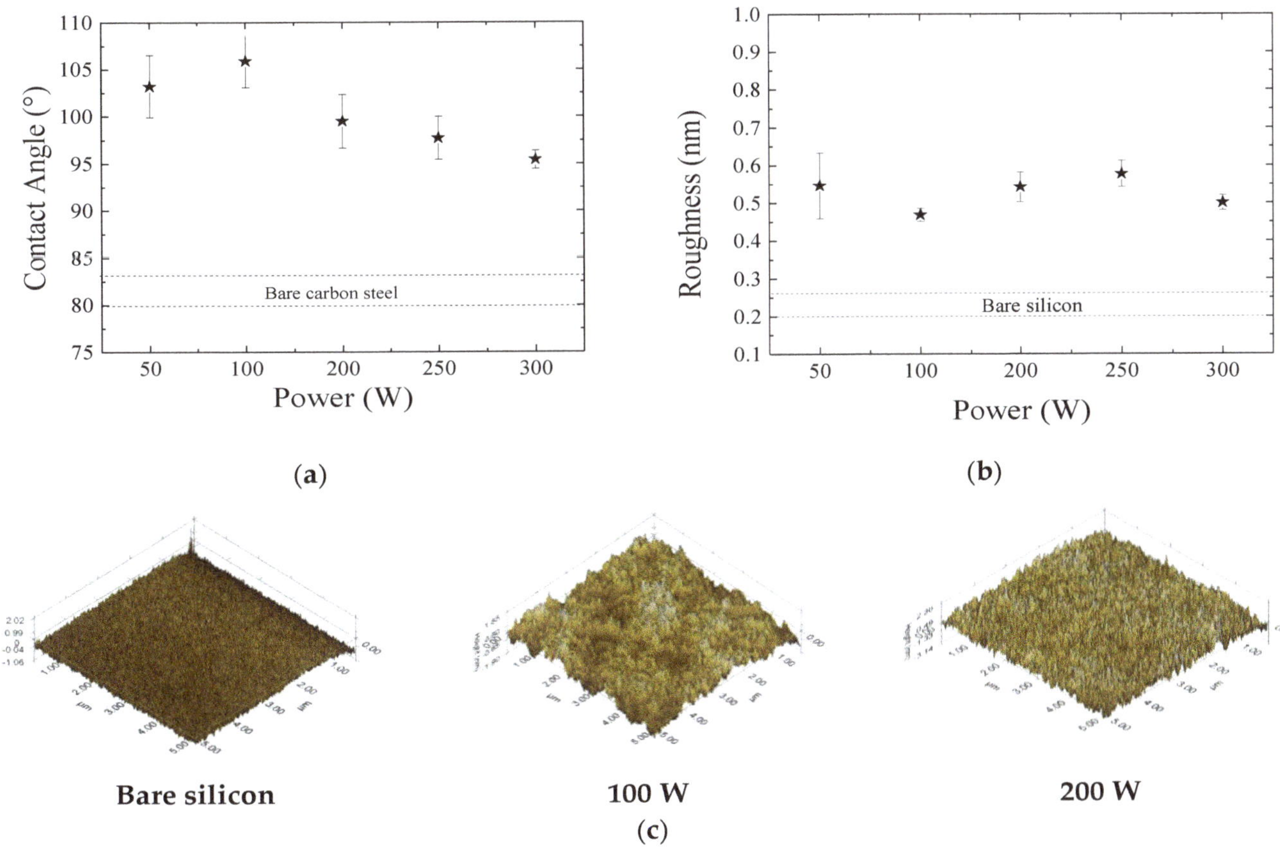

Figure 5. (**a**) Contact angle of the films as a function of the plasma excitation power. (**b**) Average quadratic roughness (RMS) of the films as a function of P. Dotted lines represent the contact angle and roughness for the untreated substrates. (**c**) Atomic force microscopy three-dimensional (3D) topographic profiles (5 μm × 5 μm) of the samples prepared in plasmas with 100 and 200 W, as well as of the bare silicon substrate.

Results in the literature show significant differences between the contact angle of silicon oxide and organosilicon films [6,14,49]. The former is hydrophilic due to high concentration of Si–O, favoring the incorporation of OH groups. The uptake of water molecules by hydroxyls is then induced by Keesom interactions. Although organosilicon films also present Si–O functionals, these bonds are shielded by nonpolar methyl groups ($–CH_3$), which rotate to the air-film interface, causing a reduction in the electrostatic forces acting on the water molecules [50]. Thus, the higher contact angle of the samples prepared with the lowest powers (50 and 100 W) is consistent with the preservation of the CH_3 groups on these surfaces, in accordance with the structure proposed in Figure 3a. On the other hand, the decreasing in θ for $P > 100$ W confirms the abstraction of such shielding groups, which affects the surface polarity.

Since roughness can affect surface wettability, the root mean square roughness (RMS) of the films was determined and is presented in Figure 5b as a function of *P*. Dotted lines in this graph represent the roughness values for the silicon substrate (0.20 to 0.23 nm). Surface roughness increases upon film deposition, attaining values (~0.55 nm) similar to those reported in the literature for organosilicon films with the same thickness [22,51]. It is readily observed in this graph that power variation does not affect significantly the roughness of the films.

These results are understood taking into account the three-dimensional (3D) topographic profiles of the samples, depicted in Figure 5c. While the bare Si plate is smooth, the surfaces of coated samples are no longer flat due to corrugations created by peaks and valleys. The asperities, formed by agglomerates with hundreds of nanometers in size, have already been detected in previous works [22,51]. According to Huan et al. [51], the geometry and concentration of post-like structures straightly affect the coating porosity. Such structures are formed by oligomerization of the HMDSO fragments in the plasma, generating high molecular weight-macromolecules that reach the substrate and stay connected to it by van der Waal's or chemical bonds. Moreover, voids, pinholes, radicals and micro-cracks [52,53], induced on the surface by ion bombardment, may act as preferential sites for the film nucleation explaining the rather uniform aspect of the surfaces with increasing bombardment intensity. Chain crosslinking, however, is pointed as the main responsible mechanism for the surface smoothening. However, despite the agglomerates, the films cover uniformly the entire inspected areas with no signal of detachments, cracks or uncoated regions. Thus, the clear transition from rougher to smoother surface, demonstrated by disappearance of the cavities (darker regions in Figure 5c) with the elevation of the power beyond 100 W, is further evidence of the structural crosslinking induced by changes in *P*. Finally, no clear influence of the roughness on the contact angle is observed with the variation of *P*. Therefore, chemical composition is pointed as the main responsible for the change of the wettability.

3.4. Mechanical and Barrier Properties

Figure 6a shows the hardness of the films as a function of depth (the distance from the film surface). Dashed line represents the average hardness for the silicon substrate, which is in good agreement with literature [54]. Surfaces softer than Si were produced in all cases, but a straight dependency of hardness with the plasma power can be observed as the curves shifted upwards with increasing *P*. Furthermore, a slight increasing in hardness is observed in all the individual curves with increasing depth, revealing the interference of the mechanical properties of the substrate on the results. To minimize such interference, hardness was determined at depths corresponding to 15% of the films thickness [55–57]. Results are shown in Figure 6b as a function of *P*.

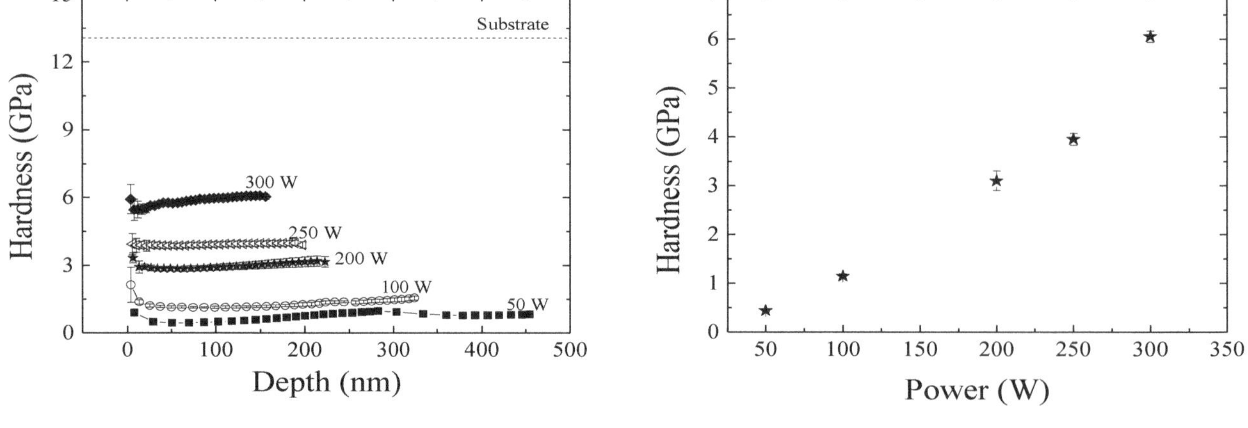

(a) (b)

Figure 6. *Cont.*

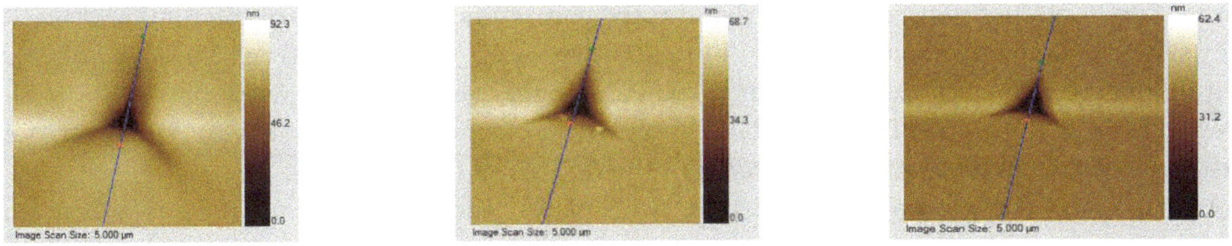

(c)

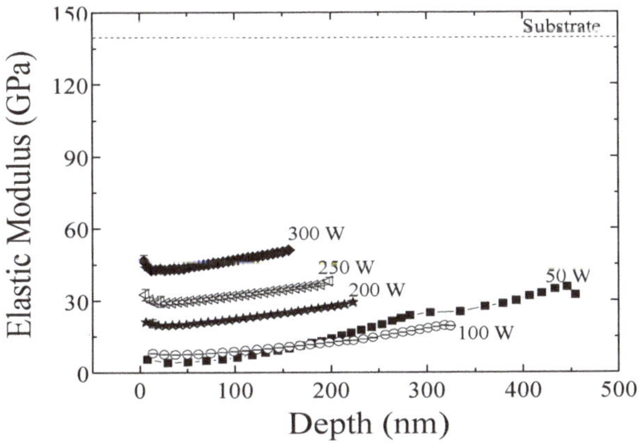

(d)

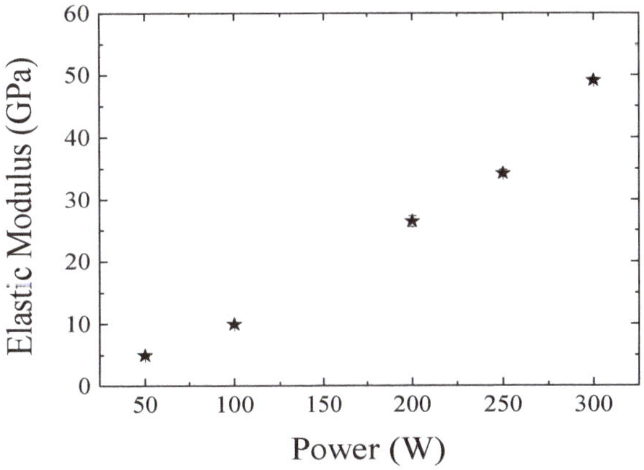

(e)

Figure 6. (**a**) Hardness as a function of depth for films deposited in plasmas of different powers. Dashed line corresponds to the average hardness for the bare silicon substrate. (**b**) Hardness of the films as a function of the plasma excitation power. (**c**) Topographic profiles in 5 μm X 5 μm area of the samples prepared in plasmas of 50, 200 and 300 W, from left to right, respectively. (**d**) Modulus of elasticity as a function of depth for films deposited in plasmas of different powers. Dashed line corresponds to the average elastic modulus for the bare silicon substrate. (**e**) Elastic modulus of the films as a function of plasma excitation power.

Soft films (0.4 GPa) were deposited under the lowest power condition confirming the organosilicon nature of the structure [58,59]. However, a continuous elevation is observed in the hardness curve with increasing *P*. For films deposited under the highest power condition, hardness reached 6.0 GPa, which is 93% higher than that of the softest film. As a reference, plain carbon steels have hardness ranging from 2.0 to 3.0 GPa [60] while the hardness of tool steels ranges from 5.0 to 12.0 GPa [57,61]. The increasing of hardness with *P* is confirmed as one analyze the dimensions of the indentations produced using the maximum indentation load (5000 μN) on the different surfaces as shown in Figure 6c, where a reduction in the strain area can be clearly observed, with increasing *P*.

Even though the organosilicon nature of the films is not altered by varying the plasma activation degree, mechanical resistance is significantly improved. This result corroborates the idea of increment in the reticulation degree and evidences the importance of this characteristic for the structure properties. Aside to decreasing the distance of neighboring chains, crosslinking generates anchoring points, which increase the connectivity of the inorganic backbone, turning the material more resistant to mechanical requirement. The anchoring points also tend to reduce the flexibility of elastomeric systems, such as rubbers or the silicone-like films deposited here under low activation powers, resulting in more rigid structures. To evaluate this aspect, the reduced elastic modulus of the films, E_r, was evaluated and is presented in Figure 6d as a function of depth. The average E_r for the Si substrate is represented by the dotted line. In general, curves shift upwards with increasing *P*, indicating the enhancement of the structure rigidity. However, the ascending trend of each individual curve with increasing depth indicates, by itself, interference of the mechanical properties of the substrate. In order to minimize such interference, E_r values were determined at depths corresponding to 15% of the total film thickness (Figure 6e). The elastic modulus increases by almost an order of magnitude (from 5.0 to 49.0 GPa) with increasing P from 50 to 300 W. Reference for these values is provided in the work of Coclite et al. [62] which prepared SiOx films with E_r = 90 GPa. That is, it is possible to prepare organosilicon structures with rigidity similar to that of silica by increasing the degree of crosslinking.

Figure 7a shows the phase angle for bare and coated carbon steel samples, as a function of polarization frequency. The curve with one concavity is characteristic of the untreated substrate [63]. The results obtained with the samples produced under 200 and 300 W are very similar to those found for the bare carbon steel. On the other hand, two concavities and displacement of the high frequency tail to higher phase angles, both revealing the presence of the coating, can be observed on the curves of the other samples.

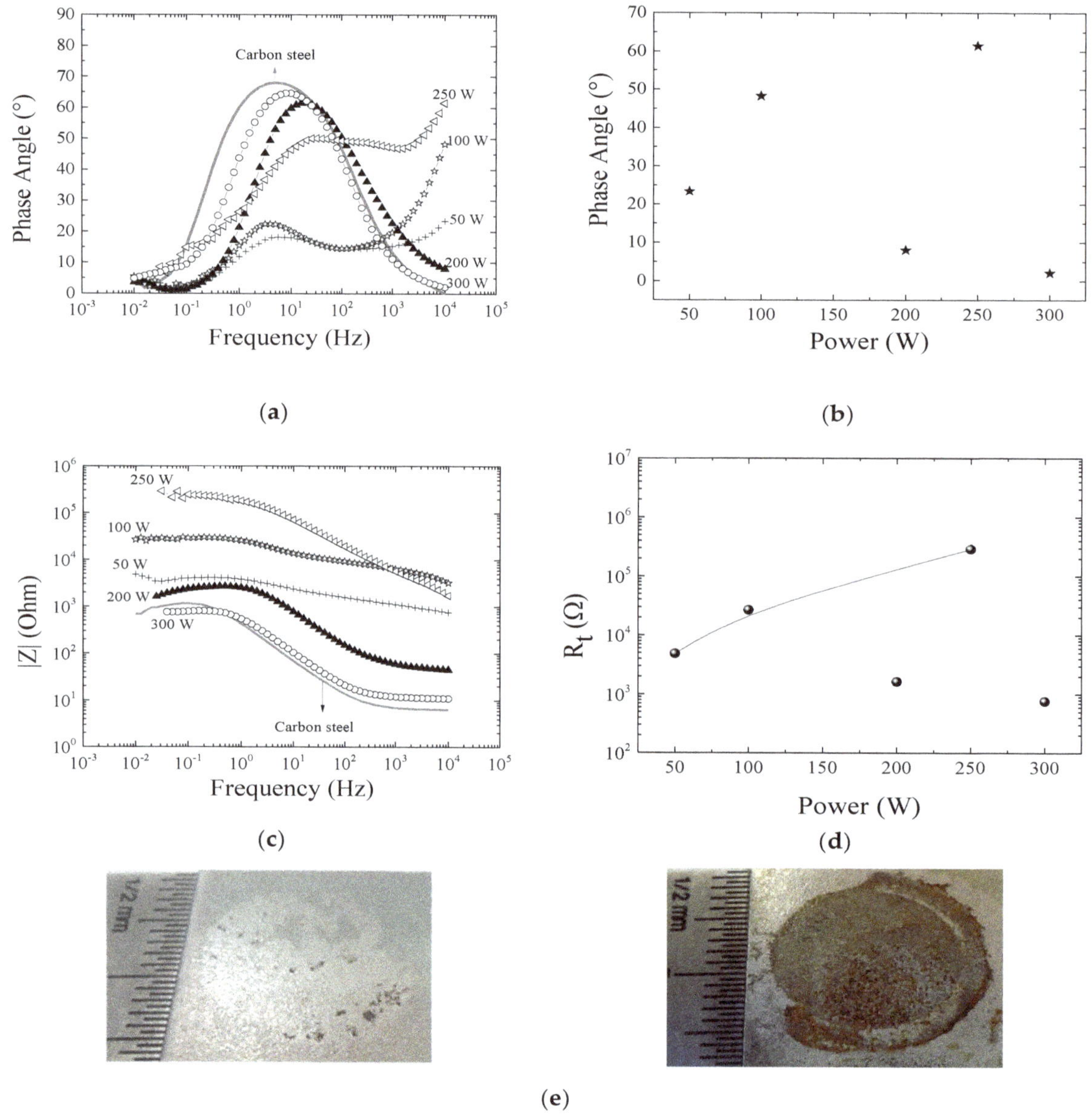

Figure 7. (**a**) Phase angle as a function of the frequency of carbon steel samples as-received and after plasma treatment with diverse applied power. (**b**) Phase angle of the films at 10^4 Hz as a function of P. At this frequency, the phase angle for the bare carbon steel is 0°. (**c**) Impedance modulus as a function of the frequency for the bare and coated carbon steel substrates. (**d**) Total resistance of the system as a function of P. (**e**) From left to right, photographs of the tested regions on the samples prepared under 100 and 300 W of power, respectively.

Capacitive layers are expected to better inhibit the transport of electrolyte species. Since in capacitive systems the phase angle between current and tension is 90°, samples with phase angles close to this value are expected to better prevent the flow of electrolyte species. The phase angle of the systems at the highest frequency edge (10^4 Hz) is presented in Figure 7b as a function of *P*. A null value of phase angle was detected for the bare substrate while reduced values are observed for the samples prepared with *P* equal to 200 and 300 W. If these two samples are not taken into account, a continuous rising in phase angle is observed with increasing *P*. Visually inspecting the tested area of these two samples in the photographs of Figure 7e, it is clearly evidenced the detachment of the films during the electrochemical experiments. This finding indicates the formation of stressed structures for the thickest (200 W) and more crosslinked (300 W) films. Although the results obtained by nanoindentation revealed well adhered films in all the cases, for the corrosion analysis the films were deposited on carbon steel instead of silicon and immersed in saline water solution, factors that may influence the adherence. The adhesion of the layer to the substrates, internal stress and mainly defects generated during the deposition process affect the chemical stability and effectiveness of the protective layers [22,49]. The general increasing of the phase angle with *P* observed on the coatings which presented physical stability to the corrosion test, can be ascribed to a reduction of the concentration and/or mobility of the species going through the film, caused by the structural densification. Thus, phase angle results corroborate the assumption of structural densification induced by chain crosslinking.

Figure 7c shows the impedance modulus, |Z|, as a function of the frequency for the samples investigated here, including the bare carbon steel (solid line). The curves for the samples prepared with 200 and 300 W of power present just slight differences with respect to that of the carbon steel, consistently with the observed peeling off of these films. For the other curves, an upward shifting is observed with increasing *P*, showing an overall improvement of the corrosion resistance. To quantify this improvement, the total resistance [64], R_t, of the systems was determined, as shown in Figure 7d as a function of *P*. If the results of the samples which peeled off are disregarded, an increase in R_t is obtained with increasing *P*, in a way very similar to that observed for the phase angle. Specifically, R_t is about 60 times higher in the system prepared in highly activated plasma (250 W) than in the less activated one (50 W), showing the importance of the densification induced by crosslinking for the current flow in the circuit.

Therefore, from the phase angle and impedance modulus results, it is possible to infer that permeation of species through the deposited layer is changed increasing the plasma power since it straightly affects the connectivity of the organosilicon network.

4. Conclusions

All the deposition conditions employed here resulted in surfaces uniformly coated independently of the substrate material. No signals of detachment, cracks or uncoated regions were observed in the inspected areas.

Under the conditions used here, the deposition rate and layer thickness are sensitive to the plasma power while molecular structure and chemical composition are just barely affected. In general, films are composed by C, H, O and Si, ordered as an organosilicon network. The loss of H and methyl is the main route of alteration, influencing the overall structure connectivity. The intensity of ion bombardment and the type of film precursor species rule such modifications. Therefore, changes in homogeneous and heterogeneous reactions of the plasma slightly alter the deposition kinetic but have important implications on the material properties.

The results of water contact angle measurements coordinate with the reduction in the concentration of non-polar groups (methylsilyl) responsible for the electrostatic shielding of the polar Si–O–Si backbone. Roughness changed after film deposition but without affecting the water contact angle, ruled majorly by chemical rather than physical aspects. Furthermore, the reduction in the coating porosity with the elevation of power was suggested by the surface smoothening.

The plasma power affects significantly the mechanical and barrier properties of the films. Harder and less flexible structures are created as connectivity increases. For treatments with excitation power higher than 200 W, the crosslinking increases substantially, making the film more stressed and, therefore, inappropriate for application as protective coatings.

The corrosion resistance of the carbon steel was improved by the film deposition in all the cases. However, with the improvement of the organosilicon crosslinking degree, the total resistance increased 60 times. All these modifications were implemented without changing the nature of the films, demonstrating the possibility of production of denser and harder organosilicon barrier layers with properties very similar to those of silica by tuning the plasma activation degree.

Finally, it should be stated that the films developed herein may be reproduced on other types of substrates (i.e., ceramics, glasses, polymers and other metallic surfaces) for different types of applications.

Author Contributions: Supervision, E.C.R.; conceptualization, E.C.R. and R.C.C.R.; investigation, R.C.C.R.; and N.C.C.; writing, R.C.C.R. and E.C.R.; resources, R.C.C.R. and E.C.R. All authors have read and agreed to the published version of the manuscript.

Acknowledgments: The authors acknowledge Lucas Araújo Proença and Gustavo Henrique de França for technical support.

References

1. Eduok, U.; Faye, O.; Szpunar, J. Recent developments and applications of protective silicone coatings: A review of PDMS functional materials. *Prog. Org. Coat.* **2017**, *111*, 124–163. [CrossRef]
2. Fuerst, T.F.; Reese, M.O.; Wolden, C.A. PECVD Synthesis of Flexible Optical Coatings for Renewable Energy Applications. *Plasma Proc. Polym.* **2016**, *13*, 184–190. [CrossRef]
3. Yu, Y. Package Structure of Flexible OLED Device and Display Device. U.S. Patent 20170352833, 12 July 2017.
4. Hermosa, C.S.; Top, M.; Dagar, J.; Fahlteich, J.; Brown, T.M. Quantifying Performance of Permeation Barrier—Encapsulation Systems for Flexible and Glass-Based Electronics and Their Application to Perovskite Solar Cells. *Adv. Electron. Mater.* **2019**, 1800978. [CrossRef]
5. Wu, C.Y.; Liao, R.M.; Lai, L.W.; Jeng, M.S.; Liu, D.S. Organosilicon/silicon oxide gas barrier structure encapsulated flexible plastic substrate by using plasma-enhanced chemical vapor deposition. *Surf. Coat. Technol.* **2012**, *206*, 4685–4691. [CrossRef]
6. Vendemiatti, C.; Hosokawa, R.S.; Rangel, R.C.C.; Bortoleto, J.R.R.; Cruz, N.C.; Rangel, E.C. Wettability and surface microstructure of polyamide 6 coated with $SiO_XC_YH_Z$ films. *Surf. Coat. Technol.* **2015**, *275*, 32–40. [CrossRef]
7. Gosar, Ž.; Kovač, J.; Mozetič, M.; Primc, G.; Vesel, A.; Zaplotnik, R. Deposition of $SiO_xC_yH_z$ Protective Coatings on Polymer Substrates in an Industrial-Scale PECVD Reactor. *Coatings* **2019**, *9*, 234. [CrossRef]
8. Cui, X.; Zhu, G.; Pan, Y.; Zhang, Y. Preparation and anticorrosion properties of Polydimethylsiloxane (PDMS). In Proceedings of the 4th Annual International Conference on Material Engineering and Application (ICMEA 2017), Wuhan, China, 15–17 December 2017; pp. 15–17.
9. Santos, N.M.; Gonçalves, T.M.; de Amorim, J.; Freire, C.M.A.; Bortoleto, J.R.R.; Durrant, S.F.; Ribeiro, R.P.; Cruz, N.C.; Rangel, E.C. Effect of the plasma excitation power on the properties of SiOxCyHz films deposited on AISI 304 steel. *Surf. Coat. Technol.* **2017**, *311*, 127–137. [CrossRef]
10. Xue, C.H.; Bai, X.; Jia, S.T. Robust, self-healing superhydrophobic fabrics prepared by one-step coating of PDMS and octadecylamine. *Sci. Rep.* **2016**, *6*, 27262. [CrossRef]
11. Liu, B.; Tian, T.; Yao, J.; Huang, C.; Tang, W.; Xiang, Z.; Xu, X.; Min, J. Superhydrophobic organosilicon-based coating system by a novel ultraviolet-curable method. *Nanomater. Nanotechnol.* **2017**, *7*, 1–10. [CrossRef]
12. Dos Reis, M.C.; Silva, V.R.M.; Sgura, R.; da Cruz, N.C.; Rangel, E.C.; Medeiros, I.S. Surface characteristics and optical properties of plasma deposited films on indirect aesthetic restorative dental materials. *Surf. Coat. Technol.* **2018**, *348*, 55–63. [CrossRef]

13. Lou, B.S.; Wang, S.B.; Hung, S.B.; Wang, C.J.; Lee, J.W. Characterization of plasma polymerized organosilicon thin films deposited on 316L stainless steel. *Thin Solid Films* **2018**, *660*, 637–645. [CrossRef]
14. Souza, J.G.S.; Cordeiro, J.M.; Bertolini, M.; de Almeida, A.B.; Cavalcante, R.C.B.R.; Junior, F.H.N.; Feres, M.; Rangel, E.C.; Barão, V.A.R. Silicon-doped superhydrophobic coating to reduce polymicrobial biofilm on titanium surface. *Biomaterials* **2019**, submitted.
15. Torabinejad, V.; Aliofkhazraei, M.; Assareh, S.; Allahyarzadeh, M.H.; Rouhaghdam, A.S. Electrodeposition of Ni-Fe alloys, composites, and nano coatings–A review. *J. Alloy. Compd.* **2017**, *691*, 841–859. [CrossRef]
16. Yang, T.C.; Chang, J.K.; Lin, C.S. The Influence of Aluminum Nitrate Pre-Treatment on High Temperature Oxidation Resistance of Dip-Coated Silica Coating on Galvanized Steel. *ECS Trans.* **2018**, *85*, 67–73. [CrossRef]
17. Dalibón, E.L.; Escalada, L.; Simison, S.; Forsich, S.; Heim, D.; Brühl, S.P. Mechanical and corrosion behavior of thick and soft DLC coatings. *Surf. Coat. Technol.* **2017**, *312*, 101–109. [CrossRef]
18. Choudhury, A.J.; Barve, S.A.; Chutia, J.; Pal, A.R.; Kishore, R.; Pande, M.; Patil, D.S. RF-PACVD of water repellent and protective HMDSO coatings o bell metal surfaces: Correlation between discharge parameters and film properties. *Appl. Surf. Sci.* **2011**, *257*, 8469–8477. [CrossRef]
19. Hegemann, D.; Korner, E.; Blanchard, N.; Drabik, M.; Guimond, S. Densification of functional plasma polymers by momentum transfer during film growth. *Appl. Phys. Lett.* **2012**, *101*, 21160. [CrossRef]
20. Hegemann, D. Plasma polymer deposition and coatings on polymers. *Compr. Mater. Process.* **2014**, *4*, 201–228.
21. Blanchard, N.E.; Hanselmann, B.; Drosten, J.; Heuberger, M.; Hegemann, D. Densification and hydration of HMDSO plasma polymers. *Plasma Process. Polym.* **2015**, *12*, 32–41. [CrossRef]
22. Zhou, L.; Lv, G.H.; Pang, H.; Zhang, G.P.; Yang, S.Z. Preparation of biomedical Ag incorporated hydroxyapatite/titania coatings on Ti6Al4V alloy by plasma electrolytic oxidation. *Surf. Coat. Technol.* **2012**, *206*, 2552–2557. [CrossRef]
23. Nagasawa, H.; Yamamoto, Y.; Tsuda, N.; Kanezashi, M.; Yoshioka, T.; Tsuru, T. Atmospheric-pressure plasma-enhanced chemical vapor deposition of microporous silica membranes for gas separation. *J. Membr. Sci.* **2017**, *524*, 644–651. [CrossRef]
24. Lee, S.; Kang, Y.J.; Jung, S.; Kim, J.; Kim, D.G. Plasma-Deposited SiOxCyHz Barrier Coatings for Organic Device Encapsulation. *Jpn. J. Appl. Phys.* **2013**, *52*, 076001. [CrossRef]
25. Nagasawa, H.; Yamamoto, Y.; Kanezashi, M.; Tsuru, T. Atmospheric-pressure plasma-enhancedchemical vapor deposition of hybridsilica membranes. *J. Chem. Eng. Jpn.* **2018**, *51*, 732–739. [CrossRef]
26. Guo, M.; Kanezashi, M.; Nagasawa, H.; Yu, L.; Yamamoto, K.; Gunji, T.; Ohshita, J.; Tsuru, T. Tailoring the Microstructure and permeation properties of bridged organosilica membranes via control of the bond angles. *J. Membr. Sci.* **2019**, *584*, 56–65. [CrossRef]
27. Supiot, P.; Vivien, C.; Granier, A.; Bousquet, A.; Mackova, A.; Escaich, D.; Clergereaux, R.; Raynaud, P.; Stryhal, Z.; Pavli, J. Growth and Modification of Organosilicon Films in PECVD and Remote Afterglow Reactors. *Plasma Process. Polym.* **2006**, *3*, 100–109. [CrossRef]
28. Van Hest, M.F.A.M.; Mitu, B.; Schram, D.C.; Van de Sanden, M.C.M. Deposition of organosilicon thin films using a remote thermal plasma. *Thin Solid Films* **2004**, *449*, 52–62. [CrossRef]
29. Martinu, L.; Poitras, D. Plasma deposition of optical films and coatings: A review. *J. Vac. Sci. Technol. A Vac. Surf. Films* **2000**, *18*, 2619–2645. [CrossRef]
30. Messier, R.; Giri, A.P.; Roy, R.A. Revised structure zone model for thin film physical structure. *J. Vac. Sci. Technol. A Vac. Surf. Films* **1984**, *2*, 500–503. [CrossRef]
31. Barni, R.; Zanini, S.; Riccardi, C. Characterization of the chemical kinetics in an O_2/HMDSO RF plasma for material processing. *Adv. Phys. Chem.* **2016**, *6*, 205380. [CrossRef]
32. Oliver, W.; Pharr, G.R. Measurement of hardness and elastic modulus by instrumented indentation: Advances in understanding and refinements to methodology. *J. Mater. Res.* **2004**, *19*, 3–20. [CrossRef]
33. Wavhal, D.S.; Zhang, J.; Steen, M.L.; Fisher, E.L. Investigation of Gas Phase Species and Deposition of SiO_2 Films from HMDSO/O_2 Plasmas. *Plasma Process. Polym.* **2006**, *3*, 276–287. [CrossRef]
34. Choudhury, A.J.; Chutia, J.; Kakati, H.; Barve, S.A.; Pal, A.R.; Sarma, N.S.; Chowdhury, D.; Patil, D.S. Studies of radiofrequency plasma deposition of hexamethyldisiloxane films and their thermal stability and corrosion resistance behavior. *Vacuum* **2010**, *84*, 1327–1333. [CrossRef]
35. Durrant, S.F.; Mota, R.P.; de Moraes, M.A.B. Fluorinated polymer films from rf plasmas containing benzene and sulfur hexafluorine. *Thin Solid Films* **1992**, *220*, 295–302. [CrossRef]
36. Yasuda, H. *Plasma Polymerization*, 1st ed.; Academic Press Inc.: Orlando, FL, USA, 1985; p. 34. ISBN 0-12-768760-2.

37. Rangel, R.C.C.; Pompeu, T.C.; Barros, J.L.S., Jr.; Antonio, C.A.; Santos, N.M.; Pelici, B.O.; Freire, C.M.A.; Cruz, N.C.; Rangel, E.C. *Improvement of the Corrosion Resistance of Carbon Steel by Plasma Deposited Thin Films*, 1st ed.; InTech: Rijeka, Croatia, 2012; p. 91. ISBN 978-953-307-920-2.
38. Fracassi, F.; d'Agostino, R.; Palumbo, F.; Angelini, E.; Grassini, S.; Rosalbino, F. Application of plasma deposited organosilicon thin films for the corrosion protection of metals. *Surf. Coat. Technol.* **2003**, *107*, 174–175. [CrossRef]
39. Lucovsky, G.; Mantini, M.J.; Srivastava, J.K.; Irene, E.A. Low-temperature growth of silicon dioxide films: A study of chemical bonding by ellipsometry and infrared spectroscopy. *J. Vac. Sci. Technol. B* **1987**, *5*, 530–537. [CrossRef]
40. Grill, A.; Neumayer, D.A. Structure of low dielectric constant to extreme low dielectric constant SiCOH films: Fourier transform infrared spectroscopy characterization. *J. Appl. Phys.* **2003**, *94*, 6697–6707. [CrossRef]
41. Chou, J.S.; Lee, S.C. Effect of porosity on infrared-absorption spectra of silicon dioxide. *J. Appl. Phys.* **1995**, *77*, 1805–1807. [CrossRef]
42. Fanelli, F.; d'Agostino, R.; Fracassi, F. GC-MS Investigation of Hexamethyldisiloxane–Oxygen Fed Cold Plasmas: Low Pressure Versus Atmospheric Pressure Operation. *Plasma Proc. Polym.* **2011**, *8*, 932–941. [CrossRef]
43. Alexander, M.R.; Jones, F.R.; Short, R.D. Radio-Frequency Hexamethyldisiloxane Plasma Deposition: A Comparison of Plasma- and Deposit-Chemistry. *Plasmas Polym.* **1997**, *2*, 277–300. [CrossRef]
44. Rao, A.P.; Rao, A.V. Modifying the surface energy and hydrophobicity of the low-density silica aerogels through the use of combinations of surface-modification agents. *J. Mater. Sci.* **2010**, *45*, 51–63.
45. Ping, Y.; Kirkpatrick, R.J.; Poe, B.; McMillan, P.F.; Cong, X. Structure of Calcium Silicate Hydrate (C-S-H): Near-, Mid-, and Far-Infrared Spectroscopy. *J. Am. Ceram. Soc.* **1999**, *82*, 742–748.
46. Lanford, W.A.; Rand, M.J. The hydrogen content of plasma-deposited silicon nitride. *J. App. Phys.* **1978**, *49*, 2473–2477. [CrossRef]
47. Montarsolo, A.; Mossottia, R.; Innocentia, R.; Vassallo, E. A study on washing resistance of pp-HMDSO films deposited on wool fabrics for anti-pilling purposes. *Surf. Coat. Technol.* **2013**, *224*, 109–113. [CrossRef]
48. Li, Y.; Huang, K.; MacGregor, J.D.; Xing, Y. The Role of PTFE in Cathode Transition Layer in Aqueous Electrolyte Li-Air Battery. *Electrochim. Acta* **2016**, *191*, 996–1000. [CrossRef]
49. Benítez, F.; Martínez, E.; Esteve, J. Improvement of hardness in plasma polymerized hexamethyldisiloxane coatings by silica-like surface modification. *Thin Solid Films* **2000**, *109*, 377–378. [CrossRef]
50. Gengenbach, T.E.; Griesser, H. Post-deposition ageing reactions differ markedly between plasma polymers deposited from siloxane and silazane monomers. *J. Polym.* **1999**, *40*, 5091–5094. [CrossRef]
51. Huan, W.; Lizhen, Y.; Qiang, C. Investigation of Microwave Surface-Wave Plasma Deposited SiO*x* Coatings on Polymeric Substrates. *Plasma Sci. Technol.* **2014**, *16*, 37–40.
52. Kondo, M.; Toyoshima, Y.; Matsuda, A. Substrate dependence of initial growth of microcrystalline silicon in plasma-enhanced chemical vapor deposition. *J. Appl. Phys.* **1996**, *80*, 6061–6063. [CrossRef]
53. Granata, S.N.; Bearda, T.; Xu, M.L.; Abdulraheem, Y.; Gordon, I.; Mertens, R.; Poortmans, J. Influence of substrate potential on a-Si: H passivation of Si foils bonded to glass. *Thin Solid Films* **2015**, *579*, 9–13. [CrossRef]
54. Guo, X.; Momota, S.; Nittac, N.; Yamaguchi, T.; Sato, M.; Tokaji, H. Modification of mechanical properties of Si crystal irradiated by Kr-beam. *Appl. Surf. Sci.* **2015**, *349*, 123–128. [CrossRef]
55. Cao, Z.; Zhang, X. Nanoindentation creep of plasma-enhanced chemical vapor deposited silicon oxide thin films. *Scr. Mater.* **2007**, *56*, 249–252. [CrossRef]
56. Zeng, K.; Chen, Z.K.; Shen, L.; Liu, B. Study of mechanical properties of light-emitting polymer films by nano-indentation technique. *Thin Solid Films* **2005**, *477*, 111–118. [CrossRef]
57. Martinatti, J.F.; Santos, L.V.; Durrant, S.F.; Cruz, N.C.; Rangel, E.C. Lubricating coating prepared by PIIID on a forming tool. *J. Phys. Conf. Series.* **2012**, *370*, 012022. [CrossRef]
58. Lopes, B.B.; Rangel, R.C.C.; Antonio, C.A.; Durrant, S.F.; Cruz, N.C.; Rangel, R.C. *Mechanical and Tribological Properties of Plasma Deposited a-C:H:Si:O Films*, 1st ed.; InTech: Rijeka, Croatia, 2012; ISBN 978-953-51-0802-3.
59. Albuquerque, M.D.F.; Santos, E., Jr.; Perdone, R.R.T.; Simao, R.A. Effect of selfbias voltage on the wettability, chemical functionality and nanomechanical properties of hexamethyldisiloxane films. *Thin Solid Films* **2014**, *564*, 73–78. [CrossRef]

60. Wang, Z.B.; Tao, N.R.; Li, S.; Wang, W.; Liu, G.; Lu, J.; Lu, K. Effect of surface nanocrystallization on friction and wear properties in low carbon steel. *Mater. Sci. Eng. A* **2003**, *352*, 144–149. [CrossRef]
61. Kostov, K.G.; Ueda, M.; Lepiensky, M.; Soares, P.C., Jr.; Gomes, G.F.; Silva, M.M.; Reuther, H. Surface modification of metal alloys by plasma immersion ion implantation and subsequent plasma nitriding. *Surf. Coat. Technol.* **2004**, *186*, 204–208. [CrossRef]
62. Coclite, A.M.; de Luca, F.; Gleason, K.K. Mechanically robust silica-like coatings deposited by microwave plasmas for barrier applications. *J. Vac. Sci. Technol. A* **2012**, *30*, 061502. [CrossRef]
63. Mello, C.B.; Ueda, M.; Oliveira, R.M.; Garcia, J.A. Corrosion effects of plasma immersion ion implantation-enhanced Cr deposition on SAE 1070 carbon steel. *Surf. Coat. Technol.* **2011**, *205*, S151–S156. [CrossRef]
64. Barranco, V.; Carpentier, J.; Grundmeier, G. Correlation of morphology and barrier properties of thin microwave plasma polymer films on metal substrate. *Electrochim. Acta* **2004**, *49*, 1999–2013. [CrossRef]

3

The Silacyclobutene Ring: An Indicator of Triplet State Baird-Aromaticity

Rabia Ayub [1,2], Kjell Jorner [1,2] and Henrik Ottosson [1,2,*]

1 Department of Chemistry—BMC, Uppsala University, Box 576, SE-751 23 Uppsala, Sweden; rabia.ayub@kemi.uu.se (R.A.); kjell.jorner@kemi.uu.se (K.J.)

2 Department of Chemistry-Ångström Laboratory Uppsala University, Box 523, SE-751 20 Uppsala, Sweden

* Correspondence: henrik.ottosson@kemi.uu.se

Abstract: Baird's rule tells that the electron counts for aromaticity and antiaromaticity in the first $\pi\pi^*$ triplet and singlet excited states (T_1 and S_1) are opposite to those in the ground state (S_0). Our hypothesis is that a silacyclobutene (SCB) ring fused with a [4*n*]annulene will remain closed in the T_1 state so as to retain T_1 aromaticity of the annulene while it will ring-open when fused to a [4*n* + 2]annulene in order to alleviate T_1 antiaromaticity. This feature should allow the SCB ring to function as an indicator for triplet state aromaticity. Quantum chemical calculations of energy and (anti)aromaticity changes along the reaction paths in the T_1 state support our hypothesis. The SCB ring should indicate T_1 aromaticity of [4*n*]annulenes by being photoinert except when fused to cyclobutadiene, where it ring-opens due to ring-strain relief.

Keywords: Baird's rule; computational chemistry; excited state aromaticity; Photostability

---

1. Introduction

Baird showed in 1972 that the rules for aromaticity and antiaromaticity of annulenes are reversed in the lowest $\pi\pi^*$ triplet state (T_1) when compared to Hückel's rule for the electronic ground state (S_0) [1–3]. The rule has subsequently been confirmed by a series of quantum chemical calculations [3,4], and it has also been shown that 4*n* π-electron species can have triplet multiplicity ground states (T_0). Interestingly, the T_0 state cyclopentadienyl cation and the isomeric vinylcyclopropenium cation (a closed-shell singlet) are nearly isoenergetic [5,6], revealing that Baird-aromatic stabilization of triplet state species can be significant [3,7]. It has also been shown through computations that Baird's rule can be extended to the lowest $\pi\pi^*$ excited singlet states (S_1) of cyclobutadiene (CBD), benzene, and cyclooctatetraene (COT) [8–13]. Thus, [4*n*]annulenes display aromatic character in both their T_1 (or T_0) and S_1 states whereas [4*n* + 2]annulenes display anti-aromaticity. With Hückel's and Baird's rules it becomes clear that benzene has a dual character and can be labelled as a molecular "Dr. Jekyll and Mr. Hyde" [3,14,15]. This is in line with the early conclusions by Baird as well as by Aihara [1,16], and the excited state antiaromaticity explains the photoreactivity of many benzene derivatives [14]. On the other hand, CBD and COT are both aromatic in the T_1 state [17–19].

In the last few years, the excited state aromaticity and antiaromaticity concepts (abbr. ES(A)A) have gained gradually more attention [20–22], even though the pioneering experimental studies were presented by Wan and co-workers already in the 80s and 90s [23–28]. We earlier stressed that Baird's rule can be used as a qualitative back-of-an-envelope tool for the design of photochemically active materials, as well as for the development of new photoreactions [29,30]. Indeed, a recent combined experimental and computational study of a chiral thiopheno-fused COT compound by Itoh and co-workers reveals that the aromatic stabilization energies in both the T_1 and S_1 states are extensive (~21 kcal/mol) [31], in agreement with previous computational estimates of T_1 state (anti)aromatic (de)stabilization [32,33]. For experimental identification of excited state aromatic cycles vs. anti- and nonaromatic ones, there is a need for suitable indicator moieties. Based on both computations and experiments we recently reported that the cyclopropyl (cPr) group can differentiate T_1 and S_1 state aromatic rings from those that are antiaromatic or nonaromatic in these states [34]. Yet, the cPr group also has a drawback in that the products formed upon ring-opening are not easily identified as they are complicated mixtures or polymeric material.

Herein, we discuss a computational study on the effects of T_1 state (anti)aromaticity of [4*n*]- and [4*n* + 2]annulenes on the photochemical ring-opening of a silacyclobutene (SCB) ring fused with an annulene. The SCB ring is interesting because of its ring-strain and high chemical reactivity [35]. One could potentially use the SCB ring as a substituent on an annulene ring, and here it is already known that 1,1-dimethyl-2-phenyl 1 silacyclobut-2-ene ring-opens photochemically [36], likely a route for excited state antiaromaticity relief. However, when used as a substituent the opening of the SCB ring will also be affected by conformational factors (Figure 1a). Instead, if fused it sits in the same arrangement regardless of annulene, and thus, should be a more unbiased indicator (Figure 1b).

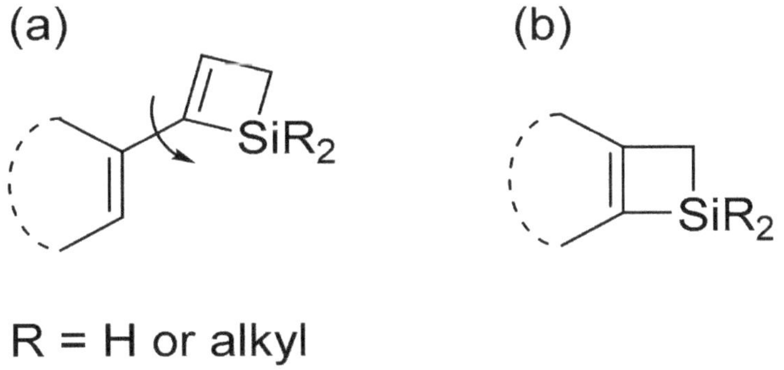

Figure 1. The silacyclobutene (SCB) ring as a substituent (**a**) and fused to an annulene ring (**b**).

We argue that the T_1 aromaticity of [4*n*]annulenoSCBs will hinder the SCB ring from opening as this will lead to loss of T_1 aromaticity, while the T_1 antiaromaticity of [4*n* + 2]annulenoSCBs instead will enhance the rate for ring-opening. We base this argument on the Bell–Evans–Polanyi principle that says that the activation energy will be proportional to the reaction energy for reactions of the same type [37]. This will lead to a photoreactivity difference which can allow the SCB ring to function as a T_1 aromaticity indicator. Noteworthily, the T_1 state potential energy surfaces (PESs) for electrocyclization reactions of compounds with $4n$ π-electrons were earlier explored by Mauksch and Tsogoeva [38], and Möbius aromatic transition states were identified in compounds with 8, 10 and 12π-electrons. Yet, the focus of our paper is not on the T_1 state electrocyclic ring-opening of SCB but instead on the explicit effect of the T_1 state (anti)aromaticity of annulenes on the SCB ring-opening when these annulenes are fused to the SCB ring. Earlier, we reported that the shapes of T_1 state PESs for twists about the C=C double bonds (cf. T_1 state *Z*/*E*-isomerizations) of [4*n*]- and [4*n* + 2]annulenyl substituted olefins are connected to changes in T_1 state (anti)aromaticity of (hetero)annulenyl substituents [39–42]. From this, one can infer that the T_1 state PES of also other photoreactions will vary in dependence of a neighboring [4*n* + 2]- or [4*n*]annulene ring. BenzoSCB (**2a**, Figure 2) should photorearrange to *o*-silaxylylene (**2b**), containing a highly reactive Si=C double bond [43], while cyclobuteno-(**1a**) and cyclooctatetraenoSCB (**3a**), based on our hypothesis, should be resistant to photochemical ring-openings. Indeed, photochemical ring-opening of **2a** in the S_1 state to the transient **2b**, trapped by alcohol solvents to yield isolable silylethers in 40–80% yield (Scheme 1), was earlier reported by Kang and co-workers [44]. Additions of alcohols to silenes, particularly naturally polarized ones ($Si^{\delta+}$=$C^{\delta-}$), proceed over very low activation barriers (a few kcal/mol) [45] due to the high oxy- and electrophilicity of the sp^2 hybridized Si atom, making these reactions highly suitable for rapid trapping of transient SCB ring-opened isomers. A potential benefit of the SCB ring over the cyclopropyl group is the persistence of the products formed upon photochemical SCB ring-opening followed by trapping [44,46].

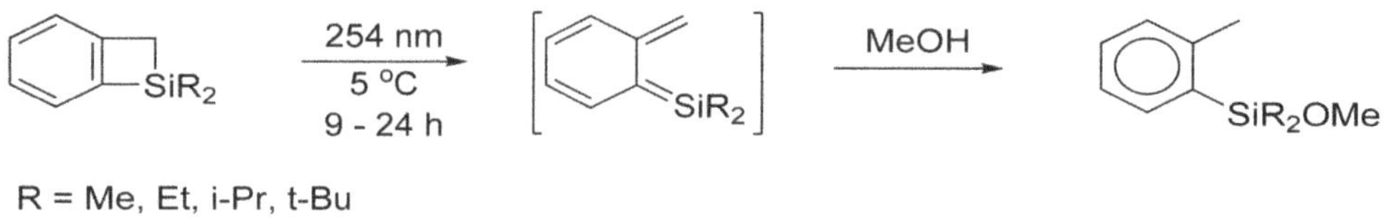

Scheme 1. The photochemical SCB ring-opening and subsequent trapping of *ortho*-silaxylylene reported by Kang, K.T. et al. [44].

Our investigation is focused on the T_1 state ring-openings rather than the S_1 state processes as the triplet states are more easily amenable to computations. We used different aromaticity indices to examine the (anti)aromatic character of ring-closed and ring-opened isomeric structures. If T_1 aromaticity of a [4*n*]annulene hinders the SCB ring from opening, then the absence of this reaction upon irradiation should indicate T_1 aromaticity. The SCB ring-opening could tentatively be connected to the bond dissociation enthalpies (BDEs) because the C–C BDE in a strain-free compound (90.4 kcal/mol) is slightly higher than the Si–C BDE (88.2 kcal/mol), which in turn is higher than the Si–Si (80.5 kcal/mol) BDE [47,48]. Although the BDE difference between the strain-free C–C and Si–C bonds is small, strain could be more important in the SCB ring than in the all-carbon cyclobutene ring (vide infra). Here it can be noted that the cyclobutene and disilacyclobut-3-ene rings are less suitable than the SCB ring because the former opens only rarely upon photolysis (e.g., benzocyclobutene does not undergo photochemical ring-opening unless further derivatized) [14,49], while the latter is unstable and readily oxidized in air to 1,3-disila-2-oxacyclopentenes [50]. It should also be noted that strained four-membered rings with heteroatoms from Groups 15 and 16 are problematic in the context of excited state aromaticity indicators as these heteroatoms provide lone-pair electrons that will interact electronically with the π-conjugated annulene. Additionally, the lowest excited states of compounds with such rings could be of $n\pi^*$ rather than of $\pi\pi^*$ character, leading to excited states for which Baird's rule is not applicable. The SCB ring when used as a substituent can also influence the annulene through π-conjugation, yet, when fused onto an annulene its C=C bond is joint with the annulene. Thus, the SCB ring could hold a unique position as a tentative excited state aromaticity indicator unit.

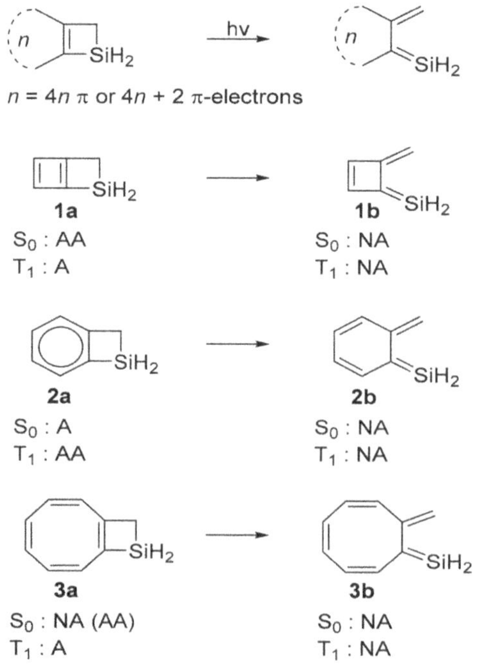

Figure 2. The annulenoSCB ring-openings and the postulated (anti)aromatic characters in the S_0 and the T_1 states of **1a**, **2a**, **3a**, **1b**, **2b**, and **3b**, respectively, with A = aromatic, AA = antiaromatic, and NA = non-aromatic.

2. Results and Discussion

We first discuss the changes in energies, geometries and (anti)aromaticities in the T_1 states during ring-opening reactions of the three molecules (Figure 2) in which a SCB ring is fused with either a [4*n*]annuleno-(cyclobuteno- and cyclooctatetraeno-) ring or a [4*n* + 2]annuleno-(benzo-) ring. We discuss reaction and activation energies, and subsequently analyze (anti)aromaticity changes through the harmonic oscillator model of aromaticity (HOMA), nucleus independent chemical shift (NICS) and isomerization stabilization energy (ISE) indices as well as anisotropy of the induced current density (ACID) plots. Finally, openings of SCB rings fused with 5- and 7-membered annulenyl cations and anions are discussed. We also explored the T_1 PES for SCB ring-opening when fused with polycyclic systems (see Scheme S1), and for comparison the all-carbon analouges of **2** and **3** were analyzed.

2.1. Energy Changes

For the first three compounds (**1–3**), three different methods, two density functional theory methods (B3LYP and OLYP), and one Coupled Cluster method (CCSD(T)) were tested to ensure that the results do not vary extensively with method. Similar results were mostly obtained at the two DFT (B3LYP and OLYP) and CCSD(T) levels, and therefore, only B3LYP energies are given for the remaining compounds unless otherwise noted. Compound **2a**(**T_1**) had a higher relative energy than that of its ring-opened isomer, **2b**(**T_1**), by 33.7–38.3 kcal/mol depending on the computational method (Figure 3). Moreover, the activation energy for Si–C bond scission in the T_1 state was merely 9.0 kcal/mol, 38.3 kcal/mol lower than in the S_0 state at the B3LYP level. This suggests that an antiaromatic destabilization of the benzene ring in the T_1 state affected **2a**(**T_1**) making it highly unstable and prone to cleave the Si–C bond.

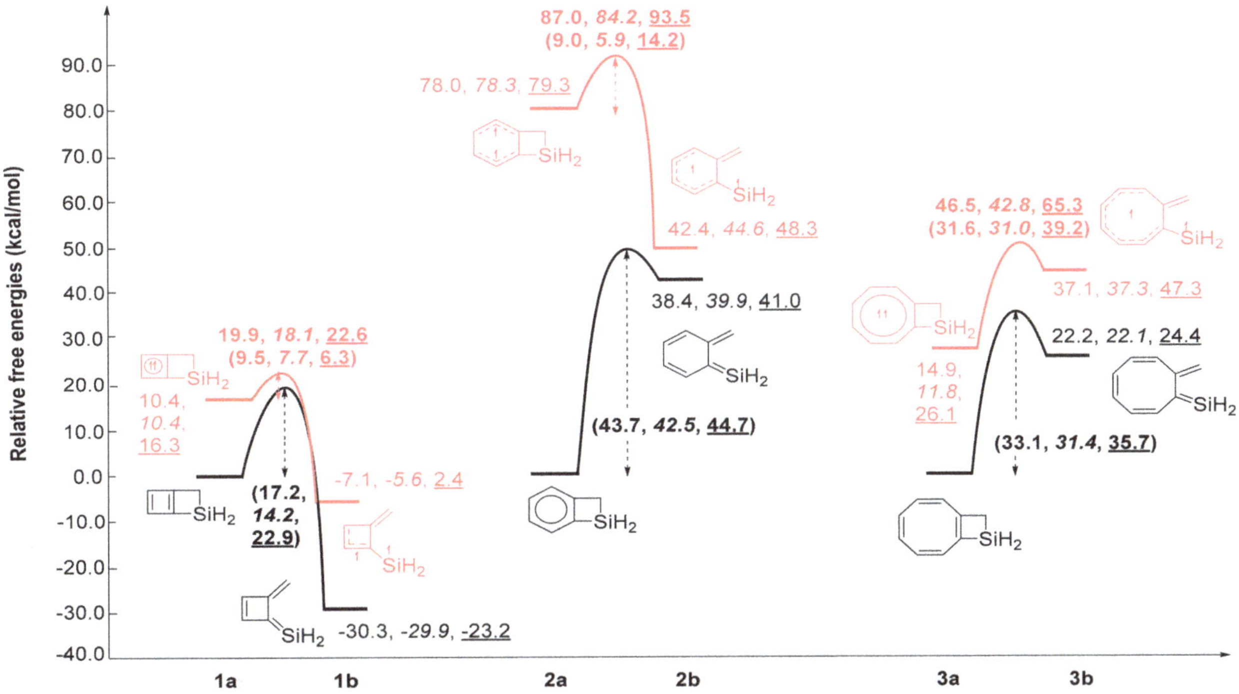

Figure 3. T_1 state potential free energy surface diagrams for ring-openings of 2,3-cyclobutadieno-1-SCB (**1**), 2,3-benzo-1-SCB (**2**) and 2,3-cyclooctatetraeno-1-SCB (**3**) at (U)B3LYP/6-311G(d,p) (normal print), (U)OLYP/6-311G(d,p) (italics), and (U)CCSD(T)/6-311G(d,p)//(U)B3LYP/6-311G(d,p) (underlined) levels. Values in parenthesis are activation energies. Values for the S_0 state in black and values for the T_1 state in red.

Interestingly, the T_1 PES of **1** displayed a similar shape as that of **2**, since **1b(T$_1$)** is of lower energy than **1a(T$_1$)** despite the fact that CBD is weakly T_1-aromatic [8,13,17,51–53]. At the two DFT levels, **1b(T$_1$)** was even lower than **1a(S$_0$)**, likely due to ring-strain relief in **1b(T$_1$)**. Furthermore, the activation energy in the T_1 state was lower than that in the S_0 state by 7.7 kcal/mol at B3LYP level. An opposite behavior was observed for **3** in its T_1 state because **3b(T$_1$)** is higher in energy than **3a(T$_1$)** by 11.8–26.1 kcal/mol. Moreover, the activation energies in the T_1 state are very high, 31.0–39.2 kcal/mol, revealing that the SCB ring will remain closed. Thus, for compounds **2** and **3**, the energy changes upon ring-opening in the T_1 states fell in line with our hypothesis; ring-opening releases energy in **2** in the T_1 state, while it requires energy in **3**. The question was, to what extent these energy changes were linked to changes in (anti)aromaticity (vide infra).

The Si–C bond lengths in **1a**, **2a**, and **3a** are 1.95 Å, 1.92 Å, and 1.90 Å, respectively, which is slightly longer than the normal Si–C bond length (1.87–1.89 Å) [54]. Thus, as one goes to gradually larger annulene rings, the SCB ring gets successively less strained when evaluated based on Si–C bond lengths.

Noteworthily, in the S_0 state, the reaction energy for ring-opening of **2a(S$_0$)** was of opposite sign to the reaction energy in the T_1 state. This reversal in endergonicity and exergonicity when going from the S_0 to the T_1 state of **2** should be a consequence of Baird's rule being the exact opposite to Hückel's rule. With regard to compounds **1** and **3** in the S_0 states, the ring-opening of **1a(S$_0$)** was exergonic, in line with relief of both S_0 antiaromaticity and ring-strain, while ring-opening in the non-aromatic isomer **3a(S$_0$)** was endergonic. These energies were a combination of factors; (i) changes in (anti)aromaticity, (ii) ring strain release, and (iii) changes in the bonding character at the Si atom as it goes from sp^3 to sp^2 hybridized. Formation of a Si=C double bond is an unfavorable process which is not sufficiently compensated by relief of ring strain in the least ring strained of the compounds (**3a(S$_0$)**). Indeed, the ring opening of the all-carbon congener (**allC-3(S$_0$)**), where a C=C double bond is formed instead, was exergonic by 6.4 kcal/mol (see Figure S13).

For **2** and **allC-2**, the S_0 electrocyclic ring-opening transition states were conrotatory, as expected for a thermal reaction with $4n$ electrons. In T_1, this was reversed to a disrotatory fashion. In contrast, for **3** and **allC-3**, both the S_0 and T_1 transition states were conrotatory. The conrotatory mode in T_1 could be explained by the fact that the spin density is delocalized both over the eight-membered and four-membered rings in the TS (Figures S44 and S45). This is consistent with a 10-electron electrocyclic ring-opening with Möbius orbital topology which would be allowed in T_1. Indeed, the ACID plots for **3** and **allC-3** supported this interpretation, as the ring current went over all 10 atoms (Figures S54 and S55). However, the difference of mechanism in T_1 with 10-electron conrotatory for **3** and 4-electron disrotatory for the other molecules was not sufficiently large to prevent application of the Bell–Evans–Polanyi principle, as they all fell on the same correlation line (Figure S40, vide infra).

In addition to the SCB ring fused to aromatic and antiaromatic annulenes, we also analyzed it when fused to the non-aromatic reference compounds cyclobutene, cyclohexene, and cyclohexadiene. When going from **4** to **6** over **5**, the reaction energies in the T_1 state became gradually less strongly exergonic, while the activation energies increased slightly. Compounds **4** and **5** both have nonconjugated C=C double bonds, yet, the cycloalkene ring was larger in **5**, leading to less ring-strain than in **4**, as well as a T_1 state SCB ring-opening reaction energy which was lower by 11.3 kcal/mol. When going from **5** to **6**, the reaction energy further decreased by 12.1 kcal/mol, revealing that the length of the conjugated path also had an impact. An indication that T_1 state antiaromaticity was alleviated in **2a(T$_1$)** was the fact that the energy released when going from **2a(T$_1$)** to **2b(T$_1$)** (Figure 3) was larger than when going from **5a(T$_1$)** to **5b(T$_1$)** (Figure 4). Indeed, the ring-closed **2a(T$_1$)** was at an even higher energy than **5a(T$_1$)** where the triplet biradical was confined to an essentially planar olefin bond. This clarified that the benzene ring in the T_1 state was strongly destabilized. The destabilized nature of T_1 state benzene became obvious when regarding compound **6**, where the ring-closed isomer has a SCB moiety with its C=C double bond being part of a conjugated, yet, nonaromatic segment because the T_1 energy of **2a(T$_1$)**, was substantially higher than that of **6a(T$_1$)** (78.0 vs. 47.5 kcal/mol, respectively). Moreover, the activation energy for SCB ring-opening of **2a(T$_1$)** was 6.6 kcal/mol lower than that of **6a(T$_1$)**, which indicated an influence of T$_1$ antiaromaticity in **2a(T$_1$)**.

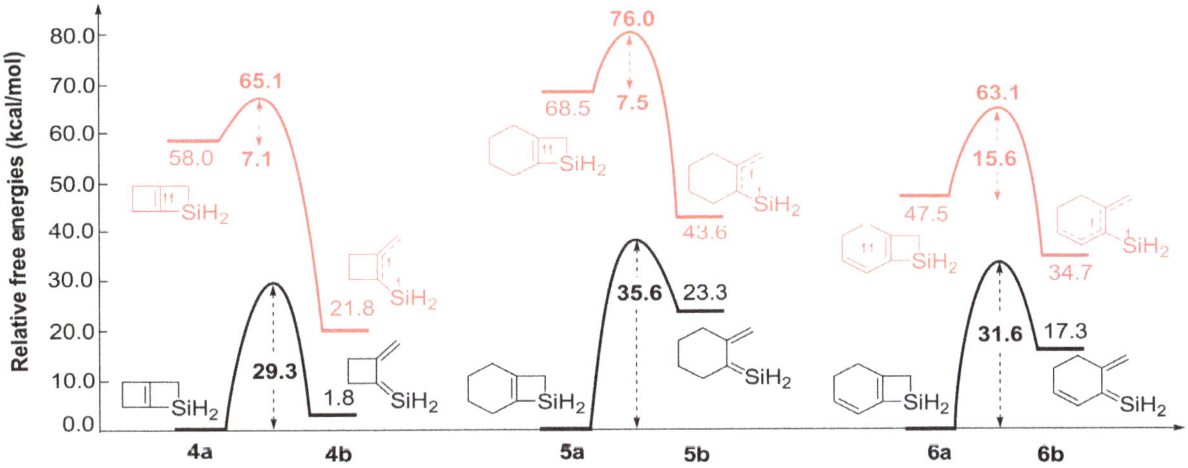

Figure 4. Relative reaction and activation free energies (kcal/mol) for the SCB ring-opening when fused to non-aromatic rings. The energies for transition states are given in bold. Black energy levels represent the S$_0$ state and red ones represent T$_1$ states at (U)B3LYP/6-311G(d,p) level.

2.2. Changes in T$_1$ State (Anti)aromaticity upon Ring-Opening

2.2.1. Harmonic Oscillator Model of Aromaticity (HOMA) Values

Bond length equalization is one indicator of aromaticity, and we chose the geometric HOMA index as one of the indices used (Table 1). The large negative HOMA values in **1a(S$_0$)**, **1b(S$_0$)** and **1b(T$_1$)** corresponded to antiaromaticity, while the small HOMA value of **1a(T$_1$)** suggested that this structure is non-aromatic. Ring-opening of **1a(T$_1$)** to **1b(T$_1$)** led to an increase in antiaromaticity (ΔHOMA(T$_1$) = −0.69).

Table 1. Harmonic oscillator model of aromaticity (HOMA) values at the (U)B3LYP/6-311G(d,p) level of compounds **1**, **2**, and **3**.

Compounds	S$_0$			T$_1$		
	a	**b**	**ΔHOMA**	**a**	**b**	**ΔHOMA**
1	−4.04	−1.11	2.93	0.10	−0.59	−0.69
2	0.97	0.07	−0.90	−0.32	0.83	1.15
3	0.08	−0.21	−0.13	0.89	0.21	−0.68

Compound **2**, in comparison, was aromatic in structures **2a(S$_0$)** and **2b(T$_1$)** while it was non-aromatic in structures **2a(T$_1$)** and **2b(S$_0$)**. Since **2a(T$_1$)** is non-aromatic, ring-opening to **2b(T$_1$)** was favored, as aromaticity was gained (ΔHOMA(T$_1$) = 1.15). Benzene in the T$_1$ state can adopt several different conformers depending on the starting geometry; (i) it can have a quinoidal structure with two unpaired electrons in the *para*-positions and two double bonds parallel to the C$_2$-axis (^{3}**Q**), (ii) it can have an anti-quinoidal structure with two allyl radical segments (^{3}**AQ**), or (iii) it can be described as a combination of a pentadienyl and a methyl radical (^{3}**PM**) [55] (Figure 5). Isomer **2a(T$_1$)** is geometrically most similar to ^{3}**AQ** since it has two long CC bonds and two allylic segments (see Figure S9).

The small HOMA values of **3a(S$_0$)**, **3b(S$_0$)** and **3b(T$_1$)** indicate that these are nonaromatic while a high positive value of **3a(T$_1$)** suggests aromatic character. Thus, ring-opening of **3** in the T$_1$ state entails an unfavorable reduction in aromaticity as ΔHOMA(T$_1$) = −0.68. Taken together, the HOMA values support our hypothesis that T_1 state aromaticity is lost in SCB ring-openings of **1a** and **3a**, while the T_1 state antiaromaticity of **2a** is alleviated in this reaction.

Figure 5. The quinoid (**Q**), anti-quinoid (**AQ**), and pentadienyl-methyl (**PM**) conformers of T_1-state benzene.

2.2.2. Nucleus Independent Chemical Shift (NICS) Scans

NICS is a magnetic indicator of aromaticity. The chemical shifts of NICS probe are scanned over a certain distance (0–5 Å) above the center of the molecular plane. The out-of-plane component obtained is then plotted against the distance. The NICS scans of **1–3** in their T_1 states in ring-closed and ring-opened isomers are shown in Figure 6, while those in the S_0 states are found in Figure S1. With regard to **1a(T$_1$)**, the out-of-plane component in the NICS scan had a negative value (−14.4 ppm; 1.1 Å) suggesting that this structure is significantly aromatic. Conversely, a high positive value of the out-of-plane component in **1b(T$_1$)** shows that this structure had antiaromatic character. For **2**, it was found instead to be **2a(T$_1$)**, as it had a high positive value for the out-of-plane component (90.9 ppm; 0 Å) revealing that this structure was T_1 antiaromatic. This T_1 antiaromaticity changed back to aromaticity when the SCB ring opened, because a value of −19.8 ppm at 1.1 Å was calculated for **2b(T$_1$)**. With regard to **3**, the NICS scan showed **3a(T$_1$)** to be aromatic with a value of −30.4 ppm at 0.9 Å, yet, **3b(T$_1$)** was non-aromatic. Because of the non-planarity of the COT of **3b(T$_1$)** a small kink was observed in its NICS scan, in contrast to that of **3a(T$_1$)** (Figure 6 and Figure S1). Discontinuities in NICS-*XY* scans due to non-planarities were earlier observed by Schaffroth and co-workers for tetraazaacenes [56]. Thus, based on NICS, we have support for our hypothesis that T_1 (anti)aromaticity influences the reaction energies for the ring-openings of **1**, **2**, and **3**. This led to loss of T_1 aromaticity in **1** and **3**, whereas it leads to alleviation of T_1 antiaromaticity in **2**. This reversal when going from **1** to **2**, and then to **3** was also viewed clearly in Figure 6, since the structures with negative (aromatic) NICS values were successively **1a(T$_1$)**, **2b(T$_1$)** and **3a(T$_1$)**.

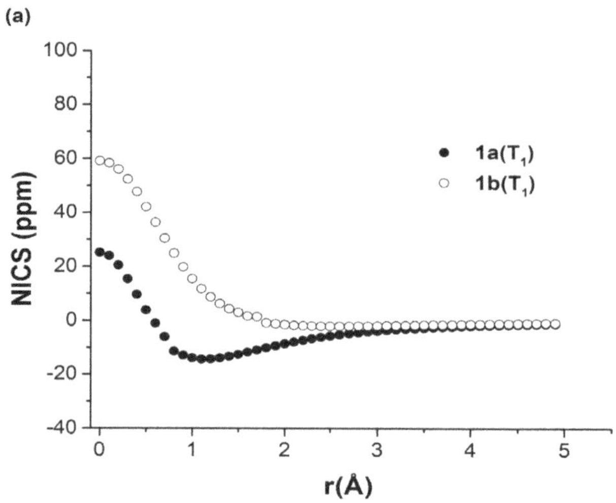

(a)

Figure 6. *Cont.*

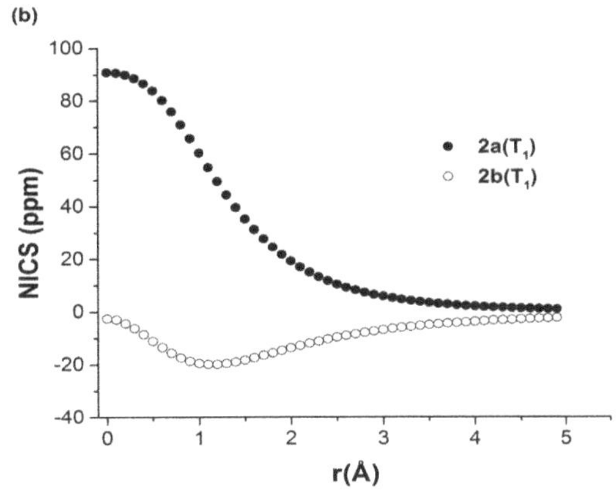

(b)

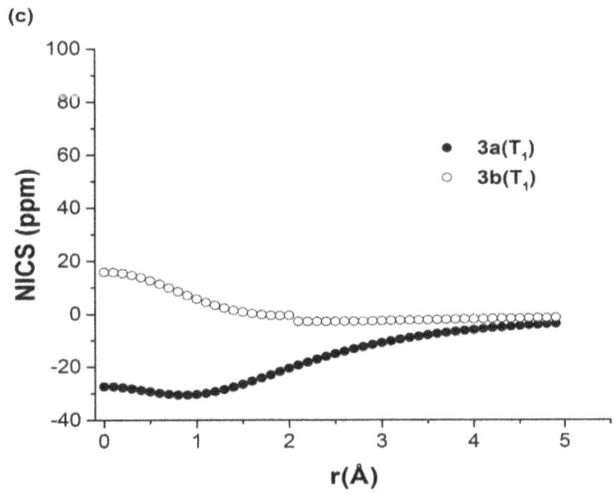

(c)

Figure 6. Nucleus independent chemical shifts (NICS) scans of (**a**) **1a**($\mathbf{T_1}$) and **1b**($\mathbf{T_1}$); (**b**) **2a**($\mathbf{T_1}$) and **2b**($\mathbf{T_1}$); as well as (**c**) **3a**($\mathbf{T_1}$) and **3b**($\mathbf{T_1}$) at the GIAO-(U)B3LYP/6-311+G(d,p)//(U)B3LYP/6-311G(d,p) level. Only the out-of-plane components are displayed.

2.2.3. Anisotropy of the Induced Current Density (ACID) Plots

ACID is a magnetic indicator of aromaticity for visualizing ring-currents and electron delocalization. The ACID plots (Figure 7) of compounds **1**, **2**, and **3** corroborated the results of the NICS scans. Clockwise ring-currents for **1a**($\mathbf{T_1}$), **2b**($\mathbf{T_1}$) and **3a**($\mathbf{T_1}$) indicated aromaticity, while counter-clockwise ring-currents in **1b**($\mathbf{T_1}$) and **2a**($\mathbf{T_1}$) represented antiaromaticity. Yet, the ring-currents in **1b**($\mathbf{T_1}$) suggested this structure to be only weakly antiaromatic. The **3b**($\mathbf{T_1}$) structure was non-aromatic. Clearly, **2a**($\mathbf{T_1}$) opened the SCB ring to alleviate T_1 antiaromaticity (a favorable process), while **1a**($\mathbf{T_1}$) and **3a**($\mathbf{T_1}$) lost aromaticity upon ring-openings (unfavorable processes).

2.2.4. Isomerization Stabilization Energy (ISE) Values

ISE is an energetic index of aromaticity and it is based on the energy difference between the calculated total energy of fully aromatic methyl isomer to that of the non-aromatic exocyclic methylene isomer. We also utilized the isomerization stabilization energy (ISE) index of Schleyer [32] to estimate either the aromaticity or antiaromaticity in ring-closed structures in the T_1 state. Here, we examined only **1** and **2**, and only in their T_1 states. With regard to the smallest compounds, the 4- and 5-methyl substituted **1a**($\mathbf{T_1}$) derivatives showed negative ISE values (ISE_{avg} −10.1 kcal/mol, Figure 8), indicative of some T_1 aromatic stabilization. With regard to **2**, the methyl-substituted **2a**($\mathbf{T_1}$) structures showed positive ISE values from 10.1 to 12.6 kcal/mol, indicative of T_1 antiaromatic destabilization. Structure **2a**($\mathbf{T_1}$) is highly destabilized, evident from the computed ISE values. On the other hand, the ISE values reported for benzene in S_0 state is −33.2 kcal/mol [57]. Thus, the ISE values support our hypothesis that **2a**(**T**$_1$) is destabilized to the same extent as **1a**(**T**$_1$) is stabilized.

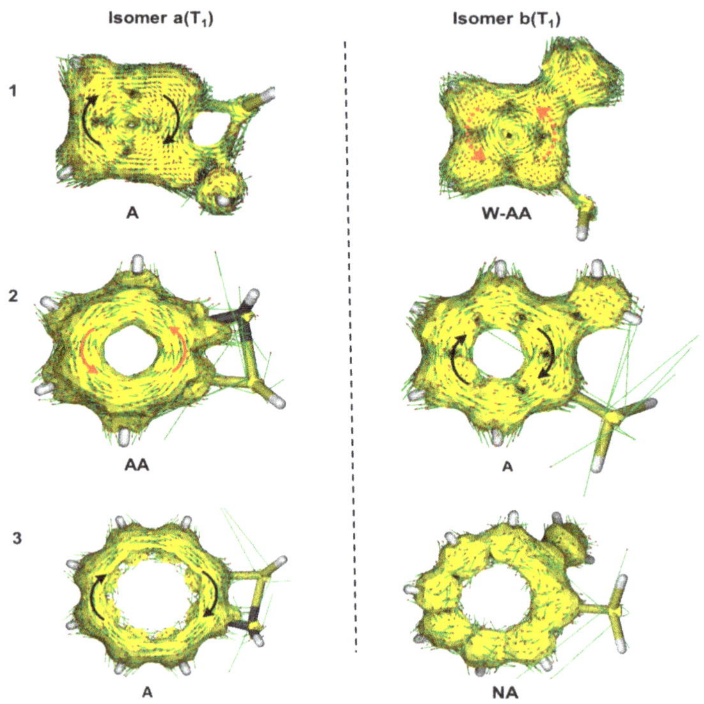

Figure 7. Anisotropy of the induced current density (ACID) plots at (U)B3LYP/6-311+G(d,p)//(U)B3LYP/6-311G(d,p) level. Broken arrows in **1b**(**T**$_1$) indicate weaker ring-currents. Aromatic = A, antiaromatic = AA, weakly antiaromatic = WAA, and non-aromatic = NA.

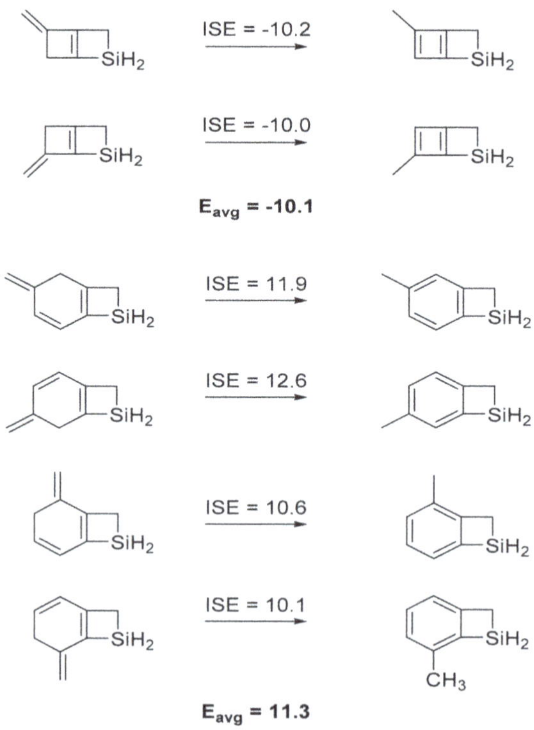

Figure 8. Isomerization stabilization energy (ISE) values (kcal/mol) of the 4- and 5-methyl substituted **1a**(**T**$_1$) derivatives, and the 5-, 6-, 7-, and 8-methyl substituted **2a**(**T**$_1$) derivatives at the UB3LYP/6-311G(d,p) level. ISE_{avg} is the average ISE value for the various methyl substitutions.

2.3. Five- and Seven-Membered Carbocyclic Anions and Cations Fused to SCB

In order to explore if the findings on **1–3** can be extended to other [4*n*]- and [4*n* + 2]annulenyl-SCBs, we examined the potential energy surfaces of 5- and 7-membered annulenyl cations and anions fused to SCB rings (**7–9**, Figure 9). This also allowed us to evaluate how the reaction energies depend on the annulene size. It should be noted that we only investigated energy and geometry changes, and we predicted **7** and **9** to resemble **1** and **3**, respectively, while **8** should resemble **2**. The SCB-fused Cp- was excluded as its calculated T_1 state was of $\pi\sigma^*$ and not of $\pi\pi^*$ character.

The **7a**($\mathbf{S_0}$) structure was a transition state; being a 4π-electron species it is strongly singlet state antiaromatic, it showed large CC bond length alternations (Figure S11), and it was unstable to SCB ring-opening, leading to singlet state antiaromaticity alleviation. On the other hand, **7a**($\mathbf{T_1}$) in the T_1 state was a minimum on the T_1 PES; its geometry met the aromaticity criterion of bond length equalization, and the spin density was uniformly distributed over the cyclopentadienyl fragment (Figure S12). Interestingly, it was 2.9 kcal/mol lower in energy than **7a**($\mathbf{S_0}$), similar to the parent cyclopentadienyl cation which has a triplet ground state [58–61]. The ring-opening of **7a**($\mathbf{T_1}$) to **7b**($\mathbf{T_1}$) was endergonic by 3.8 kcal/mol, opposite to the ring-opening of **1a**($\mathbf{T_1}$) to **1b**($\mathbf{T_1}$) which was exergonic by 17.5 kcal/mol. The reason why **1a**($\mathbf{T_1}$) does not behave similar to **7a**($\mathbf{T_1}$) and **3a**($\mathbf{T_1}$) could be explained by the ring-strain in the CBD ring.

With regard to **9a**($\mathbf{T_1}$), it was merely 4.2 kcal/mol higher in energy than **9a**($\mathbf{S_0}$), and its geometry indicated a completely delocalized cycloheptatrienyl anion. This delocalization of the triplet biradical character was also confirmed through its spin density (Figure S12). The ring-opening of **9a**($\mathbf{T_1}$) to **9b**($\mathbf{T_1}$) was energetically unfavorable, yet, not equally unfavorable as the ring-opening of **3a**($\mathbf{T_1}$) to **3b**($\mathbf{T_1}$). Thus, when going to gradually larger annulenes the SCB ring-opening energies in the T_1 state were −17.5 (**1**), 3.8 (**7**), 12.2 (**9**), and 22.2 (**3**) kcal/mol, respectively. i.e., only the most ring-strained compound (**1**) displayed an exergonic reaction energy. The activation energies in the T_1 state also increased gradually and they were 9.5 (**1**), 21.7 (**7**), 24.6 (**9**) and 31.6 (**3**) kcal/mol, respectively. Hence, the SCB ring, when fused with [4*n*]annulenes will in general not open in the T_1 state, a feature that stems from T_1 aromaticity. When an SCB ring is attached to an annulene ring, the absence of a photochemical ring-opening should therefore indicate T_1 aromaticity. Only when ring-strain is high, as in **1a**($\mathbf{T_1}$), will T_1 state ring-opening occur in such species.

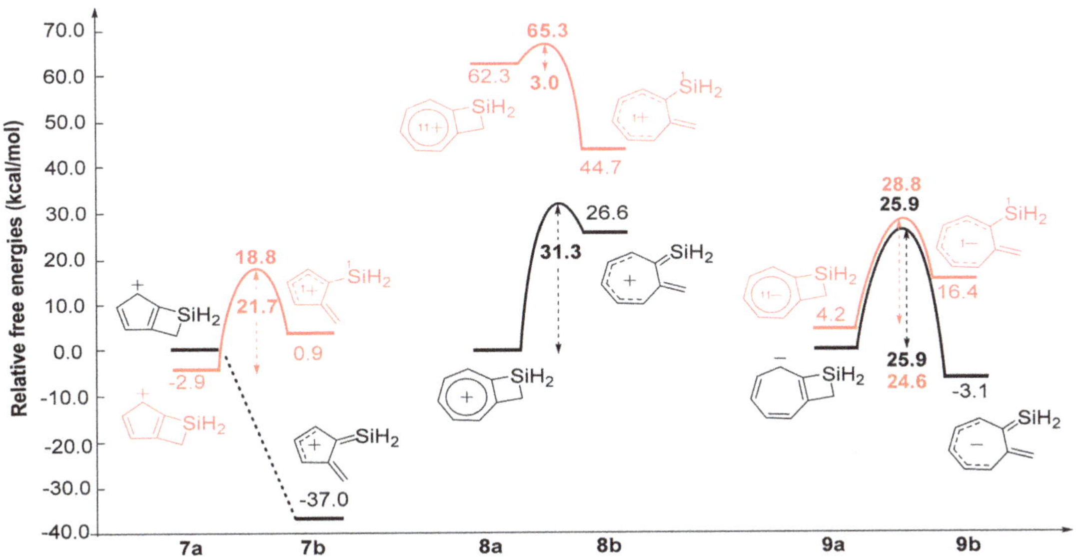

Figure 9. Free energy changes in 5- and 7-membered annulenyl cations and anions fused with SCB rings upon ring-openings at (U)B3LYP/6-311G(d,p) level. Compound **7a**($\mathbf{S_0}$) is not a minimum in the S_0 state. The black energy levels represent S_0 and those of red indicate T_1 state and energies for transition states are given in bold.

Compound **8** showed the opposite behavior to that found for **7** and **9**. Structure **8a**(**T$_1$**) was highly skewed and its ring-opening to **8b**(**T$_1$**) was exergonic by 17.6 kcal/mol. Also, the ring-opened **8b**(**T$_1$**) isomer was planar and had a CC bond delocalized structure. Thus, **8** having a [4*n* + 2]annulene moiety, displayed similar characteristics as **2**. Yet, the smaller the ring, the higher the exergonicity of the ring-opening, explained by relief of ring-strain in addition to the relief of T_1 antiaromaticity. Taken together, the T_1 state ring-opening reactions were markedly uphill for compounds **3**, **7**, and **9**, and downhill for **2** and **8**. Compounds **3**, **7**, and **9** showed T_1 aromaticity similar to [4*n*]annulenes, while compounds 2 and 8 showed T1 antiaromaticity analogous to [4*n* + 2]annulenes, suggesting that loss of T1 aromaticity was observed in ring-openings of [4*n*]annulenes while T1 antiaromaticity of [4*n* + 2]annulenes is alleviated through such reactions. Finally, the T1 state activation energies for SCB rings of the T1 aromatic compounds 3, 7, and 9 were higher than those of the non-aromatic reference compounds **4–6** (Figure 10), allowing the SCB ring to function as a T_1 state aromaticity indicator. For the (4*n* + 2)π-electron annulenoSCBs, the activation energies were similar or lower than those of the nonaromatic references. Overall, the height of the activation barriers in the T_1 state were, to a significant extent, correlated with the reaction energies (R^2 = 0.762, Figure S40), in accordance with the Bell–Evans–Polanyi principle. This correlation was even stronger for the S_0 state (R^2 = 0.872).

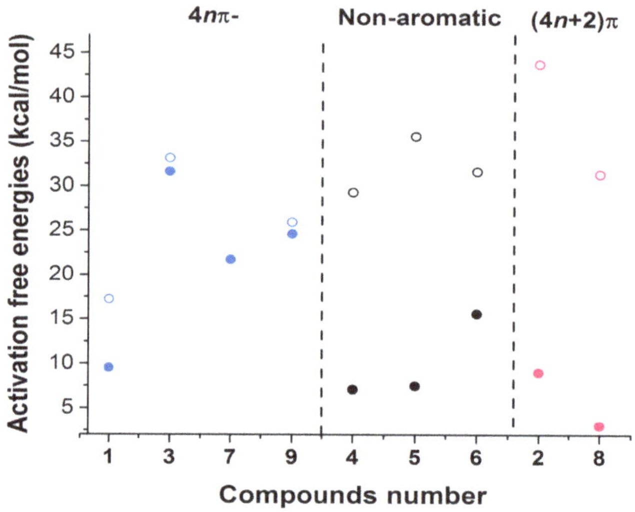

Figure 10. Activation free energies (kcal/mol) of all compounds **1–9** at the (U)B3LYP/6-311G(d,p) level in the S_0 (unfilled circles) and T_1 (filled circles) states.

2.4. Polycyclic Structural Units Fused to SCB

In order to test the generality of the hypothesis, the reaction and activation energies for the ring-opening of an SCB ring when fused to polycyclic moieties (10 and 12π-electrons) were also explored (Scheme S1). With 10 π-electrons, naphthalene is T_1 antiaromatic, and we found that the SCB ring-openings of all three isomers of naphtho-SCB (**10a–10c**) were exergonic, in line with our hypothesis. However, the activation energy for **10a** (23.3 kcal/mol) is significantly higher than observed for the cPr-naphthalenes previously studied (8–11 kcal/mol) [34], suggesting that the SCB ring will remain closed when fused to naphthalene. With regard to biphenylene, a 12π-electron compound which is T_1 state Baird-aromatic, the SCB ring-opening energies of **11a–11c** were modestly exergonic. Moreover, the activation barrier for SCB ring-opening was nearly the same as that of naphthalene (24.2 kcal/mol). Yet, it should be noted that polycyclic systems are more complex than monocycles because the SCB ring-opened products can adopt aromaticity in some of the rings leading to stabilization, a feature already observed for the ring-openings of the corresponding cPr substituted systems. Clearly, the SCB ring should remain closed for T_1 aromatic polycyclic compounds, however, it may also not open for polycyclic T_1 antiaromatic species, leading to limitations of its usage.

3. Computational Methods

All calculations were performed with Gaussian 09 revision D.01 [62]. The structures were optimized at the (U)B3LYP and (U)OLYP density functional theory levels [63–66], with the 6–311G(d,p) basis set [67,68]. Frequency calculations were carried out at the same level to confirm stationary points with real frequencies. Single-point energy calculations were performed at the (U)CCSD(T)/6-311G(d,p)//(U)B3LYP/6-311G(d,p) level and thermal corrections at the B3LYP level were added to get the free energies. Structural, magnetic and energetic indices were used to assess the extent of aromaticity [69]. The harmonic oscillator model of aromaticity (HOMA) [70] values were calculated at the (U)B3LYP/6-311G(d,p) level. Positive values approaching 1.0 correspond to aromatic compounds, negative values to antiaromatic compounds, and values close to zero indicate nonaromatic compounds. Nucleus independent chemical shift (NICS) scans along an axis perpendicular (z-axis) to the ring planes were generated with the Aroma package 1.0 [71–73], using the Gauge-Independent Atomic Orbital (GIAO) method [74] at the GIAO-(U)B3LYP/6-311+G(d,p)//(U)B3LYP/6-311G(d,p) level. Scans were performed starting at the centre of the annulene to 5.0 Å above the ring plane with increments of 0.1 Å. For aromatic compounds, the out-of-plane components show relatively deep minima. For non-aromatic compounds, the values close to the molecular plane are positive, decreases asymptotically and approach zero as the distance is increased. The antiaromatic compounds display high positive values for the out-of-plane components which go to zero with increasing distance. The anisotropy of the induced current density (ACID) calculations [75,76] were used to analyze the ring-currents with the CGST method [77] at (U)B3LYP/6-311+G(d,p) level with AICD 2.0.0 software package. The ACID plots were generated using (NMR = CGST IOp(10/93 = 1) and ultrafine grid (integral = grid = ultrafine). Clock-wise ring-currents indicates aromaticity and counter clock-wise ring-currents indicates antiaromaticity. Isomerization stabilization energies (ISE) [32,33,57] were calculated at the (U)B3LYP/6-311G(d,p) level.

4. Conclusions

We have shown that the ring-opening ability of the SCB ring in the T_1 state can be used to sense for T_1 aromaticity of a [4n]annulene to which it is fused, as its ring-opening disrupts the T_1 aromaticity of the [4n]annulene, an unfavorable (endergonic) process. Conversely, it should open regardless if fused to a T_1 non-aromatic or T_1 antiaromatic ring. By usage of a variety of (anti)aromaticity indices, we link the shapes of the T_1 PES to changes in T_1 (anti)aromaticity. Consequently, the SCB ring could be used as a T_1 aromaticity probe, in contrast to the all-carbon cyclobutene or disilacyclobutene rings which are either too photoresistent or too labile. Moreover, as the silacyclobutene ring when fused does not π-conjugate with the annulene, it has a benefit when compared to strained four-membered rings with Group 15 and 16 elements. The SCB ring also has a benefit over the cPr group examined earlier by us in the context of excited state aromaticity indicators [34] because the transient intermediate formed upon SCB ring-opening, in contrast to the ring-opened cPr intermediate, is easily trapped by alcohols to yield photostable silylethers. Yet, the SCB ring is likely of limited applicability in polycyclic systems as it may remain closed regardless of whether the ring system is T_1 aromatic or T_1 antiaromatic. Still, our study can be interesting from an applications perspective as it reveals situations when the SCB ring as a part in a larger molecule could lead to photoinstability of compounds used for various applications in organic electronics [78].

Acknowledgments: We thank the EXPERTS III (Erasmus Mundus Action II program) and the Swedish Research Council (VR) for financial support. We also thank the SNIC and UPPMAX for generous allotment of computer time.

Author Contributions: Henrik Ottosson conceived the project and Henrik Ottosson and Rabia Ayub designed the project. Rabia Ayub performed most of the Quantum Chemical calculations. Kjell Jorner performed the calculations of the transition states. Rabia Ayub, Kjell Jorner and Henrik Ottosson co-wrote the manuscript.

References

1. Baird, N.C. Quantum Organic Photochemistry. II. Resonance and Aromaticity in the Lowest $^3\pi\pi^*$ State of Cyclic Hydrocarbons. *J. Am. Chem. Soc.* **1972**, *94*, 4941–4948. [CrossRef]
2. Ottosson, H. Organic Photochemistry: Exciting Excited State Aromaticity. *Nat. Chem.* **2012**, *4*, 969–971. [CrossRef] [PubMed]
3. Rosenberg, M.; Dahlstrand, C.; Kilså, K.; Ottosson, H. Excited State Aromaticity and Antiaromaticity: Oppertunities for Photophysical and Photochemical Rationalizations. *Chem. Rev.* **2014**, *114*, 5379–5425. [CrossRef] [PubMed]
4. Gogonea, V.; Schleyer, P.R.; Schreiner, P.R. Consequences of Triplet Aromaticity in $4n\pi$-Electron Annulenes: Calculation of Magnetic Shieldings for Open-Shell Species. *Angew. Chem. Int. Ed.* **1998**, *37*, 1945–1948. [CrossRef]
5. Glukhovtsev, M.N.; Reindl, B.; Schleyer, P.R. What is the Preferred Structure of the Singlet Cyclopentadienyl Cation? *Mendeleev Commun.* **1993**, *3*, 100–102. [CrossRef]
6. Glukhovtsev, M.N.; Bach, R.D.; Laiter, S. Computational Study of the Thermochemistry of $C_5H_5^+$ Isomers: Which $C_5H_5^+$ Isomer is the Most Stable? *J. Phys. Chem.* **1996**, *100*, 10952–10955. [CrossRef]
7. Mauksch, M.; Tsogoeva, S.B. A New Architecture for High Spin Organics based on Baird's rule of $4n$ Electron Triplet Aromatics. *Phys. Chem. Chem. Phys.* **2017**, *19*, 4688–4694. [CrossRef] [PubMed]
8. Karadakov, P.B. Ground- and Excited-State Aromaticity and Antiaromaticity in Benzene and Cyclobutadiene. *J. Phys. Chem. A* **2008**, *112*, 7303–7309. [CrossRef] [PubMed]
9. Karadakov, P.B. Aromaticity and Antiaromaticity in the Low-Lying Electronic States of Cyclooctatetraene. *J. Phys. Chem. A* **2008**, *112*, 12707–12713. [CrossRef] [PubMed]
10. Karadakov, P.B.; Hearnshaw, P.; Horner, K.E. Magnetic Shielding, Aromaticity, Antiaromaticity, and Bonding in the Lowest Lying Electronic States of Benzene and Cyclobutadiene. *J. Org. Chem.* **2016**, *81*, 11346–11352. [CrossRef] [PubMed]
11. Kataoka, M. Magnetic Susceptibility and Aromaticity in the Excited States of Benzene. *J. Chem. Res.* **2004**, *2004*, 573–574. [CrossRef]
12. Haas, Y.; Zilberg, S. The $\nu_{14}(b_{2u})$ Mode of Benzene in S_0 and S_1 and the Distortive Nature of the π Electron System: Theory and Experiment. *J. Am. Chem. Soc.* **1995**, *117*, 5387–5388. [CrossRef]
13. Feixas, F.; Vandenbussche, J.; Bultinck, P.; Matito, E.; Solà, M. Electron Delocalization and Aromaticity in Low-Lying Excited States of Archetypal Organic Compounds. *Phys. Chem. Chem. Phys.* **2011**, *13*, 20690–20703. [CrossRef] [PubMed]
14. Papadakis, R.; Ottosson, H. The Excited State Antiaromatic Benzene Ring: A Molecular Mr Hyde? *Chem. Soc Rev.* **2015**, *44*, 6472–6493. [CrossRef] [PubMed]
15. Stevenson, R.L. *Strange Case of Dr. Jekyll and Mr. Hyde*; Longmans, Green and Co.: London, UK, 1886; ISBN 978-0-553-21277-8.
16. Aihara, J.-I. Aromaticity-Based Theory of Pericyclic Reactions. *Bull. Chem. Soc. Jpn.* **1978**, *51*, 1788–1792. [CrossRef]
17. Fratev, F.; Monev, V.; Janoschek, R. Ab Initio Study of Cyclobutadiene in Excited States: Optimized Geometries, Electronic Tranisitions and Aromaticities. *Tetrahedron* **1982**, *38*, 2929–2932. [CrossRef]
18. Garavelli, M.; Bernardi, F.; Cembran, A.; Castaño, O.; Frutos, L.S.; Merchán, M.; Olivucci, M. Cyclooctatetraene Computational Photo- and Thermal Chemistry: A Reactivity Model for Conjugated Hydrocarbons. *J. Am. Chem. Soc.* **2002**, *124*, 13770–13789. [CrossRef] [PubMed]
19. Villaume, S.; Fogarty, H.A.; Ottosson, H. Triplet-State Aromaticity of $4n\pi$-Electron Monocycles: Analysis of Bifurcation in the π Contribution to the Electron Localization Function. *ChemPhysChem* **2008**, *9*, 257–264. [CrossRef] [PubMed]
20. Sung, Y.M.; Yoon, M.-C.; Lim, J.M.; Rath, H.; Naoda, K.; Osuka, A.; Kim, D. Reversal of Hückel (anti)aromaticity in the Lowest Triplet States of Hexaphyrins and Spectroscopic Evidence for Baird's Rule. *Nat. Chem.* **2015**, *7*, 418–422. [CrossRef] [PubMed]
21. Sung, Y.M.; Oh, J.; Kim, W.; Mori, H.; Osuka, A.; Kim, D. Switching between Aromatic and Antiaromatic 1,3-Phenylene-Strapped[26]- and [28]Hexaphyrins upon Passage to the Singlet Excited State. *J. Am. Chem. Soc.* **2015**, *137*, 11856–11859. [CrossRef] [PubMed]

22. Sung, Y.M.; Oh, J.; Cha, W.-Y.; Kim, W.; Lim, J.M.; Yoon, M.-C.; Kim, D. Control and Switching of Aromaticity in Various All-Aza-Expanded Porphyrins: Spectroscopic and Theoretical Analyses. *Chem. Rev.* **2017**, *117*, 2257–2312. [CrossRef] [PubMed]
23. Wan, P.; Krogh, E. Evidence for the Generationo of Aromatic Cationic Systems in the Excited State. Photochemical Solvolysis of Fluoren-9-ol. *J. Chem. Soc. Chem. Commun.* **1985**, *17*, 1207–1208. [CrossRef]
24. Wan, P.; Krogh, E.; Chak, B. Enhanced Formation of $8\pi(4n)$ Conjugated Carbanions in the Excited State: First example of Photochemical C–H Bond Heterolysis in Photoexcited State. *J. Am. Chem. Soc.* **1988**, *110*, 4073–4074. [CrossRef]
25. Wan, P.; Budac, D.; Krogh, E. Excited State Carbon Acids: Base Catalysed Photoketonization of Dibenzosuberenol to Dibenzosuberone via Initial C–H Bond Heterolysis from S1. *J. Chem. Soc. Chem. Commun.* **1990**, *3*, 255–257. [CrossRef]
26. Wan, P.; Budac, D.; Earle, M.; Shukla, D. Excited-State Carbon Acid: Photochemical Carbon–Hydrogen Bond Heterolysis vs. Formal di-π-Methane Rearrangement of 5H-Dibenzo[a,c]cycloheptene and Related Compounds. *J. Am. Chem. Soc.* **1990**, *112*, 8048–8054. [CrossRef]
27. Budac, D.; Wan, P. Excited-State Carbon Acid. Facile Benzylic Carbon–Hydrogen Bond Heterolysis of Subrene on Photolysis in Aqueous Solution: A Photogenerated Cyclically Conjugated 8π-electron Carbanion. *J. Org. Chem.* **1992**, *57*, 887–894. [CrossRef]
28. Wan, P.; Shukla, D. Utility of Acid-Base Behavior of Excited States of Organic Molecules. *Chem. Rev.* **1993**, *93*, 571–584. [CrossRef]
29. Mohamed, R.K.; Mondal, S.; Jorner, K.; Delgado, T.F.; Lobodin, V.V.; Ottosson, H.; Alabugin, I.V. The Missing C1–C5 Cycloaromatization Reaction: Triplet State Antiaromaticity Relief and Self-Terminating Photorelease of Formaldehyde for Synthesis of Fulvenes from Enynes. *J. Am. Chem. Soc.* **2015**, *137*, 15441–15450. [CrossRef] [PubMed]
30. Papadakis, R.; Li, H.; Bergman, J.; Lundstedt, A.; Jorner, K.; Ayub, R.; Haldar, S.; Jahn, B.O.; Denisova, A.; Zietz, B.; et al. Metal-free Photochemical Silylations and Transfer Hydrogenations of Benzenoid Hydrocarbons and Graphene. *Nat. Commun.* **2016**, *7*, 12962. [CrossRef] [PubMed]
31. Ueda, M.; Jorner, K.; Sung, Y.M.; Mori, T.; Xiao, Q.; Kim, D.; Ottosson, H.; Aida, T.; Itoh, Y. Energetics of Baird Aromaticity Supported by Inversio of Photoexcited Chiral [4*n*]Annulene Derivatives. *Nat. Commun.* **2017**, *8*, 346. [CrossRef] [PubMed]
32. Zhu, J.; Schleyer, P.R. Evaluation of Triplet Aromaticity by the Isomerization Stabilization Energy. *Org. Lett.* **2013**, *15*, 2442–2445. [CrossRef] [PubMed]
33. An, K.; Zhu, J. Evaluation of Triplet Aromaticity by the Indene-Isoindene Isomerization Stabilization Energy Method. *Eur. J. Org. Chem.* **2014**, *13*, 2764–2769. [CrossRef]
34. Ayub, R.; Papadakis, R.; Jorner, K.; Zietz, B.; Ottosson, H. The Cyclopropyl Group: An Excited State Aromaticity Indicator? *Chem. Eur. J.* **2017**, *23*, 13684–13695. [CrossRef] [PubMed]
35. Ishikawa, M.; Naka, A.; Kobayashi, H. The Chemistry of Silacyclobutene: Synthesis, Reactions, and Theoretical Study. *Coord. Chem. Rev.* **2017**, *335*, 58–75. [CrossRef]
36. Tzeng, D.; Fong, R.H.; Dilanjan, H.S.; Weber, W.P. Evidence for the Intermediacy of 1,1-dimethyl-2-phyenyl-1-sila-1,3-butadiene in the Photochemistry and Pyrolysis of 1,1-dimethyl-2-phenyl-1-sila-2-cyclobutene. *J. Organomet. Chem.* **1981**, *219*, 153–161. [CrossRef]
37. Carey, F.A.; Sundberg, R.J. *Advanced Organic Chemistry. Part A: Structure and Mechanisms*, 5th ed.; Springer: New York, NY, USA, 2007.
38. Mauksch, M.; Tsogoeva, S.B. A Preferred Disrotatory 4*n* Electron Möbius Aromatic Transition State for a Thermal Electrocyclic Reaction. *Angew. Chem. Int. Ed.* **2009**, *48*, 2959–2963. [CrossRef] [PubMed]
39. Brink, M.; Möllerstedt, H.; Ottosson, C.-H. Characteristics of the Electronic Structure of Diabatically and Adiabatically Z/E-Isomerizing Olefins in the T_1 state. *J. Phys. Chem. A* **2001**, *105*, 4071–4083. [CrossRef]
40. Villaume, S.; Ottosson, H. Aromaticity Changes along the Lowest-Triplet State Path for C=C Bond Rotation of Annulenyl-Substituted Olefins Probed by Electron Localization Function. *J. Phys. Chem. A.* **2009**, *113*, 12304–12310. [CrossRef] [PubMed]
41. Zhu, J.; Fogarty, H.A.; Möllerstedt, H.; Brink, M.; Ottosson, H. Aromaticity Effects on the Profiles of the Lowest Triplet-State Potential-Energy Surfaces for Rotation about the C=C Bonds of Olefins with Five-Membered Ring Substituents: An Example of the Impact of Baird's Rule. *Chem. Eur. J.* **2013**, *19*, 10698–10707. [CrossRef] [PubMed]

42. Kato, H.; Brink, M.; Möllerstedt, H.; Piqueras, M.C.; Crespo, R.; Ottosson, H. *Z/E*-Photoisomerizations of Olefins with $4n\pi$-or $(4n+2)\pi$-Electron Substituents: Zigzag Variations in Olefin Properties along the T1 state Potential Energy Surfaces. *J. Org. Chem.* **2005**, *70*, 9495–9504. [CrossRef] [PubMed]
43. Ottosson, H.; Eklöf, A.M. Silenes: Connectors between Classical Alkenes and Nonclassical Heavy Alkenes. *Coord. Chem. Rev.* **2008**, *252*, 1287–1314. [CrossRef]
44. Kang, K.T.; Yoon, U.C.; Seo, H.C.; Kim, K.N.; Song, H.Y.; Lee, J.C. Thermal and Photochemical Reactions of Benzosilacyclobutenes with Alcohols. Intermediaxy of *o*-Silaquinone Methide in the Photochemical Reactions. *Bull. Korean Chem. Soc.* **1991**, *12*, 57–60.
45. Bendikov, M.; Quadt, S.R.; Rabin, O.; Apeloig, Y. Addition of Nucleophiles to Silenes. A Theoretical Study of the Effect of Substituents on Their Kinetic Stability. *Organometalllics* **2002**, *21*, 3930–3939. [CrossRef]
46. Kang, K.T.; Seo, H.C.; Kim, K.N. Thermal Reactions of Benzosilacyclobutenes with Alcohols. *Tetrahedron Lett.* **1985**, *26*, 4761–4762. [CrossRef]
47. Walsh, R. Bond Dissociation Energy Values in Silicon-Containing Compounds and Some of their Implications. *Acc. Chem. Res.* **1981**, *14*, 246–252. [CrossRef]
48. McMillen, D.F.; Golden, D.M. Hydrocarbons Bond Dissociation Energies. *Ann. Rev. Phys. Chem.* **1982**, *33*, 493–532. [CrossRef]
49. Segura, J.L.; Martín, N. *o*-Quinodimethanes: Efficient Intermediates in Organic Synthesis. *Chem. Rev.* **1999**, *99*, 3199–3246. [CrossRef] [PubMed]
50. Ishikawa, M.; Naka, A.; Yoshizawa, K. The Chemistry of Benzodisilacyclobutenes and Benzobis(disilacyclobutene)s: New Development of Transition-Metal-Catalyzed Reactions, Stereochemistry and Theoretical Studies. *Dalton Trans.* **2016**, *45*, 3210–3225. [CrossRef] [PubMed]
51. Taubert, S.; Sundholm, D.; Jusélius, J. Calculations of Spin-Current Densities Using Gauge-Including Atomic Orbitals. *J. Chem. Phys.* **2011**, *134*, 54123–54135. [CrossRef] [PubMed]
52. Soncini, A.; Fowler, P.W. Ring-Current Aromaticity in Open-Shell Systems. *Chem. Phys. Lett.* **2008**, *450*, 431–436. [CrossRef]
53. Jursic, B.S. Exploring the lowest energy triplet potential energy surface for cyclic c_4h_4 isomers with the complete basis set ab initio method. Is the transformation of triafulvene into cyclobutadiene possible in their excited states? *J. Mol. Struct. (THEOCHEM)* **1999**, *490*, 133–144. [CrossRef]
54. Brook, M.A. *Silicon in Organic, Organometallic, and Polymer Chemistry*; John Wiley and Sons: Hoboken, NJ, USA, 2000; ISBN 0-471-19658-4.
55. Zamstein, N.; Kallush, S.; Segev, B. A phase-space approach to the $T_1 \rightsquigarrow S_0$ radiationless decay in benzene: The effect of deuteration. *J. Chem. Phys.* **2005**, *123*, 074304. [CrossRef] [PubMed]
56. Schaffroth, M.; Gershoni-Poranne, R.; Stanger, A.; Bunz, U.H.F. Tetraazaacenes Containing Four-Membered Rings in Different Oxidation States. Are They Aromatic? A Computational Study. *J. Org. Chem.* **2014**, *79*, 11644–11650. [CrossRef] [PubMed]
57. Schleyer, P.R.; Pühlhofer, F. Recommendations for the Evaluation of Aromatic Stabilization Energies. *Org. Lett.* **2002**, *4*, 2873–2876. [CrossRef] [PubMed]
58. Breslow, R.; Chang, H.W.; Hill, R.; Wasserman, E. Stable Triplet States of Some Cyclopentadienyl Cations. *J. Am. Chem. Soc.* **1967**, *89*, 1112–1119. [CrossRef]
59. Saunders, M.; Berger, R.; Jaffe, A.; McBride, J.M.; O'Neill, J.; Breslow, R.; Hoffman, J.M., Jr.; Perchonock, C.; Wasserman, E.; Hutton, R.S.; et al. Unsubstituted Cyclopentadienyl Cation, a Ground-State Triplet. *J. Am. Chem. Soc.* **1973**, *95*, 3017–3018. [CrossRef]
60. Wörner, H.J.; Merkt, F. Photoelectron Spectroscopic Study of the First Singlet and Triplet States of the Cyclopentadienyl Cation. *Angew. Chem. Int. Ed.* **2006**, *45*, 293–296. [CrossRef] [PubMed]
61. Wörner, H.J.; Merkt, F. Diradicals, Antiaromaticity, and the Pseudo-Jahn–Teller Effect: Electronic and Rovibronic Structures of the Cyclopentadienyl Cation. *J. Chem. Phys.* **2007**, *127*, 34303. [CrossRef] [PubMed]
62. Frisch, M.J.; Trucks, G.W.; Schlegel, H.B.; Scuseria, G.E.; Robb, M.A.; Cheeseman, J.R.; Scalmani, G.; Barone, V.; Mennucci, B.; Petersson, G.A.; et al. *Gaussian 09, Revision D.01*; Gaussian, Inc.: Wallingford, CT, USA, 2009.
63. Stephens, P.J.; Devlin, F.J.; Chabalowski, C.F.; Frisch, M.J. Ab Initio Calculations of Vibrational Absorption and Circular Dichorism Spectra Using Density Functional Force Fields. *J. Phys. Chem.* **1994**, *98*, 11623–11627. [CrossRef]

64. Becke, A.D. A New Mixing of Hartree–Fock and Local-Density-Functional Theories. *J. Chem. Phys.* **1993**, *98*, 1372–1377. [CrossRef]
65. Handy, N.C.; Cohen, A.J. Left-Right Correlation Energy. *Mol. Phys.* **2001**, *99*, 403–412. [CrossRef]
66. Parr, R.G.; Yang, W. *Density-Functional Theory of Atoms and Molecules*; Oxford University Press: Oxford, UK, 1989.
67. Rrishnan, R.; Binkley, J.S.; Seeger, R.; Pople, J.A. Self-Consistent Molecular Orbital Methods. XX. A Basis Set for Correlated Wave Functions. *J. Chem. Phys.* **1980**, *72*, 650–654. [CrossRef]
68. McLean, A.D.; Chandler, G.S. Contracted Gaussian Basis Sets for Molecular Calculations. I. Second Row Atoms, Z = 11–18. *J. Chem. Phys.* **1980**, *72*, 5639–5648. [CrossRef]
69. Schleyer, P.R. Introduction: Aromatcitiy. *Chem. Rev.* **2001**, *101*, 1115–1118. [CrossRef] [PubMed]
70. Krygowski, T.M. Crystallographic Studies of Inter- and Intramolecular Interactions Reflected in Aromatic Character of π-Electron Systems. *J. Chem. Inf. Comput. Sci.* **1993**, *33*, 70–78. [CrossRef]
71. Stanger, A. Nucleus-Independent Chemical Shifts (NICS): Distance Dependence and Revised Criteria for Aromaticity and Antiaromaticity. *J. Org. Chem.* **2006**, *71*, 883–893. [CrossRef] [PubMed]
72. Stanger, A. Obtaining Relative Induced Ring Currents Quantitatively from NICS. *J. Org. Chem.* **2010**, *75*, 2281–2288. [CrossRef] [PubMed]
73. Rahalkar, A.; Stanger, A. Aroma. Available online: http://schulich.technion.ac.il/Amnon_Stanger.htm (accessed on 22 November 2015).
74. Cheeseman, J.R.; Trucks, G.W.; Keith, T.A.; Frisch, M.J. A Comparison of Models for Calculating Nuclear Magnetic Resonance Shielding Tensors. *J. Chem. Phys.* **1996**, *104*, 5497–5509. [CrossRef]
75. Herges, R.; Geuenich, D. Delocalization of Electrons in Molecules. *J. Phys. Chem. A* **2001**, *105*, 3214–3220. [CrossRef]
76. Geuenich, D.; Hess, K.; Kohler, F.; Herges, R. Anisotropy of the Induced Current Density (ACID), a General Method to Quantify and Visualize Electronic Delocalization. *Chem. Rev.* **2005**, *105*, 3758–3772. [CrossRef] [PubMed]
77. Keith, T.A.; Bader, R.F.W. Calculations of Magnetic Response Properties Using a Continuous Set of Gauge Transformations. *Chem. Phys. Lett.* **1993**, *210*, 223–231. [CrossRef]
78. Yan, D.; Mohsseni-Ala, J.; Auner, N.; Bolte, M.; Bats, J.W. Molecular Optical Switches: Synthesis, Structure, and Photoluminescence of Spirosila Compounds. *Chem. Eur. J.* **2007**, *13*, 7204–7214. [CrossRef] [PubMed]

4

Hybrid Disila-Crown Ethers as Hosts for Ammonium Cations: The O–Si–Si–O Linkage as an Acceptor for Hydrogen Bonding

Fabian Dankert, Kirsten Reuter, Carsten Donsbach and Carsten von Hänisch *

Fachbereich Chemie and Wissenschaftliches Zentrum für Materialwissenschaften (WZMW), Philipps-Universität Marburg, Hans-Meerwein Straße 4, D-35032 Marburg, Germany; fabian.dankert@staff.uni-marburg.de (F.D.); kirsten.reuter@staff.uni-marburg.de (K.R.); donsbach@students.uni-marburg.de (C.D.)

* Correspondence: haenisch@chemie.uni-marburg.de

Abstract: Host-guest chemistry was performed with disilane-bearing crown ethers and the ammonium cation. Equimolar reactions of 1,2-disila[18]crown-6 (**1**) or 1,2-disila-benzo[18]crown-6 (**2**) and NH_4PF_6 in dichloromethane yielded the respective compounds [NH_4(1,2-disila[18]crown-6)]PF_6 (**3**) and [NH_4(1,2-disila-benzo[18]crown-6)]PF_6 (**4**). According to X-ray crystallographic, NMR, and IR experiments, the uncommon hydrogen bonding motif $O_{(Si)}\cdots H$ could be observed and the use of cooperative effects of ethylene and disilane bridges as an effective way to incorporate guest molecules was illustrated.

Keywords: siloxanes; host-guest chemistry; supramolecular chemistry; main group coordination chemistry; hydrogen bonding

1. Introduction

Siloxane bonding has been intensely discussed for the past seventy years. However, siloxane bonding is not yet fully understood. Its discussion regarding the basicity is, to the best of our knowledge, nowadays based on two different explanatory models. Both are important in order to give insights into the Si–O bond, the associated Lewis basicity, and binding properties. As one model, negative hyperconjugation interactions are discussed especially for permethylated siloxanes [1,2]. These interactions are understood as a donation of electron density in the case $p(O) \rightarrow \sigma^*(Si–C)$, which is competing with the coordination towards a Lewis acid and vice versa. Hence, the basicity of silicon bonded oxygen atoms turns out to be lower [3–5]. The other explanatory model considers the Si–O bond as highly ionic. The electronegativity gradient in the Si–O bond is considerably larger than in the C–O bond, which causes significantly different binding properties of siloxanes in comparison to ethers. Gillespie and Robinson emphasize that the electron pairs located directly at the oxygen atoms are spatially diffused, resulting in a lower basicity [6]. Furthermore, one could argue that the partially negatively charged oxygen atoms should show strong interactions with Lewis acids. However, this argument is disproved by repulsive interactions between a positively charged silicon atom and a Lewis acid, which was recently shown via quantum chemical calculations [7,8]. Overall, this leads to an understanding of why the coordination of siloxanes turns out to be cumbersome. The whole discussion is stripped down to monosilanes, which results in a structural discrepancy regarding (cyclic) poly-silaethers. The conformation of the ligand significantly affects the coordination properties, which was shown for ring-contracted crown ethers [9,10]. Considering all those arguments, we tried to regain structural analogy towards organic (crown type) ligands with the insertion of disilane-units. Simple substitution of $–SiMe_2–$ units with $–Si_2Me_4–$ units in a residuary $–C_2H_4O–$ framework yields disilane-bearing ligands with a respectable coordination ability very close to their organic analogs. Alkali and alkaline earth metal salts could easily be coordinated by ligands of this class, so the coordination ability of siloxane compounds should be reconsidered [11–15]. However, the discussion around siloxane bonding is not restricted to the coordination of Lewis acids and includes the ability to form hydrogen bonds. Hydrogen-bonding patterns vary with the use of different substituents within a silicon-based system (see Scheme 1).

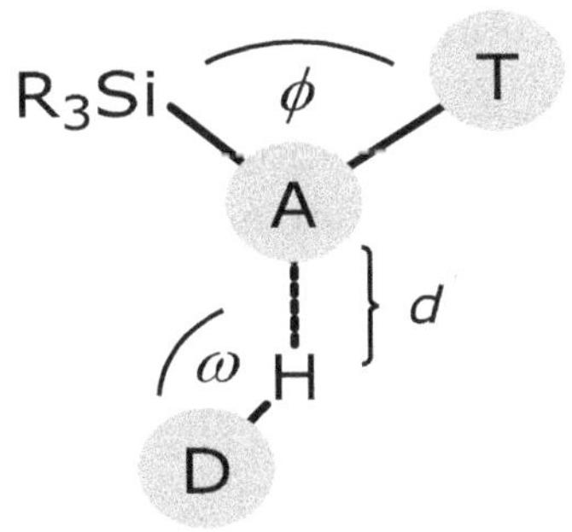

Scheme 1. Model of hydrogen-bonding involving silicon-based systems in which A = acceptor (especially O), D = donor-bearing atom (O/N), T = tetrel (C/Si), *d* = length of the respective hydrogen bonding contact, and ϕ as well as ω are relevant bond angles setting up the hydrogen bonding pattern.

The ability of these systems to form a hydrogen bond has been discussed since the early sixties, especially by the group of West. Early examinations order the affinity to form hydrogen bonds in the sequence R_3COCR_3 > R_3COSiR_3 >> $R_3SiOSiR_3$ according to IR-spectroscopic and thermodynamic studies, as well as NMR experiments [16–18]. Also, recent research confirms a low hydrogen bonding affinity of the oxygen atoms within ligands of the type $R_3SiOSiR_3$ [19]. This is also reflected by the fact that a lot more solid state structures with hydrogen bonding between D–H and R_3CDSiR_3 (D = O, N) than between D–H and $R_3SiOSiR_3$ are known to date. These results reflect the fact that a screening of the Cambridge Crystallographic Database (CCDC) reveals no more than twenty structures that exhibit hydrogen bonding between D–H and R_3CDSiR_3 (D = O, N) and just a handful of structures showing contacts in between D–H and $R_3SiOSiR_3$ in the solid state [20]. Taking all observations into account, the hydrogen bonding of siloxanes continues to be an uncommon motif and is declared as an unusual phenomenon [21]. However, it is possible to increase the ability of siloxanes to form hydrogen bonds by decreasing the ϕ-angle, which could be shown in several publications and was also supported by experimental data provided by the group of Beckmann [5,17,21–23]. The relatively small pool of experimental data motivated us to extend the coordination chemistry of hybrid disila-crown ethers to ammonium cations.

2. Results

The ability of organic crown ethers to act as host molecules is discussed regarding different systems with hydronium- and ammonium ions as the simplest hosts. Recrystallization of equimolar ratios of salt and an appropriate crown ether from organic and/or aqueous solution yields crown ether complexes with the general formula [M(CE)]A, where M = H_3O^+ or NH_4^+, CE = crown ether, and A = anion. Mainly crown ethers of the [18]crown-6 type are used, but the anion structure varies with A = Cl^-, Br_3^-, I_3^-, ClO_4^-, $[BF_4]^-$, $[PF_6]^-$ $[SbCl_6]^-$, and many more [24–33]. In the case of hybrid crown-ethers, aqueous solutions and traces of moisture lead to the entire decomposition of the ligand. Siloxane cleavage with aqueous solutions is common for this kind of ligand and has already been discussed in other publications [11,34,35]. For this reason, hydrogen bonding towards hydronium cations could not be performed. However, the incorporation of a guest turned out to be successful in the use of ammonium hexafluorophosphate as the salt and 1,2-disila[18]crown-6 (**1**), as well as 1,2-disila-benzo[18]crown-6 (**2**) as the ligands of choice. The ligands were prepared using methods described in the literature (see Scheme 2) [11,15]. Subsequent reaction of these ligands with NH_4PF_6 in anhydrous dichloromethane yielded the respective complexes $[NH_4$(1,2-disila[18]crown-6)]PF_6 (**3**) and $[NH_4$(1,2-disila-benzo[18]crown-6]PF_6 (**4**). Neat **3** is a colorless powder which can be recrystallized from dichloromethane. The resulting colorless blocks were analyzed via XRD. **3** crystallizes in the non-centrosymmetric monoclinic space group $P2_1$ as an enantiopure product very similar to the corresponding potassium complex [K(1,2-disila[18]crown-6)PF_6] according to the lattice constants [11]. As also observed for organic crown ether complexes, the ammonium cation is trapped in the cavity of the sila-crown ether beneath the anion bound to every second oxygen atom of the ligand **1**. The hydrogen bonding system of the ammonium cation now features three different binding modes due to the insertion of the O–Si–Si–O linkage and the presence of the $[PF_6]^-$ anion. Hence, etheric oxygen atoms, one silicon substituted oxygen atom, and two of the fluorine atoms of the $[PF_6]^-$ anion are participating (see Figure 1). The F···H contacts show distances that are well within the range of hydrogen bonding for this system. Freely chosen systems for comparison are [NH4([18]crown-6)]PF_6, $[NH_4$(dibenzo[18]crown-6)]PF_6, and $[NH_4$(benzo[15]crown-5)]PF_6 [32,36,37].

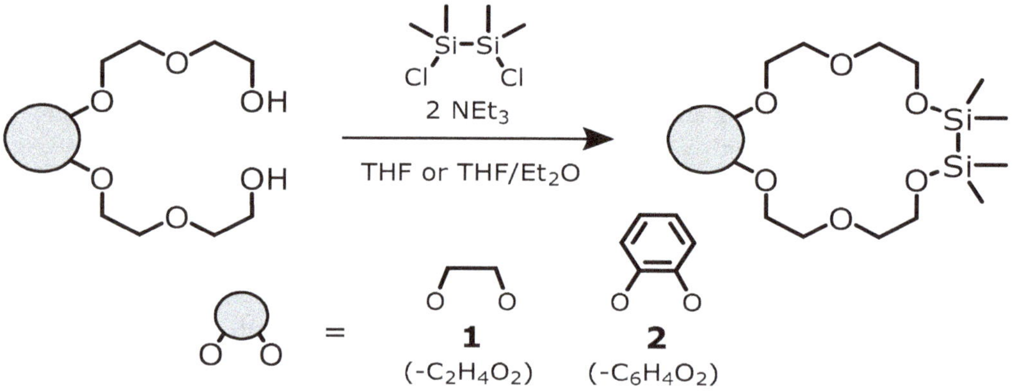

Scheme 2. Synthesis of 1,2-disila[18]crown-6 and 1,2-disila-benzo[18}crown-6 [11,15].

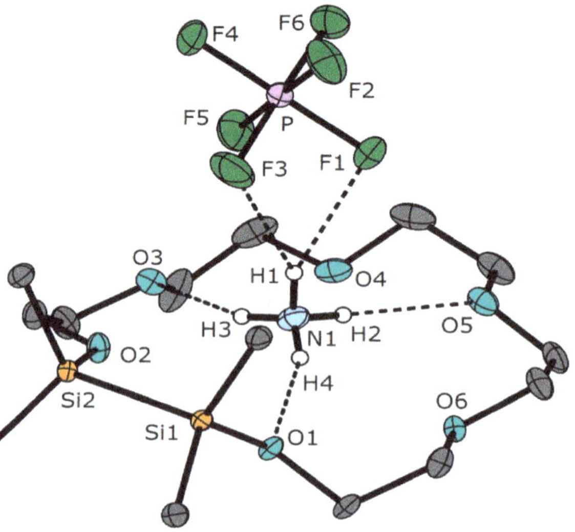

Figure 1. The molecular structure of compound **3** in the solid state. Thermal ellipsoids represent a probability level of 50%. The disordered part of the ligand with a lower occupancy and all carbon bonded hydrogen atoms are omitted for clarity, whereas hydrogen atoms of the ammonium cation are represented with arbitrary atomic radii. The latter were refined with DFIX [0.91] commands. Selected contacts and bond lengths [pm]: H1···F1 222(3), H1···F2 236(3), H2···O5 206(3), H3···O3 203(6), H4···O1 205(3), Si1–Si2 234.8(1).

Compound **3** is a rare example combining hydrogen bonding situations towards etheric oxygen atoms, as well as partially silicon substituted oxygen atoms in the same molecule. This enables a comparative analysis of the respective hydrogen bond donating properties of the different oxygen atoms. The hydrogen bonding data for **3** and related compounds-based on the representation in Scheme 1 are presented in Table 1. The respective hydrogen bonding patterns of **3** show no significant divergence between the $O_{(organic)}\cdots H$ and $O_{(Si)}\cdots H$ contacts. Hence, there is no hint of a preference of the etheric oxygen atoms to form hydrogen bonding. The N–H4···O1 contact has a rather short value of 205 pm, but is still in agreement with those in the related compounds. So, it can be assumed that the use of cooperative effects of ether and disilane bridges seems to be an effective way to incorporate guest molecules.

Table 1. Hydrogen bonding geometry in **3**, **4**, and related silicon based systems.

Compound or CSD Refcode [1]	D–H [2]	A	*d* [pm]	T [3]	ω [°]	ϕ [°]	Ref. [4]
3	N–H2	O5	206	C_E	160	113	*
	N–H3	O3	203	C_E	156	108	*
	N–H4	O1	205	C	159	122	*
4	N–H2	O4	214	C_E	165	117	*
	N–H3	O2	209	C_E	151	110	*
	N–H4	O6	200	C	172	118	*
VONMOB	N–H	O	226	C	174	128	[38]
MEQFOD	N–H	O	247	C	141	131	[39]
ITUBUI	N–H	O	198–210	C	157–160	130–132	[40]
MOLYUO	O–H	O	192	Si	167	116	[23]
TAKFOB	O–H	O	257–260	Si	148–152	139–148	[41]
ZEMXAQ	O–H	O	199	Si	156	116	[42]
EGEKAC	O–H	O	199–200	C	155–161	114–115	[43]
PERWIS	O–H	O	199	C	166	123	[44]
REXXAT	O–H	O	199	C	173	111	[45]

[1] Choice of CSD Codes is based on ref. [20]. Geometric criteria for the search were similar to those in ref [21]. CCDC ConQuest Version 1.19 was used for the search; [2] See Scheme 1 for abbreviations; [3] Subscript E denotes that the hydrogen bonding refers to an etheric oxygen atom: two carbon atoms are located next to one oxygen atom; [4] * = published within this work.

The successful incorporation of the ammonium cation in the silicon containing [18]crown-6 ether and its interesting bonding relations prompted us to synthesize another ammonium complex using the similar ligand **2**. Thereby, we obtained compound **4** as a white powder, which was further recrystallized from a mixture of dichloromethane and cyclopentane, yielding colorless rods suitable for single crystal X-ray diffraction analysis (see Figure 2 and Table 2). **4** crystallizes in the monoclinic space group *Cc* and reveals a trapped ammonium cation in the middle of the crown ether cavity bound to every second oxygen atom of the ligand **2**. The $[PF_6]^-$ anion and one molecule of co-crystalline DCM are located above and beneath the ammonium cation. In comparison to compound **3**, the ammonium cation is located closer to the calculated mean plane spanned by the donor atoms of the crown ether with 45 pm in the case of **4** and 59 pm in the case of **3**. The hydrogen bonding geometry of **4** is slightly different to that of **3** because of the rather rigid, *ortho*-bridging phenyl unit (see Table 1). Similar to the $O_{(organic)}\cdots H$ and $O_{(Si)}\cdots H$ in **3**, the respective hydrogen bonding contacts show no significant divergence. However, due to the incorporation of co-crystalline DCM, an intrinsic disorder causes problems with the crystal structure refinement. For this reason, several restraints on distances and anisotropic displacement parameters were used during the refinement. Hence, the hydrogen-bonding situation between the ligand and ammonium cation should be considered carefully and is not discussed in detail as done for **3**. Recrystallization attempts from other solvents failed. Nonetheless, the crystal structure clearly indicates the participation of the silicon bonded oxygen atom regarding hydrogen bonding.

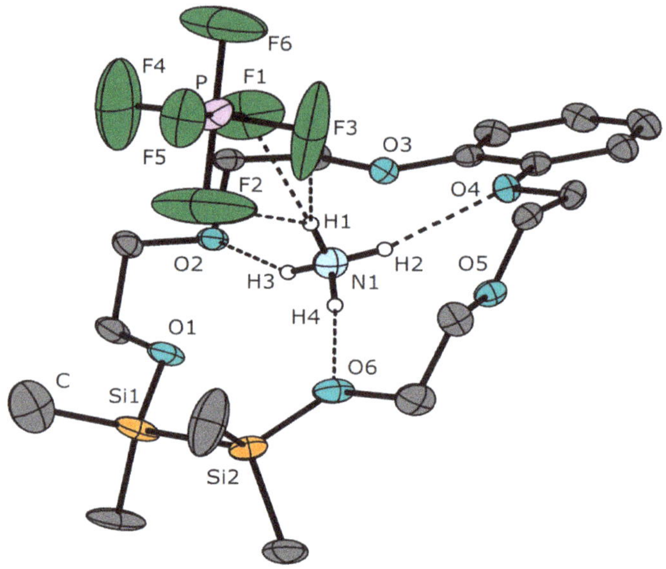

Figure 2. The molecular structure of compound **4** in the solid state. Thermal ellipsoids represent a probability level of 50%. The disordered part of the ligand with a lower occupancy and all carbon bonded hydrogen atoms, as well as the DCM molecule, are omitted for clarity. The hydrogen atoms of the ammonium cation are represented with arbitrary atomic radii. The latter were refined with DFIX [0.91] commands. Selected contacts and bond lengths [pm]: H1···F1 212(4) pm, H1···F2 247(5), H1···F3 279(5), H2···O4 213(6), H3···O2 209(5), H4···O6 200(3), Si1–Si2 233.9(3).

Table 2. X-ray crystallographic data for compounds **3** and **4**·DCM.

	3	**4·DCM**
Empirical formula	$C_{14}H_{36}F_6NO_6PSi_2$	$C_{19}H_{38}Cl_2F_6NO_6PSi_2$
Formula weight [g·mol^{-1}]	515.59	648.55
Crystal colour, shape	colourless block	colourless rod
Crystal size [mm]	0.487 × 0.432 × 0.346	0.076 × 0.106 × 0.522
Crystal system	monoclinic	monoclinic
Space group	$P2_1$	Cc
Formula units	2	4
Temperature [K]	100(2)	100(2)
Unit cell dimensions [Å, °]	a = 10.4542(5) b = 12.3869(6) c = 10.6140(5) β = 117.912(1)	a = 14.9582(7) b = 12.6454(6) c = 16.3325(8) β = 104.950(2)
Cell volume [Å^3]	1214.57(10)	2984.8(2)
ρ calc [g/cm^3]	1.410	1.443
μ [mm^{-1}]	0.286	0.422
2θ range [°]	4.342 to 50.568	4.280 to 53.570
Reflections measured	34532	68974
Independent reflections	4419	6355
R_1 (I > 2σ(I))	0.0203	0.0535
wR_2 (all data)	0.0529	0.1329
GooF	1.076	1.035
Largest diff. peak/hole [e·Å^{-3}]	0.24/−0.27	0.51/−0.44
Flack parameter	0.004(14)	0.048(16)

The interactions between the silicon affected oxygen atoms and the hydrogen atoms of the ammonium cation are further verified by NMR and IR experiments. As observed for the metal complexes of disila-crown ethers, a characteristic downfield shift of the singlet in the $^{29}Si\{^1H\}$ and the singlet of the $SiMe_2$-groups in the ^{1}H NMR spectra is observed for both compounds **3** and **4**. The ^{29}Si NMR shift is 13.7 ppm in the case of **3** and 16.2 ppm in the case of **4**. For this reason, both compounds show a dynamic process regarding the H-bonding situation. Rapid exchange results in the described equivalency of the silicon atoms. Even in VT NMR experiments, subsequently cooling the solution to 190 K did not result in an inequivalence of the $SiMe_2$ groups. The respective NMR shifts are in the range of sodium and potassium metal ion complexes of disila-crown ethers [11,14].

The ^{1}H NMR spectra represent the singlets for the $SiMe_2$ groups at 0.29 (**3**) and 0.30 (**4**) ppm, respectively. For the ammonium cations, triplets were observed at 6.44 ppm for **3** and at 6.64 ppm for **4** in the ^{1}H NMR spectra. As mentioned above, IR spectroscopic data also indicate an interaction of the ammonium cation with the silicon bonded oxygen atom. In comparison to pure NH_4PF_6, three instead of only one NH stretching vibrations are observed in both compounds. The respective signals are found at 3333 cm^{-1} in NH_4PF_6; 3317, 3188, and 3086 cm^{-1} in **3**; and 3298, 3225, and 3066 cm^{-1} in **4**, comprising a significant red-shift in the coordinated cases [46]. This is in accordance with the solid-state structures found upon single crystal X-ray diffraction analysis, as three different binding modes of the NH_4-related hydrogen atoms are revealed.

3. Materials and Methods

3.1. Laboratory Procedures and Techniques

All working procedures were performed by the use of Schlenk techniques under Ar gas. Solvents were dried and freshly distilled before use. Ammonium hexafluorophosphate was stored and handled under Ar atmosphere using a glovebox of MBRAUN-type. NMR spectra were recorded on a Bruker AV III HD 300 MHz or AV III 500 MHz spectrometer (Bruker, Ettlingen, Germany), respectively. The MestReNova package was used for analyzation [47]. Infrared (IR) spectra of the respective samples were measured using attenuated total reflectance (ATR) mode on a Bruker Model Alpha FT-IR (Bruker, Billerica, MA, USA) stored in the glove box. OPUS-software package was applied throughout [48]. ESI-MS spectrometry was performed with an LTQ-FT (Waltham, MA, USA) and LIFDI-MS with an AccuTOF-GC device (Akishima, Tokyo, Japan). Elemental analysis data cannot be provided due to the presence of fluorine in the samples, which harms the elemental analysis devices.

3.2. Crystal Structures

Single crystal X-ray experiments were carried out using a Bruker D8 Quest diffractometer (Bruker, Billerica, MA, USA) at 100(2) K with Mo Kα radiation and X-ray optics (λ = 0.71073). All structures were solved by direct methods and refinement with full-matrix-least-squares against F^2 using *SHELXT*- and *SHELXL*-2015 on the OLEX2 platform. Crystallographic data for compounds **3** and **4** are denoted as follows: CCDC Nos. 1589283 (**3**), 1589284 (**4**·DCM) [49–51]. Crystallographic information files (CIF, see Supplementary Materials) can be obtained free of charge from the Cambridge Crystallographic Data Centre (CCDC) (link: www.ccdc.cam.ac.uk/data_request/cif). Visualization of all structures was performed with Diamond software package Version 4.4.0 [52]. Thermal Ellipsoids are drawn at the 50% probability level.

3.3. Experimental Section

The sila-crown ethers 1,2-disila[18]crown-6 (**1**) and 1,2-disila-benzo[18]crown-6 (**2**) were synthesized according to methods reported in literature [11,15]. Compounds **3** and **4** were synthesized as follows.

[NH_4(1,2-disila[18]crown-6)]PF_6 (**3**): 106 mg of **1** (0.30 mmol, 1.0 eq) was dissolved in 15 mL of dichloromethane. A total of 59 mg of NH_4PF_6 (1.2 eq, 0.36 mmol) was then added. Stirring the suspension for 72 h gave a cloudy solution, which was filtered followed by the removal of the solvent under reduced pressure. The raw product was washed with 5 mL of *n*-pentane and dried in vacuo. A total of 147 mg of **3** was obtained as a pale white powder in 94% yield. For single crystal growth,

3 was dissolved in dichloromethane and the solvent was removed until saturation of the solution. Cooling to −32 °C yielded colorless blocks after a few days. ^{1}H NMR: (300 MHz, CD_2Cl_2) δ = 0.29 (s, 12H, Si(CH_3)$_2$), 3.55–3.88 (m, 4H, CH_2), 3.64 (s, 12H, CH_2), 3.74–3.76 (m, 4H, CH_2), 6.44 (t, $^1J_{NH}$ = 54 Hz, 4H, NH_4) ppm; ^{13}C{^{1}H} NMR: (75 MHz, CD_2Cl_2) δ = −1.2 (s, Si(CH_3)$_2$), 62.9 (s, CH_2), 70.7 (s, CH_2), 70.8 (s, CH_2), 70.9 (s, CH_2), 72.9 (s, CH_2) ppm; ^{19}F NMR: (283 MHz) δ = −72.5 (d, $^1J_{PF}$ = 713 Hz, PF_6) ppm; ^{29}Si{^{1}H} NMR: (99 MHz, CD_2Cl_2) δ = 13.7 (s, Si(CH_3)$_2$) ppm; ^{31}P NMR: (203 MHz, CD_2Cl_2) δ = −140.3 (hept, $^1J_{PF}$ = 713 Hz, PF_6) ppm. MS: LIFDI(+) m/z (%): 370.20784 [M–PF_6]$^+$ (100), IR (cm^{-1}): 3317 + 3188 + 3086 (m, br, $\tilde{\nu}_s$ N–H), 2891 (m), 1453 (m), 1425.98 (m), 1352 (w), 1249 (m), 1097 (s), 1075 (s), 1060 (s), 953 (s) 920 (s), 830 (vs), 791 (vs), 769 (s), 739 (s), 713 (s), 626 (m), 556 (s), 518 (w).

[NH_4(1,2-disila-benzo[18]crown-6)]PF_6 (**4**): 140 mg of **2** (0.35 mmol, 1.0 eq) was dissolved in 10 mL of Dichloromethane. Subsequent addition of 69 mg of NH_4PF_6 (0.35 mmol, 1.0 eq) and stirring of the suspension for three hours yielded a clear solution that was subsequently freed of the solvent. The raw product was well washed with 8 mL of n-pentane, followed by drying in vacuo. A total of 200 mg of **4** was obtained as a pale white powder in 95% yield. For single crystal growth, **4** was dissolved in 2 mL of dichloromethane, layered with 15 mL cyclopentane, and finally stored at −32 °C to obtain colorless rods after a few days. ^{1}H NMR: (300 MHz, CD_2Cl_2) δ = 0.30 (s, 12H, Si(CH_3)$_2$), 3.63–3.71 (m, 4H, CH_2), 3.74–3.83 (m, 4H, CH_2), 3.85–3.95 (m, 4H, CH_2), 4.15–4.23 (m, 4H, CH_2), 6.64 (t, $^1J_{NH}$ = 54 Hz, 4H, NH_4) ppm; 6.86–7.04 (m, 4H, C_6H_4). ^{13}C{^{1}H} NMR: (75 MHz, CD_2Cl_2) δ = 2.5 (s, Si(CH_3)$_2$), 62.02 (s, CH_2), 68.6 (s, CH_2), 69.6 (s, CH_2), 72.8 (s, CH_2), 113.6 (s, C_{Ar}H), 122.4 (s, C_{Ar}H), 148.3 (s, $C_{Ar,\,q}$) ppm; ^{19}F NMR: (283 MHz) δ = −72.9 (d, $^1J_{PF}$ = 712 Hz, PF_6) ppm; ^{29}Si{^{1}H} NMR: (99 MHz, CD_2Cl_2) δ = 16.2 (s, Si(CH_3)$_2$) ppm; ^{31}P NMR (203 MHz, CD_2Cl_2): −143.8 (hept, $^1J_{PF}$ = 713 Hz, PF_6) ppm. MS: ESI(+) m/z (%): 418.2080 [M–PF_6]$^+$ (100), IR (cm^{-1}): 3298, 3225, 3066 (m, br, $\tilde{\nu}_s$ N–H), 2946 (m), 2878 (m), 1594 (w), 1505 (m), 1454 (m), 1425 (m), 1249 (s), 1209 (s), 1121 (m) 1069 (s), 957 (m), 830 (vs), 791 (vs), 765 (vs), 738 (vs), 632 (w), 555 (vs).

4. Conclusions

In this work, the incorporation of guest molecules into disilane-bearing crown ethers was discussed. The complexation of ammonium cations by the ligands **1** and **2** turned out to be successful. Within the respective complexes **3** and **4**, H-bonding between a silicon affected oxygen atom and the ammonium cation was purposefully realized. So far, related H-bonding was only observed as an occasional occurrence in solid-state structures. In addition, $O_{(organic)}\cdots H$ and $O_{(Si)}\cdots H$ contacts show no significant divergence on a structural level. Hence, there is no hint for a preference of the etheric oxygen atoms to form hydrogen bonding. The interaction of the protons related to the ammonium cation with the silicon affected oxygen atoms was verified by NMR- and IR-spectroscopic experiments. It can be concluded that the use of cooperative effects of ethylene and disilane bridges is an effective way to incorporate guest molecules. We are currently aiming for the synthesis of all-disilane substituted crown ether analogs. Systems of the type SiSi–O–SiSi will help advance our understanding of the siloxane linkage in combination with hydrogen bonding.

Acknowledgments: This work was financially supported by the Deutsche Forschungsgemeinschaft (DFG). Fabian Dankert gratefully acknowledges the kind advice and meticulous data collection of Michael Marsch (X-ray department, Philipps-University Marburg). Fabian Dankert further thanks N. Mais for his help with synthetic work.

Author Contributions: Fabian Dankert performed the synthesis and analytics of compound **4**, interpretations of all the analytical data, contributed to the X-ray crystallographic refinement, and wrote the manuscript. Kirsten Reuter performed the synthesis of compound **3** and the collection of the respective analytical data. Carsten Donsbach accomplished the crystal structure solution and refinement of compound **4**. Carsten von Hänisch contributed to the interpretation and manuscript preparation and led the overarching research project.

References

1. Weinhold, F.; West, R. The nature of the silicon–oxygen bond. *Organometallics* **2011**, *30*, 5815–5824. [CrossRef]

2. Weinhold, F.; West, R. Hyperconjugative Interactions in Permethylated Siloxanes and Ethers: The Nature of the SiO Bond. *J. Am. Chem. Soc.* **2013**, *135*, 5762–5767. [CrossRef] [PubMed]
3. Ritch, J.S.; Chivers, T. Siliciumanaloga von Kronenethern und Cryptanden: Ein neues Kapitel in der Wirt-Gast-Chemie? *Angew. Chem.* **2007**, *119*, 4694–4697. [CrossRef]
4. Shambayati, S.; Blake, J.F.; Wierschke, S.G.; Jorgensen, W.L.; Schreiber, S.L. Structure and Basicity of Silyl Ethers: A Crystallographic and ab Initio Inquiry into the Nature of Silicon–Oxygen Interactions. *J. Am. Chem. Soc.* **1990**, *112*, 697–703. [CrossRef]
5. Cypryk, M.; Apeloig, Y. Ab Initio Study of Silyloxonium Ions. *Organometallics* **1997**, *16*, 5938–5949. [CrossRef]
6. Gillespie, R.J.; Robinson, E. Models of molecular geometry. *Chem. Soc. Rev.* **2005**, *34*, 396–407. [CrossRef] [PubMed]
7. Passmore, J.; Rautiainen, J.M. On The Lower Lewis Basicity of Siloxanes Compared to Ethers. *Eur. J. Inorg. Chem.* **2012**, *2012*, 6002–6010. [CrossRef]
8. Cameron, T.S.; Decken, A.; Krossing, I.; Passmore, J.; Rautiainen, J.M.; Wang, X.; Zeng, X. Reactions of a Cyclodimethylsiloxane $(Me_2SiO)_6$ with Silver Salts of Weakly Coordinating Anions; Crystal Structures of $[Ag(Me_2SiO)_6][Al]$ ($[Al] = [FAl\{OC(CF_3)_3\}_3]$, $[Al\{OC(CF_3)_3\}_4]$) and Their Comparison with $[Ag(18\text{-Crown-}6)]_2[SbF_6]_2$. *Inorg. Chem.* **2013**, *52*, 3113–3126. [CrossRef] [PubMed]
9. Ouchi, M.; Inoue, Y.; Kanzaki, T.; Hakushi, T. Ring-contracted Crown Ethers: 14-Crown-5, 17-Crown-6, and Their Sila-analogues. Drastic Decrease in Cation-binding Ability. *Bull. Chem. Soc. Jpn.* **1984**, *57*, 887–888. [CrossRef]
10. Inoue, Y.; Ouchi, M.; Hakushi, T. Molecular Design of Crown Ethers. 3. Extraction of Alkaline Earth and Heavy Metal Picrates with 14- to 17-Crown-5 and 17- to 22-Crown-6. *Bull. Chem. Soc. Jpn.* **1985**, *58*, 525–530. [CrossRef]
11. Reuter, K.; Buchner, M.R.; Thiele, G.; von Hänisch, C. Stable Alkali-Metal Complexes of Hybrid Disila-Crown Ethers. *Inorg. Chem.* **2016**, *55*, 4441–4447. [CrossRef] [PubMed]
12. Reuter, K.; Thiele, G.; Hafner, T.; Uhlig, F.; von Hänisch, C. Synthesis and coordination ability of a partially silicon based crown ether. *Chem. Commun.* **2016**, *52*, 13265–13268. [CrossRef] [PubMed]
13. Reuter, K.; Rudel, S.S.; Buchner, M.R.; Kraus, F.; von Hänisch, C. Crown ether complexes of alkali metal chlorides from SO_2. *Chem. Eur. J.* **2017**, *23*, 9607–9617. [CrossRef] [PubMed]
14. Reuter, K.; Dankert, F.; Donsbach, C.; von Hänisch, C. Structural Study of Mismatched Disila-Crown Ether Complexes. *Inorganics* **2017**, *5*, 11. [CrossRef]
15. Dankert, F.; Reuter, K.; Donsbach, C.; von Hänisch, C. A structural study of alkaline earth metal complexes with hybrid disila-crown ethers. *Dalton Trans.* **2017**, *46*, 8727–8735. [CrossRef] [PubMed]
16. West, R.; Whatley, L.S.; Lake, K.J. Hydrogen Bonding Studies. V. The Relative Basicities of Ethers, Alkoxysilanes and Siloxanes and the Nature of the Silicon–Oxygen Bond. *J. Am. Chem. Soc.* **1961**, *83*, 761–764. [CrossRef]
17. West, R.; Wilson, L.S.; Powell, D.L. Basicity of siloxanes, alkoxysilanes and ethers toward hydrogen bonding. *J. Organomet. Chem.* **1979**, *178*, 5–9. [CrossRef]
18. Popowski, E.; Schulz, J.; Feist, K.; Kelling, H.; Jancke, H. Basizität und 29Si-NMR-spektroskopische Untersuchungen von Ethoxysiloxanen. *Z. Anorg. Allg. Chem.* **1988**, *558*, 206–216. [CrossRef]
19. Yilgör, E.; Burgaz, E.; Yurtsever, E.; Yilgör, I. Comparison of hydrogen bonding in polydimethylsiloxane and polyether based urethane and urea copolymers. *Polymer* **2000**, *41*, 849–857. [CrossRef]
20. Jeffrey, G.A. *An Introduction to Hydrogen Bonding*; Oxford University Press: Oxford, UK, 1997.
21. Grabowsky, S.; Beckmann, J.; Luger, P. The Nature of Hydrogen Bonding Involving the Siloxane Group. *Aust. J. Chem.* **2012**, *65*, 785–795. [CrossRef]
22. Frolov, Y.L.; Voronkov, M.G.; Strashnikova, N.V.; Shergina, N.I. Hydrogen bonding involving the siloxane group. *J. Mol. Struct.* **1992**, *270*, 205–215. [CrossRef]
23. Grabowsky, S.; Hesse, M.F.; Paulmann, C.; Luger, P.; Beckmann, J. How to Make the Ionic Si–O Bond More Covalent and the Si–O–Si Linkage a Better Acceptor for Hydrogen Bonding. *Inorg. Chem.* **2009**, *48*, 4384–4393. [CrossRef] [PubMed]
24. Behr, J.P.; Dumas, P.; Moras, D. The H_3O^+ Cation: Molecular Structure of an Oxonium-Macrocyclic Polyether Complex. *J. Am. Chem. Soc.* **1982**, *104*, 4540–4543. [CrossRef]

25. Atwood, J.L.; Bott, S.G.; Means, C.M.; Coleman, A.W.; Zhang, H.; May, M.T. Synthesis of salts of the hydrogen dichloride anion in aromatic solvents. 2. Syntheses and crystal structures of [K·18-crown-6][Cl–H–Cl], [Mg·18-crown-6][Cl–H–Cl]$_2$, [H_3O·18-crown-6][Cl–H–Cl], and the Related [H_3O·18-crown-6][Br–H–Br]. *Inorg. Chem.* **1990**, *29*, 467–470. [CrossRef]
26. Atwood, J.L.; Bott, S.G.; Robinson, K.D.; Bishop, E.J.; May, M.T. Preparation and X-ray structure of [H_3O^+ ·18-crown-6]-[($H_5O_2^+$)(Cl^-)$_2$], a compound containing both H_3O^+ and $H_5O_2^+$ crystallized from aromatic solution. *J. Crystallogr. Spectrosc. Res.* **1991**, *21*, 459–462. [CrossRef]
27. Atwood, J.L.; Junk, P.C.; May, M.T.; Robinson, K.D. Synthesis and X-ray structure of [H_3O^+ · 18-crown-6] [Br–Br–Br^-]; a compound containing both H_3O^+ and a linear and symmetrical Br_3^- ion crystallized from aromatic solution. *J. Chem. Crystallogr.* **1994**, *24*, 243–245. [CrossRef]
28. Saleh, M.I.; Kusrini, E.; Fun, H.-K.; Teh, J.B.-J. Dicyclohexano-18-crown-6 hydroxonium tribromide. *Acta Cryst.* **2007**, *E63*, o3790–o3791. [CrossRef]
29. Kloo, L.; Svensson, P.H.; Taylor, M.J. Investigations of the polyiodides $H_3O·I_x$ (x = 3, 5 or 7) as dibenzo-18-crown-6 complexes. *J. Chem. Soc. Dalton Trans.* **2000**, 1061–1065. [CrossRef]
30. Cheng, M.; Liu, X.; Luo, Q.; Duan, X.; Pei, C. Cocrystals of ammonium perchlorate with a series of crown ethers: Preparation, structures, and properties. *CrystEngComm* **2016**, *18*, 8487–8496. [CrossRef]
31. Feinberg, H.; Columbus, I.; Cohen, S.; Rabinovitz, M.; Shoham, G.; Selig, H. Crystallographic evidence for different modes of interaction of BF_3/BF_4^- species with water molecules and 18-crown-6. *Polyhedron* **1993**, *12*, 2913–2919. [CrossRef]
32. Dapporto, P.; Paoli, P.; Matijašić, I.; Tuśek-Božić, L. Crystal structures of complexes of ammonium and potassium hexafluorophosphate with dibenzo-18-crown-6. Molecular mechanics studies on the uncomplexed macrocycle. *Inorg. Chim. Acta* **1996**, *252*, 383–389. [CrossRef]
33. Ponomarova, V.V.; Rusanova, J.A.; Rusanov, E.B.; Domasevitch, K.V. Unusual centrosymmetric structure of [M(18-crown-6)]$^+$ (M = Rb, Cs and NH_4) complexes stabilized in an environment of hexachloridoantimonate(V) anions. *Acta Crystallogr. Sect. C Struct. Chem.* **2015**, *71*, 867–872. [CrossRef] [PubMed]
34. Hurd, D.T. On the Mechanism of the Acid-catalyzed Rearrangement of Siloxane Linkages in Organopolysiloxanes. *J. Am. Chem. Soc.* **1955**, *77*, 2998–3001. [CrossRef]
35. Cypryk, M.; Apeloig, Y. Mechanism of the Acid-Catalyzed Si–O Bond Cleavage in Siloxanes and Siloxanols. A Theoretical Study. *Organometallics* **2002**, *21*, 2165–2175. [CrossRef]
36. Wu, D.H.; Wu, Q.Q. Ammonium hexafluoridophosphate-18-crown-6 (1/1). *Acta Crystallogr. Sect. E Struct. Rep. Online* **2010**, *66*, 2–9. [CrossRef] [PubMed]
37. Shephard, D.S.; Zhou, W.; Maschmeyer, T.; Matters, J.M.; Roper, C.L.; Parsons, S.; Johnson, B.F.G.; Duer, M.J. Ortsspezifische Derivatisierung von MCM-41: Molekulare Erkennung und Lokalisierung funktioneller Gruppen in mesoporösen Materialien durch hochauflösende Transmissionselektronenmikroskopie. *Angew. Chem.* **1998**, *110*, 2847–2851. [CrossRef]
38. Kociok-Köhn, G.; Molloy, K.C.; Price, G.J.; Smith, D.R.G. Structural characterisation of trimethylsilyl-protected DNA bases. *Supramol. Chem.* **2008**, *20*, 697–707. [CrossRef]
39. Ha, H.-J.; Choi, C.-J.; Ahn, Y.-G.; Yun, H.; Dong, Y.; Lee, W.K. Cycloaddition of Lewis Acid-Induced *N*-Methyleneanilines as Azadienes to 1,2-Bistrimethylsilyloxycyclobutene and Oxidative Ring Expansion to 1,2,4,5-Tetrahydro-1-benzazocine-3,6-diones. *J. Org. Chem.* **2000**, *65*, 8384–8386. [CrossRef] [PubMed]
40. Becker, B.; Baranowska, K.; Chojnacki, J.; Wojnowski, W. Cubane-like structure of a silanethiol—Primary amine assembly—A novel, unusual hydrogen bond pattern. *Chem. Commun.* **2004**, 620–621. [CrossRef] [PubMed]
41. Polishchuk, A.P.; Makarova, N.N.; Astapova, T.V. X-ray diffraction investigation of *trans*-2,8-dihydroxy-2,8-diphenyl-4,4′,6,6′,10,10′, 12,12′-octamethylcyclohexasiloxane and *trans*-2,8-dihydroxy-2,4,4′,6,6′,8,10,10′, 12,12′-decamethyl-5-carbahexacyclosiloxane. *Crystallogr. Rep.* **2002**, *47*, 798–804. [CrossRef]
42. Spielberger, A.; Gspaltl, P.; Siegl, H.; Hengge, E.; Gruber, K. Syntheses, structures and properties of dihydroxypermethylcyclosilanes and permethyloxahexasilanorbornanes. *J. Organomet. Chem.* **1995**, *499*, 241–246. [CrossRef]
43. Driver, T.G.; Franz, A.K.; Woerpel, K.A. Diastereoselective Silacyclopropanations of Functionalized Chiral Alkenes. *J. Am. Chem. Soc.* **2002**, *124*, 6524–6525. [CrossRef] [PubMed]

44. Myers, A.G.; Dragovich, P.S. A Reaction Cascade Leading to 1,6-didehydro[10]annulene → 1,5-dehydronaphthalene Cyclization Initiated by Thiol Addition. *J. Am. Chem. Soc.* **1993**, *115*, 7021–7022. [CrossRef]
45. Veith, M.; Rammo, A. Synthese und Struktur eines neuartigen molekularen λ^5-Organospirosilikats und dessen dynamisches Verhalten in Lösung. *J. Organomet. Chem.* **1996**, *521*, 429–433. [CrossRef]
46. Heyns, A.M.; van Schalkwyk, G.J. A study of the infrared and Raman spectra of ammonium hexafluorophosphate NH_4PF_6 over a wide range of temperatures. *Spectrochim. Acta Part A Mol. Spectrosc.* **1973**, *29*, 1163–1175. [CrossRef]
47. Willcott, M.R. MestRe Nova. *J. Am. Chem. Soc.* **2009**, *131*, 13180. [CrossRef]
48. *OPUS*; Version 7.2; Bruker Opt. GmbH: Ettlingen, Germany, 2012.
49. Sheldrick, G.M. *SHELXT*—Integrated space-group and crystal-structure determination. *Acta Crystallogr. Sect. A Found. Adv.* **2015**, *71*, 3–8. [CrossRef] [PubMed]
50. Sheldrick, G.M. Crystal structure refinement with *SHELXL*. *Acta Crystallogr. Sect. C Struct. Chem.* **2015**, *71*, 3–8. [CrossRef] [PubMed]
51. Dolomanov, O.V.; Bourhis, L.J.; Gildea, R.J.; Howard, J.A.K.; Puschmann, H. *OLEX2*: A complete structure solution, refinement and analysis program. *J. Appl. Crystallogr.* **2009**, *42*, 339–341. [CrossRef]
52. *Diamond*; Version 4.4.1; Crystal and Molecular Structure Visualization; Putz, H. & Brandenburg, K. GbR: Bonn, Germany, 2017.

5

(2-Pyridyloxy)silanes as Ligands in Transition Metal Coordination Chemistry

Lisa Ehrlich [1], Robert Gericke [1], Erica Brendler [2] and Jörg Wagler [1,*]

1 Institut für Anorganische Chemie, TU Bergakademie Freiberg, D-09596 Freiberg, Germany; ehrlichl@mailserver.tu-freiberg.de (L.E.); gericker.chemie@gmail.com (R.G.)
2 Institut für Analytische Chemie, TU Bergakademie Freiberg, D-09596 Freiberg, Germany; erica.brendler@chemie.tu-freiberg.de
* Correspondence: joerg.wagler@chemie.tu-freiberg.de

Abstract: Proceeding our initial studies of compounds with formally dative TM→Si bonds (TM = Ni, Pd, Pt), which feature a paddlewheel arrangement of four (*N*,*S*) or (*N*,*N*) bridging ligands around the TM–Si axis, the current study shows that the (*N*,*O*)-bidentate ligand 2-pyridyloxy (pyO) is also capable of bridging systems with TM→Si bonds (shown for TM = Pd, Cu). Reactions of $MeSi(pyO)_3$ with $[PdCl_2(NCMe)_2]$ and CuCl afforded the compounds $MeSi(\mu\text{-}pyO)_4PdCl$ (**1**) and $MeSi(\mu\text{-}pyO)_3CuCl$ (**2**), respectively. In the latter case, some crystals of the Cu(II) compound $MeSi(\mu\text{-}pyO)_4CuCl$ (**3**) were obtained as a byproduct. Analogous reactions of $Si(pyO)_4$, in the presence of HpyO, with $[PdCl_2(NCMe)_2]$ and $CuCl_2$, afforded the compounds $[(HpyO)Si(\mu\text{-}pyO)_4PdCl]Cl$ (**4**), $(HpyO)_2Si[(\mu\text{-}pyO)_2PdCl_2]_2$ (**5**), and $(HpyO)_2Si[(\mu\text{-}pyO)_2CuCl_2]_2$ (**6**), respectively. Compounds **1**–**6** and the starting silanes $MeSi(pyO)_3$ and $Si(pyO)_4$ were characterized by single-crystal X-ray diffraction analyses and, with exception of the paramagnetic compounds **3** and **6**, with NMR spectroscopy. Compound **2** features a pentacoordinate Si atom, the Si atoms of the other complexes are hexacoordinate. Whereas compounds **1**–**4** feature a TM→Si bond each, the Si atoms of compounds **5** and **6** are situated in an O_6 coordination sphere, while the $TMCl_2$ groups are coordinated to pyridine moieties in the periphery of the molecule. The TM–Si interatomic distances in compounds **1**–**4** are close to the sum of the covalent radii (**1** and **4**) or at least significantly shorter than the sum of the van-der-Waals radii (**2** and **3**). The latter indicates a noticeably weaker interaction for TM = Cu. For the series **1**, **2**, and **3**, all of which feature the Me–Si motif *trans*-disposed to the TM→Si bond, the dependence of the TM→Si interaction on the nature of TM (Pd(II), Cu(I), and Cu(II)) was analyzed using quantum chemical calculations, that is, the natural localized molecular orbitals (NLMO) analyses, the non-covalent interaction (NCI) descriptor, Wiberg bond order (WBO), and topological characteristics of the bond critical points using the atoms in molecules (AIM) approach.

Keywords: AIM; DFT; intermetallic bond; ^{29}Si NMR spectroscopy; X-ray diffraction

---

1. Introduction

The coordination number of tetravalent silicon can easily be enhanced (up to five or six) with the aid of monodentate or chelating ligands [1–4]. In some of our studies, we have also shown that late transition metals (Ni(II), Pd(II), and Pt(II)) may serve the role of a lone pair donor at hexacoordinate silicon, for example, in **I** and **II** (Chart 1) [5–8]. That kind of complexes with silicon as a lone pair acceptor in the coordination sphere of a transition metal (TM) thus complements TM silicon complexes with, e.g., silylene ligands, in which the Si atom is the formal lone pair donor, such as **III**, **IV**, and **V** [9–13], and the silyl complexes, in which the TM–Si bond is one out of four bonds to a tetravalent Si atom (e.g., **VI**, **VII**, **VIII**, and **IX**) [14–18]. In complex **IX** and some other compounds with group 9 metals [19–22], the 2-pyridyloxy (pyO) ligand was successfully utilized for stabilizing the TM–Si bond by forming two bridges over the heterodinuclear core. Some other compounds have been reported, in which one pyO ligand bridges the TM–Si bond (e.g., **X**, and some others) [23–25]. In all of these Si(μ-pyO)TM compounds, the ambidentate pyO ligand is bound to silicon via the Si–O bond, while the softer Lewis base (pyridine N atom) is TM bound. This structural motif should enable access to a new class of paddlewheel-shaped complexes with formally dative TM→Si bonds, in which the Si atom may carry a sterically more demanding group or alkyl group, because of the rather poor steric demand of the surrounding donor atoms (i.e., O atoms), whereas in the previously reported TM→Si paddlewheel complexes (such as **I** and **II**), the donor atom situation at silicon merely allowed for the presence of a small electronegative H-acceptor moiety (i.e., a halide).

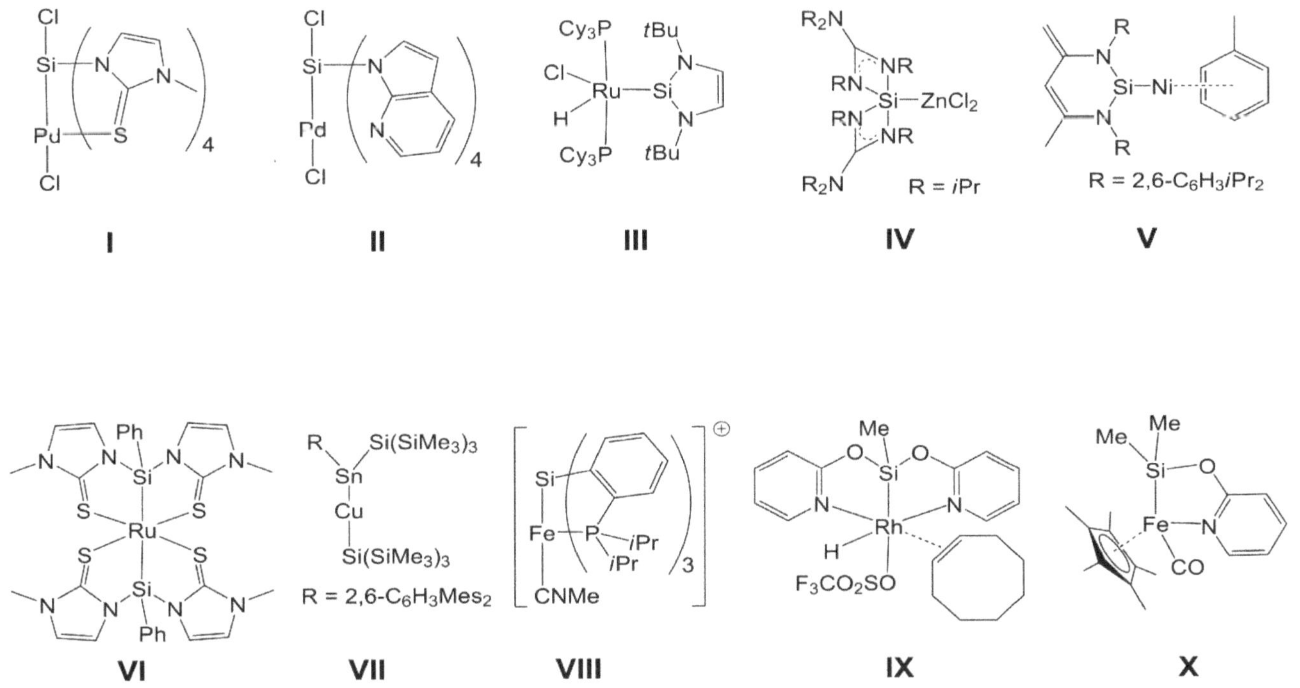

Chart 1. Selected metal–silicon complexes.

2. Results and Discussion

2.1. Syntheses and Characterization of Silanes $MeSi(pyO)_3$ and $Si(pyO)_4$

For the starting materials, we synthesized $MeSi(pyO)_3$ via triethylamine supported reaction of $MeSiCl_3$ and 2-hydroxypyridine, and $Si(pyO)_4$, via transsilylation, with the preceding synthesis of $Me_3Si(pyO)$, which is known in the literature [26] (Scheme 1). Both of the silanes formed crystals suitable for single-crystal X-ray diffraction analysis (Figure 1 and Table 1). In both of the compounds, the Si atom is essentially tetracoordinate and the coordination sphere may be described as (4 + 3) in $MeSi(pyO)_3$ and (4 + 4) in $Si(pyO)_4$, because of the pyridine N atoms, which are capping the faces of the tetrahedral coordination spheres from distances close to the sum of the van-der-Waals radii (ranging between 2.91 and 3.03 Å). This tetracoordination of Si in $Si(pyO)_4$ is in contrast to the Si hexacoordination in the thio analog $Si(pyS)_4$ [27] (and other pyS-bearing Si compounds [28]), in which two of the 2-mercaptopyridyl ligands form four-membered chelates, and in complexes of the *N*-oxide of the pyO system, which forms five-membered chelates [29,30].

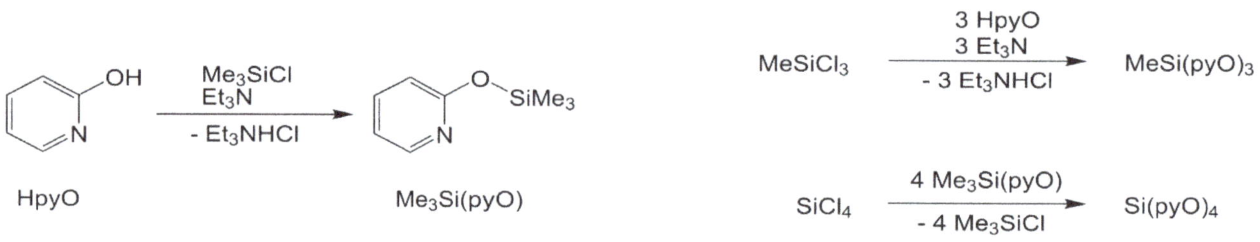

Scheme 1. Syntheses of starting materials $MeSi(pyO)_3$ and $Si(pyO)_4$.

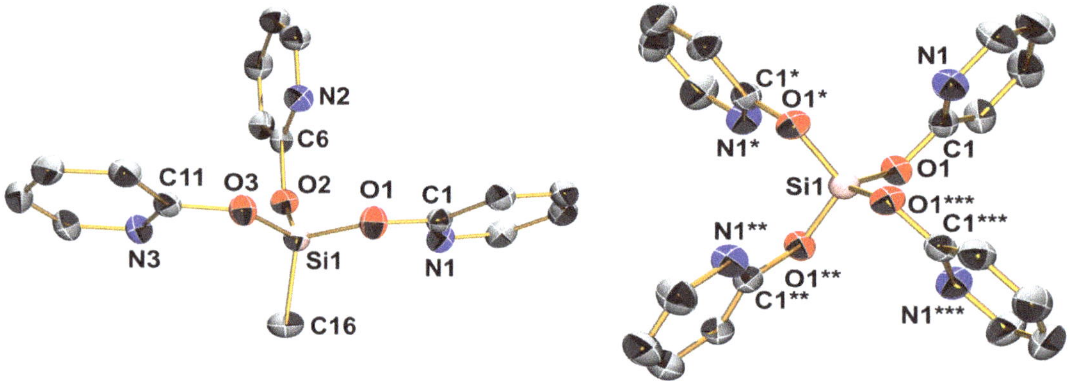

Figure 1. Molecular structures of $MeSi(pyO)_3$ and $Si(pyO)_4$ in the crystal; thermal displacement ellipsoids are drawn at the 50% probability level; H atoms are omitted for clarity and selected atoms are labeled. Because of the special crystallographic position of the Si atom of $Si(pyO)_4$ in the solid (S_4 symmetry), the asymmetric unit consists of 1/4 of the molecule. The asterisked labels indicate the symmetry equivalents. Selected interatomic distances (Å) and angles (deg) are as follows, for $MeSi(pyO)_3$: Si1–O1 1.645(1), Si1–O2 1.638(1), Si1–O3 1.652(1), Si1–C16 1.837(2), Si1···N1 2.998(1), Si1···N2 3.028(1), Si1···N3 2.920(1), O1–Si1–O2 114.68(5), O2–Si1–O3 112.74(5), O1–Si1–O3 97.77(5), O1–Si1–C16 111.96(7), O2–Si1–C16 106.46(6), O3–Si1–C16 113.72(6); for $Si(pyO)_4$: Si1–O1 1.630(1), Si1···N1 2.913(2); O1–Si1–O1* = O1*–Si1–O1** = O1**–Si1–O1*** = O1–Si1–O1*** 113.64(5), O1–Si1–O1** = O1*–Si1–O1*** 101.42(9).

The ^{29}Si NMR shifts ($MeSi(pyO)_3$ in $CDCl_3$: −46.5; $Si(pyO)_4$ in solid state: −87.9, in $CDCl_3$: −97.2) are in support of tetracoordination, as they are similar to (and even more downfield shifted than) the ^{29}Si NMR shifts of the related phenoxysilanes $MeSi(OPh)_3$ (−54.0) [31] and $Si(OPh)_4$ (−101.1) [32], respectively. Some effect of the three- or four-fold capped coordination spheres is evident from the bond angles of the Si atoms, which exhibit notable deviations from the ideal tetrahedral angle (in $MeSi(pyO)_3$, O1–Si1–O3 97.39(5)° and O2–Si1–O3 114.68(5)°; in $Si(pyO)_4$, two sets of O–Si–O angles of 101.42(9)° and 113.64(5)°). In general, the Si tetracoordination in these silanes, in combination with the vacant pyridine N atoms, should be favorable for complex formation with additional metal atoms. The same feature, tetracoordinate Si atom and vacant additional donor atoms, has also been encountered with methimazolylsilanes [5,33] and 7-azaindolylsilanes [8], which turned out to be suitable starting materials for compounds such as **I** and **II**.

2.2. Choice of Metals: Pd(II) and Cu(I)

In reactions with 7-azaindolylsilanes, Pd(II) already proved to be a suitable candidate for forming paddlewheel-complexes with TM→Si bonds and TM-bound pyridine moieties (e.g., in **II**). In addition to d^8 systems, other electron rich TMs may also be capable of forming complexes with TM→Si bonds, as shown by Bourissou et al. for the Au(I)→Si system (e.g., **XI**, Chart 2), in which the d^{10} metal is the electron pair donor [34–36]. Furthermore, for compounds **XII** (M = Cu, Ag, Au), Kameo et al. have shown that Cu(I) and Ag(I) are weaker donors [37]. In this context, we need to note that a previous study by Bourissou et al. [38] revealed an interaction between the Si–Si σ-bond electron pair of a disilane (as donor) and Cu(I) (as acceptor) (**XIII**). Similar systems (with rather weak d^{10} metal–silicon interaction) have been investigated for Ni(0), Pd(0), and Pt(0) by Grobe et al. (e.g., **XIV**) [39–41]. Whereas Au(I), Ni(0), Pd(0), and Pt(0) are more susceptible to phosphine ligands, Cu(I) represents a d^{10} system likely to bind to more than one or two N-donor ligands, and thus we included Cu(I) (as CuCl) in our investigations.

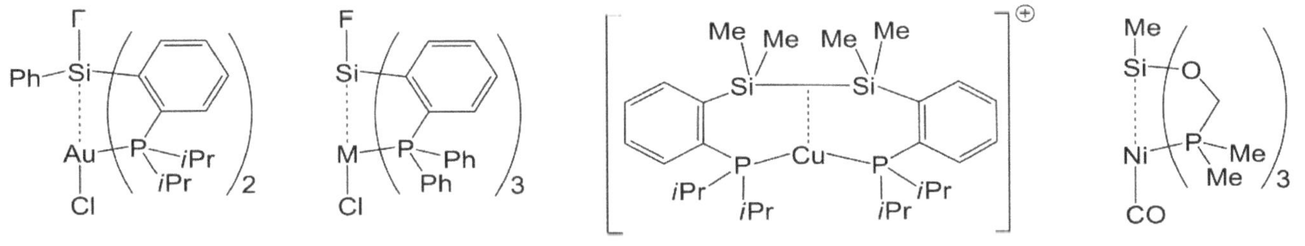

XI **XII** **XIII** **XIV**

Chart 2. Selected metal–silicon complexes with donor–acceptor-interactions between the Si atom of a tetravalent silane and a d^{10} metal atom.

2.3. *Reactions of MeSi(pyO)$_3$ with [PdCl$_2$(NCMe)$_2$] and CuCl*

The reactions of $MeSi(pyO)_3$ and $[PdCl_2(NCMe)_2]$ (in 1:1 molar ratio) in chloroform proceeded with the partial dissolution of $[PdCl_2(NCMe)_2]$ and the formation of a clear (slightly yellow, almost colorless) solution, from which the crystals of the chloroform solvate of compound **1** formed within one day (Scheme 2). The formal loss of Cl and the addition of a fourth pyO-bridge indicate ligand scrambling in the course of this reaction. The addition of excess $MeSi(pyO)_3$ (as sacrificial pyO source) eventually led to the complete dissolution of $[PdCl_2(NCMe)_2]$ and the formation of crystals of compound **1** · 2 $CHCl_3$ in good yield. From such a crystal, the molecular structure of **1** was determined using X-ray diffraction analysis (Figure 2 and Table 1). In principle, the molecule has paddlewheel architecture with four pyO ligands attached to Si (via Si–O bonds) and Pd (via Pd–N bonds) of a Me–Si–Pd–Cl axis. The idealized planes of the pyO ligands are slightly tilted against the Si–Pd axis (Pd–Si–O–C torsion angles ranging between 34.6(3)° and 26.3(3)°), and the axial angles (C21–Si1–Pd1 and Si1–Pd1–Cl1) exhibit some deviation from linearity (177.88(12)° and 177.36(3)°, respectively). The Si and Pd atoms are displaced from the O_4 and N_4 least-squares planes, respectively (into opposite directions), by 0.252(1) and 0.165(1) Å, respectively. The Pd–Si bond (2.6268(2) Å) is slightly longer than in the methimazolyl bridged paddlewheel complexes (where the Pd–Si bond lengths in the range 2.53–2.60 Å were observed) [7], and slightly shorter than in the 7-azaindolyl bridged paddlewheels [8]. The Si–C bond (1.853(4) Å) is only marginally longer than in the starting silane $MeSi(pyO)_3$ (1.837(2) Å), thus hinting at only a weak Pd→Si lone pair donor action. At hexacoordinate silicon with *trans*-disposed stronger donor (e.g., another hydrocarbyl group [42,43], 2-pyridinethiolato N atom [28] or 8-oxyquinolinyl N atom [44,45]), one would expect a Si–C bond lengthening beyond 1.90 Å. The Pd–Cl bond (2.891(1) Å), however, is unexpectedly long, thus hinting at ionic dissociation. This is supported by four H···Cl contacts with the pyO-H^6 atoms. The ^{29}Si NMR shift of compound **1** (−116.9 ppm in $CDCl_3$) is notably more upfield with respect to $MeSi(pyO)_3$ and $Si(pyO)_4$, thus indicating the hypercoordination of the Si atom. This ^{29}Si NMR shift, however, may be representative of either penta- or hexacoordinate silicon, and therefore the role of the sixth donor moiety (Pd→Si) requires further elucidation (vide infra).

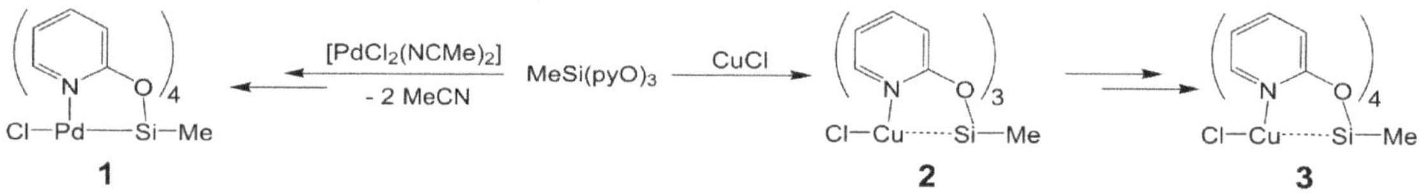

Scheme 2. Reactions of $MeSi(pyO)_3$ with $[PdCl_2(NCMe)_2]$ and CuCl.

Table 1. Crystallographic data from data collection and refinement for $MeSi(pyO)_3$, $Si(pyO)_4$, **1** · 2 $CHCl_3$, and **2**.

Parameter	$MeSi(pyO)_3$	$Si(pyO)_4$	1 · 2 $CHCl_3$	2
Formula	$C_{16}H_{15}N_3O_3Si$	$C_{20}H_{16}N_4O_4Si$	$C_{23}H_{21}Cl_7N_4O_4PdSi$	$C_{16}H_{15}ClCuN_3O_3Si$
M_r	325.40	404.46	800.08	424.39
T(K)	200(2)	200(2)	180(2)	200(2)
λ(Å)	0.71073	0.71073	0.71073	0.71073
Crystal system	triclinic	tetragonal	monoclinic	triclinic
Space group	P-1	$I4_1/a$	$C2/c$	P-1
a(Å)	9.1581(7)	9.5163(7)	14.8719(5)	8.7497(4)
b(Å)	9.3250(7)	9.5163(7)	10.5112(5)	9.2334(5)
c(Å)	11.4078(9)	21.824(2)	39.3020(13)	23.5781(13)
α(°)	92.440(6)	90	90	88.255(4)
β(°)	109.582(6)	90	95.404(3)	89.283(4)
γ(°)	116.896(6)	90	90	68.654(4)
V(Å^3)	796.41(12)	1976.4(3)	6116.4(4)	1773.36(17)
Z	2	4	8	4
ρ_{calc}(g·cm^{-1})	1.36	1.36	1.74	1.59
$\mu_{Mo\ K\alpha}$ (mm^{-1})	0.2	0.2	1.3	1.5
F(000)	340	840	3184	864
θ_{max}(°), R_{int}	28.0, 0.0263	25.0, 0.0238	25.0, 0.0355	28.0, 0.0310
Completeness	99.9%	99.8%	99.9%	99.8%
Reflections collected	12193	3724	52652	28693
Reflns unique	3836	873	5379	8555
Restraints	0	0	18	0
Parameters	209	66	403	453
GoF	1.066	1.137	1.147	1.073
$R1$, $wR2$ [$I > 2\sigma(I)$]	0.0343, 0.0871	0.0404, 0.0934	0.0339, 0.0765	0.0350, 0.0858
$R1$, $wR2$ (all data)	0.0426, 0.0925	0.0567, 0.1062	0.0423, 0.0818	0.0444, 0.0900
Largest peak/hole (e·Å^{-3})	0.22, −0.31	0.16, −0.28	0.67, −0.65	0.77, −0.28

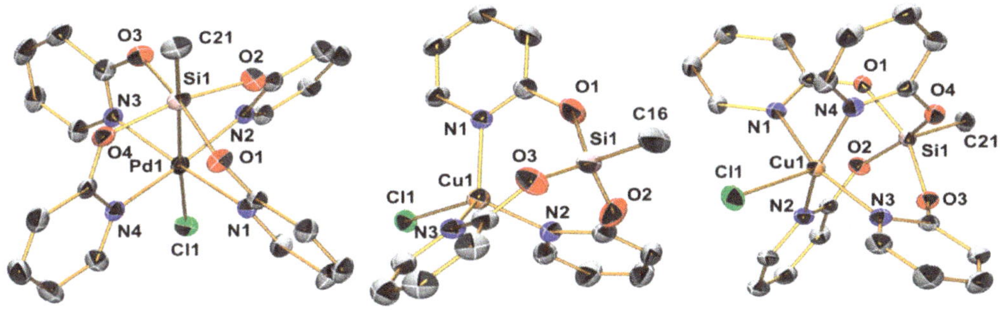

1 2 3

Figure 2. Molecular structures of compounds **1** (in **1** · 2 $CHCl_3$), **2**, and **3** in the crystal; thermal displacement ellipsoids are drawn at the 50% probability level; H atoms are omitted for clarity and selected atoms are labeled. For compound **2**, only one of the two crystallographically independent (but conformationally similar) molecules is depicted. The selected interatomic distances (Å) and angles (deg) for **1** are as follows: Pd1–Si1 2.627(1), Pd1–Cl1 2.891(1), Pd1–N1 2.023(3), Pd1–N2 2.039(3), Pd1–N3 2.028(3), Pd1–N4 2.013(3), Si1–O1 1.785(2), Si1–O2 1.775(2), Si1–O3 1.789(2), Si1–O4 1.780(2), Si1–C21 1.853(4), Pd1–Si1–C21 177.88(12), Cl1–Pd1–Si1 177.36(3), N1–Pd1–N3 170.10(11), N2–Pd1–N4 170.91(11), O1–Si1–O3 163.66(12), O2–Si1–O4 163.72(12); for **2**: Cu1–Cl1 2.361(1), Cu1–N1 2.038(2), Cu1–N2 2.039(2), Cu1–N3 2.023(2), Cu1···Si1 3.204(1), Si1–O1 1.626(2), Si1–O2 1.629(2), Si1–O3 1.618(2), Si1–C16 1.834(3); O1–Si1–C16 106.11(12), O2–Si1–C16 106.90(14), O3–Si1–C16 106.05(14), O1–Si1–O2 111.41(13), O1–Si1–O3 113.91(12), O2–Si1–O3 111.89(12); and for **3**: Cu1···Si1 2.919(1), Cu1–Cl1 2.403(1), Cu1–N1 2.023(2), Cu1–N2 2.055(2), Cu1–N3 2.025(2), Cu1–N4 2.049(2), Si1–O1 1.753(2), Si1–O2 1.757(2), Si1–O3 1.753(2), Si1–O4 1.755(2), Si1–C21 1.847(2); Cu1–Si1–C21 177.51(6), Cl1–Cu1–Si1 178.24(2), N1–Cu1–N3 159.32(5), N2–Cu1–N4 158.66(5), O1–Si1–O3 154.22(6), and O2–Si1–O4 155.83(6).

The reaction of $MeSi(pyO)_3$ with CuCl (Scheme 2) proceeds in the expected straightforward manner in a 1:1 molar ratio, i.e., CuCl dissolves in chloroform and in tetrahydrofuran (THF) in the presence of one mol equivalent of $MeSi(pyO)_3$, to afford an almost colorless (slightly greenish, by traces of Cu(II)) solution. Also, whereas the starting material produces a ^{29}Si NMR signal at −46.5 ppm (in $CDCl_3$), the resonance is shifted upfield for the solutions of $MeSi(pyO)_3$ with CuCl (−49.6 ppm for the THF solution, −64.1 ppm in $CDCl_3$). From the THF solution, some colorless crystals of the expected product **2** formed, which were suitable for single-crystal X-ray diffraction analysis (Figure 2 and Table 1). The crystal structure is comprised of two independent molecules that exhibit similar conformation, that is, a propeller with a Cl–Cu–Si–CH_3 axis and three pyO-bridges (bound to Si via Si–O bonds and to Cu via Cu–N bonds). The Cu···Si separation is rather long (ca. 3.2 Å), but it is shorter than in complex **XII** (M = Cu, Cu···Si 3.48 Å) [37], and the effect of Cu(I) on the Si coordination sphere is evident from the flattening of the SiO_3 pyramidal base (sum of angles ca. 337°). Furthermore, the ^{29}Si NMR shift of this solid is significantly upfield with respect to the starting silane (−70.0 and −71.4 ppm for the two crystallographically independent Si sites). In the $CDCl_3$ solution, compound **2** produces one set of broad ^{1}H NMR signals for the pyO moieties. Thus, this ^{1}H NMR signal broadening and the less pronounced upfield shift of the ^{29}Si NMR signal in the solution indicate conformational changes, such as coordination equilibria between isomers MeSi(μ-pyO)$_3$CuCl and Me(κO-pyO)$_2$Si(μ-pyO)CuCl, the latter with a weaker or absent Si···Cu interaction. Upon repeated opening/closing of the Schlenk flasks with solutions of **2** in THF or chloroform (e.g., for drawing NMR samples), some blue crystals of the related copper(II) compound **3** formed (Scheme 2) as solvent free crystals (from THF) and chloroform solvate (from the chloroform solution). The crystal structures of both compounds were determined using single-crystal X-ray diffraction (Table 2), and the molecular structure of **3** in the solvent free crystals (Figure 2) is included in the discussion as a representative example. This molecule combines features of both compounds **1** and **2**, as the Si···Cu separation is rather long (2.92 Å), thus being similar to compound **2**, but the axis of this paddlewheel complex is bridged by four pyO ligands. The Si–C bond length is intermediate between those of **1** and **2**, but the Si–O bonds are almost as long as in **1**, presumably as a result of the *trans*-arrangement of the Si–O bonds, but with a somewhat greater deviation of the O–Si–O axes from linearity in **3** (by ca. 25°). The Cu–N bond lengths in **3** are similar to those in **2**, and we attribute this similarity to a combination of two antagonist effects, that is, bond shortening by a higher oxidation state of the Cu atom and bond lengthening by *trans*-arrangement along the N–Cu–N axes. Deviations of the O–Si–O and N–Cu–N axes from linearity as well as the rather long Si···Cu separations are combined with displacement of Si and Cu atoms from the O_4 and N_4 least-squares planes, respectively, into opposite directions by 0.379(1) and 0.371(1) Å, respectively. Thus, the coordination spheres of both the Si and Cu atoms are best described as square–pyramidal. Unfortunately, the deliberate synthesis of larger amounts of **3** (by deliberate exposure of solutions of **2** to air) failed, and therefore we have not been able to isolate pure **3** for further spectroscopic or other investigation. Nonetheless, this compound represents a welcome link between compounds **1** and **2**, and therefore we performed computational analyses of the electronic situations in **1**, **2**, and **3**.

Table 2. Crystallographic data from data collection and refinement for **3**, **3** · $CHCl_3$, and **4** · 4 $CHCl_3$.

Parameter	3	3 · $CHCl_3$	4 · 4 $CHCl_3$
Formula	$C_{21}H_{19}ClCuN_4O_4Si$	$C_{22}H_{20}Cl_4CuN_4O_4Si$	$C_{29}H_{25}Cl_{14}N_5O_5PdSi$
M_r	518.48	637.85	1154.33
T(K)	200(2)	200(2)	200(2)
λ(Å)	0.71073	0.71073	0.71073
Crystal system	tetragonal	monoclinic	triclinic
Space group	$I4_1/a$	$P2_1$	P-1
a(Å)	18.3774(5)	9.2616(4)	11.2079(5)

Table 2. *Cont.*

Parameter	3	3 · $CHCl_3$	4 · 4 $CHCl_3$
b(Å)	18.3774(5)	15.6201(7)	13.1431(6)
c(Å)	26.4847(8)	9.2773(5)	15.1242(7)
α(°)	90	90	78.113(4)
β(°)	90	93.049(4)	87.403(4)
γ(°)	90	90	89.899(4)
V(Å^3)	8944.6(6)	1340.22(11)	2177.81(17)
Z	16	2	2
ρ_{calc}(g·cm^{-1})	1.54	1.58	1.76
$\mu_{Mo\ K\alpha}$ (mm^{-1})	1.2	1.3	1.4
F(000)	4240	646	1144
θ_{max}(°), R_{int}	28.0, 0.0425	28.0, 0.0302	27.0, 0.0271
Completeness	99.9%	99.9%	99.9%
Reflns collected	71,378	22,760	34,910
Reflns unique	5409	6449	9502
Restraints	0	1	12
Parameters	290	326	562
GoF	1.081	1.060	1.059
χ_{Flack}		−0.008(4)	
$R1$, $wR2$ [$I > 2\sigma(I)$]	0.0278, 0.0679	0.0269, 0.0624	0.0230, 0.0557
$R1$, $wR2$ (all data)	0.0355, 0.0712	0.0302, 0.0639	0.0275, 0.0577
Largest peak/hole (e·Å^{-3})	0.45, −0.25	0.44, −0.31	0.52, −0.48

2.4. Computational Analyses of the Pd→Si and Cu→Si Interactions in Compounds **1**, **2**, *and* **3**

For the following investigations, we used the crystallographically determined molecular structures of **1**, **2**, and **3** (Figure 2) as a starting point, followed by the optimization of the H atom positions for the isolated molecules in the gas phase.

Natural localized molecular orbital (NLMO) analyses were performed using DFT-(RO)B3LYP functional with an SDD basis set for Pd and Cu, 6-311+G(d) for C, Cl, H, N, O, and Si (for details see experimental section). Figure 3 shows the NLMOs identified for the TM→Si donor–acceptor σ-interaction, and Table 3 lists the selected features of these NLMOs. In all three of the cases, the NLMO analysis treated this interaction as a donation of a mainly TM localized lone pair into a vacant orbital at Si. The latter has a σ*(Si–Me) character, which reasons its rather poor acceptor qualities. Thus, in contrast to Pd→Si–Cl systems such as **I** and **II**, which feature Pd/Si contributions of 84%/12% [6] and 83%/15% [8], respectively, the corresponding σ-donor electron pair in **1** is more TM localized (91% Pd contribution). In Cu–Si complexes **2** and **3**, the corresponding lone pair is even more metal localized (ca. 98%), and has less than a 1% Si contribution. Thus, it resembles a d(z^2) orbital more closely. Regardless of the different oxidation states of the Cu atoms and the different Cu···Si separations, the characteristics of the Cu→Si NLMOs of these two compounds are surprisingly similar. As it was shown for the related Pd and Ni systems that TM→Si interactions of related complexes from first and second row TMs may be very similar (83% Ni, 13% Si contribution to the Ni→Si NLMO in the Ni-analog of compound **I**) [6], we attribute the poor donicity of Cu in compounds **2** and **3** to the enhanced effective nuclear charge (group 11 instead of group 10 element) rather than to Cu being a first row transition metal.

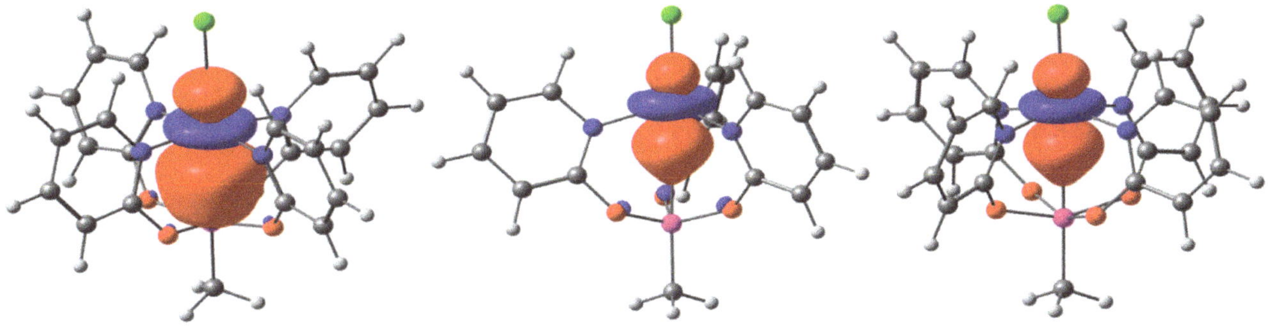

Figure 3. Natural localized molecular orbital (NLMO) representations of the TM→Si interaction in (from left) compounds **1**, **2**, and **3**. For compound **3**, the α-spin contribution is shown. NLMOs are depicted with an isosurface value of 0.02, the atom color code is consistent with Figure 2.

Table 3. Selected features of the natural localized molecular orbitals (NLMOs) of the TM→Si interaction in compounds **1**, **2**, and **3**.

Feature	1	2	3 [1]
% contribution TM	90.7	97.9	98.6
Hybrid (TM)	97.6% 4d, 2.2% 5s	99.6% 3d	99.5% 3d
% contribution Si	8.3	0.8	0.8
Hybrid (Si)	37.7% 3s, 61.7% 3p	18.2% 3s, 79.5% 3p	18.0% 3s, 79.7% 3p

[1] Contributions of α-spin and β-spin are essentially identical. TM—transition metal.

As the TM–Si interactions in compounds **1** and especially in **2** and **3** appear to be of an electrostatic/polar nature rather than covalent, the non-covalent interactions (NCI) descriptor of these compounds was analyzed (Figure 4). In compound **1**, a strong electrostatic non-covalent interaction between Pd and Si is detected, represented by a deep blue disc- or toroid-shaped area of the NCI along the Pd–Si bond. This feature is similar to the NCI along the Pd–Cl bond in this compound (and similar to the Cu–Cl bonds in compounds **2** and **3**). In sharp contrast, the light blue color of the NCI encountered along the Cu–Si paths in **2** and **3** hints at significantly weaker interactions, and their nature seems to be more closely related to the polar interactions of the pyO^6-hydrogen atoms with the metal bound chloride (i.e., weak hydrogen contacts).

In order to quantify these non-covalent interactions (TM···Si vs. C–H···Cl), topological analyses of the electron density distributions in compounds **1**, **2**, and **3** were performed using the atoms in molecules (AIM) approach. This AIM analysis of the wave function detected the bond critical points (BCPs) between Si and TM (TM = Pd, Cu) in all three of the complexes. Some of their features are listed in Table 4. For compounds **2** and **3**, the electron density $\rho(\mathbf{r_b})$ and the positive Laplacian $\nabla^2\rho(\mathbf{r_b})$ are of low magnitude, and the ratio $|V(\mathbf{r_b})|/G(\mathbf{r_b})$ slightly above 1 is indicative of an intermediate closed shell interaction with pronounced ionic contribution [46] (in accordance with the low Wiberg bond order (WBO) [47]).

The $|V(\mathbf{r_b})|/G(\mathbf{r_b})$ ratio in complex **1** is slightly greater than 2, and in combination with the negative Laplacian, it is indicative of a strong polarized shared shell interaction [48–50]. According to Espinosa et al. [51], the ratio $H(\mathbf{r_b})/\rho(\mathbf{r_b})$ can be utilized as a covalence degree parameter (for systems where $d < d_{cov}$, $|V(\mathbf{r_b})| > G(\mathbf{r_b})$, $H(\mathbf{r_b}) < 0$), the greater magnitude of which indicates the stronger atom–atom interaction (leading to an order of increasing strength **2** < **3** < **1**). For the series under investigation, the analysis of the estimated interaction energies (E_{int}) [52,53] yields an order of increasing E_{int}(TM–Si) in compounds **2** **3** < **1**. The estimated E_{int} and the ratio $H(\mathbf{r_b})/\rho(\mathbf{r_b})$ indicate a slightly stronger interaction in **3** relative to **2**. In accordance with the NCI, the Cu···Si interactions in **2** and **3** are indeed similar to the C–H···Cl contacts in these molecules in terms of energetics. Whereas the E_{int}(Cu···Si) in **2** and **3** were estimated to -2.3 and -3.7 kcal·mol^{-1}, respectively, the same approach of the analysis of $V(\mathbf{r_b})$ at the BCPs the C–H···Cl contacts in **1**, **2**, and **3** yielded an average E_{int} of -1.90, -1.65, and -2.02 kcal mol^{-1}, respectively.

Furthermore, for the paramagnetic compound **3**, this analysis afforded a Mulliken spin density distribution of 65.9% Cu-localization and 6.9–7.3%, located at each of the four nitrogen atoms. This is in accordance with the ligand field theory, which would assign the unpaired electron of a d^9 system in the square pyramidal coordination sphere to the $d(x^2-y^2)$ orbital, thus providing a $d(z^2)$-located lone pair for potential donor–acceptor interactions perpendicular to the $Cu(II)N_4$ plane in this particular case of compound **3**.

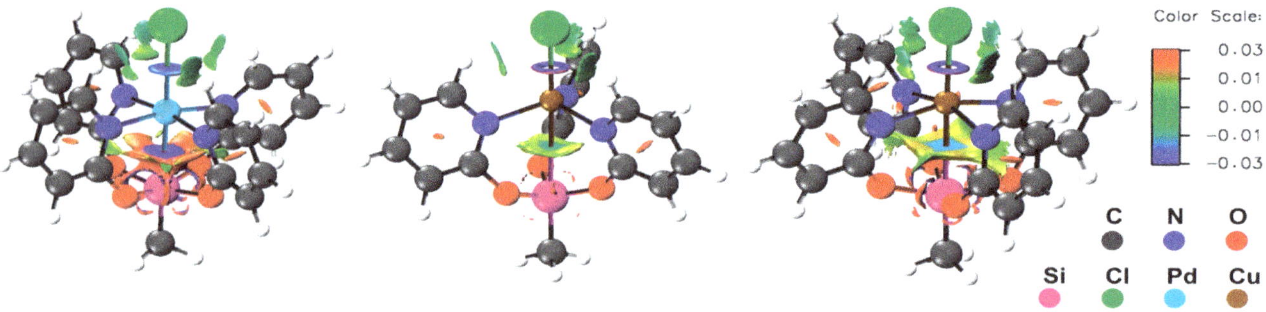

Figure 4. Non-covalent interaction (NCI) representations of (from left) compounds **1**, **2**, and **3** depicted with an isosurface value of 0.4 and a color range from -0.03 (blue, attractive) to 0.03 (red, repulsive).

Table 4. Selected features of the bond critical points (BCPs) of the TM–Si interaction in compounds **1**, **2**, and **3** derived from the topological analyses (AIM) of the wave function.

Feature [1]	1	2	3
$\rho(\mathbf{r_b})$	0.04461	0.01229	0.01742
$\nabla^2\rho(\mathbf{r_b})$	−0.00127	0.02835	0.02896
$G(\mathbf{r_b})$	0.01721	0.00721	0.00958
$V(\mathbf{r_b})$	−0.03475	−0.00734	−0.01192
$\mid V(\mathbf{r_b}) \mid / G(\mathbf{r_b})$	2.018	1.017	1.244
$G(\mathbf{r_b})/\rho(\mathbf{r_b})$	0.386	0.587	0.550
$H(\mathbf{r_b})$	−0.01753	−0.00012	−0.00234
$H(\mathbf{r_b})/\rho(\mathbf{r_b})$	−0.393	−0.010	−0.134
E_{int}	−10.9	−2.3	−3.7
WBO	0.270	0.057	0.037

[1] Electron density ($\rho(\mathbf{r_b})$ in a.u.), Laplacian of electron density ($\nabla^2\rho(\mathbf{r_b})$ in a.u.), Lagrangian kinetic energy density ($G(\mathbf{r_b})$ in a.u.), potential energy density ($V(\mathbf{r_b})$ in a.u.), ratio $\mid V(\mathbf{r_b}) \mid / G(\mathbf{r_b})$, ratio $G(\mathbf{r_b})/\rho(\mathbf{r_b})$ in a.u., electron energy density ($H(\mathbf{r_b})$ in a.u.), ratio $H(\mathbf{r_b})/\rho(\mathbf{r_b})$ in a.u., estimated interaction energy according to Lepetit et al. [53] and Espinosa et al. [52] E_{int} = 1/2 627.509469 $V(\mathbf{r_b})$ in kcal·mol^{-1}. WBO—Wiberg bond order.

2.5. Reactions of Si(pyO)$_4$ with [PdCl$_2$(NCMe)$_2$] and CuCl$_2$

In analogy to the formation of **I** and **II** from the respective silane Si(L)$_4$ (L = bridging ligand) and [PdCl$_2$(NCMe)$_2$], Si(pyO)$_4$ should be capable of forming paddlewheel complexes of the type ClSi(μ-pyO)$_4$TMCl upon reaction with TMCl$_2$, or suitable complexes thereof. Thus, we aimed at synthesizing ClSi(μ-pyO)$_4$PdCl (**4'**) through the reaction of [PdCl$_2$(NCMe)$_2$] and Si(pyO)$_4$ in chloroform (Scheme 3). As the main product, a beige solid of very poor solubility formed. In the dispersion of this fine solid in chloroform, some coarse crystals (beige, almost colorless) formed in the course of some days of storage at room temperature. Single-crystal X-ray diffraction analysis (Table 2, Figure 5) revealed the identity of this compound as the HpyO-adduct of the intended product (compound **4**, Scheme 3). Presumably, the intended product **4'** had formed initially and then reacted further with traces of free HpyO. The deliberate synthesis of **4** by reacting [PdCl$_2$(NCMe)$_2$], Si(pyO)$_4$, and HpyO in a 1:1:1 molar ratio in chloroform, eventually afforded this compound in good yield. The molecular structure of the cation [ClPd(μ-pyO)$_4$Si(HpyO)]$^+$ of **4** resembles the paddlewheel architecture of compound **1**, and notable differences that arise from the different substituent at Si (*trans* to Pd) are the following: The Pd–Si bond is noticeably shorter (2.50 Å) because of the more electronegative Si-bound substituent. As a consequence, the Si atom is less displaced from the O$_4$ least-squares plane (by 0.101(1) Å), whereas the displacement of the Pd atom from the N$_4$ plane (into the opposite direction, by 0.167(1) Å) is not altered. Interestingly, the Si1–O5 bond (*trans* to Pd) is significantly shorter than the equatorial Si–O bonds, thus hinting at still rather poor Pd→Si donor action. In spite of the NH group of the *trans*-Pd–Si located HpyO moiety, the C–O bonds of all five of the pyO and HpyO ligands are very similar, ranging between 1.313(2) and 1.325(2) Å. The ^{29}Si NMR shift of this compound (δ^{29}Si −147.9 ppm in CD$_2$Cl$_2$) is in accordance with the hexacoordination of the Si atom, and the ^{1}H and ^{13}C NMR spectra feature two sets of pyO-signals in a 4:1 intensity ratio, reflecting the four bridging and the dangling pyO moieties, respectively, and the retention of this molecular architecture in solution.

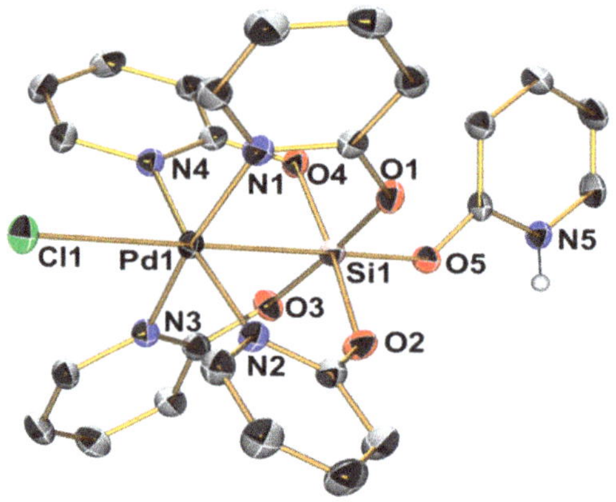

Figure 5. Molecular structure of the cation $[ClPd(\mu\text{-}pyO)_4Si(HpyO)]^+$ in the crystal structure of compound **4** · 4 $CHCl_3$; thermal displacement ellipsoids are drawn at the 50% probability level; C-bound H atoms are omitted for clarity and the selected atoms are labeled. Selected bond lengths (Å) and angles (deg) are as follows: Pd1–Si1 2.496(1), Pd1–Cl1 2.785(1), Pd1–N1 2.033(1), Pd1–N2 2.036(1), Pd1–N3 2.019(1), Pd1–N4 2.031(1), Si1–O1 1.765(1), Si1–O2 1.751(1), Si1–O3 1.758(1), Si1–O4 1.759(1), Si1–O5 1.709(1), Pd1–Si1–O5 178.57(5), Cl1–Pd1–Si1 178.87(2), N1–Pd1–N3 170.88(6), N2–Pd1–N4 170.04(5), O1–Si1–O3 173.01(6), and O2–Si1–O4 173.77(6).

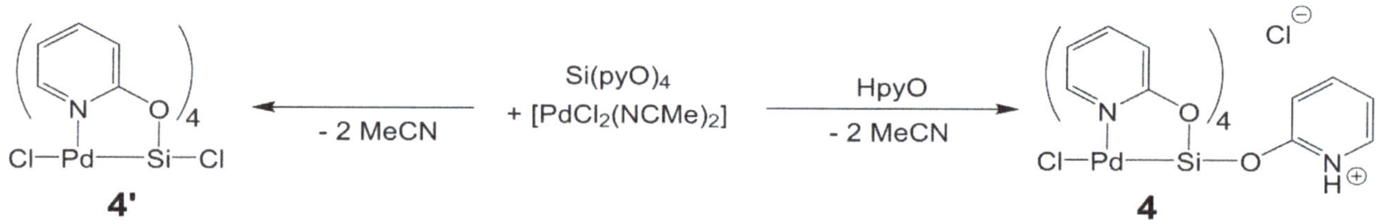

Scheme 3. Reactions of $Si(pyO)_4$ with $[PdCl_2(NCMe)_2]$.

Upon harvesting the crystalline solid **4** · 4 $CHCl_3$, some yellow crystals of another compound formed in the filtrate. Using single-crystal X-ray diffraction analysis, they were identified as the chloroform solvate **5** · 6 $CHCl_3$ of the complex **5** (Scheme 4). The deliberate synthesis of this complex, by using $[PdCl_2(NCMe)_2]$, $Si(pyO)_4$, and HpyO in a 2:1:2 molar ratio in chloroform, afforded compound **5** in good yield. Interestingly, the crystals that were initially formed during the synthesis consisted of a different solvate (**5** · 8 $CHCl_3$), and upon filtration, some more crystals of solvate **5** · 2 $CHCl_3$ formed in the filtrate. The elemental analysis of the final product upon drying was in agreement with the composition of the solvate **5** · 2 $CHCl_3$. Even though all three solvates (**5** · 8, 6, 2 $CHCl_3$, respectively) were characterized crystallographically (Table 5), only the molecular structure of **5** in the solvate **5** · 6 $CHCl_3$ is discussed as a representative example (Figure 6), as there are only marginal differences between the molecular conformations in the three different solvates.

The Si atom of compound **5** is, in the crystal structures, located on a center of inversion, thus the *trans* angles are 180°, and the very similar Si–O bond lengths and *cis* angles close to 90° (maximum deviations of ca. 1°) furnish an Si atom almost perfect octahedrally coordinated by six pyO oxygen atoms. Two sets of mutually *cis* situated anionic pyO ligands, the four O atoms of which are located in one plane, act as (*N*,*N*)-chelate donors toward $PdCl_2$, and the axial positions of the Si coordination sphere are occupied by the O atoms of HpyO, each of which establishes an N–H···O contact to an adjacent pyO oxygen atom (as indicated in Scheme 4). In spite of the very similar Si–O bond lengths, the C–O bonds of the HpyO moieties (1.303(3) Å) are significantly shorter than the C–O bonds of the bridging pyO moieties (1.328(3) Å), and thus exhibit a pronounced double bond character, as expected for HpyO. In the solvate **5** · 8 $CHCl_3$, these differences are even more pronounced with 1.296(3) vs. 1.331(3) and 1.336(3) Å. As expected, the Pd atoms are situated in a square planar coordination sphere (with a sum of *cis* angles of 360.3(1)°). Compound **5** exhibited too poor of a solubility for solution NMR characterization, but the ^{29}Si cross-polarization magic-angle-spinning (CP/MAS) NMR spectroscopy of the solid unequivocally confirmed the hexacoordination of the central Si atom (δ^{29}Si −190.2 ppm). This chemical shift is similar to other hexacoordinate Si complexes with SiO_6 coordination sphere (e.g., Si(acetylacetonate)$_2$(salicylate) δ^{29}Si −191.7 ppm) [54].

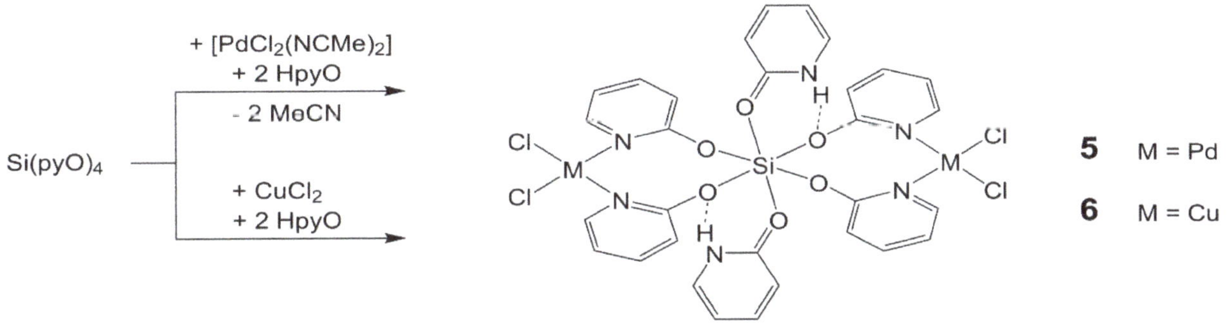

Scheme 4. Reactions of Si(pyO)$_4$ and HpyO with [PdCl$_2$(NCMe)$_2$] and CuCl$_2$.

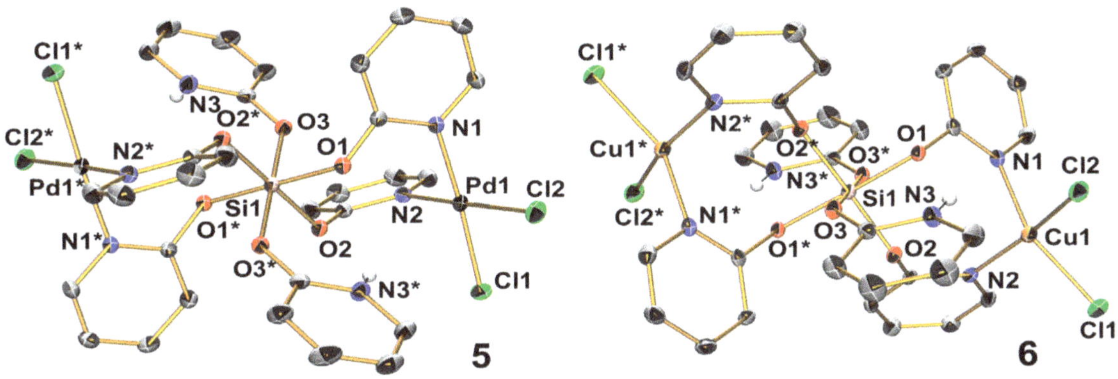

Figure 6. Molecular structures of **5** and **6** in the crystal (**5** in the structure of solvate **5** · 6 CHCl$_3$); thermal displacement ellipsoids are drawn at the 30% probability level; C-bound H atoms are omitted for clarity and selected atoms are labeled. In both cases, the Si atom is located on a crystallographically imposed center of inversion, and therefore the asymmetric unit consists of a half molecule and the symmetry related sites (e.g., O1 and O1* at Si1) are *trans* to each other. Selected bond lengths (Å) and angles (deg) for **5** are as follows: Pd1–Cl1 2.290(1), Pd1–Cl2 2.303(1), Pd1–N1 2.028(2), Pd1–N2 2.036(2), Si1–O1 1.794(2), Si1–O2 1.760(2), Si1–O3 1.770(2), N1–Pd1–Cl1 175.93(7), N2–Pd1–Cl2 176.15(7), Cl1–Pd1–Cl2 91.60(3), Cl1–Pd1–N2 88.91(7), Cl2–Pd1–N1 87.65(7), and N1–Pd1–N2 92.10(9); for **6**: Cu1–Cl1 2.238(1), Cu1–Cl2 2.256(2), Cu1–N1 2.008(4), Cu1–N2 1.990(4), Si1–O1 1.790(3), Si1–O2 1.770(3), Si1–O3 1.755(3), N1–Cu1–Cl1 153.97(12), N2–Cu1–Cl2 162.50(12), Cl1–Cu1–Cl2 93.69(5), Cl1–Cu1–N2 90.82(11), Cl2–Cu1–N1 90.31(12), and N1–Cu1–N2 93.03(16).

In an attempt at synthesizing a paddlewheel complex of the composition ClCu(μ-pyO)$_4$SiCl, in comparison to the attempted synthesis of **4'** (Scheme 3, left), anhydrous CuCl$_2$ and Si(pyO)$_4$ were dispersed in chloroform. Whereas most of the reactants remained unchanged, the solution phase became blue and some blue crystals formed upon storage within one week. The crystals were identified as compound **6** by single-crystal X-ray diffraction analysis. Apparently, traces of HpyO in the sample gave rise to the formation of this Cu-analog of compound **5**. Deliberate synthesis, by using CuCl$_2$, Si(pyO)$_4$, and HpyO in a 2:1:2 molar ratio in chloroform, afforded compound **6** in good yield (Scheme 4). Because of the very poor solubility and the paramagnetic Cu(II) sites in this compound, NMR spectroscopic characterization was no option, and therefore we only discuss the molecular structure of this complex. The central parts of the molecule, that is, the bond lengths and angles of the hexacoordinate Si atom as well as the equatorial arrangement of two Si(μ-pyO)$_2$M clamps and two axially situated HpyO ligands, which establish N–H···O hydrogen bridges to adjacent pyO oxygen atoms, are very similar to the arrangement in compound **5**. The noteworthy difference is associated with the Cu(II) coordination sphere, which is distorted and intermediate between the tetrahedral and square planar. Then sum of the *cis* angles of Cu (367.9(2)°) deviates significantly from planarity. The N–Cu–N and Cl–Cu–Cl angles in particular are wider than 90°, and the angle between the CuN_2 and $CuCl_2$ planes is 30.7(2)°. Thus, in spite of the notable distortion, this coordination sphere is still closer to the square rather than tetrahedral.

Table 5. Crystallographic data from data collection and refinement for **5** · 2 $CHCl_3$, **5** · 6 $CHCl_3$, **5** · 8 $CHCl_3$, and **6**. R_{int} for the data set of **5** · 2 $CHCl_3$ was not reported in the refinement output because of the HKLF5 format used for twin refinement.

Parameter	**5 · 2 CHCl$_3$**	**5 · 6 CHCl$_3$**	**5 · 8 CHCl$_3$**	**6**
Formula	$C_{32}H_{28}Cl_{10}N_6O_6Pd_2Si$	$C_{36}H_{32}Cl_{22}N_6O_6Pd_2Si$	$C_{38}H_{34}Cl_{28}N_6O_6Pd_2Si$	$C_{30}H_{26}Cl_4Cu_2N_6O_6Si$
M_r	1187.99	1665.46	1904.20	863.54
T(K)	200(2)	200(2)	200(2)	200(2)
λ(Å)	0.71073	0.71073	0.71073	0.71073
Crystal system	monoclinic	triclinic	monoclinic	monoclinic
Space group	$C2/c$	P-1	$P2_1/n$	$P2_1/n$
a(Å)	22.458(2)	12.1698(6)	12.3681(5)	9.1063(8)
b(Å)	11.8102(7)	12.1951(7)	13.1356(4)	11.0475(6)
c(Å)	17.4818(18)	13.1643(8)	22.2337(9)	17.1744(17)
α(°)	90	107.904(4)	90	90
β(°)	111.330(7)	103.147(4)	95.990(3)	104.705(7)
γ(°)	90	114.671(4)	90	90
V(Å^3)	4319.1(7)	1539.81(19)	3592.4(2)	1671.2(2)
Z	4	1	2	2
ρ_{calc}(g·cm^{-1})	1.83	1.80	1.76	1.72
$\mu_{Mo\ K\alpha}$ (mm^{-1})	1.5	1.6	1.6	1.7
F(000)	2344	818	1868	872
θ_{max}(°), R_{int}	25.0, /	27.0, 0.0264	27.0, 0.0396	25.0, 0.0983
Completeness	99.8%	99.9%	99.9%	99.7%
Reflns collected	16169	18872	46707	16933
Reflns unique	3789	6728	7828	2933
Restraints	9	66	166	0
Parameters	289	410	539	226
GoF	1.034	1.055	1.033	0.924
$R1$, $wR2$ [$I > 2\sigma(I)$]	0.0474, 0.1009	0.0331, 0.0720	0.0299, 0.0663	0.0441, 0.0776
$R1$, $wR2$ (all data)	0.0958, 0.1166	0.0448, 0.0776	0.0399, 0.0699	0.1057, 0.0907
Largest peak/hole (e·Å^{-3})	0.76, −0.85	0.56, −0.58	0.46, −0.35	0.46, −0.72

3. Experimental Section

3.1. General Considerations

The commercially available chemicals (2-hydroxypyridine, anhydrous $CuCl_2$, Me_3SiCl, $MeSiCl_3$, and $SiCl_4$) were used as received without further purification. Chloroform ($CDCl_3$ stabilized with silver, $CHCl_3$ stabilized with amylenes) and CD_2Cl_2 were stored over activated molecular sieves (3 Å) for at least seven days. THF, diethyl ether, toluene, and triethylamine were distilled from sodium benzophenone. All of the reactions were carried out under an atmosphere of dry argon utilizing standard Schlenk techniques. 2-Trimethylsiloxypyridine [26] was synthesized according to a literature procedure. [$PdCl_2(NCMe)_2$] [8] and CuCl [55] were available in the laboratory from previous studies. The solution NMR spectra (^1H, ^{13}C, ^{29}Si) were recorded on Bruker Avance III 500 MHz and Bruker Nanobay 400 MHz spectrometers (Bruker Biospin, Rheinstetten, Germany) and Me_4Si was used as internal standard. The ^{29}Si (CP/MAS) NMR spectra were recorded on a Bruker Avance HD 400 WB spectrometer with 7 mm zirconia (ZrO_2) rotors and KelF inserts (compound **2**) or 4 mm zirconia rotors (compounds Si(pyO)$_4$ and **5**) at an MAS frequency of ν_{spin} = 5 kHz. The elemental analyses were performed on an Elementar Vario MICRO cube (Elementar, Hanau, Germany). The single-crystal X-ray diffraction data were collected on a Stoe IPDS-2T diffractometer (Stoe, Darmstadt, Germany) using Mo

Kα-radiation. The structures were solved by direct methods using *SHELXS*-97, and were refined with the full-matrix least-squares methods of F^2 against all reflections with *SHELXL*-2014 [56–58]. All of the non-hydrogen atoms were anisotropically refined. The C-bound hydrogen atoms were isotropically refined in an idealized position (riding model), the N-bound H atoms of compounds **4** · 4 $CHCl_3$, **5** · 2 $CHCl_3$, **5** · 6 $CHCl_3$, and **5** · 8 $CHCl_3$ were refined isotropically without restraints, and in the case of compound **6**, the isotropic displacement parameter of the N-bound H atom was set at the 1.2-fold mean displacement parameter of the pivot atom for stable refinement (because of the rather poor data set). The graphics of the molecular structures were generated with ORTEP-3 [59] and POV-Ray 3.6 [60]. CCDC 1869874 (MeSi(pyO)$_3$), 1869875 (Si(pyO)$_4$), 1869876 (**3** · $CHCl_3$), 1869877 (**3**), 1869878 (**6**), 1869879 (**5** · 2 $CHCl_3$), 1869880 (**2**), 1869881 (**5** · 8 $CHCl_3$), 1869882 (**1**· 2 $CHCl_3$), 1869883 (**5** · 6 $CHCl_3$), and 1869884 (**4** · 4 $CHCl_3$) contain the supplementary crystal data for this article (see Supplementary Materials). These data can be obtained free of charge from the Cambridge Crystallographic Data Centre via www.ccdc.cam.ac.uk/data_request/cif. For the computational analyses of compounds **1**, **2**, and **3**, the atomic coordinates from the crystallographically determined molecular structures were used for the non-hydrogen atoms. The geometry optimization of the H atom positions was carried out with *Gaussian*09 [60] using a DFT-PBEPBE functional and def2tzvpp basis set for all of the atoms. Subsequently, starting from these molecular structures, the NBO (natural bond orbital) and NLMO (natural localized bond orbital) calculations were performed using *Gaussian*09 [61] with the *NBO*6.0 package [62] using DFT-(RO)B3LYP functional (for Cu, Pd with SDD basis set; for C, H, N, O, Cl, and Si, with the 6-311+G(d) basis set including Douglas–Kroll–Hess second order scalar relativistic). The NLMO graphics were generated using *ChemCraft* [63]. The NCI [64], AIM [65], and Wiberg bond order [47] calculations were carried out using *MultiWFN* [66], using the same wave function that had been used for the NBO/NLMO calculations. The graphical representations of the NCI results were created using VMD [67].

3.2. *Syntheses*

MeSi(pyO)$_3$. A Schlenk flask with magnetic stirring bar was charged with 2-hydroxypyridine (1.50 g, 15.6 mmol), evacuated, and set under Ar atmosphere prior to adding THF (50 mL) and triethylamine (1.89 g, 18.7 mmol), with stirring to afford a colorless solution. With continuous stirring at room temperature, methyltrichlorosilane (0.82 g, 5.5 mmol) was added dropwise via syringe, while the simultaneous formation of a white precipitate (Et_3NHCl) was observed. Upon the complete addition of the silane stirring at room temperature was continued for 1 h, whereupon the hydrochloride precipitate was filtered off and washed with THF (2 $\times$ 5 mL). From the combined filtrate and washings, the solvent was removed under reduced pressure (condensed into a cold trap) to afford a crystalline residue, which was dissolved in hot THF (5 mL) and filtered prior to the addition of hexane (10 mL), and was stored at 6 °C. In the course of three days, colorless crystals of the product formed, which were separated by decantation (while cold) and were dried in vacuo (yield 0.90 g (2.7 mmol, 50%)). The crystals were suitable for X-ray diffraction analysis. The elemental analysis for $C_{16}H_{15}N_3O_3Si$ (325.39 g·mol^{-1}) was as follows: C, 59.06; H, 4.65; N, 12.91; found C, 56.37; H, 5.29; and N, 12.28. The composition found indicates hydrolysis upon sample preparation. The following was calculated for $C_{16}H_{18}N_4O_{5.5}Si$ (i.e., MeSi(pyO)$_3$ · 1.5 H_2O) (352.42 g·mol^{-1}): C, 54.53; H, 5.15; N, 11.92; multiplied with a mass correction factor of 1.035, which accounts for the uptake of similar amounts of water prior to and after weighing of the sample (C, 56.44; H, 5.33; N, 12.34). ^{1}H NMR ($CDCl_3$): δ (ppm) 0.92 (s, 3H, SiCH$_3$), 6.84–6.88 (m, 6H, H^3 and H^5), 7.53–7.57 (m, 3H, H^4), 8.07–8.08 (m, 3H, H^6); ^{13}C{^{1}H} NMR ($CDCl_3$): δ (ppm) −2.6 (SiCH$_3$), 113.1 (C^5), 118.0 (C^3), 139.2 (C^4), 147.4 (C^6), 160.6 (C^2); ^{29}Si{^{1}H} NMR ($CDCl_3$): δ (ppm) −46.5.

Si(pyO)$_4$. In a Schlenk flask with magnetic stirring bar 2-trimethylsiloxypyridine (3.00 g, 18.0 mmol) was dissolved in chloroform (4 mL), whereupon $SiCl_4$ (0.82 g, 4.8 mmol) was added (via syringe) with stirring. The resultant solution was heated with stirring under reflux for 2 h, to afford a white precipitate of the product. The mixture was allowed to attain room temperature, the solid was filtered, washed with chloroform (2 × 5 mL), and dried in vacuo. The yield was 1.08 g (2.67 mmol, 57%). The single-crystals suitable for X-ray diffraction analysis were obtained by recrystallization in THF. The elemental analysis for $C_{20}H_{16}N_4O_4Si$ (404.45 g·mol^{-1}) was as follows: C, 59.39; H, 3.99; N, 13.85; found C, 57.19; H, 4.90; N, 13.36. The composition found indicates hydrolysis upon sample preparation. The following was calculated for $C_{20}H_{20}N_4O_6Si$ (i.e., Si(pyO)$_4$ · 2 H_2O) (440.48 g·mol^{-1}): C, 54.53; H, 4.58; N, 12.72; multiplied with a mass correction factor of 1.05, which accounts for uptake of similar amounts of water prior to and after weighing of the sample: C, 57.26; H, 4.81; N, 13.36. ^{1}H NMR ($CDCl_3$): δ (ppm) 6.86–6.98 (m, 8H, H^3 and H^5), 7.54–7.58 (m, 4H, H^4), 8.01 (m, 4H, H^6); ^{13}C{^{1}H} NMR ($CDCl_3$): δ (ppm) 113.2 (C^5), 118.4 (C^3), 139.2 (C^4), 147.3 (C^6), 159.7 (C^2); ^{29}Si{^{1}H} NMR ($CDCl_3$): δ (ppm) −97.2, (CP/MAS): δ (ppm) −87.9.

ClPd(μ-pyO)$_4$SiMe · 2 $CHCl_3$ (complex **1** · 2 $CHCl_3$). A Schlenk flask was charged with a magnetic stirring bar, MeSi(pyO)$_3$ (0.40 g, 1.2 mmol) and [PdCl$_2$(NCMe)$_2$] (0.16 g, 0.62 mmol), evacuated, and set under Ar atmosphere prior to adding chloroform (3 mL). Upon brief stirring at room temperature (within two minutes), the starting materials dissolved completely to afford a light-yellow solution. Immediately after dissolution, the stirring was stopped and the solution was stored undisturbed at room temperature. In the course of one day, colorless crystals of the product formed, but the solution was stored for another week in order to complete crystallization, whereupon the crystals were separated from the supernatant by decantation and then briefly dried in vacuo. A single-crystal suitable for X-ray diffraction analysis was taken out of the mother liquor. The yield was 0.31 g (0.39 mmol, 65%). The elemental analysis for $C_{23}H_{21}Cl_7N_4O_4SiPd$ (800.11 g·mol^{-1}) was as follows: C, 34.53; H, 2.65; N, 7.00; found C, 37.81; H, 2.85; N, 8.35. The composition found indicates a loss of solvent upon drying and storage, the C, H, N values found correspond to the composition ClPd(μ-pyO)$_4$SiMe · 1.2 $CHCl_3$ (for $C_{22.2}H_{20.2}Cl_{4.6}N_4O_4SiPd$ (704.61 g·mol^{-1}): C, 37.84; H, 2.89; N, 7.95). ^{1}H NMR ($CDCl_3$): δ (ppm) 0.92 (s, 3H, CH_3) 6.66–6.67 (m, 8H, H^3 and H^5), 7.47–7.51 (m, 4H, H^4), 9.22–9.24 (m, 4H, H^6); ^{13}C{^{1}H} NMR ($CDCl_3$): δ (ppm) 3.5 ($SiCH_3$), 114.2 (C^5), 116.0 (C^3), 141.4 (C^4), 148.8 (C^6), 162.7 (C^2); ^{29}Si{^{1}H} NMR ($CDCl_3$): δ (ppm) −116.9.

ClCu(μ-pyO)$_3$SiMe (complex **2**). Procedure A: A Schlenk flask was charged with a magnetic stirring bar, MeSi(pyO)$_3$ (0.44 g, 1.4 mmol) and CuCl (0.14 g, 1.4 mmol), evacuated, and set under Ar atmosphere prior to adding THF (1.5 mL). Upon brief stirring at room temperature (within five minutes), the starting materials dissolved completely to afford a light turquoise (almost colorless) solution. ^{29}Si NMR spectroscopic analysis of this solution (with D_2O capillary used as lock) revealed one signal (at −49.6 ppm). In the course of one day, colorless crystals of the product formed, whereupon the crystals were separated from the supernatant by decantation, and then briefly dried in vacuo. A single-crystal suitable for X-ray diffraction analysis was taken out of the mother liquor. The yield was 0.12 g (0.28 mmol, 20%). The elemental analysis for $C_{16}H_{15}ClCuN_3O_3Si$ (424.39 g·mol^{-1}) was as follows: C, 45.28; H, 3.56; N, 9.90; found C, 43.44; H, 3.67; N, 9.44. The composition found indicates the presence of ca. 4% of "inert" materials (free of C, H, N), such as CuCl, as C, H, and N were found in the expected ratio. ^{29}Si{^{1}H} NMR (CP/MAS): δ (ppm) −70.0, −71.4 (intensity ratio 1:1).

Procedure B: A Schlenk flask was charged with a magnetic stirring bar, MeSi(pyO)$_3$ (0.30 g, 0.92 mmol) and CuCl (0.09 g, 0.92 mmol), evacuated, and set under Ar atmosphere prior to adding $CDCl_3$ (1.0 mL). Upon brief stirring at room temperature (within five minutes), the starting materials dissolved completely to afford a light turquoise (almost colorless) solution, which was used for the ^{1}H and ^{29}Si NMR spectroscopic analysis. Compound **2** did not crystallize from this solution. ^{1}H NMR ($CDCl_3$): δ (ppm) 0.71 (s, 3H, CH_3) 6.87–6.95 (m, 6H, H^3 and H^5), 7.66 (m, 3H, H^4), 8.78 (m, 3H, H^6); ^{29}Si{^{1}H} NMR ($CDCl_3$): δ (ppm) −64.1.

ClCu(μ-pyO)$_4$SiMe (complex **3**). In the Schlenk flasks with the crude product solutions of complex **2**, upon opening, some deep blue crystals formed over the course of some days. In both cases, the crystals were suitable for single-crystal X-ray diffraction analysis. From Procedure A, the solvent free variety **3** crystallized. From Procedure B, some crystals of the mono-chloroform solvate of **3** formed. The amount of crystals of solvent free **3** obtained was sufficient for elemental analysis. The elemental analysis for $C_{21}H_{19}ClCuN_4O_4Si$ (518.48 g·mol^{-1}) was as follows: C, 48.65; H, 3.69; N, 10.81; found C, 48.29; H, 3.92; N, 10.63.

[ClPd(μ-pyO)$_4$Si(HpyO)]Cl · 4 $CHCl_3$ (complex **4** · 4 $CHCl_3$). A Schlenk flask was charged with a magnetic stirring bar; Si(pyO)$_4$ (0.30 g, 0.74 mmol), [$PdCl_2$(NCMe)$_2$] (0.19 g, 0.74 mmol), and 2-hydroxypyridine (0.07 g, 0.74 mmol); evacuated; and set under Ar atmosphere prior to adding chloroform (3 mL). Upon stirring at room temperature, a beige dispersion formed, and the supernatant was orange. Over the course of one week, the finely dispersed powder transformed into a beige coarse crystalline product. A single-crystal suitable for X-ray diffraction analysis was taken out of the mother liquor. This solid was filtered, washed with chloroform (2 × 2 mL), and briefly dried in vacuo. The yield was 0.40 g (0.30 mmol, 47%). From the filtrate, some yellow crystals of compound **5** · 6 $CHCl_3$ formed, which were suitable for single-crystal X-ray diffraction analysis. The elemental analysis for $C_{29}H_{25}Cl_{14}N_5O_5SiPd$ (1154.38 g·mol^{-1}) was as follows: C, 30.17; H, 2.18; N, 6.07; found C, 30.43; H, 2.10; N, 6.94. ^{1}H NMR (CD_2Cl_2): δ (ppm) 6.82 (ddd, 4H, H^5, 7.1 Hz, 6.1 Hz, 1.3 Hz), 6.85 (ddd, 4H, H^3, 8.4 Hz, 1.3 Hz, 0.6 Hz), 7.40 (ddd, 1H, H^5, 7.2 Hz, 6.1 Hz, 1.1 Hz) 7.62 (ddd, 4H, H^4, 8.5 Hz, 7.0 Hz, 1.8 Hz), 7.76 (ddd, 1H, H^3, 8.8 Hz, 1.0 Hz, 0.7 Hz), 8.25 (ddd, 1H, H^4, 8.8 Hz, 7.2 Hz, 2.0 Hz), 8.53 (ddd, 1H, H^6, 6.1 Hz, 2.0 Hz, 0.7 Hz), 9.29 (ddd, 4H, H^6, 6.2 Hz, 1.8 Hz, 0.6 Hz), 15.82 (s, 1H, NH); ^{13}C{^{1}H} NMR (CD_2Cl_2): δ (ppm) bridging pyO: 114.2 (C^5), 117.0 (C^3), 142.3 (C^4), 148.5 (C^6), 162.1 (C^2), Si-bound HpyO: 117.7 (C^5), 118.6 (C^3), 139.2 (C^4), 145.8 (C^6), 158.3 (C^2); ^{29}Si{^{1}H} NMR (CD_2Cl_2): δ (ppm) −147.9, (CP/MAS): δ (ppm) −146.3.

Cl_2Pd(μ-pyO)$_2$(HpyO)Si(HpyO)(μ-pyO)$_2$PdCl_2 · 2 $CHCl_3$ (complex **5** · 2 $CHCl_3$). A Schlenk flask was charged with a magnetic stirring bar; Si(pyO)$_4$ (0.40 g, 0.99 mmol), [$PdCl_2$(NCMe)$_2$] (0.51 g, 2.0 mmol), and 2-hydroxypyridine (0.19 g, 2.0 mmol); evacuated; and set under Ar atmosphere prior to adding chloroform (3 mL). Upon stirring at room temperature, an orange dispersion formed. Within three days of undisturbed storage at room temperature, some yellow crystals formed, which were identified as the solvate **5** · 8 $CHCl_3$ by single-crystal X-ray diffraction analysis. To the mixture, further chloroform (2 mL) was added, and was briefly stirred prior to filtration and washing with chloroform (2 × 3 mL). The solid was dried in vacuo to afford a yellow powdery solid. In spite of the solvent rich solvate found in the crude mixture, the elemental analysis of the dried product corresponds to the approximate composition of **5** · 2 $CHCl_3$. The yield was 0.83 g (0.70 mmol, 71%). In the filtrate some yellow crystals formed which were identified as the solvate **5** · 2 $CHCl_3$ by single-crystal X-ray diffraction analysis. The elemental analysis for $C_{32}H_{28}Cl_{10}N_6O_6SiPd_2$ (1188.05 g·mol^{-1}) was as follows: C, 32.35; H, 2.38; N, 7.07; found C, 31.68; H, 2.23; N, 7.10. The solubility of the product in various organic solvents was not sufficient for solution NMR spectroscopic characterization. ^{29}Si{^{1}H} NMR (CP/MAS): δ (ppm) −190.2.

Cl_2Cu(μ-pyO)$_2$(HpyO)Si(HpyO)(μ-pyO)$_2$CuCl_2 (complex **6**). A Schlenk flask was charged with a magnetic stirring bar; Si(pyO)$_4$ (0.38 g, 0.95 mmol), $CuCl_2$ (0.25 g, 1.9 mmol), and 2-hydroxypyridine (0.18 g, 1.9 mmol); evacuated; and set under Ar atmosphere prior to adding chloroform (2.5 mL). Upon stirring at room temperature, a turquoise dispersion formed. After three days of storage at room temperature, further chloroform (3 mL) was added and was briefly stirred prior to filtration and washing with chloroform (2 × 3 mL). The solid was dried in vacuo to afford a blue powdery solid. The yield was 0.70 g (0.74 mmol, 78%). The elemental analysis indicates the presence of solvent (0.7 $CHCl_3$) in the product. The following was calculated for $C_{30}H_{26}Cl_4N_6O_6SiCu_2$ (947.12 g·mol^{-1}): C, 38.93; H, 2.84; N, 8.87; found C, 38.70; H, 3.08; N, 8.86. Some single-crystals of compound **6** were obtained from a mixture of Si(pyO)$_4$ (0.10 g, 0.25 mmol) and $CuCl_2$ (0.03 g, 0.025 mmol), which was layered with chloroform (1 mL) and stored undisturbed at room temperature for one week.

4. Conclusions

Two different 2-pyridyloxysilanes, MeSi(pyO)$_3$ and Si(pyO)$_4$, proved to be suitable starting materials for the syntheses of heteronuclear complexes, in which the ambidentate pyO ligand binds to

Si via Si–O, and to transition metals via TM–N bonds. The chelation of the Si atom by this ligand has not been encountered, thus the pyO nitrogen atoms are readily available for complex formation.

As shown for a d^8 and a d^{10} system (with formation of $ClPd(\mu\text{-}pyO)_4SiMe$ **1** and $ClCu(\mu\text{-}pyO)_3SiMe$ **2**, respectively), pyridyloxysilanes may support the formation of different paddlewheel structures, that is, with four and three bridging ligands, respectively. The pyO buttresses do not force Si and TM into close proximity, as it is evident from the different Pd–Si and Cu–Si separations in **1** and **2**, respectively. A higher coordination of the Si atom by the transition metal site still depends on the donicity of the TM site and the acceptor qualities of Si. As shown in the current study, the Si–Me moiety *trans* to an electron rich TM (Pd(II) in this particular case) causes weaker TM–Si interactions with respect to previously reported complexes with related TM→Si–Cl motif. Furthermore, group 11 metals (Cu(I) and Cu(II)) were shown to be significantly weaker lone pair donors toward Si(IV) than group 10 metals (e.g., Pd(II) and, according to previous studies, Ni(II) [6,7]). A systematic comparison of this series was enabled by the unintended formation of the decomposition product $ClCu(\mu\text{-}pyO)_4SiMe$ **3**. To our knowledge, compound **3** represents the first complex with a Cu(II)–Si(IV) bond (or at least with such a short (2.9 Å) Cu(II)···Si(IV) separation). As shown by the quantum chemical calculations, the Cu···Si interactions in compounds **2** and **3** are slightly stronger than the C–H···Cl hydrogen contacts encountered in the same molecules.

The reactions of $Si(pyO)_4$ with $[PdCl_2(NCMe)_2]$ and $CuCl_2$, in the presence of 2-hydroxypyridine HpyO, gave rise to the formation of an entirely different class of hypercoordinate Si-complexes of the type $(HpyO)_2Si[(\mu\text{-}pyO)_2TMCl_2]_2$, **5** and **6**, respectively. As the coordination spheres about TM vary between square planar and distorted tetrahedral in compounds **5** and **6**, respectively, this kind of complex architecture may turn out to be suitable for the complexation of various further TMX_2 moieties (X = halide, pseudo-halide etc.).

Author Contributions: L.E. and J.W. conceived and designed the experiments; L.E. performed the experiments (syntheses); R.G. the computational analyses; J.W. the single-crystal X-ray diffraction analyses; E.B. the solid state NMR spectroscopic analyses; J.W. wrote the paper.

Acknowledgments: We are grateful to Ute Groß and Brunhilde Süßner for performing the elemental analyses and to Beate Kutzner and Erik Wächtler for the solution NMR spectroscopic measurements.

References

1. Tacke, R.; Ribbeck, T. Bis(amidinato)- and bis(guanidinato) silylenes and silylenes with one sterically demanding amidinato or guanidinato ligand: Synthesis and reactivity. *Dalton Trans.* **2017**, *46*, 13628–13659. [CrossRef] [PubMed]
2. Peloquin, D.M.; Schmedake, T.A. Recent advances in hexacoordinate silicon with pyridine-containing ligands: Chemistry and emerging applications. *Coord. Chem. Rev.* **2016**, *323*, 107–119. [CrossRef]
3. Wagler, J.; Böhme, U.; Kroke, E. Higher-Coordinated Molecular Silicon Compounds. In *Structure and Bonding*; Scheschkewitz, D., Ed.; Springer: Berlin, Germany, 2014; Volume 155, pp. 29–105.
4. Levason, W.; Reid, G.; Zhang, W. Coordination complexes of silicon and germanium halides with neutral ligands. *Coord. Chem. Rev.* **2011**, *255*, 1319–1341. [CrossRef]
5. Wagler, J.; Brendler, E. Metallasilatranes: Palladium(II) and Platinum(II) as Lone-Pair Donors to Silicon(IV). *Angew. Chem. Int. Ed.* **2010**, *49*, 624–627. [CrossRef] [PubMed]
6. Truflandier, L.A.; Brendler, E.; Wagler, J.; Autschbach, J. ^{29}Si DFT/NMR Observation of Spin-Orbit Effect in Metallasilatrane Sheds Some Light on the Strength of the Metal → Si Interaction. *Angew. Chem. Int. Ed.* **2011**, *50*, 255–259. [CrossRef] [PubMed]
7. Autschbach, J.; Sutter, K.; Truflandier, L.A.; Brendler, E.; Wagler, J. Atomic Contributions from Spin-Orbit Coupling to ^{29}Si NMR Chemical Shifts in Metallasilatrane Complexes. *Chem. Eur. J.* **2012**, *18*, 12803–12813. [CrossRef] [PubMed]
8. Wahlicht, S.; Brendler, E.; Heine, T.; Zhechkov, L.; Wagler, J. 7-Azaindol-1-yl(organo)silanes and Their $PdCl_2$ Complexes: Pd-Capped Tetrahedral Silicon Coordination Spheres and Paddlewheels with a Pd-Si Axis. *Organometallics* **2014**, *33*, 2479–2488. [CrossRef]

9. Cade, I.A.; Hill, A.F.; Kämpfe, A.; Wagler, J. Five-Coordinate Hydrido-Ruthenium(II) Complexes Featuring N-Heterocyclic Silylene and Carbene Ligands. *Organometallics* **2010**, *29*, 4012–4017. [CrossRef]
10. Baus, J.A.; Mück, F.M.; Schneider, H.; Tacke, R. Iron(II), Cobalt(II), Nickel(II), and Zinc(II) Silylene Complexes: Reaction of the Silylene $[iPrNC(NiPr_2)NiPr]_2Si$ with $FeBr_2$, $CoBr_2$, $NiBr_2 \cdot MeOCH_2CH_2OMe$, $ZnCl_2$, and $ZnBr_2$. *Chem. Eur. J.* **2017**, *23*, 296–303. [CrossRef] [PubMed]
11. Junold, K.; Baus, J.A.; Burschka, C.; Vent-Schmidt, T.; Riedel, S.; Tacke, R. Five-Coordinate Silicon(II) Compounds with Si—M Bonds (M = Cr, Mo, W, Fe): Bis[*N,N'*-diisopropylbenzamidinato(−)]silicon(II) as a Ligand in Transition-Metal Complexes. *Inorg. Chem.* **2013**, *52*, 11593–11599. [CrossRef] [PubMed]
12. Schäfer, S.; Köppe, R.; Roesky, P.W. Investigations of the Nature of Zn^{II}–Si^{II} Bonds. *Chem. Eur. J.* **2016**, *22*, 7127–7133. [CrossRef] [PubMed]
13. Meltzer, A.; Präsang, C.; Milsmann, C.; Driess, M. The Striking Stabilization of $Ni^0(\eta^6$-Arene) Complexes by an Ylide-Like Silylene Ligand. *Angew. Chem. Int. Ed.* **2009**, *48*, 3170–3173. [CrossRef] [PubMed]
14. Hill, A.F.; Neumann, H.; Wagler, J. Bis(methimazolyl)silyl Complexes of Ruthenium. *Organometallics* **2010**, *29*, 1026–1031. [CrossRef]
15. Klett, J.; Klinkhammer, K.W.; Niemeyer, M. Ligand Exchange between Arylcopper Compounds and Bis(hypersilyl)tin or Bis(hypersilyl)lead: Synthesis and Characterization of Hypersilylcopper and a Stannanediyl Complex with a Cu–Sn Bond. *Chem. Eur. J.* **2009**, *5*, 2531–2536. [CrossRef]
16. Rittle, J.; Peters, J.C. N—H Bond Dissociation Enthalpies and Facile H Atom Transfers for Early Intermediates of Fe—N_2 and Fe—CN Reductions. *J. Am. Chem. Soc.* **2017**, *139*, 3161–3170. [CrossRef] [PubMed]
17. Suess, D.L.M.; Tsay, C.; Peters, J.C. Dihydrogen Binding to Isostructural S = 1/2 and S = 0 Cobalt Complexes. *J. Am. Chem. Soc.* **2012**, *134*, 14158–14164. [CrossRef] [PubMed]
18. Garcés, K.; Lalrempuia, R.; Polo, V.; Fernández-Alvarez, F.J.; García-Orduña, P.; Lahoz, F.J.; Pérez-Torrente, J.J.; Oro, L.A. Rhodium-Catalyzed Dehydrogenative Silylation of Acetophenone Derivatives: Formation of Silyl Enol Ethers versus Silyl Ethers. *Chem. Eur. J.* **2016**, *22*, 14717–14729. [CrossRef] [PubMed]
19. Lalrempuia, R.; Iglesias, M.; Polo, V.; Sanz Miguel, P.J.; Fernández-Alvarez, F.J.; Pérez-Torrente, J.J.; Oro, L.A. Effective Fixation of CO_2 by Iridium-Catalyzed Hydrosilylation. *Angew. Chem. Int. Ed.* **2012**, *51*, 12824–12827. [CrossRef] [PubMed]
20. Julián, A.; Jaseer, E.A.; Garcés, K.; Fernández-Alvarez, F.J.; García-Orduña, P.; Lahoz, F.J.; Oro, L.A. Tuning the activity and selectivity of iridium-NSiN catalyzed CO_2 hydrosilylation processes. *Catal. Sci. Technol.* **2016**, *6*, 4410–4417. [CrossRef]
21. Julián, A.; Guzmán, J.; Jaseer, E.A.; Fernández-Alvarez, F.J.; Royo, R.; Polo, V.; García-Orduña, P.; Lahoz, F.J.; Oro, L.A. Mechanistic Insights on the Reduction of CO_2 to Silylformates Catalyzed by Ir-NSiN Species. *Chem. Eur. J.* **2017**, *23*, 11898–11907. [CrossRef] [PubMed]
22. Sun, J.; Ou, C.; Wang, C.; Uchiyama, M.; Deng, L. Silane-Functionalized N-Heterocyclic Carbene—Cobalt Complexes Containing a Five-Coordinate Silicon with a Covalent Co—Si Bond. *Organometallics* **2015**, *34*, 1546–1551. [CrossRef]
23. Sato, T.; Okazaki, M.; Tobita, H.; Ogino, H. Synthesis, structure, and reactivity of novel iron(II) complexes with a five-membered chelate ligand κ^2(*Si,N*)-$SiMe_2O(2-C_5H_4N)$. *J. Organomet. Chem.* **2003**, *669*, 189–199. [CrossRef]
24. Kwok, W.-H.; Lu, G.-L.; Rickard, C.E.F.; Roper, W.R.; Wright, L.J. Tethered silyl complexes from nucleophilic substitution reactions at the Si–Cl bond of the chloro(diphenyl)silyl ligand in $Ru(SiClPh_2)(\kappa^2-S_2CNMe_2)(CO)(PPh_3)_2$. *J. Organomet. Chem.* **2004**, *689*, 2979–2987. [CrossRef]
25. Kanno, Y.; Komuro, T.; Tobita, H. Direct Conversion of a Si—C(aryl) Bond to Si—Heteroatom Bonds in the Reactions of η^3-α-Silabenzyl Molybdenum and Tungsten Complexes with 2-Substituted Pyridines. *Organometallics* **2015**, *34*, 3699–3705. [CrossRef]
26. Motherwell, W.B.; Storey, L.J. Some studies on nucleophilic trifluoromethylation using the shelf-stable trifluoromethylacetophenone-*N,N*-dimethyltrimethylsilylamine adduct. *J. Fluor. Chem.* **2005**, *126*, 491–498. [CrossRef]
27. Wächtler, E.; Gericke, R.; Kutter, S.; Brendler, E.; Wagler, J. Molecular structures of pyridinethiolato complexes of Sn(II), Sn(IV), Ge(IV), and Si(IV). *Main Group Met. Chem.* **2013**, *36*, 181–191. [CrossRef]
28. Baus, J.A.; Burschka, C.; Bertermann, R.; Fonseca Guerra, C.; Bickelhaupt, F.M.; Tacke, R. Neutral Six-Coordinate and Cationic Five-Coordinate Silicon(IV) Complexes with Two Bidentate Monoanionic *N,S*-Pyridine-2-thiolato(-) Ligands. *Inorg. Chem.* **2013**, *52*, 10664–10676. [CrossRef] [PubMed]

29. Koch, J.G.; Brennessel, W.W.; Kraft, B.M. Neutral and Cationic Bis-Chelate Monoorganosilicon(IV) Complexes of 1-Hydroxy-2-pyridinone. *Organometallics* **2017**, *36*, 594–604. [CrossRef]
30. Kraft, B.M.; Brennessel, W.W. Chelation and Stereodynamic Equilibria in Neutral Hypercoordinate Organosilicon Complexes of 1-Hydroxy-2-pyridinone. *Organometallics* **2014**, *33*, 158–171. [CrossRef]
31. Schraml, J.; Chvalovsky, V.; Magi, M.; Lippmaa, E. NMR study of organosilicon compounds. XI. The role of electronic and steric effects in silicon-29 NMR spectra of compounds with a silicon-oxygen-carbon group. *Collect. Czechoslov. Chem. Commun.* **1981**, *46*, 377–390. [CrossRef]
32. Schraml, J.; Brezny, R.; Cermak, J.; Chvalovsky, V. Silicon-29 and carbon-13 NMR spectra of some substituted bis(trimethylsiloxy)benzenes. *Collect. Czechoslov. Chem. Commun.* **1990**, *55*, 2033–2037. [CrossRef]
33. Wagler, J.; Heine, T.; Hill, A.F. Poly(methimazolyl)silanes: Syntheses and Molecular Structures. *Organometallics* **2010**, *29*, 5607–5613. [CrossRef]
34. Gualco, P.; Mallet-Ladeira, S.; Kameo, H.; Nakazawa, H.; Mercy, M.; Maron, L.; Amgoune, A.; Bourissou, D. Coordination of a Triphosphine-Silane to Gold: Formation of a Trigonal Pyramidal Complex Featuring Au$\rightarrow$Si Interaction. *Organometallics* **2015**, *34*, 1449–1453. [CrossRef]
35. Gualco, P.; Mercy, M.; Ladeira, S.; Coppel, Y.; Maron, L.; Amgoune, A.; Bourissou, D. Hypervalent Silicon Compounds by Coordination of Diphosphine-Silanes to Gold. *Chem. Eur. J.* **2010**, *16*, 10808–10817. [CrossRef] [PubMed]
36. Gualco, P.; Lin, T.-P.; Sircoglou, M.; Mercy, M.; Ladeira, S.; Bouhadir, G.; Pérez, L.M.; Amgoune, A.; Maron, L.; Gabbaï, F.P.; Bourissou, D. Gold–Silane and Gold–Stannane Complexes: Saturated Molecules as σ-Acceptor Ligands. *Angew. Chem. Int. Ed.* **2009**, *48*, 9892–9895. [CrossRef] [PubMed]
37. Kameo, H.; Kawamoto, T.; Bourissou, D.; Sakaki, S.; Nakazawa, H. Evaluation of the σ-Donation from Group 11 Metals (Cu, Ag, Au) to Silane, Germane, and Stannane Based on the Experimental/Theoretical Systematic Approach. *Organometallics* **2015**, *34*, 1440–1448. [CrossRef]
38. Gualco, P.; Amgoune, A.; Miqueu, K.; Ladeira, S.; Bourissou, D. A Crystalline σ Complex of Copper. *J. Am. Chem. Soc.* **2011**, *133*, 4257–4259. [CrossRef] [PubMed]
39. Grobe, J.; Lütke-Brochtrup, K.; Krebs, B.; Läge, M.; Niemeyer, H.-H.; Würthwein, E.-U. Alternativ-Liganden XXXVIII. Neue Versuche zur Synthese von Pd(0)- und Pt(0)-Komplexen des Tripod-Phosphanliganden $FSi(CH_2CH_2PMe_2)_3$. *Z. Naturforsch.* **2007**, *62*, 55–65. [CrossRef]
40. Grobe, J.; Wehmschulte, R.; Krebs, B.; Läge, M. Alternativ-Liganden. XXXII Neue Tetraphosphan-Nickelkomplexe mit Tripod-Liganden des Typs $XM'(OCH_2PMe_2)_n(CH_2CH_2PR_2)_{3-n}$ (M′ = Si, Ge; n = 0–3). *Z. Anorg. Allg. Chem.* **1995**, *621*, 583–596. [CrossRef]
41. Grobe, J.; Krummen, N.; Wehmschulte, R.; Krebs, B.; Läge, M. Alternativ-Liganden. XXXI Nickelcarbonylkomplexe mit Tripod-Liganden des Typs $XM'(OCH_2PMe_2)_n(CH_2CH_2PR_2)_{3-n}$ (M′ = Si, Ge; n = 0–3). *Z. Anorg. Allg. Chem.* **1994**, *620*, 1645–1658. [CrossRef]
42. Wagler, J.; Roewer, G.; Gerlach, D. Photo-Driven Si–C Bond Cleavage in Hexacoordinate Silicon Complexes. *Z. Anorg. Allg. Chem.* **2009**, *635*, 1279–1287. [CrossRef]
43. Gerlach, D.; Brendler, E.; Wagler, J. Hexacoordinate Silicon Compounds with a Dianionic Tetradentate (N,N',N,N')-Chelating Ligand. *Inorganics* **2016**, *4*, 8. [CrossRef]
44. Wächtler, E.; Kämpfe, A.; Krupinski, K.; Gerlach, D.; Kroke, E.; Brendler, E.; Wagler, J. New Insights into Hexacoordinated Silicon Complexes with 8-Oxyquinolinato Ligands: 1,3-Shift of Si-Bound Hydrocarbyl Substituents and the Influence of Si-Bound Halides on the 8-Oxyquinolinate Coordination Features. *Z. Naturforsch.* **2014**, *69*, 1402–1418. [CrossRef]
45. Brendler, E.; Wächtler, E.; Wagler, J. Hypercoordinate Silacycloalkanes: Step-by-Step Tuning of N$\rightarrow$Si Interactions. *Organometallics* **2009**, *28*, 5459–5465. [CrossRef]
46. Dinda, S.; Samuelson, A.G. The Nature of Bond Critical Points in Dinuclear Copper(I) Complexes. *Chem. Eur. J.* **2012**, *18*, 3032–3042. [CrossRef] [PubMed]
47. Wiberg, K.B. Application of the Pople-Santry-Segal CNDO Method to the Cyclopropylcarbinyl and Cyclobutyl cation and to Bicyclobutane. *Tetrahedron* **1967**, *24*, 1083–1096. [CrossRef]
48. Bianchi, R.; Gervasio, G.; Marabello, D. Experimental Electron Density Analysis of $Mn_2(CO)_{10}$: Metal-Metal and Metal-Ligand Bond Characterization. *Inorg. Chem.* **2000**, *39*, 2360–2366. [CrossRef] [PubMed]
49. Bianchi, R.; Gervasio, G.; Marabello, D. The experimental charge density in transition metal compounds. *C. R. Chim.* **2005**, *8*, 1392–1399. [CrossRef]

50. Macchi, P.; Proserpio, D.M.; Sironi, A. Experimental Electron Density in a Transition Metal Dimer: Metal-Metal and Metal-Ligand Bonds. *J. Am. Chem. Soc.* **1998**, *120*, 13429–13435. [CrossRef]
51. Espinosa, E.; Alkorta, I.; Elguero, J.; Molins, E. From weak to strong interactions: A comprehensive analysis of the topological and energetic properties of the electron density distribution involving systems. *J. Chem. Phys.* **2002**, *117*, 5529–5542. [CrossRef]
52. Espinosa, E.; Molins, E.; Lecomte, C. Hydrogen bond strengths revealed by topological analyses of experimentally observed electron densities. *Chem. Phys. Lett.* **1998**, *285*, 170–173. [CrossRef]
53. Lepetit, C.; Fau, P.; Fajerwerg, K.; Kahn, M.L.; Silvi, B. Topological analysis of the metal-metal bond: A tutorial review. *Coord. Chem. Rev.* **2017**, *345*, 150–181. [CrossRef]
54. Seiler, O.; Burschka, C.; Fenske, T.; Troegel, D.; Tacke, R. Neutral Hexa- and Pentacoordinate Silicon(IV) Complexes with SiO_6 and SiO_4N Skeletons. *Inorg. Chem.* **2007**, *46*, 5419–5424. [CrossRef] [PubMed]
55. Kämpfe, A.; Brendler, E.; Kroke, E.; Wagler, J. Tp*Cu(I)–CN–SiL2–NC–Cu(I)Tp*—A hexacoordinate Si-complex as connector for redox active metals via π-conjugated ligands. *Dalton Trans.* **2015**, *44*, 4744–4750. [CrossRef] [PubMed]
56. Sheldrick, G.M. *Program for the Solution of Crystal Structures*; shelxs-97; University of Göttingen: Göttingen, Germany, 1997.
57. Sheldrick, G.M. *Program for the Refinement of Crystal Structures*; shelxl-2014/7; University of Göttingen: Göttingen, Germany, 2014.
58. Sheldrick, G.M. A short history of SHELX. *Acta Crystallogr.* **2008**, *A64*, 112–122. [CrossRef] [PubMed]
59. Farrugia, L.J. ORTEP-3 for windows—A version of ORTEP-III with a graphical user interface (GUI). *J. Appl. Crystallogr.* **1997**, *30*, 565. [CrossRef]
60. POV-RAY (Version 3.6), Trademark of Persistence of Vision Raytracer Pty. Ltd., Williamstown, Victoria (Australia). Copyright Hallam Oaks Pty. Ltd., 1994–2004. Available online: http://www.povray.org/download/ (accessed on 22 December 2011).
61. Frisch, M.J.; Trucks, G.W.; Schlegel, H.B.; Scurseria, G.E.; Robb, M.A.; Cheeseman, J.R.; Scalmani, G.; Barone, V.; Mennucci, B.; Petersson, A.; et al. *Gaussian09*; revision E.01; Gaussian, Inc.: Wallingford, CT, USA, 2009.
62. Glendening, E.D.; Badenhoop, J.K.; Reed, A.E.; Carpenter, J.E.; Bohmann, J.A.; Morales, C.M.; Landis, C.R.; Weinhold, F. *NBO 6.0*; Theoretical Chemistry Institute, University of Wisconsin: Madison, WI, USA, 2013. Available online: http://nbo6.chem.wisc.edu/ (accessed on 16 August 2016).
63. Chemcraft ver. 1.8 (Build 164). 2016. Available online: http://www.chemcraftprog.com (accessed on 8 April 2016).
64. Johnson, E.R.; Keinan, S.; Mori-Sánchez, P.; Contreras-García, J.; Cohen, A.J.; Yang, W. Revealing Noncovalent Interactions. *J. Am. Chem. Soc.* **2010**, *132*, 6498–6506. [CrossRef] [PubMed]
65. Bader, R.F.W. *Atoms in Molecules*; Clarendon Press: Oxford, UK, 1994.
66. Lu, T.; Chen, F. Multiwfn: A Multifunctional Wavefunction Analyzer. *J. Comput. Chem.* **2012**, *33*, 580–592. [CrossRef] [PubMed]
67. Humphrey, W.; Dalke, A.; Schulten, K.J. VMD: visual molecular dynamics. *J. Mol. Gr.* **1996**, *14*, 33–38. Available online: http://www.ks.uiuc.edu/Research/vmd/ (accessed on 5 April 2018). [CrossRef]

6

Synthesis and Characterization of the Germathioacid Chloride Coordinated by an *N*-Heterocyclic Carbene

Yasunobu Egawa [1,*,†], Chihiro Fukumoto [1], Koichiro Mikami [2], Nobuhiro Takeda [1] and Masafumi Unno [1,*]

1 Department of Chemistry and Chemical Biology, Graduate School of Science and Technology, Gunma University, Kiryu, Gunma 376-8515, Japan; t161a085@gunma-u.ac.jp (C.F.); ntakeda@gunma-u.ac.jp (N.T.)

2 Functional Polymer Group, Sagami Chemical Research Institute, 2743-1 Hayakawa, Ayase, Kanagawa 252-1193, Japan; koichiro.mikami@sagami.or.jp

* Correspondence: nobu_egawa@jamstec.go.jp (Y.E.); unno@gunma-u.ac.jp (M.U.)

† Present Address: Frontier Research Group, Research and Development (R&D) Center for Marine Bioscience, Japan Agency for Marine-Earth Science and Technology (JAMSTEC), 2-15 Natsushima-cho, Yokosuka, Kanagawa 237-0061, Japan.

§ Dedicated to Prof. Dr. Robert West in honors of his 90th birthday.

Abstract: Carboxylic acid chlorides are useful substrates in organic chemistry. Many germanium analogues of carboxylic acid chloride have been synthesized so far. Nevertheless, all of the reported germathioacid chlorides use bidentate nitrogen ligands and contain germanium-nitrogen bonds. Our group synthesized germathioacid chloride, Ge(S)Cl{C_6H_3-2,6-Tip_2}(Im-*i*-Pr_2Me_2), using *N*-heterocyclic carbene (Im-*i*-Pr_2Me_2). As a result of density functional theory (DFT) calculation, it was found that electrons are localized on sulfur, and the germanium-sulfur bond is a single bond with a slight double bond property.

Keywords: germanium; germanethione; germathioacid chloride; *N*-heterocyclic carbines

1. Introduction

Heavier analogues of multiple bonded organic species have attracted the interest of many chemists in terms of comparisons for structures, physical properties, and reactivities [1–3]. For a long time, multiple bonds of higher row main group elements have been thought to be unstable due to the small overlap of π-orbitals. However, a breakthrough occurred in 1981 with the achievement of the synthesis of Si=C [4], S=Si [5], and P=P [6] bonds by taking advantage of the protection by bulky substituents. They have also opened a breach in the elemental bond theory. In recent years, increasing interest in the chemistry of double-bonded species between group 14 and 16 elements has emerged, since carbonyl compounds represent one of the most important functionalities in chemistry [3,7,8]. Among them, focusing on the double bond of germanium and sulfur, that is, germanetione, there is a history of research as shown below (Figure 1).

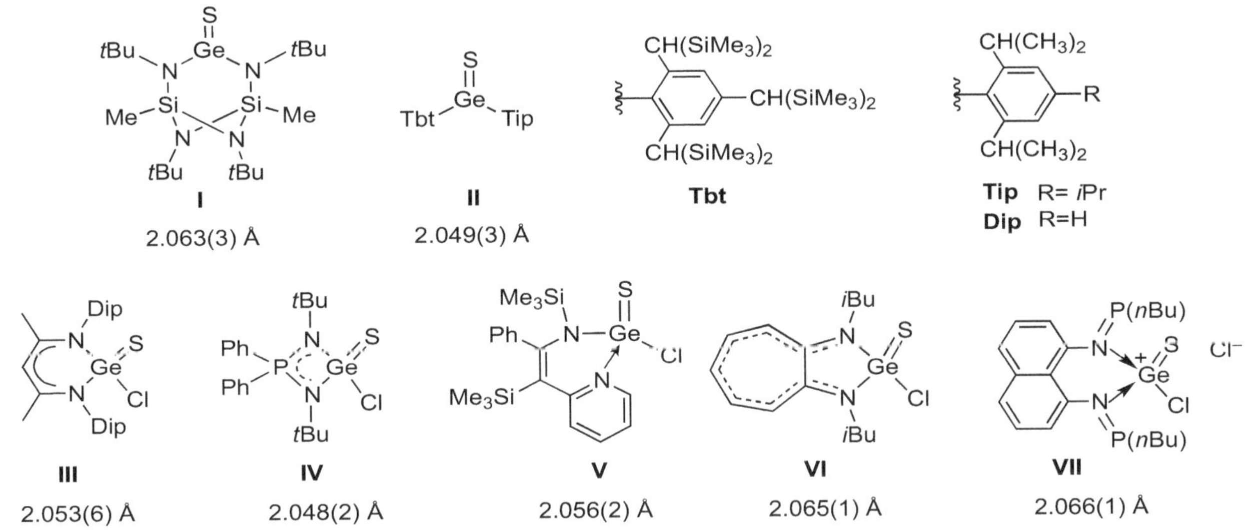

Figure 1. Ge=S-containing compounds and Ge–S bond length.

In 1989, Veith and co-workers synthesized germanethione (**I**) (germanium urea), which was stabilized by nitrogen ligand, and determined its structure by X-ray analysis [9]. After that, Okazaki, Tokitoh and co-workers isolated the first diarylgermanethione (**II**) using a very bulky protecting group and determined its structure [10,11]. Meanwhile, in organic chemistry, carboxylic acid chloride is an important compound group that serves as a substrate for the Friedel–Crafts reaction or Rosenmund reduction. The development of heavy analogues of carboxylic acid chloride was achieved by Roesky's group in 2002. They succeeded in the synthesis and isolation of the target substance (**III**) by thermodynamic stabilization using a π-diketiminate ligand [12]. Following their work, several results of germathioacid chloride syntheses using a bidentate nitrogen ligand have been reported (**IV–VII**) [13–16]. Nevertheless, due to the limitation of the use of a bidentate nitrogen ligand, all germathioacid chloride syntheses reported so far are only those in which nitrogen atoms and germanium are bonded; thus, all are heavy analogues of carbamoyl chloride.

In this paper, we report the synthesis and structure of germathioacid chloride stabilized with an NHC ligand. Among germathioacid chlorides, this is the first example containing a germanium carbon bond. We have clarified the bonding state of the germathioacid chloride by density functional theory (DFT) calculation.

2. Results and Discussion

2.1. *Synthesis and Structure of Germathioacid Chloride* 3

As shown in Scheme 1, 1,3-diisopropyl-4,5-dimethylimidazol-2-ylidene (Im-*i*-Pr_2Me_2)-substituted chlorogermylene (**2**) was prepared by the reaction of the Ge(Cl){C_6H_3-2,6-Tip_2} (**1**) with an equivalent amount of Im-*i*-Pr_2Me_2 in dry toluene. This synthesis method was based on Filippou's method [17]. Compound **2** was treated with sulfur element in dry benzene-d_6 at ambient temperature for 3 days, and after work up, the germathioacid chloride coordinated by Im-*i*-Pr_2Me_2 (**3**), Ge(S)Cl{C_6H_3-2,6-Tip_2}(Im-*i*-Pr_2Me_2) was obtained in 31% yield. The germathioacid chloride (**3**) was characterized by NMR spectroscopy together with elemental analysis, and the structure of **3** was finally determined by single-crystal X-ray analysis (Figure 2).

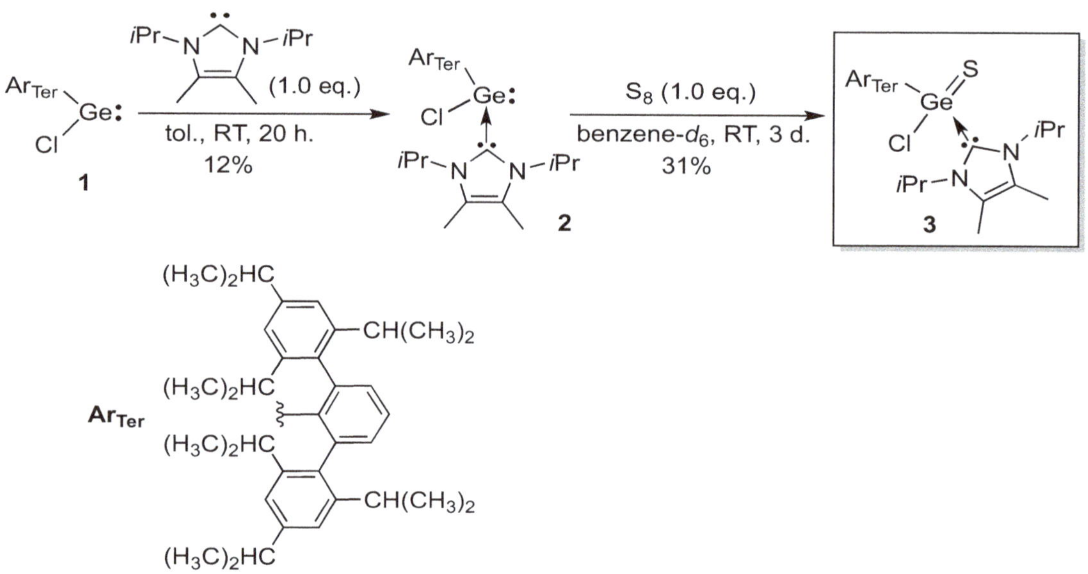

Scheme 1. Synthesis of Ge(S)Cl{C_6H_3-2,6-Tip_2}(Im-*i*-Pr_2Me_2) **3**.

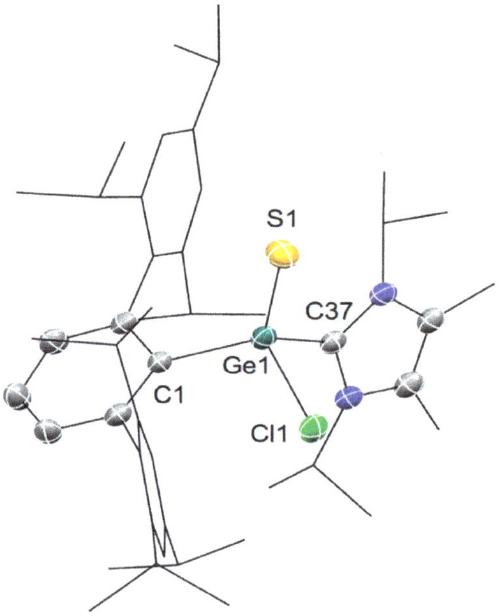

Figure 2. Molecular structure of Ge(S)Cl{C_6H_3-2,6-Tip_2}(Im-*i*-Pr_2Me_2) **3**, with thermal ellipsoids shown at the 50% probability level. Hydrogen atoms are omitted for clarity. Ge: Green, S: Yellow, Cl: Light green, N: Blue, C: gray. Selected bond distances (Å) and angles (°): Ge1–S1 2.0846(8), Ge1–Cl1 1.991(3), Ge1–C37 2.060(3), Ge1–Cl 2.2261(7); C1–Ge1–C37 108.65(10), C37–Ge1–Cl1 95.26(7), C1–Ge1–Cl1 106.70(7), C1–Ge1–S1 102.04(8), C37–Ge1–S1 104.82(8), S1–Ge1–Cl1 108.17(3).

Crystals suitable for X-ray crystallographic analysis of **3** were obtained from benzene-d_6. Compound (**3**) crystallized in the monoclinic crystal system with $P2_1/n$ space group. Compound (**3**) is the first example of germathioacid chloride in which the Ge atom is not coordinated to an N atom. The germanium center is bonded to the terphenyl ligand, chlorine, and sulfur atoms, and the other site is occupied by the NHC (Im-*i*-Pr_2Me_2), resulting in a tetrahedral geometry at the germanium center. The Ge–S bond of **3** (2.0846(8) Å) is slightly longer than the Ge=S bonds stabilized by N ligands (ranging from 2.048(2) Å to 2.066(1) Å [12–16]), but shorter than the Ge–S single bond (2.239(1) Å [18]) (Figure 1). This result suggests that the Ge–S bond has a partial double bond character.

2.2. Density Functional Theory Studies on the Germathioacid Chloride **3**

In order to better understand the characteristics of germathioacid chloride **3**, a density functional theory (DFT) calculation (B3LYP/6-31G(d,p)) was carried out [19]. The optimized structure of the germathioacid chloride **3** well-reproduced the structure experimentally observed in single-crystal X-ray analysis (Figure 2 and Figure S3). The natural population analysis (NPA) showed that a large positive charge was located on Ge (+1.240), and the S atom and Cl atom had negative charges of −0.762 and −0.456, respectively. The calculated IR stretching frequency of G=S bond was 469.64 cm^{-1}.

As shown in Figure 3, the highest occupied molecular orbital (HOMO) and HOMO − 1 were predominantly localized on the S atom, which should correspond to non-bonding pair electrons (Figure 3a,b), and an π–bond between the Ge atom and the S atom was not observed. The molecular orbitals (MOs), which should correspond to the σ-bonds of the Ge–S/Ge–C(NHC), and the σ^*-bonds of the Ge–S/Ge–Cl were also observed in HOMO − 1/HOMO − 29 (Figure 3c,d) and LUMO + 8, respectively (Figure 3e). These results suggest that the Ge–S bond would have a single bond character. On the other hand, the value of the Wiberg Bond Index (WBI) of Ge–S is 1.367, implying that the Ge–S bond has a partial double bond character. Although the value of WBI seemingly contradicted the analysis of the MOs, the second-order perturbation theory analysis rationalized the Ge–S bond feature; both non-bonding pair electrons localized on the S atom donated stabilization energy to the σ^*-orbitals of the Ge–Cl (23.39 kcal/mol), Ge–C(Ar_{Tip}) (11.23 kcal/mol), and Ge–C(NHC) (16.51 kcal/mol), respectively, and the secondary orbital interaction should be attributed to the partial π-bonding nature between the Ge–S bond in the WBI analysis. These results indicate that the Ge–S bond is a single bond with a partial double bond character, which is in good agreement with observation that the bond length (2.0846(8) Å) in the germathioacid chloride **3** (Figure 2) is shorter than that of the single bond of Ge–S (2.239(1)Å) [18], and longer than that of the Ge–S double bond (2.048(2) Å to 2.066(1) Å) [12–16].

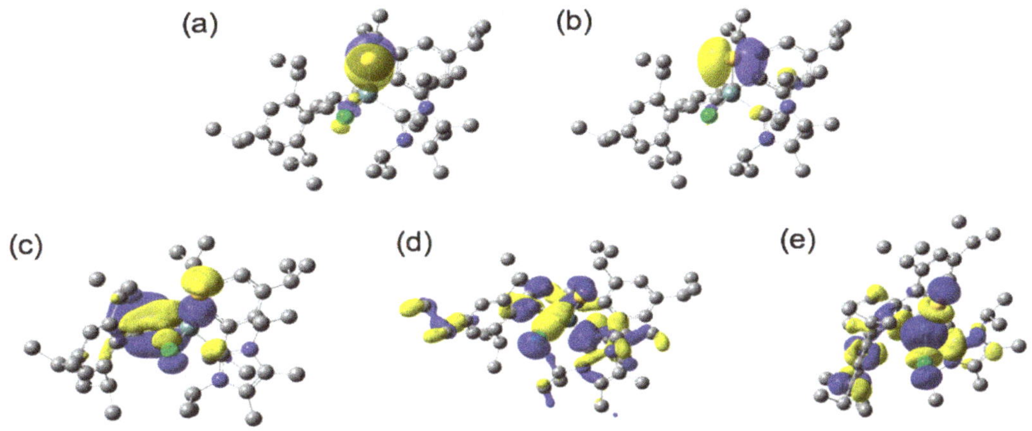

Figure 3. Selected molecular orbitals and their energy levels (eV): (**a**) HOMO (−4.815 eV), (**b**) HOMO − 1 (−4.925 eV), (**c**) HOMO − 7 (−6.6529 eV), (**d**) HOMO − 29 (−9.1961 eV), (**e**) LUMO + 8 (−0.6615 eV).

Based on experiments and calculation results, compound **3** could be an intermediate property between **3′** and **3″** (Figure 4). Further investigation will be carried out in the future.

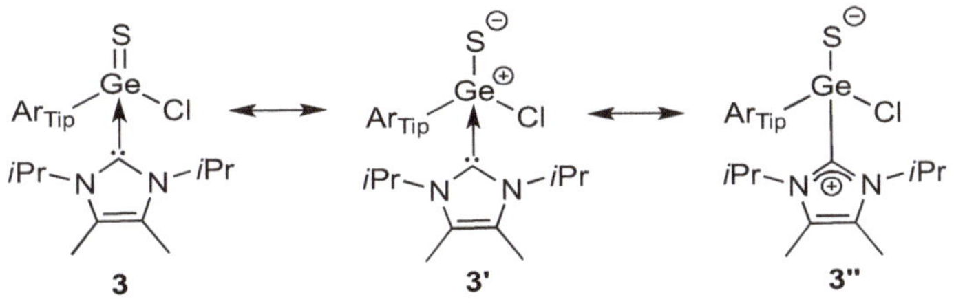

Figure 4. Resonant structural formula of **3**.

3. Materials and Methods

3.1. General Information

Materials: All manipulations were carried out under an argon atmosphere using standard Schlenk-line or glove-box techniques. Hexane, ether, and THF were dried by being passed through columns of activated alumina and a supported copper catalyst supplied by Nikko Hansen & Co., Ltd. (Osaka, Japan). Toluene was refluxed over sodium and benzophenone, distilled, and degassed prior to use. Deuterated benzene (C_6D_6, benzene-d_6) was dried and degassed over a potassium mirror in vacuo prior to use. $(Et_2O)LiC_6H_3$-2,6-Tip_2 [20], Ge(Cl){C_6H_3-2,6-Tip_2} [21] and 1,3-diisopropyl-4,5-dimethylimidazol-2-ylidene (Im-*i*-Pr_2Me_2) [22] were prepared according to procedures published in the literature.

The Fourier transformation nuclear magnetic resonance (NMR) spectra were obtained using JEOL JNM-ECS 300 (^{1}H at 300 MHz, ^{13}C at 75 MHz) and JEOL JNM-ECA 600 (^{1}H at 600 MHz, ^{13}C at 151 MHz) NMR instruments (JEOL, Tokyo, Japan). For ^{1}H-NMR, chemical shifts are reported as δ units (ppm) relative to $SiMe_4$ and the residual solvents peaks were used as standards. Analysis by electron impact mass spectrometry (EI-MS) was performed on a SHIMADZU GCMS-QP2010SE/DI2010 (SHIMADZU, Kyoto, Japan). Elemental analyses were performed by the Center for Material Research by Instrumental Analysis (CIA), Gunma University, Japan.

3.2. Synthesis

3.2.1. Synthesis of Ge(Cl){C_6H_3-2,6-Tip_2}(Im-*i*-Pr_2Me_2) **2**

To a solution of Ge(Cl){C_6H_3-2,6-Tip_2} (249 mg, 0.423 mmol) in dry toluene (3 mL) was added 1,3-diisopropyl-4,5-dimethlylimidazol-2-ylidene (76.3 mg, 0.423 mmol) in dry toluene (6 mL) at room temperature under an argon atmosphere. The solution was stirred for 20 h. The resulting solution was concentrated under reduced pressure. Then, toluene was added for crystallization to afford **2** as colorless crystals (yield: 24.4 mg, 12%).

2: colorless crystals; ^{1}H-NMR (300 MHz, C_6D_6, 70 °C, δ in ppm) δ 0.98 (broad, 6H) 1.04 (broad, 6H), 1.11 (*d*, *J* = 6.8 Hz, 6H), 1.16 (*d*, *J* = 6.8 Hz, 6H), 1.26 (*d*, *J* = 6.8 Hz, 6H), 1.31 (*d*, *J* = 6.8 Hz, 6H), 1.33 (*d*, *J* = 6.8 Hz, 6H), 1.58 (s, 6H), 1.64 (*d*, *J* = 6.8 Hz, 6H), 2.91 (sept, *J* = 6.8 Hz, 2H), 3.19 (sept, *J* = 6.8 Hz, 2H), 3.38 (sept, *J* = 6.8 Hz, 2H), 5.39 (broad, 2H), 6.96–7.12 (m, 3H), 7.23 (s, 4H); ^{13}C{^{1}H}-NMR (151 MHz, C_6D_6, RT, δ in ppm) δ 10.2 (CH_3), 20.6 (CH_3), 21.7 (CH_3), 22.6 (CH_3), 23.4 (CH_3), 24.4 (CH_3), 24.5 (CH_3), 24.6 (CH_3), 26.2 (CH_3), 26.5 (CH_3), 30.9 (CH), 31.2 (CH), 31.4 (CH), 31.5 (CH), 34.6 (CH), 34.9 (CH), 52.2 (CH), 52.8 (CH), 120.8 (CH), 121.0 (CH), 121.2 (CH), 125.0 (CH), 125.7 (CH), 126.3 (CH), 128.3 (CH), 129.3 (C), 131.8 (C), 140.0(C), 147.1 (C), 147.6 (C), 156.9 (C), 172.3 (C); Anal. Calcd. for $C_{47}H_{69}ClGeN_2$: C 73.30, H 9.03, N 3.64; found: C 73.01, H 8.94, N 3.61.

3.2.2. Synthesis of Ge(S)Cl{C_6H_3-2,6-Tip_2}(Im-*i*-Pr_2Me_2) **3**

To a solution of Ge(Cl){C_6H_3-2,6-Tip_2}(Im-*i*-Pr_2Me_2) (90.6 mg, 0.118 mmol) (**2**) in benzene-d_6 (0.5 mL) in an NMR tube equipped with a Young stopcock at room temperature, S_8 (3.77 mg, 0.118 mmol) was added. The solution was reacted for 3 day in the NMR tube. The resulting solution was concentrated under reduced pressure to yield colorless crystals of **3** (yield: 28.9 mg, 31%).

3: colorless crystals; ^{1}H-NMR (600 MHz, C_6D_6, RT, δ in ppm) δ 0.34 (broad, 3H), 0.50 (broad, 3H), 0.94 (broad, 3H), 0.98 (broad, 3H), 1.04 (broad, 3H), 1.06 (broad, 3H), 1.16 (broad, 3H), 1.20 (broad, 3H), 1.30 (broad, 6H), 1.36 (broad, 12H), 1.49 (broad, 3H), 1.55 (broad, 3H), 1.97 (broad, 3H), 2.02 (broad, 3H), 2.64 (broad, 1H), 2.86 (broad, 1H), 2.97 (broad, 1H), 3.06 (broad, 1H), 3.73 (broad, 1H), 3.97 (broad, 1H), 5.38 (broad, 1H), 6.92–7.02 (m, 4H), 7.35–7.44 (m, 3H), 7.75 (broad, 1H); ^{13}C{^{1}H}-NMR (151 MHz, C_6D_6, RT, δ in ppm) δ 10.2 (CH_3), 10.4 (CH_3), 19.9 (CH_3), 19.9 (CH_3), 20.5 (CH_3), 20.6 (CH_3), 22.8 (CH_3), 22.9 (CH_3), 23.0 (CH_3), 23.1 (CH_3), 23.8 (CH_3), 24.2 (CH_3), 24.5 (CH_3), 25.6 (CH_3), 25.9 (CH_3), 26.1 (CH_3),

26.9 (CH_3), 27.4 (CH_3), 31.3 (CH), 31.4 (CH), 31.7 (CH), 32.1 (CH), 34.7 (CH), 34.8 (CH), 49.4 (CH), 52.7 (CH), 120.6 (CH), 120.8 (CH), 121.5 (CH), 122.1 (CH), 126.8 (CH), 128.4 (CH), 134.1 (CH), 138.9 (C), 139.7 (C), 142.0(C × 2), 145.2 (C), 145.3 (C), 145.4 (C), 146.3 (C), 148.3 (C), 148.4 (C), 148.5 (C), 148.4 (C), 148.5 (C), 148.8 (C), 152.0 (C), 153.3 (C); Anal. Calcd. for $C_{47}H_{69}ClGeN_2S_2$: C 70.28, H 8.78, N 3.49; found: C 69.99, H 8.72, N 3.88.

3.3. *Single-Crystal X-ray Analysis of Ge(S)Cl{C_6H_3-2,6-Tip$_2$}(Im-i-Pr$_2$Me$_2$)* **3**

Crystals suitable for the X-ray crystallographic analysis of the germathioacid chloride (**3**) were obtained from benzene-d_6. The intensity data were collected on a Rigaku XtaLab P200 diffractometer (Rigaku, Tokyo, Japan) with multi layer mirror mono chromated Mo Kα radiation (λ = 0.71075 Å). The structures were determined by direct methods (*SHELXS*-97) [23], and refined by full-matrix least-squares procedures on *F*2 for all reflections (*SHELXL*-97) [23]. All of the non-hydrogen atoms were refined anisotropically. All hydrogens were placed using AFIX instructions. All calculations were carried out using *Yadokari*-XG2009 [24]. Crystallographic data have been deposited into the Cambridge Crystallographic Data Centre, with deposition numbers CCDC 1851711 for compound **3**. Copies of the data can be obtained free of charge via http://www.ccdc.cam.ac.uk/conts/retrieving.html (or from the Cambridge Crystallographic Data Centre, 12, Union Road, Cambridge, CB2 1EZ, U.K.

Crystal data for Ge(S)Cl{C_6H_3-2,6-Tip$_2$}(Im-*i*-Pr$_2$Me$_2$) (**3**) (123 K): $C_{47}H_{69}ClGeN_2S$, Fw 802.14, Monoclinic, Space group $P2_1/n$, colorless crystals, a = 9.3421(10), b = 19.570(2), c = 23.901(3) Å, β = 90.387(3)°, V = 4369.6(8) Å^3, Z = 2, D_{calcd} = 1.219 Mg/m^3, $R1$ = 0.0486 ($I > 2\sigma$), $wR2$ = 0.1243 (all data), Goodness of Fit (GOF)= 0.906.

3.4. *DFT Calculations*

Density functional theory (DFT) calculations were performed using the *Gaussian* 16 program package [25]. Geometry optimizations and vibrational analyses of all local equilibriums were performed using the B3LYP functional with a basis set of 6-31G(d,p) for all atoms. The natural population analysis (NPA) [26], charge distribution, Wiberg bond index (WBI) [27], and second-order perturbation analysis were calculated with the natural bond orbital (NBO) program package at the B3LYP/6-31G(d,p) level of theory. Cartesian coordinates and energies of the computed structures are listed in the Supplementary Materials.

4. Conclusions

In conclusion, we have achieved the synthesis of germathioacid chloride without using nitrogen ligands. The nature of the bonds between the germanium atom and sulfur atom was investigated by X-ray analysis and DFT calculation. These experimental and theoretical results suggest that the Ge–S bond is a single bond with a partial double bond character. We hope that our research will help us clarify the nature of the Ge–S bond.

Author Contributions: Y.E., N.T. and M.U. conceived and designed the experiments; C.F. and Y.E. performed the experiments; Y.E. and N.T. performed the XRD analysis; K.M. performed the theoretical calculation; Y.E., K.M. and M.U. wrote the paper.

Acknowledgments: The computations were performed by the Research Center for Computational Science, Okazaki, Japan.

References

1. Power, P.P. Main-group elements as transition metal. *Nature* **2010**, *463*, 171–177. [CrossRef] [PubMed]
2. Lee, V.Y.; Sekiguchi, A. *Organometallic Compounds of Low Coordinate Si, Ge, Sn and Pb: From Phantom Species to Stable Compounds*; Wiley: Chichester, UK, 2010.
3. Tokitoh, N.; Okazaki, R. Recent advances in the chemistry of group 14–group 16 double bond compound. *Adv. Organomet. Chem.* **2001**, *47*, 121–166.

4. Brook, A.G.; Abdesaken, F.; Gutekunst, B.; Gutekunst, G.; Kallury, R.K. A solid silaethene: Isolation and characterization. *J. Chem. Soc. Chem. Commun.* **1981**, *4*, 191–192. [CrossRef]
5. West, R.; Fink, M.J.; Michl, J. Tetramesityldisilene, a stable compoundd containing a silicon-silicon double bond. *Science* **1981**, *214*, 1343–1344. [CrossRef] [PubMed]
6. Yoshifuji, M.; Shima, I.; Inamoto, N.; Hirotsu, K.; Higuchi, T. Synthesis and structure of bis(2,4,6-tri-tert-buthylphenyl)diphosphene: Isolation of a true phosphobenzene. *J. Am. Chem. Soc.* **1981**, *103*, 4587–4589. [CrossRef]
7. Okazaki, R.; Tokitoh, N. Heavy Ketones, the Heavier Element Congeners of a Ketone. *Acc. Chem. Res.* **2000**, *33*, 625–630. [CrossRef] [PubMed]
8. Fischer, R.C.; Power, P.P. π-Bonding and the Lone Pair Effect in Multiple Bonds involving Heavier Main Group Elements: Developments in the New Millennium. *Chem. Rev.* **2010**, *110*, 3877–3923. [CrossRef] [PubMed]
9. Veith, M.; Becker, S.; Huch, V. A Base-Stabilized Ge–S Double Bond. *Angew. Chem. Int. Ed. Engl.* **1989**, *28*, 1237–1238. [CrossRef]
10. Tokitoh, N.; Matsumoto, T.; Manmaru, K.; Okazaki, R. Synthesis and Crystal Structure of the First Stable Diarylgermanetione. *J. Am. Chem. Soc.* **1993**, *115*, 8855–8856. [CrossRef]
11. Matsumoto, T.; Tokitoh, N.; Okazaki, R. The First Kinetically Stabilized Germanethiones and Germaneselones: Syntheses, Structure, and Reactivities. *J. Am. Chem. Soc.* **1999**, *121*, 8811–8824. [CrossRef]
12. Ding, Y.; Ma, Q.; Usón, I.; Roesky, H.W.; Noltemeyer, M.; Schmidt, H.-G. Synthesis and Structures of [{HC(CMeNAr)$_2$}Ge(S)X] (Ar = 2,6-iPr$_2$C$_6$H$_3$, X = F, Cl, Me): Structurally Characterized Examples with a Formal Double Bond between Group 14 and 16 Elements Bearing a Halide. *J. Am. Chem. Soc.* **2002**, *124*, 8542–8543. [CrossRef] [PubMed]
13. Prashanth, B.; Singh, S. Concise access to iminophosphonamide stabilized heteroleptic germylenes: Chemical reactivity and structural investigation. *Dalton Trans.* **2016**, *45*, 6079–6087. [CrossRef] [PubMed]
14. Leung, W.-P.; Chong, K.-H.; Wu, Y.-S.; So, C.-W.; Chan, H.-S.; Mak, T.C.W. Synthesis of Chalcogeno[3-(pyrid-2-yl)-1-azaallyl]germanium Complexes. *Eur. J. Inorg. Chem.* **2006**, 808–812. [CrossRef]
15. Sinhababu, S.; Siwatch, R.K.; Mukherjee, G.; Rajaraman, G.; Nagendran, S. Aminotroponiminatogermaacid Halides with a Ge(E)X Moiety (E = S, Se; X = F, Cl). *Inorg. Chem.* **2012**, *51*, 9240–9248. [CrossRef] [PubMed]
16. Xiong, Y.; Yao, S.; Inoue, S.; Berkefeld, A.; Driess, M. Taming the germyliumylidene [ClGe:]$^+$ and germathionium [ClGe=S]$^+$ ions by donor–acceptor stabilization using 1,8-bis(tributylphosphazenyl) naphthalene. *Chem. Commun.* **2012**, *48*, 12198–12200. [CrossRef] [PubMed]
17. Filippou, A.C.; Chernov, O.; Blom, B.; Stumpf, K.W.; Schnakenburg, G. Stable N-Heterocyclic Carbene Adducts of Arylchlorosilylenes and Their Germanium Homologues. *Chem. Eur. J.* **2010**, *16*, 2866–2872. [CrossRef] [PubMed]
18. Ossig, G.; Meller, A.; Brönneke, C.; Müller, O.; Schäfer, M.; Herbst-Irmer, R. Bis[(2-pyridyl)bis(trimethylsilyl) methyl-*C,N*]germanium(II): A Base-Stabilized Germylene and the Corresponding Germanethione, Germaneselenone, and Germanetellurone. *Organometallics* **1997**, *16*, 2116–2120. [CrossRef]
19. Li, L.; Fukawa, T.; Matsuo, T.; Hashizume, D.; Fueno, H.; Tanaka, K.; Tamao, K. A stable germanone as the first isolated heavy ketone with a terminal oxygen atom. *Nat. Chem.* **2012**, *4*, 361–365. [CrossRef] [PubMed]
20. Schiemenz, B.; Power, P.P. Synthesis of Sterically Encumbered Terphenyls and Characterization of Their Metal Derivatives Et$_2$OLiC$_6$H$_3$-2,6-Trip$_2$ and Me$_2$SCuC$_6$H$_3$-2,6-Trip$_2$ (Trip = 2,4,6-i-Pr$_3$C$_6$H$_2$$^-$). *Organometallics* **1996**, *15*, 958–964. [CrossRef]
21. Pu, L.; Olmstead, M.M.; Power, P.P. Synthesis and Characterization of the Monomeric Terphenyl-Metal Halides Ge(Cl){C$_6$H$_3$-2,6-Trip$_2$}(Trip = C$_6$H$_2$-2,4,6-i-Pr$_3$) and Sn(I){C$_6$H$_3$-2,6-Trip$_2$} and the Terphenyl-Metal Amide Sn{N(SiMe$_3$)$_2$}{C$_6$H$_3$-2,6-Trip$_2$}. *Organometallics* **1998**, *17*, 5602–5606. [CrossRef]
22. Kuhn, N.; Kratz, Y. Synthesis of Imidazol-2-ylidenes by Reduction of Imidazole-2-(3*H*)-thiones. *Synthesis* **1993**, 561–562. [CrossRef]
23. Sheldrick, G.M. *SHELXS-97 and SHELXL-97*; Program for the Solution of Crystal Structures; University of Göttingen: Göttingen, Germany, 1997.
24. Kabuto, C.; Akine, S.; Nemoto, T.; Kwon, E. Release of Software (Yadokari-XG 2009) for Crystal Structure Analyses. *J. Cryst. Soc. Jpn.* **2009** *51*, 218–224. [CrossRef]
25. Frisch, M.J.; Trucks, G.W.; Schlegel, H.B.; Scuseria, G.E.; Robb, M.A.; Cheeseman, J.R.; Scalmani, G.; Barone, V.; Petersson, G.A.; Nakatsuji, H.; et al. *Gaussian 16*; Revision A.03; Gaussian, Inc.: Wallingford, CT, USA, 2016.

26. Reed, A.E.; Curtiss, L.A.; Weinhold, F. Intermolecular interactions from a natural bond orbital, donor-acceptor viewpoint. *Chem. Rev.* **1988**, *88*, 899–926. [CrossRef]
27. Sizova, O.V.; Skripnikov, L.V.; Sokolv, A.Y. Symmetry decomposition of quantum chemical bond orders. *THEOCHEM* **2008**, *870*, 1–9. [CrossRef]

7

Si–H Bond Activation of a Primary Silane with a Pt(0) Complex: Synthesis and Structures of Mononuclear (Hydrido)(dihydrosilyl) Platinum(II) Complexes

Norio Nakata *, Nanami Kato, Noriko Sekizawa and Akihiko Ishii *

Department of Chemistry, Graduate School of Science and Engineering, Saitama University, 255 Shimo-okubo, Sakura-ku, Saitama 338-8570, Japan; rdhns681@yahoo.co.jp (N.K.); jjhs61ns@yahoo.co.jp (N.S.)

* Correspondence: nakata@chem.saitama-u.ac.jp (N.N.); ishiiaki@chem.saitama-u.ac.jp (A.I.)

Abstract: A hydrido platinum(II) complex with a dihydrosilyl ligand, [*cis*-PtH(SiH_2Trip)(PPh_3)$_2$] (**2**) was prepared by oxidative addition of an overcrowded primary silane, TripSiH_3 (**1**, Trip = 9-triptycyl) with [Pt(η^2-C_2H_4)(PPh_3)$_2$] in toluene. The ligand-exchange reactions of complex **2** with free phosphine ligands resulted in the formation of a series of (hydrido)(dihydrosilyl) complexes (**3**–**5**). Thus, the replacement of two PPh_3 ligands in **2** with a bidentate bis(phosphine) ligand such as DPPF [1,2-bis(diphenylphosphino)ferrocene] or DCPE [1,2-bis(dicyclohexylphosphino)ethane] gave the corresponding complexes [PtH(SiH_2Trip)(L-L)] (**3**: L-L = dppf, **4**: L-L = dcpe). In contrast, the ligand-exchange reaction of **2** with an excess amount of PMe_3 in toluene quantitatively produced [PtH(SiH_2Trip)(PMe_3)(PPh_3)] (**5**), where the PMe_3 ligand is adopting *trans* to the hydrido ligand. The structures of complexes **2**–**5** were fully determined on the basis of their NMR and IR spectra, and elemental analyses. Moreover, the low-temperature X-ray crystallography of **2**, **3**, and **5** revealed that the platinum center has a distorted square planar environment, which is probably due to the steric requirement of the *cis*-coordinated phosphine ligands and the bulky 9-triptycyl group on the silicon atom.

Keywords: platinum; primary silane; hydrido complex; oxidative addition; ligand-exchange reaction; X-ray crystallography

---

1. Introduction

The transition metal catalyzed synthesis of functionalized organosilicon compounds gained substantial momentum during the past few decades [1]. Among these catalytic conversions, the oxidative addition of hydrosilanes with platinum(0) complexes is an efficient method for the generation of the platinum(II) hydride species, which has been proposed as a key intermediate in platinum-catalyzed hydrosilylations [2–6] and bis-silylations [7,8], as well as the dehydrogenative couplings of hydrosilanes [9–13]. While a number of reactions of hydrosilanes with platinum(0) complexes affording mononuclear bis(silyl) [14–18] and silyl-bridged multinuclear complexes [19–29] have been described so far, the isolation of mononuclear hydrido(silyl) complexes has been less well studied due to the high reactivity of a Pt–H bond [30–34]. In particular, the synthesis of hydrido(dihydrosilyl) platinum(II) complexes, which are anticipated as the initial products in the Si–H bond activation reactions of primary silanes with platinum(0) complexes, is quite rare. Indeed, only two publications have previously reported the characterization of hydrido(dihydrosilyl) platinum(II) complexes. In 2000, Tessier et al. reported that the reaction of a primary silane with a bulky *m*-terphenyl group with [Pt(PPr_3)$_3$] produced the first example of a stable hydrido(dihydrosilyl) complex [*cis*-PtH(SiH_2Ar)(PPr_3)$_2$] (Ar = 2,6-MesC_6H_3) [35]. Quite recently, Lai et al. also described the synthesis of a bis(phosphine) hydrido(dihydrosilyl) complex [$PtH(SiH_2Si^tBu_2Me)(dcpe)$] (dcpe = 1,2-bis(dicyclohexylphosphino)ethane) containing a Si–Si bond [36]. Meanwhile, we succeeded in the first isolation of a series of hydrido(dihydrogermyl) platinum(II) complexes [$PtH(GeH_2Trip)(L)$] (Trip = 9-triptycyl) using a bulky substituent, 9-triptycyl group [37]. In addition, we reported the first syntheses and structural characterizations of hydrido palladium(II) complexes with a dihydrosilyl- or dihydrogermyl ligand, [$PdH(EH_2Trip)(dcpe)$] (E = Si, Ge) [38]. Very recently, we also found that hydride-abstraction reactions of [$MH(EH_2Trip)(dcpe)$] (M = Pt, Pd, E = Si, Ge) with $B(C_6F_5)_3$ led to the formations of new cationic dinuclear complexes with bridging hydrogermylene and hydrido ligands, [$\{M(dcpe)\}_2(\mu\text{-}GeHTrip)(\mu\text{-}H)$]$^+$ [39]. As an extension of our previous work and taking into account the interest devoted to hydrido platinum(II) complexes, we present here the synthesis and characterization of a series of mononuclear (hydrido)(dihydrosilyl) complexes [$PtH(SiH_2Trip)(L)_2$].

2. Results

2.1. Synthesis and Characterization of [cis-$PtH(SiH_2Trip)(PPh_3)_2$] (2)

The reaction of $TripSiH_3$ **1** with [$Pt(\eta^2\text{-}C_2H_4)(PPh_3)_2$] in toluene proceeded efficiently at room temperature under inert atmosphere to form the corresponding complex [*cis*-$PtH(SiH_2Trip)(PPh_3)_2$] (**2**) in 91% yield as colorless crystals (Scheme 1). In the ^{1}H NMR spectrum of **2**, the characteristic signals of the platinum hydride were observed at $\delta = -2.15$, which were split by 19 and 157 Hz of ^{31}P–^1H couplings accompanying 958 Hz of satellite signals from the ^{195}Pt isotope. This chemical shift is comparable with those of the related (hydrido)(dihydrosilyl) complexes, [*cis*-$PtH(SiH_2Ar)(PPr_3)_2$] (Ar = 2,6-$MesC_6H_3$) ($\delta = -3.40$) [35] and [*cis*-$PtH(SiH_2Si^tBu_2Me)(dcpe)$] ($\delta = -0.89$) [36]. The SiH_2 resonance appeared as a multiplet at $\delta = 4.68$ ppm. The $^{31}P\{^1H\}$ NMR spectrum of **2** exhibited two doublets ($^2J_{P\text{–}P} = 15$ Hz) at $\delta = 33.8$ and 34.5 with ^{195}Pt–^{31}P coupling constants, 2183 and 1963 Hz, which were assigned to the phosphorus atoms lying *trans* to the hydrido and dihydrosilyl ligands, respectively, in agreement with the NMR data for reported germanium congener [*cis*-$PtH(GeH_2Trip)(PPh_3)_2$] [$\delta = 31.2$ ($^1J_{Pt\text{–}P} = 2317$ Hz) and 31.6 ($^1J_{Pt\text{–}P} = 2252$ Hz)] [37]. The silicon atom of **2** gave rise to a resonance around $\delta = -40.6$ with splitting due to ^{31}P–^{29}Si couplings ($^2J_{P(trans)\text{–}Si} = 161$, $^2J_{P(cis)\text{–}Si} = 15$ Hz) and ^{195}Pt satellites ($^1J_{Pt\text{–}Si} = 1220$ Hz) in the $^{29}Si\{^1H\}$ NMR spectrum. In the solid state IR spectrum for **2**, Pt–H and Si–H stretching vibrations were observed at 2041 and 2080 cm^{-1}, respectively. Complex **2** is thermally and air stable in the solid state (melting point: 123 °C (dec.)) or in solution, and no dimerization or dissociation of phosphine ligands was observed.

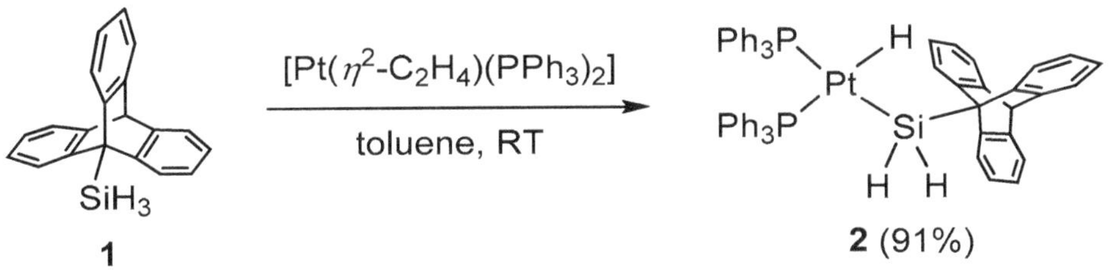

Scheme 1. Synthesis of [*cis*-$PtH(SiH_2Trip)(PPh_3)_2$] **2**.

The molecular structure of **2** was determined unambiguously by X-ray crystallographic analysis, as depicted in Figure 1. The X-ray crystallographic analysis of **2** revealed that the platinum center attains a distorted square-planar environment, which was probably due to the steric requirement of the *cis*-coordinated PPh_3 ligands and the bulky 9-triptycyl group on the silicon atom. The P1–Pt1–P2 angle of 101.63(3)° and P1–Pt1–Si1 angle of 96.17(3)° deviated considerably from the ideal 90° of square-planar geometry. The Pt–Si bond length is 2.3458(9) Å, which is comparable to those ranging from 2.321 to 2.406 Å observed in the related mononuclear platinum(II) complexes bearing silyl ligands [1]. The hydrogen atom on the platinum atom was located in the electron density map and has a Pt–H distance of 1.59(4) Å. The Pt1–P1 bond length [2.2945(8) Å] is slightly shorter than the Pt1–P2 bond length [2.3401(8) Å], which indicates the stronger *trans* influence of the silicon atom compared with that of the hydride in this complex. This result is consistent with the ^{195}Pt–^{31}P coupling constants (2183 and 1963 Hz) observed in the ^{31}P{^{1}H} NMR spectrum.

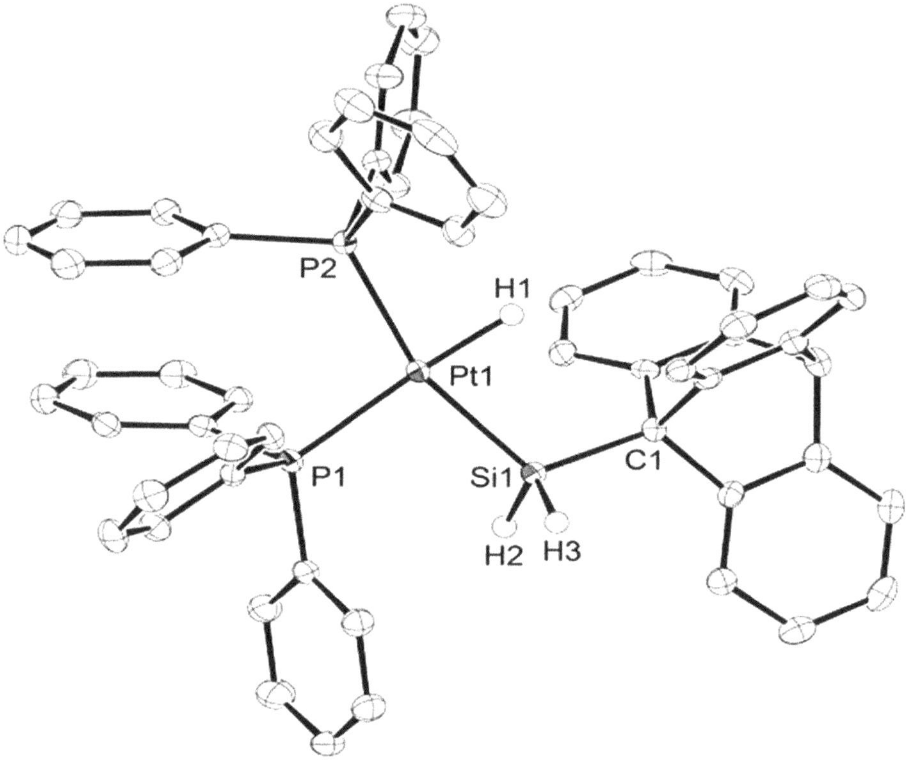

Figure 1. ORTEP of [*cis*-PtH(SiH_2Trip)(PPh_3)$_2$] **2** (50% thermal ellipsoids, a solvation toluene molecule, and hydrogen atoms, except H1, H2, and H3 were omitted for clarity). Selected bond lengths (Å) and bond angles (°): Pt1–Si1 = 2.3458(9), Pt1–P1 = 2.2945(8), Pt1–P2 = 2.3401(8), Pt1–H1 = 1.59(4), Si1–C1 = 1.918(3), P1–Pt1–P2 = 101.63(3), Si1–Pt1–P1 = 96.17(3), Si1–Pt1–H1 = 79.7(16), P2–Pt1–H1 = 82.5(16), Si1–Pt1–P2 = 162.13(3), P1–Pt1–H1 = 175.8(16).

2.2. Ligand Exchange Reactions of **2** *with Free Phosphine Ligands*

We next examined the ligand-exchange reactions of complex **2** with free phosphine ligands. The replacement of two PPh_3 ligands in **2** with a bidentate bis(phosphine) ligand such as DPPF (1,2-bis(diphenylphosphino)ferrocene) or DCPE gave the corresponding complexes [PtH(SiH_2Trip)(L-L)] (**3**: L-L = dppf, **4**: L-L = dcpe) in 87% and 80% yields, respectively (Scheme 2). In the ^{1}H NMR spectra of **3** and **4**, the hydride resonated as a doublet of doublets at δ = –1.62 ($^2J_{P\text{–}H}$ = 20, 164, $^1J_{Pt\text{–}H}$ = 995 Hz) for **3** and –0.46 ($^2J_{P\text{–}H}$ = 13, 166, $^1J_{Pt\text{–}H}$ = 1004 Hz) for **4**. These chemical shifts are shifted downfield in comparison with that of the starting complex **2** (δ = −2.15), which is probably due to the stronger electron-donating ability of chelating phosphines compared with PPh_3. The spectrum for **3** also displayed a multiplet signal centering at δ = 4.61 corresponding to the SiH_2 protons, which is shifted upfield by 0.89 ppm in comparison with that of **4**. The ^{31}P{^{1}H} NMR spectrum of **3** showed two doublets ($^2J_{P\text{–}P}$ = 21 Hz) with ^{195}Pt satellites at δ = 30.5 ($^1J_{Pt\text{–}P}$ = 2247 Hz) and 34.5 ($^1J_{Pt\text{–}P}$ = 1837 Hz), which are close to those of **2** [δ = 33.8 ($^1J_{Pt\text{–}P}$ = 2183 Hz) and 34.5 ($^1J_{Pt\text{–}P}$ = 1963 Hz)]. The observation of P–P coupling indicates a large deviation of the P–Pt–P angle from 90° of the ideal square planar geometry around the Pt(II) center (vide infra). In contrast, the ^{31}P{^{1}H} resonances for **4** were observed as two singlets at δ = 69.2 ($^1J_{Pt\text{–}P}$ = 1809 Hz) and 85.3 ($^1J_{Pt\text{–}P}$ = 1678 Hz), respectively, which are relatively shifted downfield relative to those of **2** and **3**. The larger $^1J_{Pt\text{–}P}$ values

(2183 Hz for **3**, 1809 Hz for **4**) are assigned to the phosphorus atom *trans* to the hydrido ligand, as in the case of **2**. Furthermore, the $^{29}Si\{^1H\}$ NMR spectra of **3** and **4** showed a doublet of doublets signals at $\delta = -39.0$ ($^2J_{Si-P} = 167, 12$ Hz) for **3**, and -44.6 ($^2J_{Si-P} = 173, 11$ Hz) for **4**, which were accompanied by ^{195}Pt satellites of 1207 Hz for **3** and 1253 Hz for **4**, respectively.

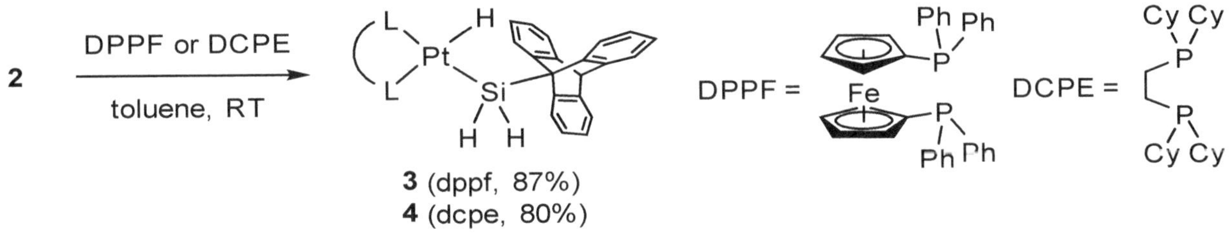

Scheme 2. Ligand-exchange reaction of [*cis*-PtH(SiH$_2$Trip)(PPh$_3$)$_2$] **2** with chelating bis(phosphine)s.

The molecular structure of DPPF-derivative **3** in the crystalline state was confirmed by X-ray crystallography (Figure 2). The platinum atom lies in a distorted square-planar geometry; the sum of the bond angles around the platinum atom is 360.48°. The P1–Pt1–P2 angle is 102.29(9)°, and other angles around the platinum atom are less than 90°, except the P1–Pt1–Si1 angle [95.19(9)°]. The Pt–Si [2.331(3) Å] and two Pt–P bond lengths [2.286(2), 2.319(2) Å] are comparable to those of the corresponding DPPF-ligated hydrido complex [PtH(SiHPh$_2$)(dppf)] [2.3366(4), 2.2830(4), and 2.3192(4) Å, respectively] [40].

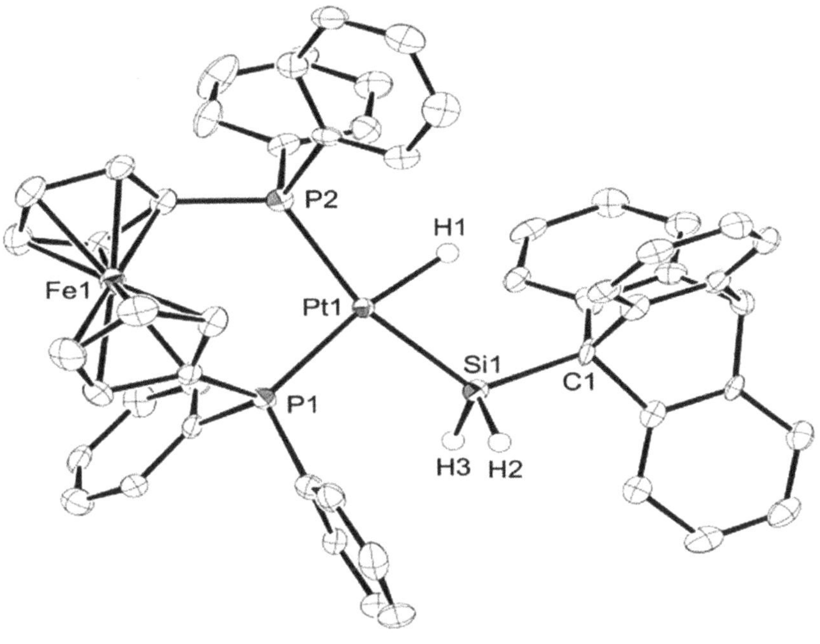

Figure 2. ORTEP of [PtH(SiH$_2$Trip)(dppf)] **3** 50% thermal ellipsoids, a solvation CH_2Cl_2 molecule, and hydrogen atoms, except H1, H2, and H3 were omitted for clarity. Selected bond lengths (Å) and bond angles (°): Pt1–Si1 = 2.331(3), Pt1–P1 = 2.286(2), Pt1–P2 = 2.319(2), Pt1–H1 = 1.687(10), Si1–C1 = 1.920(10), P1–Pt1–P2 = 102.29(9), Si1–Pt1–P1 = 95.19(9), Si1–Pt1–H1 = 76(5), P2–Pt1–H1 = 87(5), Si1–Pt1–P2 = 162.10(9), P1–Pt1–H1 = 171(5).

It is well known that trimethylphosphine (PMe$_3$) is a strong σ-donating ligand for a wide variety of transition-metal complexes. Therefore, one can reasonably expect the formation of [PtH(SiH$_2$Trip)(PMe$_3$)$_2$] in a similar ligand-exchange reaction of **2** with PMe$_3$. However, we found that the reaction of **2** with 2.2 equivalents of PMe$_3$ in toluene at room temperature did not proceed completely, which resulted in the formation of [PtH(SiH$_2$Trip)(PMe$_3$)(PPh$_3$)] (**5**), where the PPh$_3$ ligand *trans* to hydrido in **2** is exchanged with a PMe$_3$ ligand (Scheme 3). Complex **5** was isolated as colorless crystals in quantitative yield after workup. In the ^{1}H NMR spectrum of **5** at 298 K, the characteristic broad doublet signal due to the platinum hydride was observed centering at $\delta = -1.91$ with splitting by ^{31}P–^{1}H ($^{2}J_{P(trans)\text{–}H}$ = 157 Hz) and ^{195}Pt–^{1}H ($^{1}J_{Pt\text{–}H}$ = 899 Hz) couplings. The SiH_2 protons also appeared as a broad multiplet at δ = 5.95, which is shifted downfield relative to those of the above complexes **2**–**4** (δ = 4.68–5.50). The $^{31}P\{^{1}H\}$ NMR spectrum of **5** at 298 K exhibited two nonequivalent broad singlet signals at $\delta = -22.9$ and 37.4, with two sets of ^{195}Pt satellites of 2115 and 1833 Hz. The former signal was assigned to the PMe_3 *trans* to the hydrido ligand using a non ^{1}H-decoupled ^{31}P NMR technique at 253 K. The $^{29}Si\{^{1}H\}$ NMR spectrum of **5** at 223 K featured one ^{29}Si resonance as a broad doublet signal at $\delta = -43.9$ ($^{2}J_{Si\text{–}P}$ = 151 Hz). The broadening of NMR signals possibly implies the existence of the Si–H σ-complex intermediate **6** in the NMR time scale [38,41,42]. Unfortunately, attempts to probe further the fluxional behavior of **5** by VT (variable temperature)-NMR in solution revealed no appreciable changes in spectroscopic features by ^{1}H and ^{31}P NMR spectroscopy. The molecular structure of **5** is also determined by X-ray analysis, as shown in Figure 3. Distortions from square planar geometry at the platinum center were observed, similar to the cases of **2** and **4**. The Pt–Si bond length of **5** [2.3414(13) Å] is almost equal to those of **2** [2.3458(9) Å] and **4** [2.331(3) Å]. As expected, the Pt1–P1 bond length for the PMe_3 ligand of **5** [2.2978(12) Å] is shortened compared with the Pt1–P2 bond length for the PPh_3 ligand of **5** [2.3203(11) Å] due to the different ligands in the *trans* positions of the phosphorus atoms. While only a few cationic platinum complexes containing different phosphine ligands have been reported [43–45], complex **5** is the first example of a neutral platinum complex bearing a weakly electron-donating PPh_3 and strongly electron-donating PMe_3.

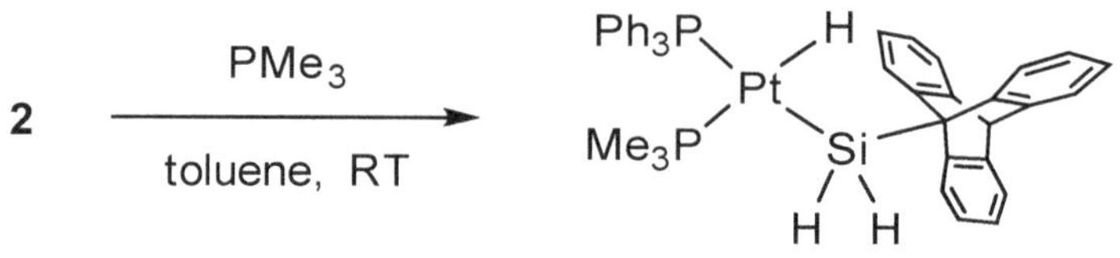

5 (94%)

Scheme 3. Ligand-exchange reaction of [*cis*-$PtH(SiH_2Trip)(PPh_3)_2$] **2** with trimethylphosphine (PMe_3).

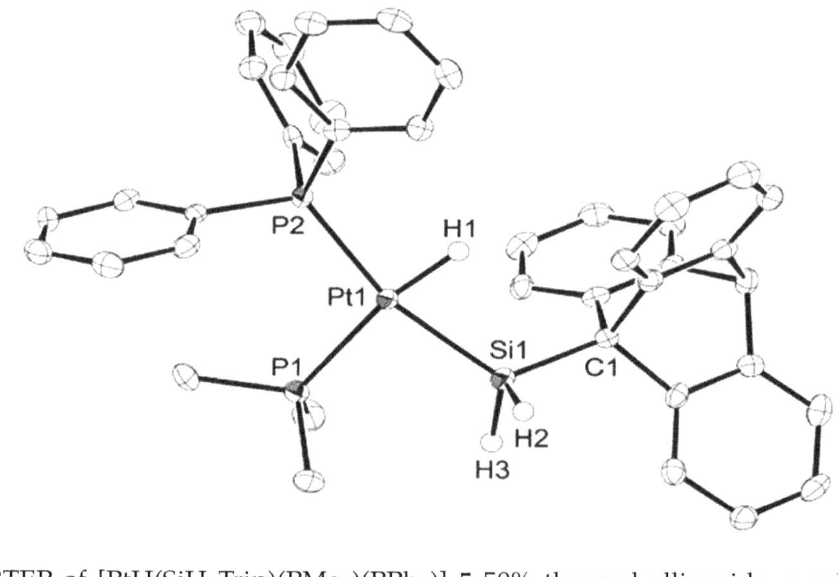

Figure 3. ORTEP of [$PtH(SiH_2Trip)(PMe_3)(PPh_3)$] **5** 50% thermal ellipsoids, a solvation CH_2Cl_2 molecule, and hydrogen atoms, except H1, H2, and H3 were omitted for clarity. Selected bond lengths (Å) and bond angles (°): Pt1–Si1 = 2.3414(13), Pt1–P1 = 2.2978(12), Pt1–P2 = 2.3203(11), Pt1–H1 = 1.44(6), Si1–C1 = 1.930(5), P1–Pt1–P2 = 103.47(4), Si1–Pt1–P1 = 92.34(4), Si1–Pt1–H1 = 75(2), P2–Pt1–H1 = 89(2), Si1–Pt1–P2 = 164.19(4), P1–Pt1–H1 = 167(2).

A plausible formation mechanism for **5** is shown in Scheme 4. According to the stronger *trans* influence of the silyl ligand than that of the hydrido ligand, the ligand-exchange of a PPh_3 *trans* to silyl ligand takes place to yield the intermediate **5′** in the first step, while **5** might be formed directly from the corresponding coordinatively unsaturated intermediate (3-coordinated 14-electron complexes) [46,47]. Then, the intramolecular interchange of coordination environments between the silyl and hydrido ligands through the Si–H σ-complex intermediate **6** would occur, probably due to the steric repulsion between the bulky 9-triptycyl group on the silicon atom and the *cis*-PPh_3 ligand in **5′**. Finally, the corresponding complex **5** was obtained as the thermodynamic product. As another pathway, it is likely that the direct formation of **5** is caused by an initial dissociation of the PPh_3 ligand at the *trans* position of the hydrido ligand in **2** due to steric reason.

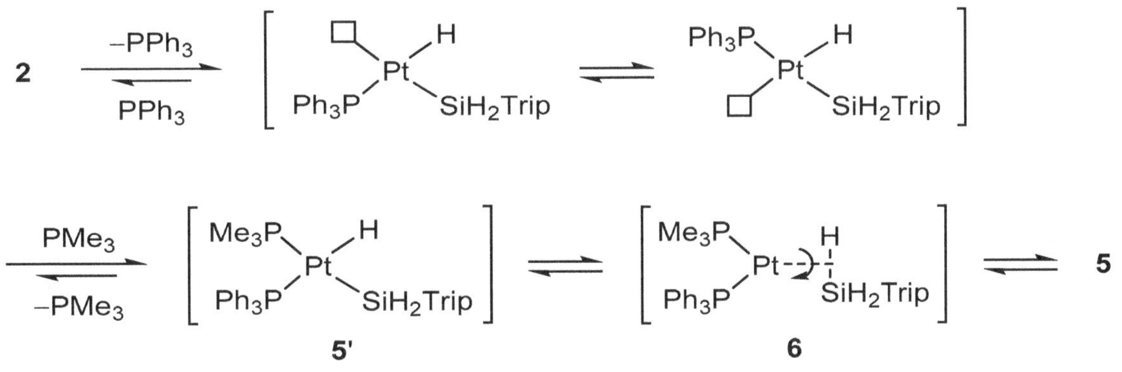

Scheme 4. Plausible reaction pathway for the formation of [PtH(SiH_2Trip)(PMe_3)(PPh_3)] **5**.

3. Materials and Methods

3.1. General Procedures

All of the experiments were performed under an argon atmosphere unless otherwise noted. Solvents were dried by standard methods and freshly distilled prior to use. ^{1}H, ^{13}C, and ^{31}P NMR spectra were recorded on Bruker DPX-400 or DRX-400 (400, 101 and 162 MHz, respectively), Avance-500 (500, 126 and 202 MHz, respectively) (Karlsruhe, Germany) spectrometers using $CDCl_3$ or C_6D_6 as the solvent at room temperature. ^{29}Si NMR spectra were recorded on Bruker Avance-500 (Karlsruhe, Germany) or JEOL EX-400 (Tokyo, Japan) (99.4 and 79.3 MHz, respectively) spectrometers using $CDCl_3$, CD_2Cl_2, C_6D_6, or THF-d_8 as the solvent at room temperature, unless otherwise noted. IR spectra were obtained on a Perkin-Elmer System 2000 FT-IR spectrometer (Walham, MA, USA). Elemental analyses were carried out at the Molecular Analysis and Life Science Center of Saitama University. All of the melting points were determined on a Mel-Temp capillary tube apparatus (Stafford, UK) and are uncorrected. 9-Triptycylsilane (TripSiH_3, **1**) [48] and [Pt(η^2-C_2H_4)(PPh_3)$_2$] [49] were prepared according to the reported procedures.

3.1.1. [*cis*-PtH(SiH_2Trip)(PPh_3)$_2$] (**2**)

A solution of TripSiH_3 **1** (50.9 mg, 0.179 mmol) and [Pt(η^2-C_2H_4)(PPh_3)$_2$] (149.3 mg, 0.199 mmol) in toluene (3 mL) was stirred at room temperature for 1 h to form a pale yellow solution. After the removal of the solvent in vacuo, the residual colorless solid was purified by washing with hexane to give [*cis*-PtH(SiH_2Trip)(PPh_3)$_2$] (**2**) (164.3 mg, 91%) as colorless crystals.

^{1}H NMR (400 MHz, $CDCl_3$): δ = −2.15 (dd, $^2J_{H\text{–}P(trans)}$ = 157, $^2J_{H\text{–}P(cis)}$ = 19, $^1J_{H\text{–}Pt}$ = 958 Hz, 1H, PtH), 4.68 (m, 2H, SiH_2), 5.28 (s, 1H, TripCH), 6.83–6.89 (m, 6H, Ar), 7.02–7.06 (m, 6H, Ar), 7.15–7.31 (m, 21H, Ar), 7.49 (t, J = 7 Hz, 6H, Ar), 7.81 (d, J = 7 Hz, 3H, Ar). ^{13}C{^{1}H} NMR (101 MHz, $CDCl_3$): δ = 46.6 (TripC), 55.3 (TripCH), 122.7 (Ar(CH)), 123.7 (Ar(CH) × 2), 126.2 (Ar(CH)), 127.8 (Ar(CH)), 127.9 (Ar(CH)), 129.4 (Ar(CH)), 129.6 (Ar(CH)), 134.1 (Ar(CH)), 134.2 (Ar(CH)),

133.7–134.8 (m, Ar(C)), 135.5 (d, $^{1}J_{C\text{–}P}$ = 38 Hz, Ar(C)), 148.4 (Ar(C)), 149.5 (Ar(C)). $^{31}P\{^{1}H\}$ NMR (162.0 MHz, $CDCl_3$): δ = 33.8 (d, $^{2}J_{P\text{–}P}$ = 15, $^{1}J_{P\text{–}Pt}$ = 2183 Hz), 34.5 (d, $^{2}J_{P\text{–}P}$ = 15, $^{1}J_{P\text{–}Pt}$ = 1963 Hz). $^{29}Si\{^{1}H\}$ NMR (79.3 MHz, CD_2Cl_2): δ = −40.6 (dd, $^{2}J_{Si\text{–}P(trans)}$ = 161, $^{2}J_{Si\text{–}P(cis)}$ = 15, $^{1}J_{Si\text{–}Pt}$ = 1220 Hz). IR (KBr, cm^{-1}): ν = 2041 (Pt–H), 2080 (Si–H). Anal. Calcd. for $C_{56}H_{46}P_2PtSi$: C, 66.99; H, 4.62. Found: C, 66.55; H, 4.61. Melting point: 123 °C (dec.).

3.1.2. [PtH(SiH_2Trip)(dppf)] (**3**)

A solution of **2** (40.5 mg, 0.040 mmol) and DPPF (28.2 mg, 0.048 mmol) in toluene (3 mL) was stirred at room temperature for 5 h. After the removal of the solvent in vacuo, the residual colorless solid was purified by washing with Et_2O and hexane to give [PtH(SiH_2Trip)(dppf)] (**3**) (37.1 mg, 0.035 mmol, 87%) as yellow crystals.

^{1}H NMR (400 MHz, $CDCl_3$): δ = −1.62 (dd, $^{2}J_{H\text{–}P(trans)}$ = 164, $^{2}J_{H\text{–}P(cis)}$ = 20, $^{1}J_{H\text{–}Pt}$ = 995 Hz, 1H, PtH), 3.86 (s, 2H, Cp), 4.20 (s, 2H, Cp), 4.37 (s, 2H, Cp), 4.58–4.64 (m, 4H, Cp and SiH_2), 5.30 (s, 1H, TripCH), 6.87–6.89 (m, 6H, Ar), 7.26 (m, 6H, Ar), 7.41 (br, 6 H, Ar), 7.66–7.81 (m, 14H, Ar). $^{13}C\{^{1}H\}$ NMR (101 Hz, $CDCl_3$): δ = 46.5 (m, TripC), 55.2 (TripCH), 71.3 (d, $^{3}J_{C\text{–}P}$ = 5 Hz, Cp(CH)), 72.1 (d, $^{3}J_{C\text{–}P}$ = 6 Hz, Cp(CH)), 74.5 (d, $^{2}J_{C\text{–}P}$ = 7 Hz, Cp(CH)), 75.6 (d, $^{2}J_{C\text{–}P}$ = 8 Hz, Cp(CH)), 79.3 (dd, $^{1}J_{C\text{–}P}$ = 42, $^{3}J_{C\text{–}P}$ = 5 Hz, Cp(C)), 80.7 (dd, $^{1}J_{C\text{–}P}$ = 47, $^{3}J_{C\text{–}P}$ = 6 Hz, Cp(C)), 122.7 (Ar(CH)), 123.71 (Ar(CH)), 123.68 (Ar(CH)), 126.4 (Ar(CH)), 127.7 (d, $^{3}J_{C\text{–}P}$ = 11 Hz, Ar(CH)), 128.0 (d, $^{3}J_{C\text{–}P}$ = 10 Hz, Ar(CH)), 129.9 (Ar(CH)), 130.1 (Ar(CH)), 134.3 (d, $^{2}J_{C\text{–}P}$ = 13 Hz, Ar(CH)), 134.6 (d, $^{2}J_{C\text{–}P}$ = 14 Hz, Ar(CH)), 134.8–135.5 (m, Ar(C)), 136.2 (d, $^{1}J_{C\text{–}P}$ = 43 Hz, Ar(C)), 148.4 (Ar(C)), 149.7 (Ar(C)). $^{31}P\{^{1}H\}$ NMR (202 MHz, $CDCl_3$): δ = 30.5 (d, $^{2}J_{P\text{–}P}$ = 21 Hz, $^{1}J_{Pt\text{–}P}$ = 2247 Hz), 34.5 (d, $^{2}J_{P\text{–}P}$ = 21, $^{1}J_{Pt\text{–}P}$ = 1837 Hz). $^{29}Si\{^{1}H\}$ NMR (79.3 MHz, $CDCl_3$): δ = −39.0 (dd, $^{2}J_{Si\text{–}P(trans)}$ = 167, $^{2}J_{Si\text{–}P(cis)}$ = 12, $^{1}J_{Si\text{–}Pt}$ = 1207 Hz). IR (KBr, cm^{-1}): ν = 2055 (Pt–H), 2081 (Si–H). Anal. Calcd. for $C_{54}H_{44}FeP_2PtSi$: C, 62.73; H, 4.29. Found: C, 62.70; H, 4.27. Melting point: 183 °C (dec.).

3.1.3. [PtH(SiH_2Trip)(dcpe)] (**4**)

A solution of **2** (35.6 mg, 0.035 mmol) and DCPE (18.8 mg, 0.044 mmol) in toluene (3 mL) was stirred at room temperature for 5 h. After the removal of the solvent in vacuo, the residual colorless solid was purified by washing with Et_2O and hexane to give [PtH(SiH_2Trip)(dcpe)] (**4**) (25.5 mg, 0.028 mmol, 80%) as colorless crystals.

^{1}H NMR (500 MHz, $CDCl_3$): δ = −0.45 (dd, $^{2}J_{H\text{–}P(trans)}$ = 165, $^{2}J_{H\text{–}P(cis)}$ = 13, $^{1}J_{Pt\text{–}H}$ = 1004 Hz, 1H, PtH), 1.16–1.51 (m, 20H, Cy), 1.65–1.86 (m, 24H, Cy), 2.16–2.19 (m, 2H, Cy), 2.30–2.37 (m, 2H, Cy), 5.32 (s, 1H, TripCH), 5.50 (dd, $^{2}J_{H\text{–}H}$ = 15, $^{3}J_{H\text{–}P(trans)}$ = 6, $^{2}J_{Pt\text{–}H}$ = 31 Hz, 2H, SiH_2), 6.84–6.90 (m, 6H, Ar), 7.30–7.32 (d, J = 7 Hz, 3H, Ar), 7.94–7.96 (d, J = 7 Hz, 3H, Ar). $^{13}C\{^{1}H\}$ NMR (101 Hz, $CDCl_3$): δ = 23.2 (dd, $^{3}J_{C\text{–}P}$ = 21, 16 Hz, PCH_2), 26.2 (dd, $^{3}J_{C\text{–}P}$ = 23, 21 Hz, PCH_2), 26.9 (d, $^{3}J_{C\text{–}P}$ = 14 Hz, PCy(CH_2)), 26.4 (d, $^{3}J_{C\text{–}P}$ = 13 Hz, PCy(CH_2)), 26.8 (d, $^{3}J_{C\text{–}P}$ = 10 Hz, PCy(CH_2)), 27.0 (d, $^{3}J_{C\text{–}P}$ = 12 Hz, PCy(CH_2)), 28.9 (m, PCy(CH_2) × 2), 29.7 (m, PCy(CH_2)), 35.3–35.8 (m, PCy(CH) × 2), 46.8 (TripC), 55.4 (TripCH), 122.7 (Ar(CH)), 123.6 (Ar(CH)), 126.9 (Ar(CH)), 148.6 (Ar(C)), 159.2 (Ar(C)). $^{31}P\{^{1}H\}$ NMR (162 Hz, $CDCl_3$): δ = 69.2 (s, $^{1}J_{Pt\text{–}P}$ = 1809 Hz), 85.3 (s, $^{1}J_{Pt\text{–}P}$ = 1678 Hz). $^{29}Si\{^{1}H\}$ NMR (79.3 MHz, CD_2Cl_2): δ = −44.6 (dd, $^{2}J_{Si\text{–}P(trans)}$ = 173, $^{2}J_{Si\text{–}P(cis)}$ = 11, $^{1}J_{Si\text{–}Pt}$ = 1253 Hz). IR (KBr, cm^{-1}): ν = 2057 (Pt–H), 2081 (Si–H). Anal. Calcd for $C_{46}H_{64}P_2PtSi$: C, 61.24; H, 7.15. Found: C, 61.10; H, 7.10. Melting point: 134 °C (dec.).

3.1.4. [PtH(SiH_2Trip)(PMe_3)(PPh_3)] (**5**)

A toluene solution of PMe_3 (1.0 M, 0.2 mL, 0.200 mmol) was added to a solution of **2** (98.0 mg, 0.098 mmol) in toluene (3.5 mL) at room temperature. The reaction mixture was stirred at room temperature for 30 min. After the removal of the solvent in vacuo, the residual colorless solid was purified by washing with Et_2O and hexane to give [PtH(SiH_2Trip)(PMe_3)(PPh_3)] **5** (75.6 mg, 94%) as colorless crystals.

^{1}H NMR (400 MHz, C_6D_6): δ = −1.91 (d, $^2J_{H-P(trans)}$ = 157, $^1J_{H-Pt}$ = 899 Hz, 1H, PtH), 1.03–1.11 (m, 9H, PMe), 5.34 (s, 1H, TripCH), 5.95–5.97 (m, 2H, SiH_2), 6.82–6.97 (m, 15H, Ar), 7.31 (d, J = 7 Hz, 3H, Ar), 7.57 (br, 6H, Ar), 8.50 (d, J = 7 Hz, 3H, Ar). $^{13}C\{^1H\}$-NMR (101 Hz, $CDCl_3$): δ = 16.4–17.1 (m, PMe), 53.6 (TripC), 55.4 (TripCH), 123.0 (Ar(CH)), 123.9 (Ar(CH)), 124.1 (Ar(CH)), 126.5 (Ar(CH)), 128.5 (d, $^3J_{C-P}$ = 10 Hz, Ar(CH)), 130.0 (Ar(CH)), 134.4 (d, $^2J_{C-P}$ = 13 Hz, Ar(CH)), 135.6 (d, $^1J_{C-P}$ = 36 Hz, Ar(CH), 148.5 (Ar(C)), 149.7 (Ar(C)). $^{31}P\{^1H\}$-NMR (162 Hz, C_6D_6): δ = −22.9 (s, $^1J_{Pt-P}$ = 2115 Hz), 37.4 (br, $^1J_{Pt-P}$ = 1833 Hz). $^{29}Si\{^1H\}$-NMR (79.3 MHz, THF-d_8, 223 K) δ −43.9 (d, $^2J_{Si-P(trans)}$ = 151 Hz). IR (KBr, cm^{-1}) ν = 2029 (Pt–H), 2054 (Si–H). Anal. Calcd for $C_{41}H_{40}P_2PtSi$: C, 60.21; H, 4.93. Found: C, 60.57; H, 5.00. Melting point: 115 °C (dec.).

3.2. *X-ray Crystallographic Studies of* **2**, **3**, *and* **5**

Colorless single crystals of **2** were grown by the slow evaporation of its saturated toluene solution, and single crystals of **3** and **5** were grown by the slow evaporation of its saturated CH_2Cl_2 and hexane solution. The intensity data were collected at 103 K on a Bruker AXS SMART diffractometer (Karlsruhe, Germany) employing graphite-monochromatized Mo Kα radiation (λ = 0.71073 Å). The structures were solved by direct methods and refined by full-matrix least-squares procedures on F^2 for all reflections (*SHELX*-97) [50]. Hydrogen atoms, except for the PtH and SiH hydrogens of **2**, **3**, and **5**, were located by assuming ideal geometry, and were included in the structure calculations without further refinement of the parameters. Full details of the crystallographic analysis and accompanying cif files (see Supplementary Materials) may be obtained free of charge from the Cambridge Crystallographic Data Centre (CCDC numbers 1577277, 1577278, and 1577279) via http://www.ccdc.cam.ac.uk/conts/retrieving.html (or from the CCDC, 12 Union Road, Cambridge CB2 1EZ, UK.

3.2.1. [*cis*-PtH(SiH_2Trip)(PPh_3)$_2$] (**2**)

$C_{56}H_{46}P_2PtSi$, C_7H_8, M_W = 1096.18, triclinic, space group *P*-1, a = 12.7971(6) Å, b = 13.4847(6) Å, c = 14.8861(7) Å, α = 99.2791(10)°, β = 99.2791(10)°, γ = 90.3020(10)°, V = 2472.4(2) Å^3, Z = 2, $D_{calc.}$ = 1.472 g·cm^{-3}, R_1 ($I > 2\sigma I$) = 0.0310, wR_2 (all data) = 0.0718 for 11639 reflections and 617 parameters, *GOF* = 1.025.

3.2.2. [PtH(SiH_2Trip)(dppf)] (**3**)

$C_{54}H_{44}FeP_2PtSi$, CH_2Cl_2, MW = 1118.79, monoclinic, space group $P2_1$, a = 12.2585(6) Å, b = 15.5003(7) Å, c = 12.9367(6) Å, β = 110.8060(10)°, V = 2297.81(19) Å^3, Z = 2, D_{calc} = 1.617 g·cm^{-3}, R_1 ($I > 2\sigma I$) = 0.0484, wR_2 (all data) = 0.1171 for 8350 reflections, 571 parameters, and 2 restraints, *GOF* = 1.018.

3.2.3. [PtH(SiH_2Trip)(PMe_3)(PPh_3)] (**5**)

$C_{41}H_{40}P_2PtSi$, CH_2Cl_2, M_W = 902.78, orthorhombic, space group *Pbca*, a = 15.9074(6) Å, b = 20.9224(8) Å, c = 22.6113(9) Å, V = 7525.5(5) Å^3, Z = 8, D_{calc} = 1.594 g·cm^{-3}, R_1 ($I > 2\sigma I$) = 0.0325, wR_2 (all data) = 0.0706 for 7011 reflections and 448 parameters, *GOF* = 1.026.

4. Conclusions

We have demonstrated that the oxidative addition of the sterically bulky primary silane, $TripSiH_3$ **1** with [Pt(η^2-C_2H_4)(PPh_3)$_2$] in toluene, resulted in the formation of the mononuclear (hydrido)(dihydrosilyl) complex [*cis*-PtH(SiH_2Trip)(PPh_3)$_2$] **2**. The ligand-exchange reactions of **2** with free chelating bis(phosphine)s such as DPPF or DCPE resulted in the formations of a series of (hydrido)(dihydrosilyl) complexes [PtH(SiH_2Trip)(L)] (**3**: L = dppf, **4**: L = dcpe). In contrast, the reaction of **2** with an excess amount of PMe_3 in toluene quantitatively produced [PtH(SiH_2Trip)(PMe_3)(PPh_3)] **5**. The latter is of particular interest, as it represents the first platinum complex having different simple phosphine ligands such as a weakly electron-donating PPh_3 and a strongly electron-donating PMe_3. Further investigations on the reactivity of these complexes are currently in progress.

Acknowledgments: This work was partially supported by JSPS KAKENHI Grant Number T17K05771 (to Norio Nakata). We are grateful to Kohtaro Osakada and Makoto Tanabe (Tokyo Institute of Technology, Japan) for their assistance in measurement of ^{29}Si NMR spectroscopy.

Author Contributions: Norio Nakata, Nanami Kato, and Norio Sekizawa contributed to the synthesis and data analysis; Norio Nakata performed X-ray crystallography and wrote the manuscript; Norio Nakata and Akihiko Ishii proposed idea on the design of the experiment, reviewed and approved the final manuscript.

References

1. Corey, J.Y. Reactions of hydrosilanes with transition metal complexes. *Chem. Rev.* **2016**, *116*, 11291–11435. [CrossRef] [PubMed]
2. Marciniec, B. *Comprehensive Handbook on Hydrosilylation*; Pergamon Press: Oxford, UK, 1992; ISBN 0-08-040272-0.
3. Ojima, I.; Li, Z.; Zhu, J. Recent Advances in Hydrosilylation and Related Reactions. In *The Chemistry of Organic Silicon Compounds*; Patai, S., Rappoport, Z., Eds.; John Wiley and Sons: New York, NY, USA, 1998; pp. 1687–1792.
4. Marciniec, B.; Maciejewski, H.; Pietraszuk, C.; Pawluc, P. *Hydrosilylation: A Comprehensive Review on Recent Advances*; Marciniec, B., Ed.; Advances in Silicon Science; Springer: Berlin, Germany, 2009; Volume 1, ISBN 978-1-4020-8172-9.
5. Chalk, A.J.; Harrod, J.F. Homogeneous catalysis. II. The mechanism of the hydrosilation of olefins catalyzed by group VIII metal complexes. *J. Am. Chem. Soc.* **1965**, *87*, 16–21. [CrossRef]
6. Ozawa, F. The chemistry of organo(silyl)platinum(II) complexes relevant to catalysis. *J. Organomet. Chem.* **2000**, *611*, 332–342. [CrossRef]
7. Sharma, H.; Pannell, K.H. Activation of the Si–Si bond by transition metal complexes. *Chem. Rev.* **1995**, *95*, 1351–1374. [CrossRef]
8. Suginome, M.; Ito, Y. Transition-metal-catalyzed additions of silicon−silicon and silicon−heteroatom bonds to unsaturated organic molecules. *Chem. Rev.* **2000**, *100*, 3221–3256. [CrossRef] [PubMed]
9. Ojima, I.; Inaba, S.-I.; Kogure, T.; Nagai, Y. The action of tris(triphenylphosphine)chlororhodium on polyhydromonosilanes. *J. Organomet. Chem.* **1973**, *55*, C7–C8. [CrossRef]
10. Chang, L.S.; Johnson, M.P.; Fink, J. Polysilyl complexes of platinum—Synthesis and thermochemistry. *Organometallics* **1989**, *8*, 1369–1371. [CrossRef]
11. Harrod, J.F.; Mu, Y.; Samuel, E. Catalytic dehydrocoupling: A general method for the formation of element-element bonds. *Polyhedron* **1991**, *10*, 1239–1245. [CrossRef]
12. Woo, H.G.; Walzer, J.F.; Tilley, T.D. A σ-bond metathesis mechanism for dehydropolymerization of silanes to polysilanes by d^0 metal catalysts. *J. Am. Chem. Soc.* **1992**, *114*, 7047–7055. [CrossRef]
13. Grimmond, B.J.; Corey, J.Y. Amino-functionalized zirconocene $(C_5H_4CH(Me)NMe_2)_2ZrCl_2$, a catalyst for the dehydropolymerization of $PhSiH_3$: Probing of peripheral substituent effects on catalyst dehydropolymerization activity. *Inorg. Chim. Acta* **2002**, *330*, 89–94. [CrossRef]
14. Kim, Y.-J.; Park, J.-I.; Lee, S.-C.; Osakada, K.; Tanabe, M.; Choi, J.-C.; Koizumi, T.; Yamamoto, T. *Cis* and *trans* isomers of $Pt(SiHAr_2)_2(PR_3)_2$ (R = Me, Et) in the solid state and in solutions. *Organometallics* **1999**, *18*, 1349–1352. [CrossRef]
15. Braddock-Wilking, J.; Corey, J.Y.; Trankler, K.A.; Xu, H.; French, L.M.; Praingam, N.; White, C.; Rath, N.P. Spectroscopic and reactivity studies of platinum−silicon monomers and dimers. *Organometallics* **2006**, *25*, 2859–2871. [CrossRef]
16. Arii, H.; Takahashi, M.; Noda, A.; Nanjo, M.; Mochida, K. Spectroscopic and structural studies of thermally unstable intermediates generated in the reaction of $[Pt(PPh_3)_2(\eta^2\text{-}C_2H_4)]$ with dihydrodisilanes. *Organometallics* **2008**, *27*, 1929–1935. [CrossRef]
17. Ahrens, T.; Braun, T.; Braun, B. Activation of 1,2-dihydrodisilanes at platinum(0) phosphine complexes: Syntheses and structures of bissilyl platinum(II) complexes. *Z. Anorg. Allg. Chem.* **2014**, *640*, 93–99. [CrossRef]

18. Li, Y.-H.; Huang, Z.-F.; Li, X.-A.; Lai, W.-Y.; Wang, L.-H.; Ye, S.-H.; Cui, L.-F.; Wang, S. Synthesis and structural characterization of a novel bis(silyl) platinum(II) complex bearing SiH_3 ligand. *J. Orgnomet. Chem.* **2014**, *749*, 246–250. [CrossRef]
19. Auburn, M.; Ciriano, M.; Howard, J.A.K.; Murray, M.; Pugh, N.J.; Spencer, J.L.; Stone, F.G.A.; Woodward, P. Synthesis of bis-μ-diorganosilanediyl-*af*-dihydridobis(triorganophosphine)diplatinum complexes: Crystal and molecular structure of $[\{PtH(\mu\text{-}SiMe_2)[P(C_6H_{11})_3]\}_2]$. *J. Chem. Soc. Dalton Trans.* **1980**, 659–666. [CrossRef]
20. Zarate, E.A.; Tessier-Youngs, C.A.; Young, W.J. Synthesis and structural characterization of platinum–silicon dimers with unusually short cross-ring silicon–silicon interactions. *J. Am. Chem. Soc.* **1988**, *110*, 4068–4070. [CrossRef]
21. Levchinsky, Y.; Rath, N.P.; Braddock-Wilking, J. Reaction of a symmetrical diplatinum complex containing bridging μ-η^2-H–SiH(IMP) ligands (IMP = 2-isopropyl-6-methylphenyl) with PMe_2Ph. Formation and characterization of $\{(PhMe_2P)_2Pt[\mu\text{-}SiH(IMP)]\}_2$. *Organometallics* **1999**, *18*, 2583–2586. [CrossRef]
22. Sanow, L.M.; Chai, M.; McConnville, D.B.; Galat, K.J.; Simons, R.S.; Rinaldi, P.L.; Young, W.; Tessier, C.A. Platinum–silicon four-membered rings of two different structural types. *Organometallics* **2000**, *19*, 192–205. [CrossRef]
23. Braddock-Wilking, J.; Levchinsky, Y.; Rath, N.P. Synthesis and characterization of diplatinum complexes containing bridging μ-η^2-H–SiHAr ligands. X-ray crystal structure determination of $\{(Ph_3P)Pt[\mu\text{-}\eta^2\text{-}H\text{–}SiHAr]\}_2$ (Ar = 2,4,6-$(CF_3)_3C_6H_2$, C_6Ph_5). *Organometallics* **2000**, *19*, 5500–5510. [CrossRef]
24. Osakada, K.; Tanabe, M.; Tanase, T. A triangular triplatinum complex with electron-releasing $SiPh_2$ and PMe_3 Ligands: $[\{Pt(\mu\text{-}SiPh_2)(PMe_3)\}_3]$. *Angew. Chem. Int. Ed.* **2000**, *39*, 4053–4054. [CrossRef]
25. Braddock-Wilking, J.; Corey, J.Y.; Trankler, K.A.; Dill, K.M.; French, L.M.; Rath, N.P. Reaction of silafluorenes with $(Ph_3P)_2Pt(\eta_2\text{-}C_2H_4)$: Generation and characterization of Pt–Si monomers, dimers, and trimers. *Organometallics* **2004**, *23*, 4576–4584. [CrossRef]
26. Braddock-Wilking, J.; Corey, J.Y.; French, L.M.; Choi, E.; Speedie, V.J.; Rutherford, M.F.; Yao, S.; Xu, H.; Rath, N.P. Si–H bond activation by $(Ph_3P)_2Pt(\eta_2\text{-}C_2H_4)$ in dihydrosilicon tricycles that also contain O and N heteroatoms. *Organometallics* **2006**, *25*, 3974–3988. [CrossRef]
27. Tanabe, M.; Ito, D.; Osakada, K. Diplatinum complexes with bridging silyl ligands. Si–H bond activation of μ-silyl ligand leading to a new platinum complex with bridging silylene and silane ligands. *Organometallics* **2007**, *26*, 459–462. [CrossRef]
28. Tanabe, M.; Ito, D.; Osakada, K. Ligand exchange of diplatinum complexes with bridging silyl ligands involving Si–H bond cleavage and formation. *Organometallics* **2008**, *27*, 2258–2267. [CrossRef]
29. Tanabe, M.; Kamono, M.; Tanaka, K.; Osakada, K. Triangular triplatinum complex with four bridging Si ligands: Dynamic behavior of the molecule and catalysis. *Organometallics* **2017**, *36*, 1929–1935. [CrossRef]
30. Mullica, D.F.; Sappenfield, E.L. Synthesis and X-ray structural investigation of *cis*-hydrido(triphenylsilyl)-1,4-butanediyl-*bis*-(dicyclohexylphosphine)platinum(II). *Polyhedron* **1991**, *10*, 867–872. [CrossRef]
31. Latif, L.A.; Eaborn, C.; Pidcock, A.P. Square planar platinum(II) complexes. Crystal structures of *cis*- bis(triphenylphosphine)hydro (triphenylstannyl)platinum(II) and *cis*-bis(triphenylphosphine)-hydro(triphenylsilyl)platinum(II). *J. Organomet. Chem.* **1994**, *474*, 217–221. [CrossRef]
32. Koizumi, T.; Osakada, K.; Yamamoto, T. Intermolecular transfer of triarylsilane from $RhCl(H)(SiAr_3)[P(i\text{-}Pr)_3]_2$ to a platinum(0) complex, giving *cis*-$PtH(SiAr_3)(PEt_3)_2$ (Ar = C_6H_5, C_6H_4F-*p*, C_6H_4Cl-*p*). *Organometallics* **1997**, *16*, 6014–6016. [CrossRef]
33. Chan, D.; Duckett, S.B.; Heath, S.L.; Khazal, I.G.; Perutz, R.N.; Sabo-Etienne, S.; Timmins, P.L. Platinum bis(tricyclohexylphosphine) silyl hydride complexes. *Organometallics* **2004**, *23*, 5744–5756. [CrossRef]
34. Arii, H.; Takahashi, M.; Nanjo, M.; Mochida, K. Syntheses of mono- and dinuclear silylplatinum complexes bearing a diphosphino ligand via stepwise bond activation of unsymmetric disilanes. *Dalton Trans.* **2010**, *39*, 6434–6440. [CrossRef] [PubMed]
35. Simons, R.S.; Sanow, L.M.; Galat, K.J.; Tessier, C.A.; Young, W.J. Synthesis and Structural Characterization of the Mono(silyl)platinum(II) Complex *cis*-$(2,6\text{-}Mes_2C_6H_3(H)_2Si)Pt(H)(PPr_3)_2$. *Organometallics* **2000**, *19*, 3994–3996. [CrossRef]

36. Wang, S.; Yu, X.-F.; Li, N.; Yang, T.; Lai, W.-Y.; Mi, B.-X.; Li, J.-F.; Li, Y.-H.; Wang, L.-H.; Huang, W. Synthesis and structural studies of a rare bis(phosphine) (hydrido) (silyl) platinum(II) complex containing a Si–Si single bond. *J. Orgnomet. Chem.* **2015**, *776*, 113–116. [CrossRef]
37. Nakata, N.; Fukazawa, S.; Ishii, A. Synthesis and crystal structures of the first stable mononuclear dihydrogermyl(hydrido) platinum(II) complexes. *Organometallics* **2009**, *28*, 534–538. [CrossRef]
38. Nakata, N.; Fukazawa, S.; Kato, N.; Ishii, A. Palladium(II) hydrido complexes having a primary silyl or germyl ligand: Synthesis, crystal structures, and dynamic behavior. *Organometallics* **2011**, *30*, 4490–4493. [CrossRef]
39. Nakata, T.; Sekizawa, N.; Ishii, A. Cationic dinuclear platinum and palladium complexes with bridging hydrogermylene and hydrido ligands. *Chem. Commun.* **2015**, *51*, 10111–10114. [CrossRef] [PubMed]
40. Kalläne, S.; Laubenstein, R.; Braun, T.; Dietrich, M. Activation of Si–Si and Si–H bonds at a platinum bis(diphenylphosphanyl)ferrocene (dppf) complex: Key steps for the catalytic hydrogenolysis of disilanes. *Eur. J. Inorg. Chem.* **2016**, 530–537. [CrossRef]
41. Boyle, R.C.; Mague, J.T.; Fink, M.J. The first stable mononuclear silyl palladium hydrides. *J. Am. Chem. Soc.* **2003**, *125*, 3228–3229. [CrossRef] [PubMed]
42. Boyle, R.C.; Pool, D.; Jacobsen, H.; Fink, M.J. Dynamic processes in silyl palladium complexes: Evidence for intermediate Si−H and Si−Si σ-complexes. *J. Am. Chem. Soc.* **2006**, *128*, 9054–9055. [CrossRef] [PubMed]
43. Feducia, J.A.; Campbell, A.N.; Doherty, M.Q.; Gagné, M.R. Modular catalysts for diene cycloisomerization: Rapid and enantioselective variants for bicyclopropane synthesis. *J. Am. Chem. Soc.* **2006**, *128*, 13290–13297. [CrossRef] [PubMed]
44. Campbell, A.N.; Gagné, M.R. Room-temperature β-H elimination in $(P_2P)Pt(OR)$ cations: Convenient synthesis of a platinum hydride. *Organometallics* **2007**, *26*, 2788–2790. [CrossRef]
45. Pan, B.; Xu, Z.; Bezpalko, M.W.; Foxman, B.M.; Thomas, C.M. *N*-Heterocyclic phosphenium ligands as sterically and electronically-tunable isolobal analogues of nitrosyls. *Inorg. Chem.* **2012**, *51*, 4170–4179. [CrossRef] [PubMed]
46. Frey, U.; Helm, L.; Merbach, A.E.; Romeo, R. High-pressure NMR kinetics. Part 41. Dissociative substitution in four-coordinate planar platinum(II) complexes as evidenced by variable-pressure high-resolution proton NMR magnetization transfer experiments. *J. Am. Chem. Soc.* **1989**, *111*, 8161–8165. [CrossRef]
47. Romeo, R.; Grassi, A.; Scolaro, L.M. Factors affecting reaction pathways in nucleophilic substitution reactions on platinum(II) complexes: A comparative kinetic and theoretical study. *Inorg. Chem.* **1992**, *31*, 4383–4390. [CrossRef]
48. Brynda, M.; Bernardinelli, G.; Dutan, C.; Geoffroy, M. Kinetic stabilization of primary hydrides of main group elements. The synthesis of an air-stable, crystalline arsine and silane. *Inorg. Chem.* **2003**, *42*, 6586–6588. [CrossRef] [PubMed]
49. Cook, C.D.; Jauhal, G.S. Chemistry of low-valent complexes. II. Cyclic azo derivatives of platinum. *J. Am. Chem. Soc.* **1968**, *90*, 1464–1467. [CrossRef]
50. Sheldrick, G.M. A short history of *SHELX*. *Acta Cryst. A* **2008**, *64*, 112–122. [CrossRef] [PubMed]

8

Synthesis of a α-Chlorosilyl Functionalized Donor-Stabilized Chlorogermylene

Debabrata Dhara [1], Volker Huch [2], David Scheschkewitz [2,*] and Anukul Jana [1,*]

1 Tata Institute for Fundamental Research Hyderabad Gopanpally, Hyderabad-500107, Telangana, India; debabratad@tifrh.res.in

2 Krupp-Chair of General and Inorganic Chemistry, Saarland University, 66123 Saarbrücken, Germany; huch@mx.uni-saarland.de

* Correspondence: scheschkewitz@mx.uni-saarland.de (D.S.); ajana@tifrh.res.in (A.J.)

Abstract: Peripherally functionalized low-valent main group species allow for the introduction/interconversion of functional groups without increasing the formal oxidation state of the main group center. Herein, we report a straightforward method for the incorporation of a α-chlorosilyl moiety adjacent to the NHC-coordinated germanium(II) center.

Keywords: silylene; germylene; *N*-heterocyclic carbene; oxidative addition

1. Introduction

In recent years, the chemistry of the heavier analogues of carbenes (tetrylenes) has been expanded beyond mere synthetic curiosity [1–3] towards application in synthesis. Heavier carbene analogues are applied as a donor ligands in low-valent main group species [4,5] as well as in transition metal complexes [6,7]. An increasing number of examples show competitive catalytic activity in different organic transformations [8]. As the complexity of the tetrylenes increases with more intricate ligand architectures [9], functionalization protocols in the presence of uncompromised low-valent Group 14 centers conveniently allow for a comparatively straightforward diversification in the final stages of ligand synthesis. While the interconversion of functional groups is just beginning to emerge in the case of heavier multiple bonds [10], numerous examples have been reported for the heavier tetrylenes [11–15]. Recently, Scheschkewitz et al. have taken a similar approach with the synthesis of the multiply functional NHC-coordinated silagermenylidenes, **I** and **II** (Scheme 1) [16,17], which serve as precursors for cyclic NHC-coordinated germylenes of type **III**, **IV**, and **V** under consumption of the Si=Ge bond, but retention of the low-valent germanium center (Scheme 1) [16,18,19]. The leaving group characteristics of the peripheral chloro functionality of **II** can be exploited for the incorporation of different organic substituents by treatment with organolithium reagents in order to fine-tune the steric requirements of the ligand scaffold of cyclic germylenes of type **IV** [20,21]. Directly chloro-functionalized silylenes and germylenes are readily converted to a variety of novel low-valent group 14 compounds by functional group interconversion at the low-valent tetrel center [22].

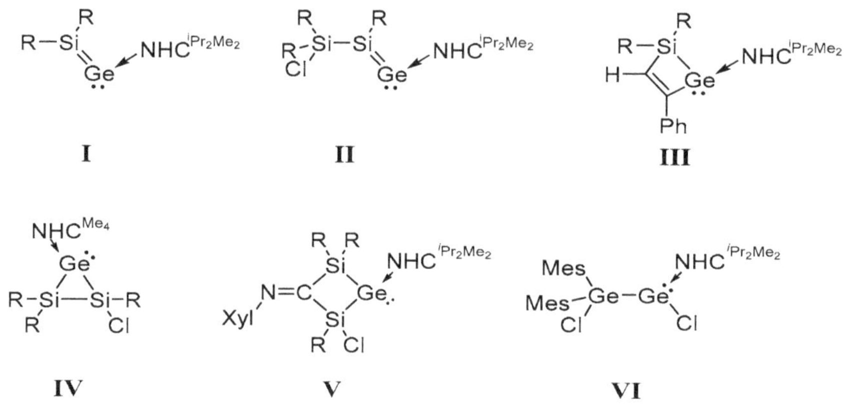

Scheme 1. Chemical structures of α-chlorosilyl-functionalized silagermenyledene **I**, and α-chloro-functionalized germylenes **II–IV** (R = Tip = 2,4,6-$iPr_3C_6H_2$, $NHC^{iPr_2Me_2}$ = 1,3-diisopropyl-4,5-dimethylimidazol-2-ylidene, NHC^{Me_4} = 1,3,4,5-tetramethylimidazol-2-ylidene, Xyl = 2,6-$Me_2C_6H_3$, and Mes = 2,6-$Me_2C_6H_3$).

An additional leaving group adjacent to a chloro-functionalized heavier carbene center would in principle provide a precursor for the synthesis of further examples of heavier vinylidene such as **I** and **II**. The synthesis of an NHC-coordinated (chlorogermyl)chlorogermylene, **IV**, from NHC-coordinated diaryl germylene and NHC- or 1,4-dioxane coordinated dichlorogermylene has been reported by the groups of Baines and Tobitah [23,24]. Herein, we now report the synthesis of NHC-stabilized (chlorosilyl)chlorogermylene, **1**.

2. Results and Discussions

We anticipated that West's *N*-heterocyclic silylene, **2** [25] would insert into the Ge–Cl bond of the NHC-germanium(II)dichloride adduct **3** [26] as it is well known for the oxidative addition of different types of bonds e.g., C–Cl [27] and Ge–N [28]. Indeed, the reaction of **2** and **3** in a 1:1 ratio in toluene at room temperature afforded the NHC-stabilized (chlorosilyl)chlorogermylene **1** which was isolated as a crystalline compound (Scheme 2). We did not obtain any indication for the formation of donor–acceptor adducts between **2** and **3** or rearrangement products as often described for reactions of silylenes and germylenes [29,30].

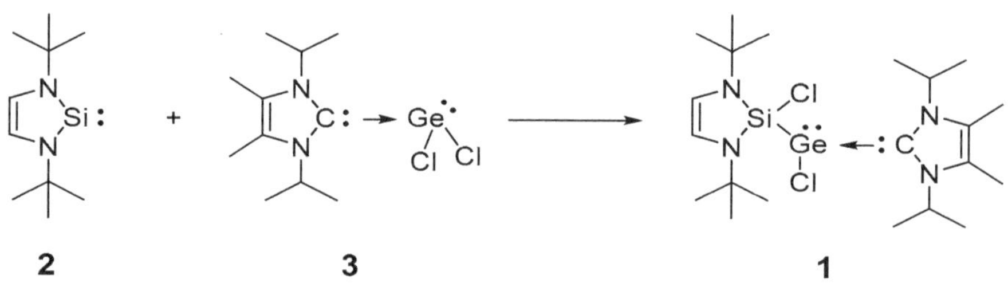

Scheme 2. Synthesis of **1**.

Compound **1** was characterized in solution state by NMR spectroscopy as well as in solid state by single crystal X-ray molecular structure determination. The insertion of the silylene into the Ge–Cl bond turns the germanium atom into a center of chirality. As a result the two diastereotopic C–*H* protons of C_2N_2Si-moiety give rise to two doublets at δ = 5.87 and 6.03 ppm ($^1J_{(H, H)}$ = 3.92 Hz) in the ^1H NMR. Similarly, the two *t*Bu groups show ^1H NMR resonances at δ = 1.29 and 1.65 ppm. Hindered rotation can be excluded as the explanation for the doubling of these resonances. Despite the increased congestion about the germanium center, the NHC retains the local rotational C_2-symmetry in solution: the ^1H NMR shows only a single septet for the two C*H* moieties of the isopropyl groups. In contrast, there are again two signals for the adjacent diastereotopic methyl groups of the *N*-isopropyl moiety. In $^{13}C\{^1H\}$ NMR, the carbenic carbon shows a resonance at δ = 171.01 ppm, which is similar to the chemical shifts observed for other NHC-coordinated Ge(II) compounds [18,19]. In $^{29}Si\{^1H\}$ NMR, the singlet at δ = −3.39 ppm is strongly highfield shifted compared with that of the free *N*-heterocyclic silylene (δ = +78.3 ppm) [25]. Notably, even repeated crystallization of **1** did not yield NMR spectra uncontaminated by residual **2**, which led us to speculate about the reversibility of the oxidative Ge–Cl addition to the silylene. NMR at variable temperatures, however, did not show any temperature dependence of the sample composition.

Nonetheless, single crystals of **1** suitable for a X-ray diffraction study were obtained from saturated toluene solution at −20 °C after one day. Compound **1** crystallizes in the monoclinic $P2_1/c$ space group. However in the obtained single crystal X-ray diffraction data we did not see any residual electron density for the cocrystalization of **2** along with **1**. Analysis of molecular structure determination revels the presence of α-chlorosilyl moiety adjacent to the NHC-coordinated germanium(II) center; which was anticipated from the solution state structure (Figure 1). The Si–Ge bond length is 2.4969(7) Å which is close to reported Si(IV)–Ge(II) bond length [19]. The distance between carbenic carbon and germanium(II) center is 2.081(2) Å, which is slightly shorter than that of the corresponding NHC-coordinated germanium(II)dichloride (2.106(3) Å) [26].

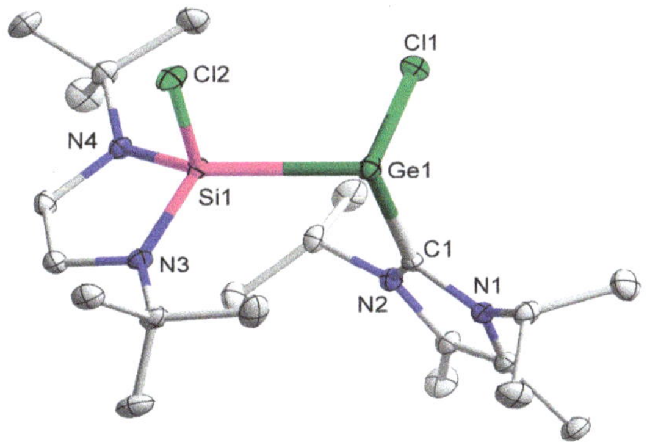

Figure 1. Molecular structure of **1** at 30% probability level, all H-atoms were deleted for clarity. Selected bond lengths (Å) and bond angles (deg.): Ge1–C1 2.081(2), Ge1–Cl1 2.2891(8), Ge1–Si1 2.4969(7); C1–Ge1–Cl1 92.58(7), C1–Ge1–Si1 102.56(7), Cl–Ge1–Si1 101.83(3).

3. Materials and Methods

3.1. General Information

All manipulation were carried out under an argon atmosphere using either a Schlenk line technique or inside a GloveBox. All solvents were dried by Innovative Technology solvent purification system. Compounds **2** [25] and **3** [26] were prepared according to literature procedures. Benzene-d6 was dried and distilled over potassium under argon. NMR spectra were recorded on a Bruker Avance III 300 MHz NanoBay NMR spectrometer (Bruker, Switzerland). ^1H and $^{13}C\{^1H\}$ NMR spectra were referenced to the peaks of residual protons of the deuterated solvent (^1H) or the deuterated solvent itself ($^{13}C\{^1H\}$). $^{29}Si\{^1H\}$ NMR spectra were referenced to external $SiMe_4$.

3.2. Experimental Details

Synthesis of compound **1**: 25-mL dry and degassed toluene were added to a Schlenk flask containing **2** (0.242 g, 1.23 mmol) and **3** (0.4 g, 1.23 mmol) at −78 °C. The mixture is brought to room temperature within one hour and stirred continuously for another two hours. Removal of the solvent in vacuum and washing of the solid residue with *n*-hexane was followed by extraction with 20 mL warm toluene. The resulting yellow solution was concentrated to about 15 mL and kept at −20 °C for one day to get bright yellow crystals of the desired compound, **1**, suitable for single crystal X-ray diffraction study. Despite apparently uniform crystals, a pure sample of compound **1** without free silylene **2** could not be obtained. Yield: 0.360 g (56% which include 13% of compound **2**). **^{1}H NMR** (300 MHz, C_6D_6, 298 K): δ = 1.13 (d, $^1J_{(H,H)}$ = 7.02 Hz, 6H, CH(CH_3)$_2$), 1.24 (d, $^1J_{(H,H)}$ = 6.9 Hz, 6H, CH(CH_3)$_2$), 1.29 (s, 9H, N(CH_3)$_3$), 1.51 (s, 6H, CCH_3), 1.63 (s, 9H, N(CH_3)$_3$), 5.74 (sept, $^1J_{(H,H)}$ = 7.02 Hz, 2H, *CH*(CH_3)$_2$), 5.87 (d, $^1J_{(H,H)}$ = 3.92 Hz, 1H, CHC*H*) 6.03 (d, 1H, $^1J_{(H,H)}$ = 3.92 Hz, *C*HCH) (1.42 and 6.76 ppm refer to the resonances for compound **2**) ppm. **^{13}C{^{1}H} NMR** (75.4 MHz, C_6D_6, 298 K): δ = 10.35 (2C, CCH_3), 22.06(4C, CH(CH_3)$_2$), 31.97 (3C, C(CH_3)$_3$), 32.15 (3C, C(CH_3)$_3$), 51.97 (1C, *C*(CH_3)$_3$), 52.49 (1C, *C*(CH_3)$_3$), 54.07 (2C, CH(CH_3)$_2$), 113.70 (1C, CHCH), 115.26 (1C, CHCH), 127.13 (2C, CCH_3), 171.01 (1C, NCN). **^{29}Si{^{1}H} NMR** (59.6 MHz, C_6D_6, 298 K): δ = −3.39 ppm.

3.3. X-ray Crystallographic Analysis

Single crystals of **1** were obtained from saturated toluene solution at −20 °C. Intensity data were collected on a Bruker SMART APEX CCD diffractometer (Bruker, Germany with a Mo Kα radiation (λ = 0.71073 Å) at T = 182(2) K. The structures were solved by a direct method (*SHELXS* [31]) and refined by a full-matrix least square method on F^2 for all reflections (*SHELXL*-2014 [32]). All hydrogen atoms were placed using AFIX instructions, while all other atoms were refined anisotropically. Crystallographic data (Supplementary Materials) were deposited at the Cambridge Crystallographic Data Center (CCDC; under reference number: CCDC-1587144) and can be obtained free of charge via https://www.ccdc.cam.ac.uk/structures/. X-ray crystallographic data for **1**: M = 520.15, monoclinic, $P2_1/c$, a = 11.3853(3) Å, b = 13.1071(3) Å, c = 17.5390(5) Å, β = 96.2860(10)°, V = 2601.58(12) Å^3, Z = 4, $D_{calc.}$ = 1.328 gcm^{-3}, m = 1.444 mm^{-1}, $2\theta_{max}$ = 54.20°, measd./unique refls. = 48784/5747 ($R_{int.}$ = 0.0292), *GOF* = 1.075, R_1 = 0.0408/0.0471 [I>2σ(I)/all data], wR_2 = 0.1077/0.1112 [I>2σ(I)/all data], largest diff. peak and hole 3.152 and −0.604 e.Å^{-3}.

4. Conclusions

We have demonstrated a proof of principle study for the straightforward incorporation of a α-chlorosilyl moiety adjacent to the donor-stabilized germanium(II) center. The resulting product features a 1,2-dicholoro functionality and should therefore in principle be suitable as precursor for the targeted synthesis of NHC-coordinated silagermenylidenes, the heavier analogues of vinylidenes. So far, attempts to eliminate the two chloro substituents reductively were not met with success.

Acknowledgments: This work is supported by the Research Group Linkage Program of the Alexander von Humboldt Foundation, Germany; the CSIR (01(2863)/16/EMR-II), India; SERB-DST (EMR/2014/001237), India, the TIFR Center for Interdisciplinary Science Hyderabad, India; and Saarland University.

Author Contributions: David Scheschkewitz and Anukul Jana conceived and designed the experiments and wrote the paper; Debabrata Dhara performed the experiments; Volker Huch performed the XRD analysis.

References

1. Kim, S.B.; Sinsermsuksakul, P.; Hock, A.S.; Pike, R.D.; Gordon, R.G. Synthesis of *N*-heterocyclic stannylene (Sn(II)) and germylene (Ge(II)) and a Sn(II) amidinate and their application as precursors for atomic layer deposition. *Chem. Mater.* **2014**, *26*, 3065–3073. [CrossRef]
2. Purkait, T.K.; Swarnakar, A.K.; De Los Reyes, G.B.; Hegmann, F.A.; Rivard, E.; Veinot, J.G.C. One-pot synthesis of functionalized germanium nanocrystals from a single source precursor. *Nanoscale* **2015**, *7*, 2241–2244. [CrossRef] [PubMed]
3. Blom, B.; Said, A.; Szilvási, T.; Menezes, P.W.; Tan, G.; Baumgartner, J.; Driess, M. Alkaline-earth-metal-induced liberation of rare allotropes of elemental silicon and germanium from *N*-heterocyclic metallylenes. *Inorg. Chem.* **2015**, *54*, 8840–8848. [CrossRef] [PubMed]
4. Shan, Y.-L.; Yim, W.-L.; So, C.-W. An *N*-heterocyclic silylene-stabilized digermanium(0) complex. *Angew. Chem. Int. Ed.* **2014**, *53*, 13155–13158. [CrossRef] [PubMed]

5. Shan, Y.-L.; Leong, B.-X.; Xi, H.-W.; Ganguly, R.; Li, Y.; Limb, K.H.; So, C.-W. Reactivity of an amidinato silylene and germylene toward germanium(II), tin(II) and lead(II) halides. *Dalton Trans.* **2017**, *46*, 3642–3648. [CrossRef] [PubMed]
6. Blom, B.; Stoelzel, M.; Driess, M. New vistas in *N*-heterocyclic silylene (NHSi) transition-metal coordination chemistry: Syntheses, structures and reactivity towards activation of small molecules. *Chem. Eur. J.* **2013**, *19*, 40–62. [CrossRef] [PubMed]
7. Parvin, N.; Dasgupta, R.; Pal, S.; Sen, S.S.; Khan, S. Strikingly diverse reactivity of structurally identical silylene and stannylene. *Dalton Trans.* **2017**, *46*, 6528–6532. [CrossRef] [PubMed]
8. Brck, A.; Gallego, D.; Wang, W.; Irran, E.; Driess, M.; Hartwig, J.F. Pushing the s-donor strength in iridium pincer complexes: Bis-(silylene) and bis(germylene) ligands are stronger donors than bis(phosphorus(III)) ligands. *Angew. Chem. Int. Ed.* **2012**, *51*, 11478–11482. [CrossRef] [PubMed]
9. Driess, M.; Yao, S.; Brym, M.; van Wüllen, C.; Lentz, D. A new type of *N*-heterocyclic silylene with ambivalent reactivity. *J. Am. Chem. Soc.* **2006**, *128*, 9628–9629. [CrossRef] [PubMed]
10. Präsang, C.; Scheschkewitz, D. Reactivity in the periphery of functionalised multiple bonds of heavier group 14 elements. *Chem. Soc. Rev.* **2016**, *45*, 900–921. [CrossRef] [PubMed]
11. Azhakar, R.; Ghadwal, R.S.; Roesky, H.W.; Wolf, H.; Stalke, D. A début for base stabilized monoalkylsilylenes. *Chem. Commun.* **2012**, *48*, 4561–4563. [CrossRef] [PubMed]
12. Gallego, D.; Brück, A.; Irran, E.; Meier, F.; Kaupp, M.; Driess, M.; Hartwig, J.F. From bis(silylene) and bis(germylene) pincer-type nickel(II) complexes to isolable intermediates of the nickel-catalyzed sonogashira cross-coupling reaction. *J. Am. Chem. Soc.* **2013**, *135*, 15617–15626. [CrossRef] [PubMed]
13. Baus, J.A.; Poater, J.; Bickelhaupt, F.M.; Tacke, R. Silylene-induced reduction of [$Mn_2(CO)_{10}$]: Formation of a five-coordinate silicon(IV) complex with an O-bound [$(OC)_4Mn{=}Mn(CO)_4$]$^{2-}$ ligand. *Eur. J. Inorg. Chem.* **2017**, 186–191. [CrossRef]
14. Cabeza, J.A.; García-Álvarez, P.; Gobetto, R.; González-Álvarez, L.; Nervi, C.; Pérez-Carreño, E.; Polo, D. [$MnBrL(CO)_4$] (L = Amidinatogermylene): Reductive dimerization, carbonyl substitution, and hydrolysis reactions. *Organometallics* **2016**, *35*, 1761–1770. [CrossRef]
15. Rivard, E. Group 14 inorganic hydrocarbon analogues. *Chem. Soc. Rev.* **2016**, *45*, 989–1003. [CrossRef] [PubMed]
16. Jana, A.; Huch, V.; Scheschkewitz, D. NHC-stabilized silagermenylidene: A heavier analogue of vinylidene. *Angew. Chem. Int. Ed.* **2013**, *52*, 12179–12182. [CrossRef] [PubMed]
17. Jana, A.; Majumdar, M.; Huch, V.; Zimmer, M.; Scheschkewitz, D. NHC-coordinated silagermenylidene functionalized in allylic position and its behaviour as a ligand. *Dalton Trans.* **2014**, *43*, 5175–5181. [CrossRef] [PubMed]
18. Jana, A.; Omlor, I.; Huch, V.; Rzepa, H.S.; Scheschkewitz, D. *N*-heterocyclic carbene coordinated neutral and cationic heavier cyclopropylidenes. *Angew. Chem. Int. Ed.* **2014**, *53*, 9953–9956. [CrossRef] [PubMed]
19. Jana, A.; Huch, V.; Rzepa, H.S.; Scheschkewitz, D. A multiply functionalized base-coordinated GeII compound and its reversible dimerization to the digermene. *Angew. Chem. Int. Ed.* **2015**, *54*, 289–292. [CrossRef] [PubMed]
20. Nieder, D.; Yildiz, C.B.; Jana, A.; Zimmer, M.; Huch, V.; Scheschkewitz, D. Dimerization of a marginally stable disilenyl germylene to tricyclic systems: Evidence for reversible NHC-coordination. *Chem. Commun.* **2016**, *52*, 2799–2802. [CrossRef] [PubMed]
21. Nieder, D.; Huch, V.; Yildiz, C.B.; Scheschkewitz, D. Regiodiscriminating reactivity of isolable NHC-coordinated disilenyl germylene and its cyclic isomer. *J. Am. Chem. Soc.* **2016**, *138*, 13996–14005. [CrossRef] [PubMed]
22. Nagendran, S.; Roesky, H.W. The chemistry of aluminum(I), silicon(II), and germanium(II). *Organometallics* **2008**, *27*, 457–492. [CrossRef]
23. Rupar, P.A.; Jennings, M.C.; Baines, K.M. Synthesis and structure of *N*-heterocyclic carbene complexes of germanium(II). *Organometallics* **2008**, *27*, 5043–5051. [CrossRef]
24. Tashita, S.; Watanabe, T.; Tobita, H. Synthesis of a base-stabilized (Chlorogermyl)metallogermylene and Its photochemical conversion to a (Chlorogermyl)germylyne complex. *Chem. Lett.* **2013**, *42*, 43–44. [CrossRef]
25. Denk, M.; Lennon, J.R.; Hayashi, R.; West, R.; Belyakov, A.V.; Verne, H.P.; Haaland, A.; Wagner, M.; Metzler, N. Synthesis and structure of a stable silylene. *J. Am. Chem. Soc.* **1994**, *116*, 2691–2692. [CrossRef]

26. Rupar, P.A.; Staroverov, V.N.; Ragogna, P.J.; Baines, K.M. A germanium(II)-centered dication. *J. Am. Chem. Soc.* **2007**, *129*, 15138–15139. [CrossRef] [PubMed]
27. Moser, D.F.; Naka, A.; Guzei, I.A.; Müller, T.; West, R. Formation of disilanes in the reaction of stable silylenes with halocarbons. *J. Am. Chem. Soc.* **2005**, *127*, 14730–14738. [CrossRef] [PubMed]
28. Gehrhus, B.; Hitchcock, P.B.; Lappert, M.F. New reactions of a silylene: Insertion into M–N bonds of $M[N(SiMe_3)_2]_2$ (M = Ge, Sn, or Pb). *Angew. Chem. Int. Ed. Engl.* **1997**, *36*, 2514–2516. [CrossRef]
29. Al-Rafia, S.M.I.; Momeni, M.R.; McDonald, R.; Ferguson, M.J.; Brown, A.; Rivard, E. Controlled growth of dichlorogermanium oligomers from Lewis basic hosts. *Angew. Chem. Int. Ed.* **2013**, *52*, 6390–6395. [CrossRef] [PubMed]
30. Schafer, A.; Saak, W.; Weidenbruch, M.; Marsmann, H.; Henkel, G. Reactions of a silylene with a germylene and a stannylene: Formation of a digermene with an unusual arrangement of the substituents and of a stannane. *Chem. Ber.* **1997**, *130*, 1733–1737. [CrossRef]
31. Sheldrick, G.M. A short history of *SHELX*. *Acta Cryst.* **2008**, *A64*, 112–122. [CrossRef] [PubMed]
32. Sheldrick, G.M. Crystal Structure Refinement with *SHELXL*. *Acta Cryst.* **2015**, *C71*, 3–8. [CrossRef]

9

Bond Insertion at Distorted Si(001) Subsurface Atoms

Lisa Pecher and Ralf Tonner *

Faculty of Chemistry and Material Sciences Center, Philipps-Universität Marburg, Hans-Meerwein-Straße 4, 35032 Marburg, Germany; lisa.pecher@chemie.uni-marburg.de
* Correspondence: tonner@chemie.uni-marburg.de

Abstract: Using density functional theory (DFT) methods, we analyze the adsorption of acetylene and ethylene on the Si(001) surface in an unusual bond insertion mode. The insertion takes place at a saturated tetravalent silicon atom and the insight gained can thus be transferred to other saturated silicon compounds in molecular and surface chemistry. Molecular orbital analysis reveals that the distorted and symmetry-reduced coordination of the silicon atoms involved due to surface reconstruction raises the electrophilicity and, additionally, makes certain σ bond orbitals more accessible. The affinity towards bond insertion is, therefore, caused by the structural constraints of the surface. Additionally, periodic energy decomposition analysis (pEDA) is used to explain why the bond insertion structure is much more stable for acetylene than for ethylene. The increased acceptor abilities of acetylene due to the presence of two π^*-orbitals (instead of one π^*-orbital and a set of σ^*(C–H) orbitals for ethylene), as well as the lower number of hydrogen atoms, which leads to reduced Pauli repulsion with the surface, are identified as the main causes. While our findings imply that this structure might be an intermediate in the adsorption of acetylene on Si(001), the predicted product distributions are in contradiction to the experimental findings. This is critically discussed and suggestions to resolve this issue are given.

Keywords: adsorption; bond activation; bonding analysis; density functional theory; distorted coordination; molecular orbital analysis; silicon surfaces

---

1. Introduction

The discovery of stable molecules containing silicon–silicon double bonds four decades ago [1] has sparked the research interest in the chemistry of disilenes [2–5]. Unlike in alkenes, the substituents at these double bonds are not arranged in a planar fashion, but tilted, a result of the bonding situation which has been described as a double dative bond [6,7]. The most stable conformation is a *trans* arrangement of the four substituents [6,8–10], which is favored not only from orbital overlap aspects at the Si–Si bond, but also Pauli repulsion within the molecule [11]. While these unusual double bonds can undergo similar reactions as alkenes, e.g., 1,2-addition, cycloaddition, or coordination in metal complexes [3], the reactivity is increased and sterically demanding substituents are needed to stabilize the molecules [5]. Additionally, the tilted geometry leads to changes in reaction mechanisms and stereochemistry, as a recent study on the addition of molecular hydrogen to a disilene showed [12]: here, the system favors *anti*-addition to *syn*-addition or a stepwise mechanism.

A *cis* arrangement of substituents in disilenes is rarely observed. Two examples include silicon atoms as part of a borane cluster [13] and a base-stabilized adduct [14]. In both cases, the coordination of the two silicon atoms in the most stable conformation is not symmetric (formally, C_{2v} symmetry, in the case of four identical substituents), but asymmetric (formally, C_s symmetry). This distortion changes the electronic structure and, as a consequence, the reactivity: in the base-stabilized adduct [14], one silicon atom acts as an acceptor in a dative bond, a situation that would not be possible in carbon–carbon double bonds. Further investigation of *cis*-substituted disilenes could, therefore, reveal new aspects of silicon chemistry.

The Si(001) surface, a widely-used substrate in surface science with a high relevance for application [15], features a structure similar to *cis*-substituted disilenes (Figure 1): in the surface reconstruction process of the bulk (a), which crystalizes in a diamond structure, two adjacent atoms form a covalent bond and tilt, yielding the characteristic buckled dimers of this surface (b). Although the bond could be, in a local picture, formally described as a double bond, the buckling leads to a localization of the occupied π-type orbital at the upper Si_{up} atom, while the empty π^*-type orbital is mostly localized at the lower Si_{down} atom (c) [16]. Therefore, the surface possesses both nucleophilic (at Si_{up}) and electrophilic (at Si_{down}) characters. This enhances the reactivity with organic molecules: In addition to typical double bond reactions like 1,2-addition and cycloaddition, the surface can form dative bonds and act as a reagent in nucleophilic substitution [17,18]. This underlines that, in many cases, it can be treated as a molecular reagent [19]. The study of reactions on Si(001) can, therefore, aid in the understanding of molecular disilenes.

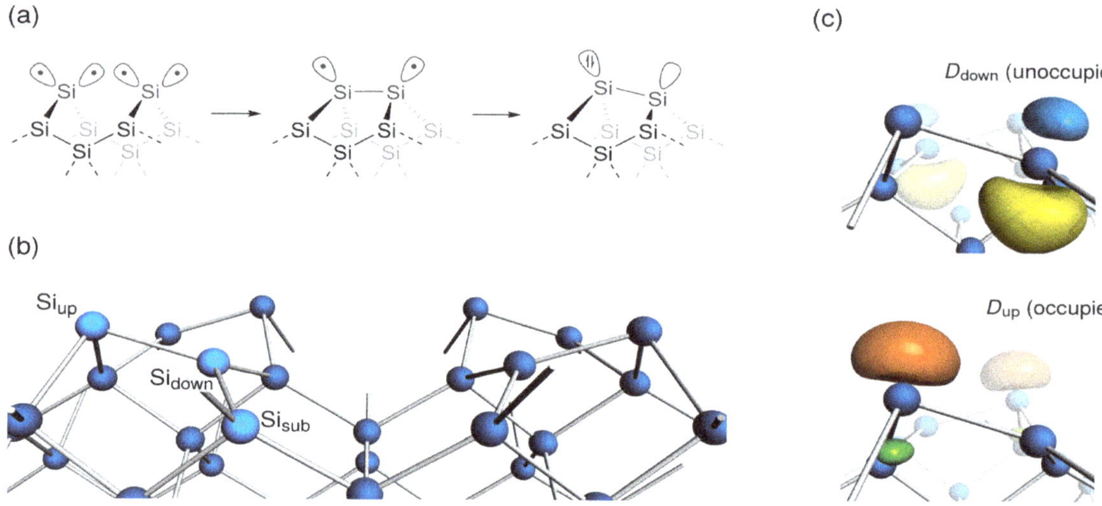

Figure 1. (**a**) Schematic depiction of the Si(001) surface reconstruction process. Dots indicate unpaired electrons. (**b**) Structure of Si(001) in the most stable reconstruction, $c(4 \times 2)$, with nomenclature used subsequently. (**c**) Crystal orbitals (at the Γ point in *k* space) of a Si(001) slab corresponding to the dimer states, calculated at PBE-D3/TZ2P.

Previously, a new reaction type on Si(001) was reported in a theoretical study [20]: the bond insertion of acetylene into a Si_{down}–Si_{sub} bond (compare Figure 1b), called the sublayer mode from now on. This behavior is highly unusual, since, until now, the Si_{up} and Si_{down} atoms were considered reactive centers in most cases. Si_{sub} atoms were disregarded because they are tetravalent, therefore fully saturated and missing the "dangling bond" orbitals of Si_{up} and Si_{down} (Figure 1c). The observation is also in stark contrast to the adsorption of alkenes like ethylene, which undergo a cycloaddition with one or two surface dimers mediated by a coordination of the molecular π system to a Si_{down} atom (π complex) [21–27]. For the adsorption of acetylene, cycloadducts are the most stable adsorption modes as well, as confirmed by theoretical calculations and experimental measurements at room temperature [20,21,28–36]. While there is experimental evidence for the presence of an intermediate state (called precursor in surface science) [37], there is no consensus on the nature of this state. Statements that have been made include a mobile precursor [37], a π complex [28,30], no stable precursor at all [21,35], and the bond insertion structure [20].

Using molecular orbital analysis and bonding analysis, we will now provide a description of acetylene adsorption on Si(001) that advocates the insertion mechanism (Scheme 1). Even though this mechanism has not been reported for other systems so far, we will explain how the electronic structure of acetylene benefits this reaction by comparing it to the adsorption of ethylene. Additionally, we will provide an explanation for the reactivity of Si_{sub} atoms by molecular orbital analysis of distorted silane. This will show that the calculated reactivity is a reasonable alternative to existing hypotheses.

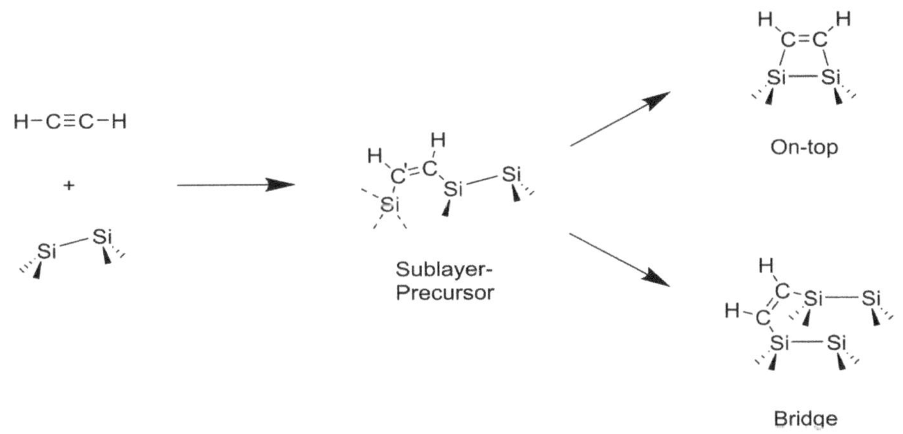

Scheme 1. Possible adsorption mechanism of acetylene on Si(001) with the sublayer mode as an intermediate (precursor) in the formation of cycloaddition modes on-top and bridge as advocated in [20].

2. Computational Details

2.1. Molecules

Structural optimization was performed using the PBE functional [38,39], the D3(BJ) dispersion correction [40,41] and the def2-TZVPP basis set [42] in ORCA 3.0.2 [43] (standard convergence criteria and integration grids). Molecular orbitals were calculated and visualized using PBE-D3(BJ)/TZ2P in ADF 2016 [44–46] (standard convergence criteria and integration grids).

2.2. Surfaces

All energies and structures were calculated with the Vienna Ab Initio Simulation Package (VASP) [47–50] version 5.3.5 using the PBE and HSE06 [51] functionals (structures and frequencies: PBE only), the D3(BJ) dispersion correction and the PAW formalism [52,53] with a basis set cutoff of 400 eV. Electronic k space was sampled using a $\Gamma(221)$ grid. Self-consistent field (SCF) calculations were converged to an accuracy of 10^{-6} eV, and structural optimizations to 10^{-2} eV·Å^{-1}. Reaction paths were calculated using the Climbing-Image Nudged Elastic Band method [54], while the transition state structures were refined using the Dimer method [55]. Harmonic vibrational frequencies were calculated from structures converged to 10^{-3} eV·Å^{-1} by a finite differences approach and construction of the Hessian using Cartesian displacements of 0.01 Å (SCF convergence: 10^{-8} eV). Gibbs energies were calculated at $T = 300$ K, $p = 1$ bar in an approach described elsewhere [27]. All minimum and transition state geometries and their respective total energies are given in the Supplementary Materials.

The Si(001) surface was modeled as a six-layer slab in $c(4 \times 2)$ reconstruction and frozen double layer approximation. The bottom was saturated with hydrogen atoms in a tetrahedral arrangement at a distance of d(Si–H) = 1.480 Å, the experimental equilibrium distance in silane [56]. Cell parameters a and b were set to 15.324 Å, corresponding to a 4×4 cell with an optimized bulk lattice parameter of 5.418 Å. In the c direction, a vacuum layer of at least 10 Å was ensured. Convergence studies on these parameters can be found in a previous work [57]. The bonding energy E_{bond} was defined as the difference between the energy E_{tot} of the relaxed total system and the energies E_{surf} and E_{mol} of the relaxed and isolated surface and molecule:

$$E_{bond} = E_{tot} - E_{surf} - E_{mol} \quad (1)$$

Please note that surface science convention is the use of the adsorption energy E_{ads} with inverse sign convention ($E_{ads} = -E_{bond}$). Scanning tunneling microscopy (STM) topographies were calculated in the Tersoff-Hamann approximation [58,59] using bSKAN [60,61]. The approach outlined has delivered accurate and reliable results for organic/semiconductor systems in the past [18,26,27,57,62,63].

2.3. Periodic Energy Decomposition Analysis

Bonding analysis was performed at PBE-D3(BJ)/TZ2P, Γ only k sampling, using pEDA [64] in ADF-BAND 2016 [65,66]. The pEDA method allows to dissect E_{bond} into well-defined quantities that allow interpreting the bonding between two fragments in a system (here: molecule and surface) in a chemically meaningful way. In the first step, E_{bond} is partitioned into the intrinsic interaction energy ΔE_{int} and the respective preparation energies ΔE_{prep} of the molecule (M) and surface (S):

$$E_{bond} = \Delta E_{int} + \Delta E_{prep}(M) + \Delta E_{prep}(S) \tag{2}$$

Since a dispersion correction is applied to the DFT calculations, ΔE_{int} can be divided into a dispersion term (disp) and an electronic term (elec):

$$\Delta E_{int} = \Delta E_{int}(disp) + \Delta E_{int}(elec) \tag{3}$$

The actual pEDA procedure then decomposes $\Delta E_{int}(elec)$ into contributions from Pauli repulsion (ΔE_{Pauli}), electrostatics (ΔE_{elstat}), and orbital interaction (ΔE_{orb}):

$$\Delta E_{int}(elec) = \Delta E_{Pauli} + \Delta E_{elstat} + \Delta E_{orb} \tag{4}$$

Additionally, the NOCV extension [67,68] was applied to the pEDA calculation in this work. This allows expressing ΔE_{orb} as the sum of individual energy contributions $\Delta E_{orb}(i)$ of different character (e.g., σ/π bonding, donation and back donation). The assignment to a character was done by visual inspection of the deformation densities $\Delta\rho(i)$ which show the corresponding charge transfer.

3. Results

3.1. Distorted Coordination of Tetravalent Silicon

The enhanced reactivity of Si_{sub} atoms can be understood from the structural distortion in the reconstruction process (Figure 1a): since the position of the surface atoms Si_{up} and Si_{down} is changed with respect to the bulk, the coordination of the Si_{sub} atoms changes from tetrahedral to a symmetry-reduced arrangement (Figure 1b). This affects the electronic structure, as can be seen from a molecular orbital analysis of a similarly distorted silane molecule: Figure 2a shows a Walsh diagram [69] depicting the Kohn-Sham energies of selected molecular orbitals as a function of the distortion angle α (Figure 2b), i.e., the tilting of two Si–H bonds from tetrahedral position without a change in bond length. The distortion reduces the symmetry of the molecule from point group T_d to C_s and causes degenerate orbitals to split in energy. This is most evident in the triply-degenerate $1t_2$ orbital, the HOMO of the tetrahedral molecule: Orbital $4a'$ rises approximately linearly in energy with increasing α, while at the same time, orbital $1a''$ is unaffected and $3a'$ decreases in energy. This is caused by changes in overlap of the involved $3p$(Si) and $1s$(H) atomic orbitals. The unoccupied and anti-bonding $2t_2$ orbitals are affected in a similar manner and the LUMO of the distorted molecule ($5a'$) gets significantly lower in energy with increasing α. At 50° distortion, it features a large lobe at the "empty coordination site", i.e., the location of a fifth bonding partner in a five-fold coordinated structure. Since this orbital becomes spatially and energetically more accessible, the distortion makes the system more electrophilic. The ability of silicon to accommodate more than four substituents is well-known in molecular silicon chemistry.

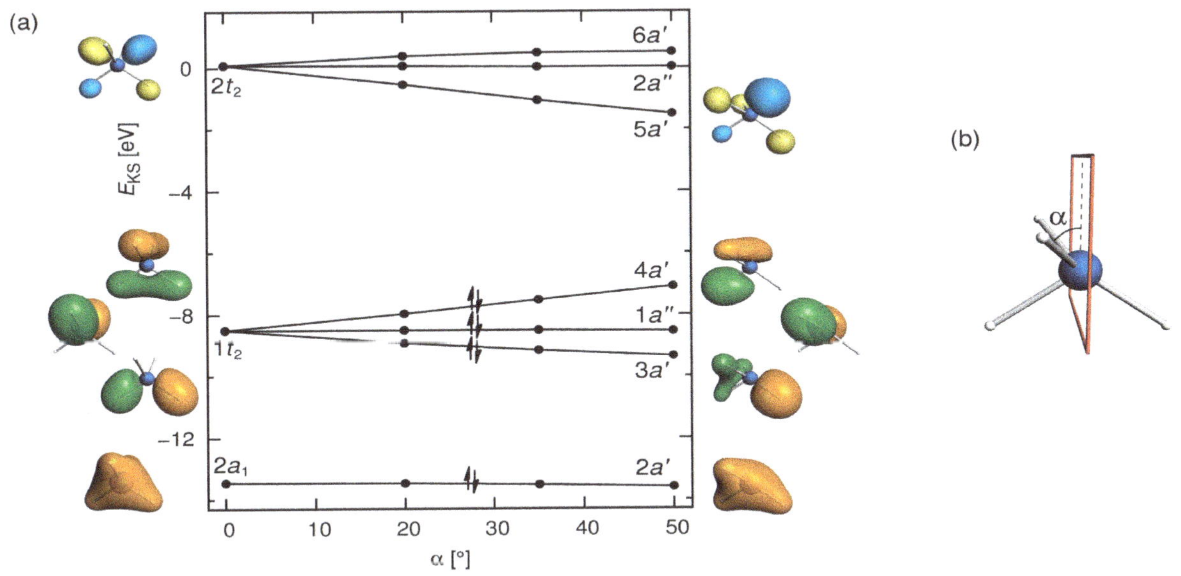

Figure 2. (**a**) Walsh diagram of the silane molecule: Kohn-Sham energies (calculated at PBE-D3/TZ2P) of selected valence orbitals as a function of the tilting angle α, i.e., the deflection of two Si–H bonds from the ideal tetrahedral arrangement. At $\alpha = 0°$, only the $2t_2$ orbital that transforms into the LUMO of the distorted molecule ($5a'$) is shown. (**b**) Definition of α.

Still, this does not explain the activation of the Si_{down}–Si_{sub} bonds, since the HOMO ($4a'$), while raised in energy, is located at three Si–H bonds. However, on Si(001) the structure is even further reduced in symmetry and the two distorted bonds are not tilted by the same amount. This is because every Si_{sub} atom is connected to one Si_{down} and one Si_{up} in the minimum configuration (Figure 1b) and the corresponding bonds are tilted by a different α: Approximately 30° (towards Si_{up}) and 50° (towards Si_{down}). When this geometry is reproduced in silane, an additional symmetry reduction takes place (C_s to C_1), the HOMO (Figure 3a) becomes a mixture of orbitals $4a'$ and $1a''$ of the C_s-symmetric molecule and features a large lobe at the Si–H bond that is tilted by 50°, i.e., the one representing the Si_{down}–Si_{sub} bond on Si(001). At the same time, the LUMO (Figure 3b) is essentially unchanged with respect to the C_s-symmetric molecule. This asymmetric distortion provides a much better explanation of the bond activation, since the HOMO could interact with an empty orbital of another molecule (adsorbate) and form a covalent bond by breaking the Si–H (Si_{down}–Si_{sub}) bond, whereas the LUMO could interact with an occupied orbital and form a second covalent bond, reinforcing the bond insertion reactivity.

Two selected crystal orbitals (COs) of a Si(001) slab at the Γ point in k space show that the electronic structure at Si_{sub} atoms is similar: the HOCO (Figure 3c, −4.85 eV) is localized primarily at the Si_{down}–Si_{sub} bonds, while the LUCO+5 (Figure 3d, −3.47 eV) features large lobes at the Si_{sub} atoms. It is surprising to see the Si_{down}–Si_{sub} bond orbital as the HOCO, since the non-bonding electron pairs at Si_{up} (Figure 1c) are usually considered as frontier orbitals in the surface reactivity of Si(001). However, at the Γ point, the highest energy orbital representing the Si_{up} states (HOCO–1) is very close in energy (−5.02 eV) to the HOCO. Additionally, this orbital becomes the HOCO in other parts of k space, as a recent theoretical determination of the band structure using hybrid functionals has shown (Figure 5c in [70]). Hence, both the non-bonding electron pairs at Si_{up} and the activated Si_{down}–Si_{sub} bonds are high enough in energy to be considered reactive towards adsorbates. The energetic location of the LUCO+5, the other crystal orbital involved in bond insertion reactivity, is more in line with previous considerations, since the orbitals representing the empty states at Si_{down} (Figure 1c) are still lower in energy (−4.43 to −3.61 eV) and, therefore, more accessible. All of this shows that the effects described for silane also apply to Si_{sub} atoms on the Si(001) surface and that the enhanced reactivity of these saturated atoms is caused by the distorted geometric arrangement.

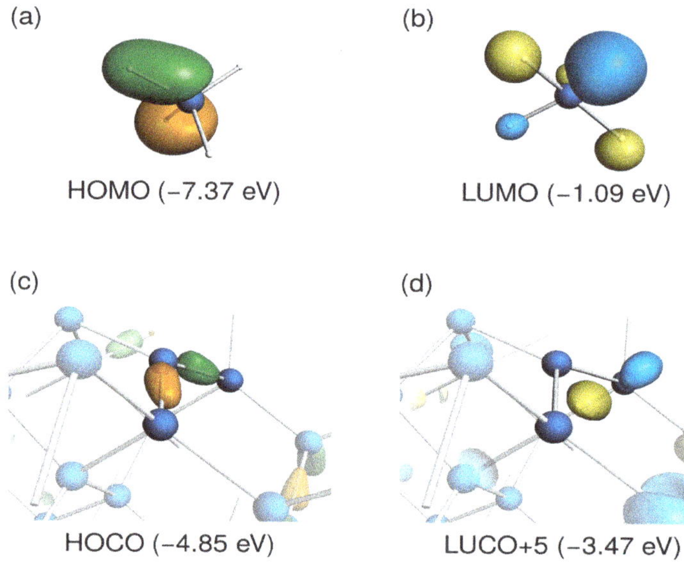

Figure 3. Top: HOMO (**a**) and LUMO (**b**) of an asymmetrically-distorted silane molecule (two Si–H bonds deflected by α = 30 and 50°, respectively), representing the coordination of a Si_{sub} atom on a Si(001) surface (see Figure 1b). Bottom: HOCO (**c**) and LUCO+5 (**d**) of a Si(001) surface slab at the Γ point in k space, showing a similar electronic structure at the Si_{sub} atoms.

3.2. Bond Insertion of Acetylene and Ethylene on Si(001)

The optimized structures and bonding energies of acetylene and ethylene on Si(001) in the π complex and sublayer adsorption modes are given in Figure 4. As previously reported [20], the π complex mode of acetylene is not a minimum for the computational approach chosen here, but a first-order saddle point ($\nu_{imag} = 71i$). This is also evident in the energy profile connecting the two stationary points (c) which shows no local energy maximum. The acetylene π complex structure can therefore be seen as a transition state (TS) between two symmetry-equivalent sublayer minima. In contrast to this, the ethylene π complex structure is a minimum, the sublayer mode is higher in energy, and the energy profile features a maximum. Furthermore, a local minimum is apparent: this represents the lowest energy rotamer of the ethylene π complex, in which the C–C bond is rotated by 30–45° approximately in the xy plane [27,71]. Since the structure shown in Figure 4b is only 3 $kJ{\cdot}mol^{-1}$ higher in energy and allows a better comparability with the corresponding acetylene TS structure due to a similar molecular orientation, it is analyzed here instead of the minimum rotamer.

The two π complex structures show several similarities, beginning with a C–C bond elongation by 3–4% compared to the gas phase value (acetylene: 1.208 → 1.239 Å, ethylene: 1.333 → 1.381 Å). This is already an indication of electron donation from the π system to the surface. An indicator of back donation is the Si_{down}–Si_{sub} bond length, which is elongated by 2% compared to its minimum value in the isolated surface slab (2.340 → 2.396 (acetylene), 2.340 → 2.389 Å (ethylene)). The nature of this interaction would be a donation of electron density from the $\sigma(Si_{down}\text{–}Si_{sub})$ crystal orbital (Figure 3c) to the molecular π^* orbital. Carbon-silicon distances $d(C\text{–}Si_{down})$ are in a similar range for both systems as well but slightly shorter for acetylene (2.175 Å, ethylene: 2.234 Å). While this could indicate a stronger interaction between molecule and surface in comparison to ethylene, the E_{bond} value of acetylene ($-64\ kJ{\cdot}mol^{-1}$) implies a weaker interaction (ethylene: $-71\ kJ{\cdot}mol^{-1}$). Bonding analysis will provide an explanation for this apparent contradiction later on. In comparison with [20], where this reaction was reported first, our E_{bond} values are lower in energy by ~25 $kJ{\cdot}mol^{-1}$. This difference can be mainly attributed to the dispersion correction to the DFT energies, which was not applied in the literature study, and other differences in the computational setup. Our G_{bond} values (Figure 4) are higher in energy by 47 (acetylene) and 58 $kJ{\cdot}mol^{-1}$ (ethylene) than E_{bond} values, highlighting the importance of molecular entropy loss upon adsorption described previously [27,72,73]. However, it should be noted that, in the case of the acetylene TS, the partition function used to calculate G_{bond} has one fewer degree of freedom, the imaginary mode. Hence, this G_{bond} value should not be taken as a bonding energy of this state in thermodynamic equilibrium, which is not possible for a saddle point, but rather the effective energy in a reaction where the structure appears as a TS, i.e., conversion between two sublayer minima.

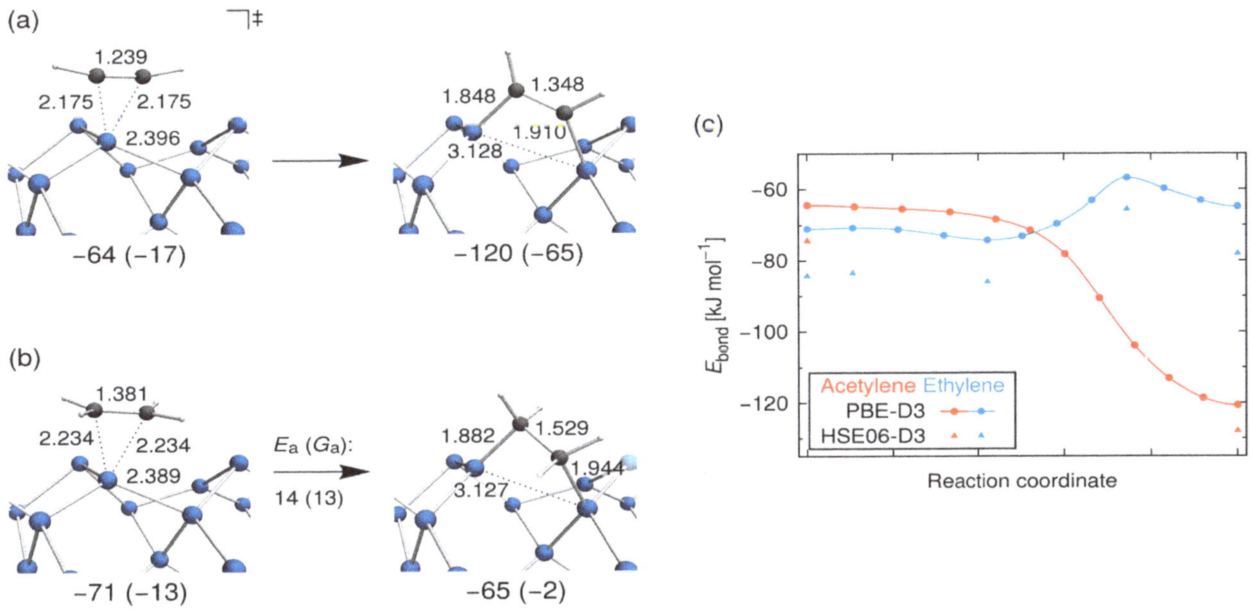

Figure 4. (**a**,**b**) Optimized structures and PBE-D3 E_{bond} (G_{bond}) values of the π complex (left) and sublayer (right) adsorption modes of acetylene (**a**) and ethylene (**b**) on Si(001). d(C–C′), d(C/C′–Si_{down}), d(C′–Si_{sub}) and d(Si_{down}–Si_{sub}) given in Å. (**c**) Corresponding reaction energy profiles (PBE-D3) and HSE06-D3 energies at the stationary points.

The sublayer mode is easily characterized by a drastically increased Si_{down}–Si_{sub} bond length (acetylene: 3.128 Å, ethylene: 3.127 Å) with respect to the isolated slab value (2.340 Å), a strong indication that this covalent bond has been broken. At the same time, two C–Si covalent bonds are formed (d(C–Si_{down}/C′–Si_{sub}) = 1.848/1.910 (acetylene), 1.882/1.944 Å (ethylene)) and the C–C′ bond elongates to 1.348 (acetylene) and 1.529 Å (ethylene), typical values for double and single bonds, respectively. This implies that one π bond is broken in the insertion process and the C–C′ bond order is reduced by one. The slightly larger d(C–Si) values for ethylene are caused by Pauli repulsion between the CH_2 groups and the surface, as bonding analysis will show later. While the acetylene sublayer mode is markedly lower in energy than the π complex TS at values of -120 (E_{bond}) and -65 kJ·mol^{-1} (G_{bond}), the ethylene structure is higher in energy than the π complex at -65 (E_{bond}) and -2 kJ·mol^{-1} (G_{bond}). This was reported in the previous study [20] as well (E_{bond} = -97 (acetylene) and -22 kJ·mol^{-1} (ethylene)) and used as an argument on why this mode was not observed in any adsorption experiments of ethylene. The differences of the literature energies to our values can again be attributed to the missing dispersion correction and the different computational setup. In particular, the smaller cell size of 2×2 atoms per layer (this study: 4×4) appears to destabilize the sublayer modes. A similar effect was previously found for the *bridge* mode of ethylene [27]. This destabilization is also apparent at the TS for the π complex $\rightarrow$ sublayer conversion in the ethylene system, since our values for the energy barrier (E_a: 14, G_a: 13 kJ·mol^{-1}) are considerably lower than the literature value of E_a = 41 kJ·mol^{-1}.

Considering that both this study and [20] were performed using GGA functionals, one could argue that the stability of the acetylene sublayer mode might be an artifact of the methodology. To address this issue, we performed single point calculations at the optimized structures using the HSE06 hybrid functional. The resulting energies of the stationary points (Figure 4c) show only a minor decrease in E_{bond} values of up to 15 kJ·mol^{-1} and barely any changes in energy differences (HSE06-D3 energies of the full reaction paths in Figure 4c are given in the Supplementary Materials). This confirms that the PBE functional is an appropriate choice for the description of these systems. An independent test of the DFT energy values, e.g., by wavefunction-based methods, would nevertheless be desirable.

Bonding analysis results obtained from pEDA calculations (see Section 2.3) are given in Table 1. By comparing the π complex values, it becomes apparent that the lower E_{bond} value of ethylene is caused by a reduced molecular preparation energy (ΔE_{prep}(M) = 10 vs. acetylene: 15 kJ·mol^{-1}) and an increased dispersion interaction (ΔE_{int}(disp) = −27 vs. acetylene: −23 kJ·mol^{-1}). The electronic interaction energy ΔE_{int}(elec), as well as the stabilizing pEDA terms electrostatic (ΔE_{elstat}) and orbital interaction (ΔE_{orb}) are all weaker in case of ethylene. These results resolve the apparent contradiction of shorter $C^{1/2}$–Si_{down} bonds, but less negative E_{bond} values in the case of acetylene: The shorter distance is caused by a larger stabilization through electrostatic and orbital interaction, which are overcompensated by the reduced dispersion interaction and higher preparation energy. However, the character of the chemical bond as given by pEDA is comparable in both structures: a dative bond between the molecule and surface, apparent from the $\Delta E_{elstat}/\Delta E_{orb}$ ratio (47:53), a typical value in covalent bonds [74], and a dominating molecule-to-surface orbital contribution $\Delta E_{orb}(M \rightarrow S)$ (acetylene: 70% of ΔE_{orb}, ethylene: 74%).

Table 1. pEDA bonding analysis of acetylene and ethylene adsorbed on Si(001) in the π complex and sublayer modes [1].

	π Complex				**Sublayer**			
	Acetylene (TS)		**Ethylene**		**Acetylene**		**Ethylene**	
ΔE_{int}	−105		−106		−656		−583	
ΔE_{int}(disp) [2]	−23	(22%)	−27	(25%)	−22	(3%)	−30	(5%)
ΔE_{int}(elec) [2]	−82	(78%)	−79	(75%)	−634	(97%)	−553	(95%)
ΔE_{Pauli}	701		653		1587		1628	
ΔE_{elstat} [3]	−367	(47%)	−342	(47%)	−912	(41%)	−942	(43%)
ΔE_{orb} [3]	−416	(53%)	−390	(53%)	−1310	(59%)	−1240	(57%)
$\Delta E_{orb}(M \rightarrow S)$ [4]	−292	(70%)	−289	(74%)	−551	(42%)	−536	(43%)
$\Delta E_{orb}(S \rightarrow M)$ [4]	−88	(21%)	−76	(19%)	−714	(55%)	−641	(52%)
ΔE_{prep}(M)	15		10		364		341	
ΔE_{prep}(S)	17		16		156		161	
E_{bond} [5]	−73	(−64)	−80	(−71)	−136	(−120)	−81	(−65)

[1] All values in kJ·mol^{-1}, calculated using PBE-D3/TZ2P. Fragments: Molecule and surface. Fragmentation: Closed-shell singlet (π complex), triplet (sublayer). M: Molecule, S: Surface. [2] Percentage values give relative contributions of dispersion and electronic effects to the interaction energy ΔE_{int}. [3] Percentage values give relative contributions of the attractive pEDA terms ΔE_{elstat} and ΔE_{orb}. [4] Percentage values give relative contributions to the total orbital interaction energy ΔE_{orb}. [5] PAW values (in parentheses) given for comparison.

In the sublayer mode of acetylene, ΔE_{int} is drastically lower than in the π complex TS, at −656 kJ·mol^{-1}. Since ΔE_{int}(disp) barely changes, this is an electronic effect that can be attributed to the formation of the two shared-electron bonds between molecule and surface. ΔE_{Pauli}, ΔE_{elstat}, and ΔE_{orb} rise significantly in absolute value, and orbital interaction is now even more predominant in stabilization at 59% vs. 41% for ΔE_{elstat}. This is mainly caused by a large increase in back bonding (surface to molecule) contributions $\Delta E_{orb}(S \rightarrow M)$, now the dominating orbital term (55% of ΔE_{orb}). The increase occurs because the nature and number of the covalent bonds has changed from one dative to two shared-electron bonds. These major contributions to stabilization, however, are partially compensated by large preparation energies in both molecule and surface (364 and 156 kJ·mol^{-1}, respectively), so that the resulting E_{bond} value is only ~60 kJ·mol^{-1} lower than the energy of the π complex TS. The ethylene sublayer analysis shows that the destabilization with respect to acetylene is not a result of increased molecular deformation, as previously suggested [20], since the ΔE_{prep}(M) value for ethylene is lower than the corresponding value for acetylene. Furthermore, ΔE_{int}(disp) is lower for ethylene as well, so the effect is entirely contained in ΔE_{int}(elec), which is 81 kJ·mol^{-1} less stabilizing than for acetylene. The pEDA decomposition reveals that the destabilization is contained in ΔE_{Pauli} and ΔE_{orb}, but not ΔE_{elstat}, which is more stabilizing for ethylene. This is in contrast to the π complex structures, where the weaker ΔE_{int} was reflected in a lower absolute value of all three pEDA terms. Here, the ΔE_{orb} destabilization is mainly caused by a reduction of back bonding contributions $\Delta E_{orb}(S \rightarrow M)$, which are weaker by 73 kJ·mol^{-1} for ethylene, whereas $\Delta E_{orb}(M \rightarrow S)$ contributions are only reduced by 15 kJ·mol^{-1}. This can be understood from the electronic structure of the molecules: Whereas acetylene has two low-lying π*-orbitals that can act as acceptors for electron density from the surface, in ethylene, only one π*-orbital is available. The back donation into σ*(C–H) bonds (negative hyperconjugation) that is possible in ethylene yields much less stabilization since the orbitals are higher in energy. Therefore, the acceptor ability of ethylene is reduced (lower $\Delta E_{orb}(S \rightarrow M)$ value) and the amount of ΔE_{orb} stabilization decreases, while, at the same time, Pauli repulsion increases since the two additional C–H bonds are interacting with the surface. The only stabilizing effects that arise from the additional atoms and bonds in ethylene are in ΔE_{elstat} and ΔE_{int}(disp), but these are not able to compensate the energy loss from the two other terms. It can therefore be concluded that the enhanced acceptor abilities of acetylene combined with the lower number of C–H bonds leading to the reduced Pauli repulsion are the determining factors that benefit a stabilization of the sublayer mode for this molecule.

3.3. Comparison with Experiment

As already mentioned in the Introduction, these findings imply that the sublayer mode could be a precursor intermediate in the adsorption mechanism of acetylene on Si(001). Up to now, scanning tunneling microscopy (STM) and spectroscopic measurements were performed at room temperature or higher [29,31–33], where the more stable on-top and bridge modes (Scheme 1) dominate and all molecules that could have possibly been trapped in a precursor have already reacted to these states. However, if one assumes that these reactions are under kinetic control, as observed for many reactions of organic molecules on Si(001) [19], the distribution of on-top and bridge structures at room temperature can be estimated from calculated energy barriers G_a (Equation (5)) and compared with experimental findings.

$$\frac{N_{on\text{-}top}}{N_{bridge}} = \exp\left(-\frac{G_{a,on\text{-}top} - G_{a,bridge}}{k_B T}\right) \tag{5}$$

For ethylene, this approach showed excellent agreement with experiment [27]. Here, the calculated barriers $G_{a,on\text{-}top}/G_{a,bridge}$, 55/35 (PBE-D3) and 62/48 kJ·mol^{-1} (HSE06-D3), yield distributions of $N_{on\text{-}top}/N_{bridge}$ = 0.0003 (PBE-D3) and 0.004 (HSE06-D3) at 300 K. Unfortunately, an experimental STM measurement at the same temperature showed a much higher abundance of on-top structures (28%) at low coverage and the corresponding $N_{on\text{-}top}/N_{bridge}$ value of 0.39 would indicate a difference in G_a values of less than 3 kJ·mol^{-1}. Hence, our proposed model for the adsorption mechanism (Scheme 1) is not able to reproduce the experimental findings with the calculated energies and is therefore either incorrect or incomplete. The alternative of formulating a π complex as a precursor from which the system converts to on-top and bridge is not supported by DFT calculations as well: previous studies were unable to determine a transition state on the π complex $\rightarrow$ on-top path, whereas for the path towards bridge, a transition state could be found [21,35]. Hence, conversion to on-top should be highly preferred, which is again not in line with the experimental observations. To minimize the possibility that we missed other precursor structures by our structural optimization approach, we carried out ten ab initio molecular dynamics simulations of the gas-surface dynamics. Eight trajectories ended up in the sublayer mode while the other two ended up directly in on-top and bridge, respectively (see the Supplementary Materials). Thermal effects are also an unlikely source of the disagreement, since molecule and surface were initially put in random states of thermal excitation corresponding to T = 300 K in these simulations. Additionally, our DFT approach provided reliable results in the description of adsorption kinetics for ethylene and other organic molecules on Si(001) [18,27,62]. Thus, we assume that the disagreement between theory and experiment is not due to a failure of the approximations in our density functional approach, but rather a result of an incomplete model of the adsorption dynamics. An alternative idea would be to assume that the reaction is under thermodynamic control. This also does not deliver a coherent picture, since all theoretical studies predict the on-top mode to be more stable than bridge at low coverage [20,21,30,35,57]. Therefore, a new, presumably more complex model of the adsorption dynamics must be devised, which we cannot provide at this moment.

As an aid to identify different possible precursor species in experimental measurements, Figure 5 shows simulated STM images of the adsorption modes sublayer, on-top, and bridge in comparison with the clean Si(001) surface. The bias voltages were chosen as −2.0, +0.8 and +1.9 V, typical experimental values. While at negative voltage, occupied orbitals are probed and the non-bonding electron pairs at Si_{up} atoms are visible on the clean surface (compare Figure 1), unoccupied orbitals are probed at a positive voltage and the empty orbitals mostly localized at Si_{down} atoms are visible. At a voltage of +1.9 V, the nodal planes of the π^* type orbitals become apparent. Adsorption modes on-top and bridge are easily identifiable at all voltages by the quenched signals of surface atoms, thereby appearing as dark spots of different orientation [31]. In contrast, the sublayer mode is invisible at negative bias voltage, as is the ethylene π complex structure [26]. However, in contrast to the π complex, which is evident at positive bias voltage from a quenched surface atom signal, this mode might be identified from a displaced Si_{down} signal, which is due to the change in coordination and electronic structure of this surface atom. However, if the STM resolution does not allow the identification of the displacement of the Si_{down} signal, the structure would appear invisible at all commonly-used voltages. If the π complex of acetylene is, in contrast to all theoretical investigations, the precursor, it should appear in STM measurements with the same features as the equivalent ethylene structure described above. It should be noted that in order to identify this mode, experimentally demanding low-temperature measurements at both positive and negative bias voltages would be ideal, since the signals at positive voltages could also stem from on-top structures, whereas the molecule is invisible at negative voltages. On the basis of our results, signals that are visible at positive and invisible at negative voltages would be strong indications for the π complex.

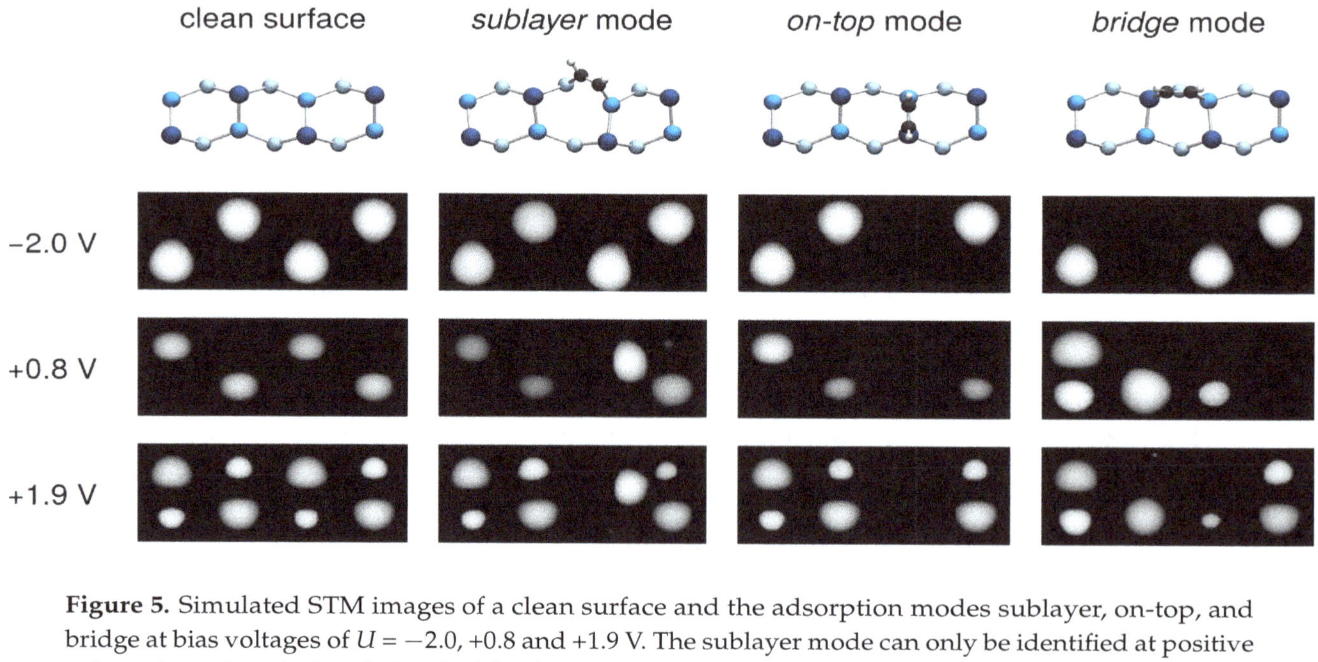

Figure 5. Simulated STM images of a clean surface and the adsorption modes sublayer, on-top, and bridge at bias voltages of $U = -2.0$, +0.8 and +1.9 V. The sublayer mode can only be identified at positive voltage from the displaced signal of the Si_{down} atom to which the molecule is bonded. Color coding of Si atoms: dark blue (Si_{up}), medium blue (Si_{down}), light blue (Si_{sub}).

Alternatively, vibrational spectroscopy measurements at low temperatures could provide indications about the nature of the precursor. The π complex of ethylene is evident from hindered vibrations at ~200 and 600 cm^{-1} [22,26] and it can be expected that an equivalent acetylene structure would show a similar vibrational fingerprint. The sublayer mode, on the contrary, could be identified from a large ~100 cm^{-1} splitting of the C–H stretching vibration band whereas on-top and bridge show a more narrow splitting of ~20–30 cm^{-1}.

An alternative to the precursor-based models discussed up to now could be the existence of direct or pseudo-direct adsorption pathways as present in the adsorption of cylooctyne on Si(001) [62,75]. Future experiments or more accurate computations might resolve the remaining open questions and help to address the difficulties of established DFT methods in describing the adsorption behavior of acetylene on Si(001).

4. Conclusions

Using molecular orbital analysis of silane, we have shown that the geometric distortion of saturated silicon atoms leads to enhanced reactivity. In particular, the electrophilic character of the silicon atom is increased while the distorted σ bonds are activated. On Si(001), this distortion is enforced by the surface reconstruction and leads to an unusually high reactivity of the saturated subsurface atoms. Similar behavior might be observed in molecular systems if steric or electronic effects of substituents could be designed to lead to similar structures. The results therefore present a type of bond activation not driven by orbital interaction, as is often the case, but rather the constrained geometry of the system.

Furthermore, pEDA analysis has shown that the increased acceptor ability and smaller number of C–H bonds of acetylene lead to a stabilization of the sublayer bond insertion mode on Si(001), whereas in the case of ethylene adsorption, this bonding pattern leads to a rather unstable structure. While this mode is a possible candidate for a precursor intermediate in the adsorption of acetylene, the results are not in agreement with the experimental findings. In particular, the ratio of product states on-top and bridge at low coverage deviates by two to three orders of magnitude. However, alternative pathways also do not lead to a better agreement and further experiments or more accurate computations are needed to settle the adsorption mechanism. We provided STM simulations of the sublayer adsorption mode in comparison with the product states which might help in identifying molecules that might be trapped in this intermediate state at low temperatures. In any case, this system provides an opportunity to improve and revise either established adsorption models and/or the theoretical methods used to model them.

Acknowledgments: This work was supported by Deutsche Forschungsgemeinschaft (DFG) within SFB 1083. Computational resources were provided by HRZ Marburg, CSC-LOEWE Frankfurt and HLR Stuttgart.

Author Contributions: Lisa Pecher performed all of the calculations, evaluated the data, and wrote the manuscript. Ralf Tonner contributed to the interpretation and wrote the manuscript.

References

1. West, R.; Fink, M.J.; Michl, J. Tetramesityldisilene, a Stable Compound Containing a Silicon–Silicon Double Bond. *Science* **1981**, *214*, 1343–1344. [CrossRef] [PubMed]
2. Raabe, G.; Michl, J. Multiple Bonding to Silicon. *Chem. Rev.* **1985**, *85*, 419–509. [CrossRef]
3. West, R. Chemistry of the Silicon–Silicon Double Bond. *Angew. Chem. Int. Ed.* **1987**, *26*, 1201–1302. [CrossRef]
4. Kira, M.; Iwamoto, T. Progress in the Chemistry of Stable Disilenes. *Adv. Organomet. Chem.* **2006**, *54*, 73–148. [CrossRef]
5. Präsang, C.; Scheschkewitz, D. Reactivity in the Periphery of Functionalised Multiple Bonds of Heavier Group 14 Elements. *Chem. Soc. Rev.* **2016**, *45*, 900–921. [CrossRef] [PubMed]
6. Malrieu, J.-P.; Trinquier, G. Trans-Bending at Double Bonds. Occurrence and Extent. *J. Am. Chem. Soc.* **1989**, *111*, 5916–5921. [CrossRef]

7. Power, P.P. π-Bonding and the Lone Pair Effect in Multiple Bonds between Heavier Main Group Elements. *Chem. Rev.* **1999**, *99*, 3463–3504. [CrossRef] [PubMed]
8. Olbrich, G. On the Structure and Stability of Si_2H_4. *Chem. Phys. Lett.* **1986**, *130*, 115–119. [CrossRef]
9. Dolgonos, G. Relative Stability and Thermodynamic Properties of Si_2H_4 Isomers. *Chem. Phys. Lett.* **2008**, *466*, 11–15. [CrossRef]
10. Kira, M. Bonding and Structure of Disilenes and Related Unsaturated Group-14 Element Compounds. *Proc. Jpn. Acad. Ser. B* **2012**, *88*, 167–191. [CrossRef]
11. Jacobsen, H.; Ziegler, T. Nonclassical Double Bonds in Ethylene Analogues: Influence of Pauli Repulsion on Trans Bending and π–Bond Strength. A Density Functional Study. *J. Am. Chem. Soc.* **1994**, *116*, 3667–3679. [CrossRef]
12. Wendel, D.; Szilvási, T.; Jandl, C.; Inoue, S.; Rieger, B. Twist of a Silicon–Silicon Double Bond: Selective Anti-Addition of Hydrogen to an Iminodisilene. *J. Am. Chem. Soc.* **2017**, *139*, 9156–9159. [CrossRef] [PubMed]
13. Jemmis, E.D.; Kiran, B. Structure and Bonding in $B_{10}X_2H_{10}$ (X = C and Si). The Kinky Surface of 1,2-Dehydro-*o*-disilaborane. *J. Am. Chem. Soc.* **1997**, *119*, 4076–4077. [CrossRef]
14. Schweizer, J.I.; Scheibel, M.G.; Diefenbach, M.; Neumeyer, F.; Würtele, C.; Kulminskaya, N.; Linser, R.; Auner, N.; Schneider, S.; Holthausen, M.C. A Disilene Base Adduct with a Dative Si–Si Single Bond. *Angew. Chem. Int. Ed.* **2016**, *55*, 1782–1786. [CrossRef] [PubMed]
15. Teplyakov, A.V.; Bent, S.F. Semiconductor Surface Functionalization for Advances in Electronics, Energy Conversion, and Dynamic Systems. *J. Vac. Sci. Technol. A* **2013**, *31*, 50810. [CrossRef]
16. Yoshinobu, J. Physical Properties and Chemical Reactivity of the Buckled Dimer on Si(100). *Prog. Surf. Sci.* **2004**, *77*, 37–70. [CrossRef]
17. Leftwich, T.R.; Teplyakov, A.V. Chemical Manipulation of Multifunctional Hydrocarbons on Silicon Surfaces. *Surf. Sci. Rep.* **2007**, *63*, 1–71. [CrossRef]
18. Pecher, L.; Laref, S.; Raupach, M.; Tonner, R. Ethers on Si(001): A Prime Example for the Common Ground between Surface Science and Molecular Organic Chemistry. *Angew. Chem. Int. Ed.* **2017**, *56*, 15150–15154. [CrossRef] [PubMed]
19. Filler, M.A.; Bent, S.F. The Surface as Molecular Reagent: Organic Chemistry at the Semiconductor Interface. *Prog. Surf. Sci.* **2003**, *73*, 1–56. [CrossRef]
20. Zhang, Q.J.; Fan, X.L.; Lau, W.M.; Liu, Z.F. Sublayer Si Atoms as Reactive Centers in the Chemisorption on Si(100): Adsorption of C_2H_2 and C_2H_4. *Phys. Rev. B* **2009**, *79*, 195303. [CrossRef]
21. Cho, J.-H.; Kleinman, L. Adsorption Kinetics of Acetylene and Ethylene on Si(001). *Phys. Rev. B* **2004**, *69*, 75303. [CrossRef]
22. Nagao, M.; Umeyama, H.; Mukai, K.; Yamashita, Y.; Yoshinobu, J. Precursor Mediated Cycloaddition Reaction of Ethylene to the Si(100)c(4 × 2) Surface. *J. Am. Chem. Soc.* **2004**, *126*, 9922–9923. [CrossRef] [PubMed]
23. Fan, X.L.; Zhang, Y.F.; Lau, W.M.; Liu, Z.F. Violation of the Symmetry Rule for the [2 + 2] Addition in the Chemisorption of C_2H_4 on Si(100). *Phys. Rev. B* **2005**, *72*, 165305. [CrossRef]
24. Mette, G.; Schwalb, C.H.; Dürr, M.; Höfer, U. Site-Selective Reactivity of Ethylene on Clean and Hydrogen Precovered Si(001). *Chem. Phys. Lett.* **2009**, *483*, 209–213. [CrossRef]
25. Lipponer, M.A.; Armbrust, N.; Dürr, M.; Höfer, U. Adsorption Dynamics of Ethylene on Si(001). *J. Chem. Phys.* **2012**, *136*, 144703. [CrossRef] [PubMed]
26. Pecher, J.; Tonner, R. Precursor States of Organic Adsorbates on Semiconductor Surfaces are Chemisorbed and Immobile. *ChemPhysChem* **2017**, *18*, 34–38. [CrossRef] [PubMed]
27. Pecher, J.; Mette, G.; Dürr, M.; Tonner, R. Site-Specific Reactivity of Ethylene at Distorted Dangling-Bond Configurations on Si(001). *ChemPhysChem* **2017**, *18*, 357–365. [CrossRef] [PubMed]
28. Liu, Q.; Hoffmann, R. The Bare and Acetylene Chemisorbed Si(001) Surface, and the Mechanism of Acetylene Chemisorption. *J. Am. Chem. Soc.* **1995**, *117*, 4082–4092. [CrossRef]
29. Matsui, F.; Yeom, H.W.; Imanishi, A.; Isawa, K.; Matsuda, I.; Ohta, T. Adsorption of Acetylene and Ethylene on the Si(001)2×1 Surface Studied by NEXAFS and UPS. *Surf. Sci.* **1998**, *401*, L413–L419. [CrossRef]
30. Sorescu, D.C.; Jordan, K.D. Theoretical Study of the Adsorption of Acetylene on the Si(001) Surface. *J. Phys. Chem. B* **2000**, *104*, 8259–8267. [CrossRef]
31. Mezhenny, S.; Lyubinetsky, I.; Choyke, W.J.; Wolkow, R.A.; Yates, J.T. Multiple Bonding Structures of C_2H_2 Chemisorbed on Si(100). *Chem. Phys. Lett.* **2001**, *344*, 7–12. [CrossRef]

32. Kim, W.; Kim, H.; Lee, G.; Hong, Y.-K.; Lee, K.; Hwang, C.; Kim, D.-H.; Koo, J.-Y. Initial Adsorption Configurations of Acetylene Molecules on the Si(001) Surface. *Surf. Sci.* **2001**, *64*, 193313. [CrossRef]
33. Kim, W.; Kim, H.; Lee, G.; Chung, J.; You, S.Y.; Hong, Y.K.; Koo, J.Y. Acetylene Molecules on the Si(001) Surface: Room-Temperature Adsorption and Structural Modification upon Annealing. *Surf. Sci.* **2002**, *514*, 376–382. [CrossRef]
34. Silvestrelli, P.L.; Pulci, O.; Palummo, M.; Del Sole, R.; Ancilotto, F. First-Principles Study of Acetylene Adsorption on Si(100): The End-Bridge Structure. *Surf. Sci.* **2003**, *68*, 235306. [CrossRef]
35. Takeuchi, N. First Principles Calculations of the Adsorption of Acetylene on the Si(001) Surface at Low and Full Coverage. *Surf. Sci.* **2007**, *601*, 3361–3365. [CrossRef]
36. Czekala, P.T.; Lin, H.; Hofer, W.A.; Gulans, A. Acetylene Adsorption on Silicon (100)-(4 × 2) Revisited. *Surf. Sci.* **2011**, *605*, 1341–1346. [CrossRef]
37. Taylor, P.A.; Wallace, R.M.; Cheng, C.C.; Weinberg, W.H.; Dresser, M.J.; Choyke, W.J.; Yates, J.J.T. Adsorption and Decomposition of Acetylene on Si(100)-(2 × 1). *J. Am. Chem. Soc.* **1992**, *114*, 6754–6760. [CrossRef]
38. Perdew, J.P.; Burke, K.; Ernzerhof, M. Generalized Gradient Approximation Made Simple. *Phys. Rev. Lett.* **1996**, *77*, 3865–3868. [CrossRef] [PubMed]
39. Perdew, J.P.; Burke, K.; Enzerhof, M. Errata: Generalized Gradient Approximation Made Simple. *Phys. Rev. Lett.* **1997**, *78*, 1396. [CrossRef]
40. Grimme, S.; Antony, J.; Ehrlich, S.; Krieg, S. A Consistent and Accurate Ab Initio Parametrization of Density Functional Dispersion Correction (DFT-D) for the 94 Elements H-Pu. *J. Chem. Phys.* **2010**, *132*, 154104. [CrossRef] [PubMed]
41. Grimme, S.; Ehrlich, S.; Goerigk, L. Effect of the Damping Function in Dispersion Corrected Density Functional Theory. *J. Comput. Chem.* **2011**, *32*, 1456–1465. [CrossRef] [PubMed]
42. Weigend, F.; Ahlrichs, R. Balanced Basis Sets of Split Valence, Triple Zeta Valence and Quadruple Zeta Valence Quality for H to Rn: Design and Assessment of Accuracy. *Phys. Chem. Chem. Phys.* **2005**, *7*, 3297–3305. [CrossRef] [PubMed]
43. Neese, F. The ORCA Program System. *WIREs Comput. Mol. Sci.* **2012**, *2*, 73–78. [CrossRef]
44. Fonseca Guerra, C.; Snijders, J.G.; te Velde, G.; Baerends, E.J. Towards an Order-*N* DFT Method. *Theor. Chem. Acc.* **1998**, *99*, 391–403. [CrossRef]
45. Te Velde, G.; Bickelhaupt, F.M.; Baerends, E.J.; Fonseca Guerra, C.; van Gisbergen, S.J.A.; Snijders, J.G.; Ziegler, T. Chemistry with ADF. *J. Comput. Chem.* **2001**, *22*, 931–967. [CrossRef]
46. Software for Chemistry & Material (SCM). *ADF2016*; Theoretical Chemistry, Vrije Universiteit: Amsterdam, The Netherlands, 2016; Available online: http://www.scm.com (accessed on 21 December 2017).
47. Kresse, G.; Hafner, J. Ab Initio Molecular Dynamics for Liquid Metals. *Phys. Rev. B* **1993**, *47*, 558–561. [CrossRef]
48. Kresse, G.; Hafner, J. *Ab Initio* Molecular-Dynamics Simulation of the Liquid-Metal–Amorphous-Semiconductor Transition in Germanium. *Phys. Rev. B* **1994**, *49*, 14251–14269. [CrossRef]
49. Kresse, G.; Furthmüller, J. Efficient Iterative Schemes for *Ab Initio* Total-Energy Calculations Using a Plane-Wave Basis Set. *Phys. Rev. B* **1996**, *54*, 11169–11186. [CrossRef]
50. Kresse, G.; Furthmüller, J. Efficiency of *Ab-Initio* Total Energy Calculations for Metals and Semiconductors Using a Plane-Wave Basis Set. *Comput. Mater. Sci.* **1996**, *6*, 15–50. [CrossRef]
51. Krukau, A.V.; Vydrov, O.A.; Izmaylov, A.F.; Scuseria, G.E. Influence of the Exchange Screening Parameter on the Performance of Screened Hybrid Functionals. *J. Chem. Phys.* **2006**, *125*, 224106. [CrossRef] [PubMed]
52. Blöchl, P. Projector Augmented-Wave Method. *Phys. Rev. B* **1994**, *50*, 17953–17979. [CrossRef]
53. Kresse, G.; Joubert, D. From Ultrasoft Pseudopotentials to the Projector Augmented-Wave Method. *Phys. Rev. B* **1999**, *59*, 1758–1775. [CrossRef]
54. Henkelman, G.; Uberuaga, B.P.; Jónsson, H. A Climbing Image Nudged Elastic Band Method for Finding Saddle Points and Minimum Energy Paths. *J. Chem. Phys.* **2000**, *113*, 9901–9904. [CrossRef]
55. Henkelman, G.; Jónsson, H. A Dimer Method for Finding Saddle Points on High Dimensional Potential Surfaces Using Only First Derivatives. *J. Chem. Phys.* **1999**, *111*, 7010–7022. [CrossRef]
56. Boyd, D.R.J. Infrared Spectrum of Trideuterosilane and the Structure of the Silane Molecule. *J. Chem. Phys.* **1955**, *23*, 922–926. [CrossRef]
57. Pecher, J.; Schober, C.; Tonner, R. Chemisorption of a Strained but Flexible Molecule: Cyclooctyne on Si(001). *Chem. Eur. J.* **2017**, *23*, 5459–5466. [CrossRef] [PubMed]

58. Tersoff, J.; Hamann, D.R. Theory and Application for the Scanning Tunneling Microscope. *Phys. Rev. Lett.* **1983**, *50*, 1998–2001. [CrossRef]
59. Tersoff, J.; Hamann, D.R. Theory of the Scanning Tunneling Microscope. *Phys. Rev. B* **1985**, *31*, 805–813. [CrossRef]
60. Hofer, W.A. Challenges and Errors: Interpreting High Resolution Images in Scanning Tunneling Microscopy. *Prog. Surf. Sci.* **2003**, *71*, 147–183. [CrossRef]
61. Palotás, K.; Hofer, W.A. Multiple Scattering in a Vacuum Barrier Obtained from Real-Space Wavefunctions. *J. Phys. Condens. Matter* **2005**, *17*, 2705–2713. [CrossRef]
62. Pecher, L.; Schmidt, S.; Tonner, R. Modeling the Complex Adsorption Dynamics of Large Organic Molecules: Cyclooctyne on Si(001). *J. Phys. Chem. C* **2017**, *121*, 26840–26850. [CrossRef]
63. Pecher, L.; Tonner, R. Computational analysis of the competitive bonding and reactivity pattern of a bifunctional cyclooctyne on Si(001). *Theor. Chem. Acc.* **2018**. in print.
64. Raupach, M.; Tonner, R. A Periodic Energy Decomposition Analysis (pEDA) Method for the Investigation of Chemical Bonding in Extended Systems. *J. Chem. Phys.* **2015**, *142*, 194105. [CrossRef] [PubMed]
65. Te Velde, G.; Baerends, E.J. Precise Density-Functional Method for Periodic Structures. *Phys. Rev. B* **1991**, *44*, 7888–7903. [CrossRef]
66. Software for Chemistry & Material (SCM). *BAND2016*; Theoretical Chemistry, Vrije Universiteit: Amsterdam, The Netherlands, 2016; Available online: http://www.scm.com (accessed on 21 December 2017).
67. Mitoraj, M.P.; Michalak, A.; Ziegler, T. A Combined Charge and Energy Decomposition Scheme for Bond Analysis. *J. Chem. Theory Comput.* **2009**, *5*, 962–975. [CrossRef] [PubMed]
68. Raupach, M. Quantum Chemical Investigation of Chemical Bonding at Surfaces—Development and Application of an Energy Decomposition Based Method. Ph.D. Thesis, Philipps-Universität Marburg, Marburg, Germany, 2015. [CrossRef]
69. Walsh, A.D. The Electronic Orbitals, Shapes, and Spectra of Polyatomic Molecules. Part I. AH_2 Molecules. *J. Chem. Soc.* **1953**, 2260–2266. [CrossRef]
70. Sagisaka, K.; Nara, J.; Bowler, D. Importance of Bulk States for the Electronic Structure of Semiconductor Surfaces: Implications for Finite Slabs. *J. Phys. Condens. Matter* **2017**, *29*, 145502. [CrossRef] [PubMed]
71. Lee, Y.T.; Lin, J.S. Ab Initio Molecular Dynamics Study of Ethylene Adsorption onto Si(001) Surface: Short-Time Fourier Transform Analysis of Structural Coordinate Autocorrelation Function. *J. Comput. Chem.* **2013**, *34*, 2697–2706. [CrossRef] [PubMed]
72. Gaberle, J.; Gao, D.Z.; Watkins, M.B.; Shluger, A.L. Calculating the Entropy Loss on Adsorption of Organic Molecules at Insulating Surfaces. *J. Phys. Chem. C* **2016**, *120*, 3913–3921. [CrossRef]
73. Campbell, C.T.; Sellers, J.R.V. The Entropies of Adsorbed Molecules. *J. Am. Chem. Soc.* **2012**, *34*, 18109–18115. [CrossRef] [PubMed]
74. Von Hopffgarten, M.; Frenking, G. Energy Decomposition Analysis. *WIREs Comput. Mol. Sci.* **2012**, *2*, 43–62. [CrossRef]
75. Reutzel, M.; Münster, N.; Lipponer, M.A.; Länger, C.; Höfer, U.; Koert, U.; Dürr, M. Chemoselective Reactivity of Bifunctional Cyclooctynes on Si(001). *J. Phys. Chem. C* **2016**, *120*, 26284–26289. [CrossRef]

10

Predicted Siliconoids by Bridging Si_9 Clusters through sp^3-Si Linkers

Laura-Alice Jantke and Thomas F. Fässler *

Department of Chemistry, Technical University of Munich, Lichtenbergstr. 4, 85747 Garching, Germany; Laura.Jantke@lrz.tu-muenchen.de

* Correspondence: thomas.faessler@lrz.tum.de

Abstract: Charged and neutral silicon clusters comprising Si atoms that are exclusively connected to atoms of the same type serve as models for bulk silicon surfaces. The experimentally known *nido*-$[Si_9]^{4-}$ Zintl cluster is investigated as a building block and allows for a theoretical prediction of novel silicon-rich oligomers and polymers by interconnection of such building units to larger aggregates. The stability and electronic properties of the polymers $\{^1_\infty([Si_9]{-}(SiCl_2)_2)_n\}$ and $\{^1_\infty([Si_9]{-}(SiH_2)_2)_n\}$, as well as of related oligomers are presented.

Keywords: silicon cluster; siliconoid; nanoparticle; computation

1. Introduction

Silicon is the element of choice for fulfilling the desire for novel materials with promising properties. Even though, silicon is an indirect band gap semiconductor resulting in poor efficiency of light emission, the observation of visible photoluminescence from porous silicon or silicon nanoparticles at room temperature reported in the early 90s [1,2] triggered the investigations of low-dimensional silicon quantum structures and had been a subject of extensive investigations due to the potential usage of nano-sized silicon in photonic and optoelectronic devices [3,4].

In recent years, some experimental molecular approaches successfully showed that molecules with low-valent Si atoms can be synthesized. So-called siliconoids [5] are stable unsaturated neutral silicon clusters that show the characteristic structural features of silicon nanoparticles and surfaces in the molecular regime generally realized through the occurrence of one or more unsubstituted Si atoms [5–8].

Computationally, many novel well-ordered Si allotropes of various dimensionality have been proposed [9,10], but only a few of those were experimentally verified, such as the low-density Si allotropes [11] with clathrate-type structures [12–19]. The focus of the research lies on the description, understanding, and discovery of well-performing photovoltaic materials as well as models for bulk silicon surfaces [20,21]. A few such predicted materials are realizable in laboratory to date also applying physical methods such as ultrafast laser-induced confined microexplosion [22].

Currently, one of the most investigated two-dimensional materials is silicene [23], the higher homologue of graphene [24]. It is reported as a buckled sheet of sp^3/sp^2-hybridized Si atoms connected to wrinkled six membered rings, which is stabilized through intrinsic van-der-Waals interactions [25]. Silicene is experimentally accessible only on metal surfaces [26–28]. Theoretical investigations on this material are reviewed in several articles [29–31]. Another two-dimensional Si modification results from adding ad-atoms to silicene. Si in this MoS_2-structure type shows a lower relative energy compared to silicene [32]. In our recent work, we introduced two-dimensional materials, which may overcome surface reconstruction problems by using $\{Si_9\}$ Zintl cages (Figure 1a) that are stable in solution [33].

Our aim is the computational investigation using substructures or molecular building units that are experimentally known, which we call the *chemi-inspired* attempt [9,10]. We found that Zintl clusters of Group 14 elements, as they occur in neat binary intermetallic solids (Zintl phases) [34], represent ideal candidates for constructing tailor-made materials. They have well-defined structures, which are retained upon solvation, and can thus be functionalized, oligomerized, or even polymerized [35,36]. Best suited for such chemistry is the nine-atomic $[E_9]^{(2-4)-}$ cluster (E = Si, Ge, Sn, and Pb). For silicon, only a few examples of following-up reactions of Si_9 clusters are known whereas the pure clusters are structurally characterized in solids [37] and in salts obtained from solution [38–42]. In all examples, reactions occur via atoms of the open square of the mono-capped square antiprism of the *nido*-$[Si_9]^{4-}$ cluster (Figure 1a); either capped in a η^4-coordination with phenylzinc (Figure 1b) [39] as well as Cu-NHC (NHC = *N*-heterocyclic carbene) [43] or two clusters are bridged by $Ni(CO)_2$ fragments via two neighboring atoms of the open square, each involved in a η^1-coordination (Figure 1c) [40,44].

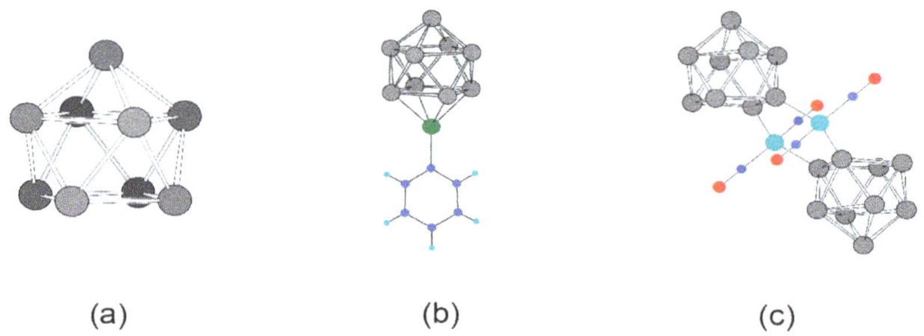

Figure 1. The building block *nido*-$[Si_9]^{4-}$ and its experimentally accessible complexes from solution. (**a**) *nido*-$[Si_9]^{4-}$ optimized on a DFT-PBE0/def2-TZVPP/PCM level of theory; (**b**) closo-$[Si_9Zn(C_6H_5)]^{3-}$ [39]; (**c**) $[\{Ni(CO)_2\}_2(\mu\text{-}Si_9)_2]^{8-}$ [40,44]. Si, C, H, Zn, Ni, and O are shown as grey, blue, light blue, green, turquoise, and red spheres.

Further, it was shown that such clusters can serve as seed crystals for the synthesis of Si nanoparticles and nanoscale materials [45,46]. In analogy to Ge_9 clusters for which covalently bonded Ge_9 oligomers and polymers are experimentally known, Si-based materials that contain Si_9 units linked via the atoms of the open square are feasible [39,41,44]. The direct linkage has already been discussed and also the stability and properties of two-dimensional Si materials containing sp^3-hybrized Si atoms as linkers between Si_9 clusters were computationally investigated [33]. In this work, we explore the possibilities of introducing tetrahedrally connecting sp^3-Si atoms, which might occur from a formal reaction of the anions with $SiCl_4$ and SiH_2Cl_2 and *nido*-$[Si_9]^{4-}$ clusters. Thus, we investigate the complexes $[Si(Si_9)Cl_3]^{3-}$ and $[Si(Si_9)Me_3]^{3-}$ (Me = $-CH_3$), $[Si(Si_9)_2Cl_2]^{6-}$, $[Si(Si_9)_3Cl]^{9-}$, and $[Si(Si_9)_4]^{12-}$. Further, we analyze the hypothetical insertion of *nido*-$[Si_9]^{4-}$ clusters into polysilanes resulting in the neutral polymers $\{^1_\infty([Si_9]-(SiCl_2)_2)_n\}$ and $\{^1_\infty([Si_9]-(SiH_2)_2)_n\}$.

2. Computational Details

Neutral polymers containing silane fragments and Si_9 clusters as well as the polysilanes themselves were optimized starting from manually drawn structures pre-optimized using UFF [47,48]. The lattice and atomic positions were allowed to relax during optimization within the constraints given by rod group symmetry [49]. All quantum chemical calculations for the polymers were carried out using the CRYSTAL09 program package [50] with a hybrid DFT functional after Perdew, Burke, and Ernzerhof (DFT-PBE0) [51,52]. For silicon, a modified split-valence + polarization (SVP) basis set [53] was applied. The shrinking factor (SHRINK) for generating the Monkhorst-Pack-type grid of k points in the reciprocal space was set to 4, resulting in three k-points in the irreducible Brillouin zone. For the evaluation of the Coulomb and exchange integrals tight tolerance factors (TOLINTEG) of 8, 8, 8, 8, and 16 were chosen. Default optimization convergence thresholds and an extra-large integration grid (XLGRID) for the density-functional part were applied in all calculations. Harmonic vibrational frequencies were calculated numerically to confirm the stationary point on the potential-energy surface as a true minimum.

The investigations on charged complexes containing *nido*-$[Si_9]^{4-}$ as a building block are done with the DFT-PBE0 [51,52] hybrid DFT functional and def2-TVZPP level basis sets for the elements Si, Cl, H, and C using the Gaussian09 program package [54]. For modelling a solvation effect, a solvation model (polarizable continuum model, PCM) with standard settings was applied [55]. For structure optimizations, extremely tight optimization convergence criteria were combined with a large and ultrafine DFT integration grid. The systems were allowed to relax without symmetry restrictions. Harmonic vibrational frequencies were calculated analytically to confirm the stationary point on the potential-energy surface as a true minimum (except for **M6**).

The position parameters for all optimized compounds are listed in the Supplementary Materials.

3. Results and Discussion

3.1. Neutral Polymers with $SiCl_2$ or SiH_2 Linkers between the Clusters

Inspired by the $[(Ni(CO)_2)_2(\mu\text{-}Si_9)_2]^{8-}$ complex [40,44] (Figure 1c) and in continuation of our work on 2D structures containing Si_9 clusters [33], we derived two polymers $\{^1_\infty([Si_9]-(SiCl_2)_2)_n\}$ (**P1**) and $\{^1_\infty([Si_9]-(SiH_2)_2)_n\}$ (**P2**) by a formal replacement of $Ni(CO)_2$ with isolobal $SiCl_2$ or SiH_2. The polymers can also be regarded as silanes with inserting Si_9 units. Pauling electronegativity of Ni and Si are almost identical with a value of $\chi_{Ni} = 1.91$ and $\chi_{Si} = 1.90$, respectively, and the electron withdrawing effect of the CO ligands is considered by choosing ligands whose electronegativity is higher if compared to Si ($\chi_{Cl} = 3.16$ and $\chi_H = 2.20$).

The energetically optimized polymers **P1** and **P2** possess nearly ideal C_{4v} symmetric Si_9 units in the rod group *pmm*2 (relevant bond lengths are listed in Table S1, Supplementary Materials). The two bridging Si atoms and two adjacent atoms of each of the two open squares form a planar hexagon which is compressed along the propagation direction of the polymer. The bond angles at the bridging Si atom consequently lie in the range of tetrahedral angles 103.4°–111.0° for both compounds. The torsional angles between these hexagons and the open square of the *nido*-clusters are 165.8° and 160.3° for **P1** and **P2**, respectively. In order to compare the total energy with respect to α-Si and since the compounds contain also X = H and Cl atoms we include polysilanes with comparable bond situations according to Equation (1):

$$\Delta E_{rel}(^1_\infty[Si_9-(SiX_2)_2]) = \Delta E_{tot}(^1_\infty[Si_9-(SiX_2)_2]) - [9\cdot E_{tot}(\alpha\text{–}Si) + 2\cdot E_{tot}(^1_\infty[SiX_2]) \quad (1)$$

The resulting values of 7.28 eV and 7.23 eV for X = Cl and H, respectively refer to one formula unit $(Si_9)(SiX_2)_2$ (X = Cl, H) for **P1** and for **P2**, respectively (Figure 2). Scaled to one Si atom per formula unit these values result in 0.56 eV for both compounds. On a first glance, the values seem high if compared to α-Si, however the polymers must rather be compared to Si nanoparticles than to bulk materials. In this context the values are rather close to the relative energy of the carbon fullerenes of which for example C_{60} which is 0.48 eV per atom higher in total energy than diamond [56]. The Si bridged clusters are also much lower in energy than the directly connected clusters (0.96 eV) [33]. Energetically, the differences between Si–Cl and Si–H bonds a negligible.

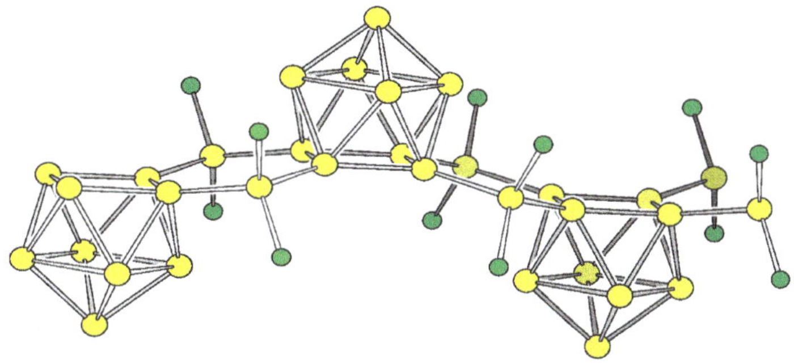

Figure 2. Cut-out of polymer **P1**. Si and Cl are shown as yellow and green spheres, respectively.

3.2. Charged Molecules—The Stepwise Ligand Exchange in $SiCl_4$ to $[Si_{17}]^{12-}$

Siliconoid molecules are formed by a stepwise substitution of heteroatomic bonds between ligands bound to a central Si atom with Si terminated groups. We computationally substituted ligands in SiX_4 by $[Si_9]^{4-}$ for X = Cl, H, and CH_3. The resulting molecular ions $[Si(Si_9)Cl_3]^{3-}$ (**M1**), $[Si(Si_9)H_3]^{3-}$ (**M2**) and $[Si(Si_9)Me_3]^{3-}$ (**M3**) differ in the ligands at the sp^3-Si atom (Table 1). We started the optimizations for all compounds from a η^1-connected isomer with an ideal C_{4v} symmetric *nido*-$[Si_9]^{4-}$ cluster.

As the energetically most favorite structure of **M1** we found an arrangement with an almost perfect D_{3h} symmetric trigonal prismatic Si_9 cluster with the $SiCl_3$ group symmetrically capping a triangular face of the Si_9 cluster. The three distances between the Si atoms of the $SiCl_3$ group and the Si atoms of the clusters are with a value of 2.58 Å identical to the lengths of the triangular face of Si_9, consequently forming an almost undistorted tetrahedron. The three Si–Cl distances are 2.20 Å. The shape of Si_9 cluster distorts to a tricapped trigonal prism with prism heights of 2.69 Å that are elongated with respect to the triangular faces. The opposing trigonal face, which is not capped, has considerable shorter lengths edges of 2.46 Å and 2.49 Å. Thus the coordination of the $SiCl_3$ group leads to an elongation of edges of the capped triangle.

Interestingly, no local minimum structure with η^1-coordination of the $SiCl_3$ groups were found. In contrast, for $[Si(Si_9)H_3]^{3-}$ (**M2**) and $[Si(Si_9)Me_3]^{3-}$ (**M3**) the energetically most favored structures remain with η^1-coordinated Si_9 clusters. For both anions, the Si_9 clusters distort to (pseudo-)C_{2v} symmetric cages as an intermediate of the mono-capped square anti prism and the tricapped trigonal prism. With the ratio of diagonal lengths of the open square of $d_1/d_2 = 1.22$ is **M2** more distorted than **M3** ($d_1/d_2 = 1.16$). The bond length between the clusters and their ligands are 2.33 Å and 2.35 Å, respectively. The prism heights are 2.63 Å, 2.64 Å, and 3.08 Å for **M2** and 2.62 Å, 2.63 Å, and 3.18 Å for **M3**.

Bridging two clusters by a single $SiCl_2$ fragment, e.g. attaching two clusters to one central Si atom leads to the anion of composition $[SiCl_2(Si_9)_2]^{6-}$ (**M4**). Both Si_9 clusters in **M4** form a η^1-coordination with distances between the cluster and the bridge atoms of 2.34 Å. The Si–Cl bonds of 2.13 Å and 2.14 Å are shorter than for **M1**. Both clusters within **M4** deform after optimization to (pseudo-)C_{2v} symmetric compounds with $d_1/d_2 = 1.21$ and 1.22. (The distortion of C_{4v} symmetric monocapped square anti prism can be characterized by the ratio of the diagonal lengths d_1 and d_2 of the open square. For an undistorted cage the ratio is equal to 1. Deviation from this ratio leads to cages with C_{2v} symmetry.) The bond angle between the central Si atom and the two bonded Si atoms of the clusters is 121.7° and thus larger than the tetrahedral angel, which is most probably due to the steric demands of the Si_9 ligands.

Substitution of another Cl ligand by a third $[Si_9]^{4-}$ results in the highly charged cluster $[SiCl(Si_9)_3]^{9-}$ (**M5**). In **M5** the three (pseudo-)C_{2v} symmetric clusters have diagonal length ratio in the narrow range of $d_1/d_2 = 1.15$–1.17 and bond lengths of 2.37 Å between cluster atoms and the central sp^3-Si atom for each bond. The distance Si–Cl is 1.18 Å.

A replacement of all Cl ligands by Si_9 units results in $[Si(Si_9)_4]^{12-}$ (**M6**). Such highly charged clusters are not unrealistic, since a salt containing a $[Ge_{45}]^{12-}$ cluster unit, which is attached to three Au^+ ions ($[Au_3Ge_{45}]^{9-}$) had been isolated and structurally characterized [57]. The distances for all four Si–Si bonds at the central Si atom are 2.39 Å. The angles range between 105.1° and 117.3°. The clusters are again distorted towards C_{2v} symmetric cages, but stay close to the C_{4v} symmetric input with ratios of the diagonal lengths of the open square between $d_1/d_2 = 1.11$–1.15. A table listening all bond distances and relevant cluster parameters is located in the Supplementary Materials.

The building unit **M6** can be extended by continuation the building scheme with more $SiCl_4$. Successive linking $[Si_9]^{4-}$ clusters with $SiCl_4$. under Cl substitution results in a neutral two-dimensional ${}^2_\infty\{[(Si_9)\text{–}Si]_n\}$ sheet [33].

Table 1. Structural analysis of molecular anions (Figure 3). All values are averaged, if more bonds or angles of the same type exist.

Molecular Anions	Distances/Å	Distances/Å	Angles/°	Angles/°	Angles/°
	X_3Si–$Si(Si_9)$	Si–Cl	X–Si–X	(Si_9)–Si–X	(Si_9)–Si–(Si_9)
M1	2.58 (η^3)	2.20	94.6	-	-
M2	2.33	-	105.3	113.3	-
M3	2.35	-	107.8	111.1	-
M4	2.34	2.14	101.2	107.9	121.7
M5	2.37	2.18	-	100.6	116.7
M6	2.39	-	-	-	109.4

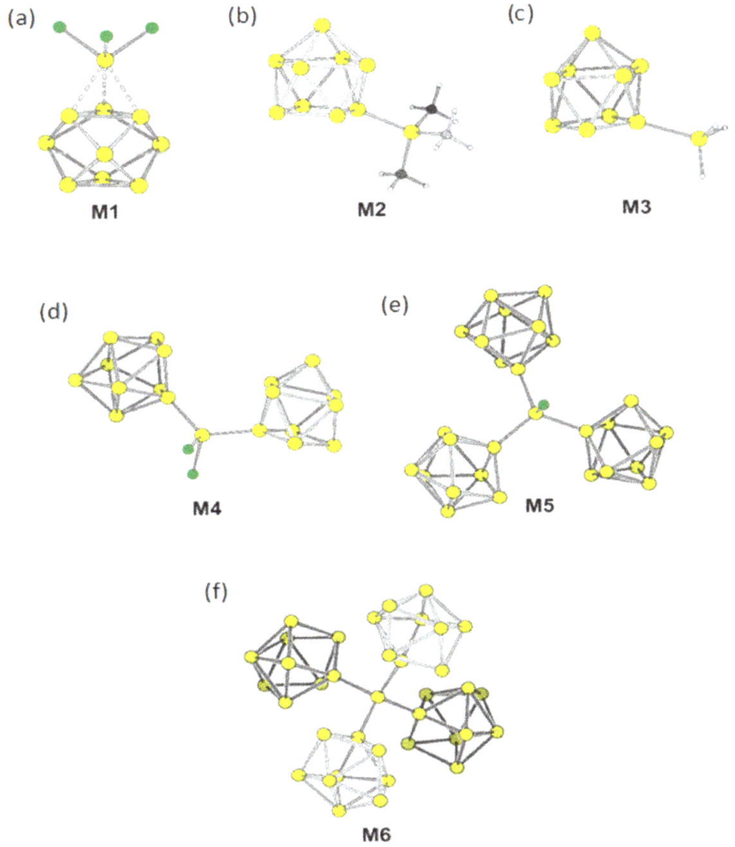

Figure 3. Optimized structures of molecular anions **M1–M6**. **(a)** η^1-$[Si(Si_9)Cl_3]^{3-}$ (**M1**), **(b)** η^1-$[Si(Si_9)Me_3]^{3-}$ (**M2**), **(c)** η^1-$[Si(Si_9)H_3]^{3-}$ (**M3**), **(d)** $[SiCl_2(Si_9)_2]^{6-}$ (**M4**), **(e)** $[SiCl(Si_9)_3]^{9-}$ (**M5**), **(f)** $[Si(Si_9)_4]^{12-}$ (**M6**). Si and Cl atoms are shown as yellow and green spheres, respectively.

4. Conclusion

The modelling of polymeric Si_9 chains gave interesting insights to the chemistry of this up to now hard to functionalize Si_9 cluster. It showed that a replacement of $Ni(CO)_2$ as a ligand by others bridging sp^3-Si units with comparable electron withdrawing effects, namely SiH_2 and $SiCl_2$, leads to polymers that are energetically in a range for realizable materials. The formation of charged oligomers by the formal substitution of X ligands in SiX_4 by $[Si_9]^{4-}$ clusters leads to stable anions $[SiCl_{4-x}(Si_9)_x]^{3x-}$. For all derivatives, stable anions with covalent Si–Si bonds to the central Si atom form, except for x = 1 and X = Cl. Here, a η^3-coordination of the cluster to the Si atom of the $SiCl_3$ group is observed. Whereas the monomeric unit $[Si(Si_9)_4]^{12-}$ (**M6**) possesses a relative high charge, further combination with bridging Si atoms leads to a relatively stable two-dimensional neutral Si allotrope.

Acknowledgments: The authors are grateful to the SolTech (Solar Technologies go Hybrid) program of the Bavarian State Ministry of Education, Science and the Arts for financial support.

Author Contributions: Laura-Alice Jantke and Thomas F. Fässler conceived and designed the structures. Laura-Alice Jantke performed all quantum chemical calculations, Laura-Alice Jantke and Thomas F. Fässler wrote and corrected the paper, respectively.

References

1. Cullis, A.G.; Canham, L.T. Visible light emission due to quantum size effects in highly porous crystalline silicon. *Nature* **1991**, *353*, 335–338. [CrossRef]
2. Canham, L.T. Silicon quantum wire array fabrication by electrochemical and chemical dissolution of wafers. *Appl. Phys. Lett.* **1990**, *57*, 1046–1048. [CrossRef]
3. Bisi, O.; Ossicini, S.; Pavesi, L. Porous silicon: A quantum sponge structure for silicon based optoelectronics. *Surf. Sci. Rep.* **2000**, *38*, 1–126. [CrossRef]
4. Geppert, T.; Schiling, J.; Wehrspohn, R.; Gösele, U. Silicon-based photonic crystals. In *Silicon Photonics*; Pavesi, L., Lockwood, D.J., Eds.; Springer: Berlin, Germany, 2004; Volume 94, pp. 295–322. ISBN 978-3-540-21022-1.
5. Abersfelder, K.; Russell, A.; Rzepa, H.S.; White, A.J.; Haycock, P.R.; Scheschkewitz, D. Contraction and expansion of the silicon scaffold of stable Si_6R_6 isomers. *J. Am. Chem. Soc.* **2012**, *134*, 16008–16016. [CrossRef] [PubMed]
6. Iwamoto, T.; Akasaka, N.; Ishida, S. A heavy analogue of the smallest bridgehead alkene stabilized by a base. *Nat. Commun.* **2014**, *5*, 5353. [CrossRef] [PubMed]
7. Abersfelder, K.; White, A.J.; Rzepa, H.S.; Scheschkewitz, D. A tricyclic aromatic isomer of hexasilabenzene. *Science* **2010**, *327*, 564–566. [CrossRef] [PubMed]
8. Scheschkewitz, D. A molecular silicon cluster with a "naked" vertex atom. *Angew. Chem. Int. Ed.* **2005**, *44*, 2954–2956. [CrossRef] [PubMed]
9. Jantke, L.A.; Stegmaier, S.; Karttunen, A.J.; Fässler, T.F. Slicing Diamond—A Guide to Deriving sp^3-Si Allotropes. *Chem. Eur J.* **2017**, *23*, 2734–2747. [CrossRef] [PubMed]
10. Jantke, L.A.; Karttunen, A.J.; Fassler, T.F. Slicing Diamond for More sp^3-Group 14 Allotropes Ranging from Direct Bandgaps to Poor Metals. *ChemPhysChem* **2017**, *18*, 1992–2006. [CrossRef] [PubMed]
11. Beekman, M.; Wei, K.; Nolas, G.S. Clathrates and beyond: Low-density allotropy in crystalline silicon. *Appl. Phys. Rev.* **2016**, *3*, 040804. [CrossRef]
12. O'Keeffe, M.; Adams, G.B.; Sankey, O.F. Duals of Frank-Kasper structures as C, Si and Ge clathrates: Energetics and structure. *Philosoph. Mag. Lett.* **1998**, *78*, 21–28. [CrossRef]
13. Krishna, L.; Martinez, A.D.; Baranowski, L.L.; Brawand, N.P.; Koh, C.A.; Stevanovic, V.; Lusk, M.T.; Toberer, E.S.; Tamboli, A.C. Group IV clathrates: Synthesis, optoelectronic properties, and photovoltaic applications. *Proceedings* **2014**, *8981*, 898108.
14. Norouzzadeh, P.; Myles, C.W.; Vashaee, D. Prediction of Giant Thermoelectric Power Factor in Type-VIII Clathrate Si_{46}. *Sci. Rep.* **2014**, *4*, 7028. [CrossRef] [PubMed]
15. Härkönen, V.J.; Karttunen, A.J. Ab initio studies on the lattice thermal conductivity of silicon clathrate frameworks II and VIII. *Phys. Rev. B* **2016**, *93*, 024307. [CrossRef]
16. Ammar, A.; Cros, C.; Pouchard, M.; Jaussaud, N.; Bassat, J.-M.; Villeneuve, G.; Duttine, M.; Ménétrier, M.; Reny, E. On the clathrate form of elemental silicon, Si_{136}: Preparation and characterisation of Na_xSi_{136} ($x \rightarrow 0$). *Solid State Sci.* **2004**, *6*, 393–400. [CrossRef]
17. Ohashi, F.; Hattori, M.; Ogura, T.; Koketsu, Y.; Himeno, R.; Kume, T.; Ban, T.; Iida, T.; Habuchi, H.; Natsuhara, H.; et al. High-yield synthesis of semiconductive type-II Si clathrates with low Na content. *J. Non-Cryst. Solids* **2012**, *358*, 2134–2137. [CrossRef]
18. Baranowski, L.L.; Krishna, L.; Martinez, A.D.; Raharjo, T.; Stevanovic, V.; Tamboli, A.C.; Toberer, E.S. Synthesis and optical band gaps of alloyed Si–Ge type II clathrates. *J. Mater. Chem. C* **2014**, *2*, 3231–3237. [CrossRef]
19. Krishna, L.; Baranowski, L.L.; Martinez, A.D.; Koh, C.A.; Taylor, P.C.; Tamboli, A.C.; Toberer, E.S. Efficient route to phase selective synthesis of type II silicon clathrates with low sodium occupancy. *Cryst. Eng. Comm* **2014**, *16*, 3940–3949. [CrossRef]

20. Buriak, J.M. Organometallic Chemistry on Silicon and Germanium Surfaces. *Chem. Rev.* **2002**, *102*, 1271–1308. [CrossRef] [PubMed]
21. Wippermann, S.; He, Y.P.; Voros, M.; Galli, G. Novel silicon phases and nanostructures for solar energy conversion. *Appl. Phys. Rev.* **2016**, *3*, 040807. [CrossRef]
22. Rapp, L.; Haberl, B.; Pickard, C.J.; Bradby, J.E.; Gamaly, E.G.; Williams, J.S.; Rode, A.V. Experimental evidence of new tetragonal polymorphs of silicon formed through ultrafast laser-induced confined microexplosion. *Nat. Commun.* **2015**, *6*, 7555. [CrossRef] [PubMed]
23. Guzmán-Verri, G.G.; Lew Yan Voon, L.C. Electronic structure of silicon-based nanostructures. *Phys. Rev. B* **2007**, *76*, 075131. [CrossRef]
24. Novoselov, K.S.; Geim, A.K.; Morozov, S.V.; Jiang, D.; Zhang, Y.; Dubonos, S.V.; Grigorieva, I.V.; Firsov, A.A. Electric Field Effect in Atomically Thin Carbon Films. *Science* **2004**, *306*, 666–669. [CrossRef] [PubMed]
25. Cahangirov, S.; Topsakal, M.; Akturk, E.; Sahin, H.; Ciraci, S. Two- and one-dimensional honeycomb structures of silicon and germanium. *Phys. Rev. Lett.* **2009**, *102*, 236804. [CrossRef] [PubMed]
26. Lin, C.-L.; Arafune, R.; Kawahara, K.; Tsukahara, N.; Minamitani, E.; Kim, Y.; Takagi, N.; Kawai, M. Structure of Silicene Grown on Ag(111). *Appl. Phys. Express* **2012**, *5*, 045802. [CrossRef]
27. Fleurence, A.; Friedlein, R.; Ozaki, T.; Kawai, H.; Wang, Y.; Yamada-Takamura, Y. Experimental evidence for epitaxial silicene on diboride thin films. *Phys. Rev. Lett.* **2012**, *108*, 245501. [CrossRef] [PubMed]
28. Meng, L.; Wang, Y.; Zhang, L.; Du, S.; Wu, R.; Li, L.; Zhang, Y.; Li, G.; Zhou, H.; Hofer, W.A.; et al. Buckled Silicene Formation on Ir(111). *Nano Lett.* **2013**, *13*, 685–690. [CrossRef] [PubMed]
29. Zhuang, J.; Xu, X.; Feng, H.; Li, Z.; Wang, X.; Du, Y. Honeycomb silicon: A review of silicene. *Sci. Bull.* **2015**, *60*, 1551–1562. [CrossRef]
30. Kaloni, T.P.; Schreckenbach, G.; Freund, M.S.; Schwingenschlögl, U. Current developments in silicene and germanene. *Phys. Status Solidi RRL* **2016**, *10*, 133–142. [CrossRef]
31. Voon, L.C.L.Y.; Zhu, J.; Schwingenschlögl, U. Silicene: Recent theoretical advances. *Appl. Phys. Rev.* **2016**, *3*, 040802. [CrossRef]
32. Gimbert, F.; Lee, C.-C.; Friedlein, R.; Fleurence, A.; Yamada-Takamura, Y.; Ozaki, T. Diverse forms of bonding in two-dimensional Si allotropes: Nematic orbitals in the MoS_2 structure. *Phys. Rev. B* **2014**, *90*, 165423. [CrossRef]
33. Jantke, L.-A.; Karttunen, A.J.; Fässler, T.F. Chemi-inspired silicon allotropes—Experimentally accessible Si9 cages as building block for 1D polymers, 2D sheets, single-walled nanotubes, and nanoparticles, submitted.
34. Fässler, T.F. Zintl phases: Principles and recent developments. In *Structure and Bonding*; Mingos, D.M.P., Ed.; Springer: Heidelberg, Germany, 2011; Volume 139, ISBN 978-3-642-21150-8.
35. Scharfe, S.; Kraus, F.; Stegmaier, S.; Schier, A.; Fässler, T.F. Zintl Ions, Cage Compounds, and Intermetalloid Clusters of Group 14 and Group 15 Elements. *Angew. Chem. Int. Ed.* **2011**, *50*, 3630–3670. [CrossRef] [PubMed]
36. Fässler, T.F. Zintl ions: Principles and recent developments. In *Structure and Bonding*; Mingos, D.M.P., Ed.; Springer: Heidelberg, Germany, 2011; Volume 140, ISBN 978-3-642-21181-2.
37. Queneau, V.; Todorov, E.; Sevov, S.C. Synthesis and structure of isolated silicon clusters of nine atoms. *J. Am. Chem. Soc.* **1998**, *120*, 3263–3264. [CrossRef]
38. Goicoechea, J.M.; Sevov, S.C. Ligand-free deltahedral clusters of silicon in solution: Synthesis, structure, and electrochemistry of $[Si_9]^{2-}$. *Inorganic Chemistry* **2005**, *44*, 2654–2658. [CrossRef] [PubMed]
39. Goicoechea, J.M.; Sevov, S.C. Organozinc derivatives of deltahedral zintl ions: Synthesis and characterization of *closo*-$[E_9Zn(C_6H_5)]^{3-}$ (E = Si, Ge, Sn, Pb). *Organometallics* **2006**, *25*, 4530–4536. [CrossRef]
40. Joseph, S.; Hamberger, M.; Mutzbauer, F.; Härtl, O.; Meier, M.; Korber, N. Chemistry with Bare Silicon Clusters in Solution: A Transition-Metal Complex of a Polysilicide. *Anion Angew. Chem. Int. Ed.* **2009**, *48*, 8770–8772. [CrossRef] [PubMed]
41. Joseph, S.; Suchentrunk, C.; Kraus, F.; Korber, N. $[Si_9]^{4-}$ Anions in Solution—Structures of the Solvates $Rb_4Si_9\cdot 4.75NH_3$ and [Rb(18-crown-6)]$Rb_3Si_9\cdot 4NH_3$, and Chemical Bonding in $[Si_9]^{4-}$. *Eur. J. Inorg. Chem.* **2009**, *2009*, 4641–4647. [CrossRef]
42. Waibel, M.; Kraus, F.; Scharfe, S.; Wahl, B.; Fässler, T.F. $[(MesCu)_2(\eta^3\text{-}Si_4)]^{4-}$: A Mesitylcopper-Stabilized Tetrasilicide Tetraanion. *Angew. Chem. Int. Ed.* **2010**, *49*, 6611–6615. [CrossRef] [PubMed]
43. Geitner, F.S.; Fässler, T.F. Low Oxidation State Silicon Clusters—Synthesis and Structure of $[NHCDippCu(\eta^4\text{-}Si_9)]^{2-}$. *Chem. Commun.* **2017**, *53*, 12974–12977. [CrossRef] [PubMed]

44. Gärtner, S.; Hamberger, M.; Korber, N. The First Chelate-Free Crystal Structure of a Silicide Transition Metal Complex $K_{0.28}Rb_{7.72}Si_9(Ni(CO)_2)_2 \cdot 16NH_3$. *Crystals* **2015**, *5*, 275–282. [CrossRef]
45. Riley, A.E.; Korlann, S.D.; Richman, E.K.; Tolbert, S.H. Synthesis of semiconducting thin films with nanometer-scale periodicity by solution-phase coassembly of zintl clusters with surfactants. *Angew. Chem. Int. Ed.* **2005**, *45*, 235–241. [CrossRef] [PubMed]
46. Nolan, B.M.; Henneberger, T.; Waibel, M.; Fässler, T.F.; Kauzlarich, S.M. Silicon Nanoparticles by the Oxidation of $[Si_4]^{4-}$- and $[Si_9]^{4-}$- Containing Zintl Phases and Their Corresponding Yield. *Inorg. Chem.* **2015**, *54*, 396–401. [CrossRef] [PubMed]
47. Hanson, B. *Jmol: An open-source Java viewer for chemical structures in 3D*; Northfield, MN, USA, 2016.
48. Rappe, A.M.; Joannopoulos, J.D.; Bash, P.A. A test of the utility of plane-waves for the study of molecules from first principles. *J. Am. Chem. Soc.* **1992**, *114*, 6466–6469. [CrossRef]
49. Dovesi, R.; Saunders, V.R.; Roetti, R.; Orlando, R.; Zicovich-Wilson, C.M.; Pascale, F.; Civalleri, B.; Doll, K.; Harrison, N.M.; Bush, I.J.; et al. *CRYSTAL09 User's Manual*; University of Turino: Turino, Italy, 2009.
50. Dovesi, R.; Orlando, R.; Civalleri, B.; Roetti, C.; Saunders, V.R.; Zicovich-Wilson, C.M. CRYSTAL: A computational tool for the ab initio study of the electronic properties of crystals. *Z. Kristallogr.* **2005**, *220*, 571–573. [CrossRef]
51. Perdew, J.P.; Burke, K.; Ernzerhof, M. Generalized gradient approximation made simple. *Phys. Rev. Lett.* **1996**, *77*, 3865–3868. [CrossRef] [PubMed]
52. Adamo, C.; Barone, V. Toward reliable density functional methods without adjustable parameters: The PBE0 model. *J. Chem. Phys.* **1999**, *110*, 6158–6170. [CrossRef]
53. Karttunen, A.J.; Fässler, T.F.; Linnolahti, M.; Pakkanen, T.A. Structural Principles of Semiconducting Group 14 Clathrate Frameworks. *Inorg. Chem.* **2011**, *50*, 1733–1742. [CrossRef] [PubMed]
54. Frisch, M.J.; Trucks, G.W.; Schlegel, H.B.; Scuseria, G.E.; Robb, M.A.; Cheeseman, J.R.; Scalmani, G.; Barone, V.; Mennucci, B.; Petersson, G.A.; et al. *Gaussian 09*; Gaussian, Inc.: Wallingford, CT, USA, 2009.
55. Scalmani, G.; Frisch, M.J. Continuous surface charge polarizable continuum models of solvation. I. General formalism. *J. Chem. Phys.* **2010**, *132*, 114110. [CrossRef] [PubMed]
56. Karttunen, A.J.; Fässler, T.F.; Linnolahti, M.; Pakkanen, T.A. Two-, One-, and Zero-Dimensional Elemental Nanostructures Based on Ge_9-Clusters. *ChemPhysChem* **2010**, *11*, 1944–1950. [CrossRef] [PubMed]
57. Spiekermann, A.; Hoffmann, S.D.; Kraus, F.; Fässler, T.F. $[Au_3Ge_{18}]^{5-}$—A gold-germanium cluster with remarkable Au–Au interactions. *Angew. Chem. Int. Ed.* **2007**, *46*, 1638–1640. [CrossRef] [PubMed]

11

Synthesis of a Dichlorodigermasilane: Double Si–Cl Activation by a Ge=Ge Unit

Tomohiro Sugahara [1], Norihiro Tokitoh [1] and Takahiro Sasamori [2,*]

1 Institute for Chemical Research, Kyoto University, Gokasho, Uji, Kyoto 611-0011, Japan; sugahara@boc.kuicr.kyoto-u.ac.jp (T.S.); tokitoh@boc.kuicr.kyoto-u.ac.jp (N.T.)
2 Graduate School of Natural Sciences, Nagoya City University, Yamanohata 1, Mizuho-cho, Mizuho-ku, Nagoya, Aichi 467-8501, Japan
* Correspondence: sasamori@nsc.nagoya-cu.ac.jp

Abstract: Halogenated oligosilanes and oligogermanes are interesting compounds in oligosilane chemistry from the viewpoint of silicon-based-materials. Herein, it was demonstrated that a 1,2-digermacyclobutadiene derivative could work as a bis-germylene building block towards double Si–Cl activation to give a halogenated oligometallane, a bis(chlorogermyl)dichlorosilane derivative.

Keywords: Si–Cl activation; germylene; digermene; digermacyclobutadiene

1. Introduction

Halogenated oligosilanes and oligogermanes are attractive compounds as functionalized oligometallanes from the standpoint of oligosilane material chemistry [1–7]. In this regard, Si(II) or Ge(II) species should be an important building block for creating such halogenated oligosilanes/germanes because silylenes (divalent Si(II) species) or germylenes (divalent Ge(II) species) have been known to undergo ready Si–Cl insertion reactions, i.e., Si–Cl activation reactions [8–11]. For example, Iwamoto and Kira reported the facile Si–Cl insertion of the isolable dialkylmetallylenes towards $SiCl_4$ under mild conditions [10]. However, especially in the germanium cases, it is difficult to isolate the insertion products, >Ge(Cl)SiR_3, because the insertion reaction of a germylene toward a Si–Cl bond would be reversible in some cases [12]. Thus, the substituents on the Si atom (R of the R_3Si–Cl species) should be bulky and/or electropositive to avoid the α-elimination of R_3Si–Cl from the Ge moiety [8,12–22]. The requirement for the bulkiness on the Si–Cl moiety could make it difficult to create halogenated oligometallanes, such as >Ge(Cl)–$SiCl_2$–Ge(Cl)< with utilizing the double Si–Cl activation of the Ge(II) species towards $SiCl_4$, because of two unfavorable factors: (i) entropy, two Ge(II) moieties should react with one $SiCl_4$ species; and (ii) the stability of the product, the Cl atoms on the Si atom should promote the α-Si–Cl elimination, i.e., the retro reaction (Scheme 1). In this paper, we chose a 1,2-digermacyclobutadiene derivative [23] as a suitable Ge(II) building block for the double Si–Cl activation of $SiCl_4$ to yield >Ge(Cl)–$SiCl_2$–Ge(Cl)< species, because the rigid cyclic skeleton should overcome the entropy-disadvantage, and the rigidness of the cyclic skeleton should suppress the α-Si–Cl elimination. Finally, it was found that the stable 1,2-digermacyclobutadiene **1** (1,2-Tbb_2-3,4-Ph_2-digermacyclobutadiene, Tbb = 2,6-$[CH(SiMe_3)_2]_2$-4-*t*-Bu-C_6H_2, Scheme 2) [24] with $SiCl_4$ afforded the corresponding 1,3-digerma-2-sila-cyclopent-4-ene derivative, the cyclic >Ge(Cl)–$SiCl_2$–Ge(Cl)< compound. 1,2-Digermacyclobut-1-ene derivative was reacted with $SiCl_4$ to give the double-Si–Cl-insertion product, and the following reduction reaction gave the corresponding > Ge = Si = Ge < species [11]. Although **1** could undergo facile double Si–Cl activation toward $SiCl_4$, neither double Ge–Cl nor C–Cl activation could occur in the reaction of **1** with $GeCl_4$/CCl_4.

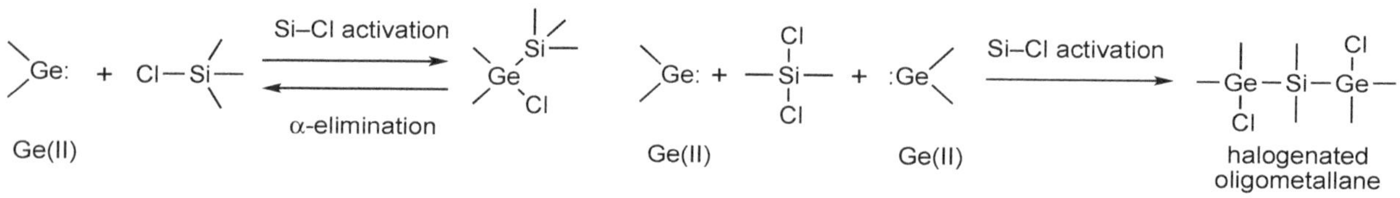

Scheme 1. Depictions for Si–Cl activations of Ge(II) species and Si–Cl α-elimination from a chlorosilylgermane.

2. Results and Discussions

When the stable digermyne **2** bearing bulky aryl substituents, Tbb groups (2,6-$[CH(SiMe_3)_2]_2$-4-*t*-Bu-C_6H_2), was treated with PhC≡CPh (tolan) at room temperature, 1,2-digermacyclobutadiene **1** was isolated as a stable crystalline compound [23–26] via formal [2+2] cycloaddition (Scheme 2). As one can see from the structure of **1**, it is a cyclic 4π-electron conjugated, anti-aromatic compound incorporating Ge(II) moieties. On the basis of theoretical calculations, **1** has considerable –Ge=Ge–C=C– character rather than =Ge–Ge=C–C– [24]. Accodingly, as expected, **1** could work as a building block of the bis-Ge(II) moiety. Reaction of **1** with $SiCl_4$ afforded digermadichlorosilane **3** quantitatively, which could be formed via double Si–Cl insertion reactions of the Ge(II) moieties of the 1,2-digermacyclobutadiene skeleton in **1**. This reaction has been performed under the neat condition at 55 °C because the addition of small amount of $SiCl_4$ or reaction at r.t. afforded very slow conversion of **3**. The obtained dichlorosilane **3** has the >Ge(Cl)–$SiCl_2$–Ge(Cl)< moiety in its 1,3-digerma-2-sila-cyclopent-4-ene skeleton, i.e., **2** should be one of a unique class of compounds of oligohalo-oligometallanes. Thus, **1** was found to work as a bis-germylene building block (>Ge: + :Ge<) towards a double Si–Cl activation.

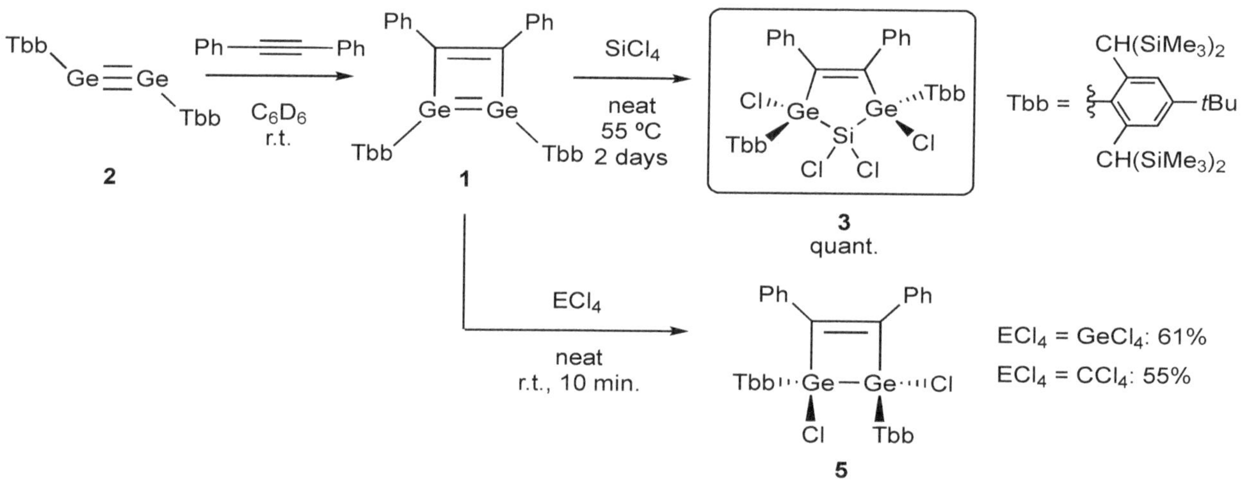

Scheme 2. Preparation of 1,2-digermacyclobutadiene **1**, and its reaction with ECl_4 giving digermadichlorosilane **3** (E = Si) and dichlorodigermacyclobutene **5** (E = C, Ge), respectively.

The molecular structure of digermadichlorosilane **3** was definitively determined by X-ray crystallographic analysis (Figure 1). The two Tbb/Cl groups are oriented in (*E*)-geometry probably due to steric reasons. The five-membered ring skeleton in **3** exhibits the envelope geometry with a deviation of the Si atom from the Ge–C=C–Ge plane by *ca.* 1.27 Å. While the two Ge–Cl bond lengths are almost the same (Ge1–Cl1: 2.2094(14) Å, Ge2–Cl4: 2.2011(15) Å) within a range of standard deviations, the orientation of the two Cl atoms are slightly different to each other. That is, one of the Cl atom (Cl4) is oriented to outside of the five-membered ring skeleton, but another one (Cl1) is approaching to the central Si atom with the Cl1···Si distance of 3.25 Å, which is far from the other one (Cl4···Si = 3.66 Å) [27]. In addition, the two Cl–Ge–Si angles are considerably different from each other, (Cl1–Ge1–Si = 90.20(8), Cl4–Ge2–Si = 105.40(8)). These asymmetrical structural features indicate weak n(Cl1)$\cdots\sigma^*$(Si–Cl3) interaction. These structural features were reasonably reproduced by the theoretical structural optimization at B3PW91/6-311G(2d) [28]. The theoretically-optimized structure of the less hindered model **3′**, which has Me groups instead of Tbb groups, exhibits a completely planar five-membered skeleton with C_2 symmetry. Thus, these structural features observed in **3** could be due to the steric congestion.

In the expectation of obtaining the Ge analogue of **3**, digermadichlorogermane **4**, the reaction of **1** with $GeCl_4$ was attempted. As a result, the expected product, **4**, was not obtained, but the 1,2-dichloro-1,2-digermacyclobut-3-ene **5** was obtained as a predominant product even under the conditions of using only a small amount of $GeCl_4$ in the dark [29]. In addition, the reaction of **1** with CCl_4 also furnished the formation of **5** without any formation of the CCl_2-insertion product **6**. 1,2-Dichloro-1,2-digermacyclobutene **5** showed considerable stability in the air, and it can object to further purification by silica gel column chromatography. Although the reaction mechanism for the formation of **5** by the reaction of **1** with $GeCl_4$ or CCl_4 was not clear at present, the formation of **5** is most likely interpreted in terms of the double-chlorination of **1** with the elimination of ECl_2 (E = Ge or C) moiety.

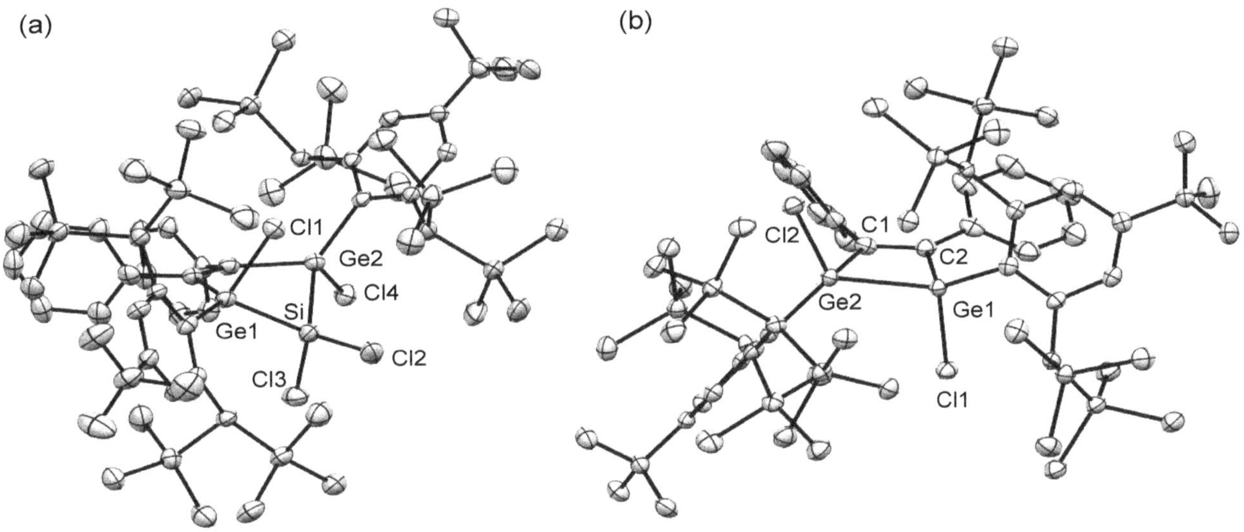

Figure 1. Molecular structures of (**a**) digermadichlorosilane **3** and (**b**) dichlorodigermacyclobutene **5** with atomic displacement parameters set at 50% probability. All hydrogen atoms and solvent molecules (THF and benzene) were omitted for clarity and only selected atoms are labeled. Selected bond lengths (Å) and angles (deg.): (**a**) **3**: Ge1–Si, 2.3734(16); Ge2–Si, 2.3938(15); Ge1–Cl1, 2.2094(14); Ge2–Cl4: 2.2011(15); Si–Cl2, 2.052(2); Si–Cl3, 2.053(2); Ge1–Si–Ge2, 91.15(5); Cl1–Ge1–Si, 90.29(5); Cl4–Ge2–Si, 105.40(6); Cl2–Si–Cl3, 106.04(9); (**b**) **5**: Ge1–Ge2, 2.4694(6); Ge1–Cl1, 2.2098(11); Ge2–Cl2, 2.2049(11); Ge1–C2, 1.984(4); Ge2–C1, 1.996(4); C2–Ge1–Ge2, 74.24(13); Ge1–Ge2–C1, 73.02(12); Ge2–C1–C2, 106.8(3); C1–C2–Ge1, 105.6(3).

The difference of the products in the reaction of **1** with ECl_4 (E = Si, Ge, and C) between E = Si and E = Ge, C cases should be of great interest. Although we could not draw a definitive conclusion, we performed the thermodynamic energy calculations (free energies) on the reaction of **1** with ECl_4 (E = Si, Ge, C) to give the insertion products, **3**, **4**, and **6**, or the chlorination products, **5** and Cl_2E: (calculated as 1/2 $Cl_2E{=}ECl_2$) at the B3PW91/6-311G(2d) level of theory (Scheme 3) [28]. In the case of E = Si, the formation of **3** should be exothermic by 2.3 kcal/mol, and that of **5** with $Cl_2Si{=}SiCl_2$ was estimated as an endothermic reaction by 4.5 kcal/mol. However, in the case of E = Ge or C, the formation of **5** with $Cl_2E{=}ECl_2$ was thermodynamically favorable (E = Ge: $\Delta G = -27$ kcal/mol, E = C: $\Delta G = -81$ kcal/mol) relative to the formation of **4** or **6** (E = Ge (**4**): $\Delta G = -24$ kcal/mol, E = C

(**6**): $\Delta G = -63$ kcal/mol). Thus, thermodynamic energy difference between cases of E = Si, Ge, and C could give us some hints on the difference of the reaction products, though the reasonable reaction mechanisms are not clear at present.

The structure of 1,2-dichloro-1,2-digermacyclobut-3-ene **5** was revealed by the X-ray crystallographic analysis. The two Tbb/Cl moieties are oriented in (*E*)-geometries, in the digermacyclobutene skeleton in **5**. The Ge–Ge bond length is 2.4694(6) Å, which is within a range of singly-bonded Ge–Ge distances. The lengths of the two Ge–Cl bonds are almost identical as Ge1–Cl1 = 2.2098(11) Å and Ge2–Cl2 = 2.2049(11) Å, which are similar to those of **3**. The Ge1–C2 and Ge2–C1 (1.984(4), 1.996(4) Å) bond lengths in the digermacyclobutene skeleton of **5** are slightly longer, and shorter relative to those of the only example of the previously reported chlorinated 1,2-digerma-3-cyclobutadiene derivative **7** (Ge–Cl: 2.145(2)–2.150(2), Ge–C: 1.998(6), 2.002(6) Å) (Scheme 4) [30]. Interestingly, reduction of the isolated **5** with lithium naphthalenide was found to reproduce 1,2-digermacyclobutadiene **1** quantitatively, as evidenced by the ^{1}H NMR spectra.

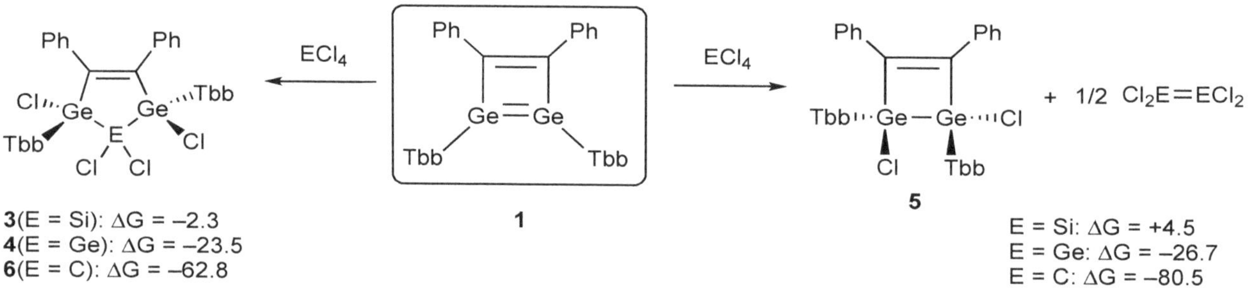

Scheme 3. Theoretical calculations on ΔG values (in kcal/mol) in the reactions of **1** with ECl_4 (E = Si, Ge, C) to give insertion products (**3**, **4**, **6**) or chlorinated product **5**.

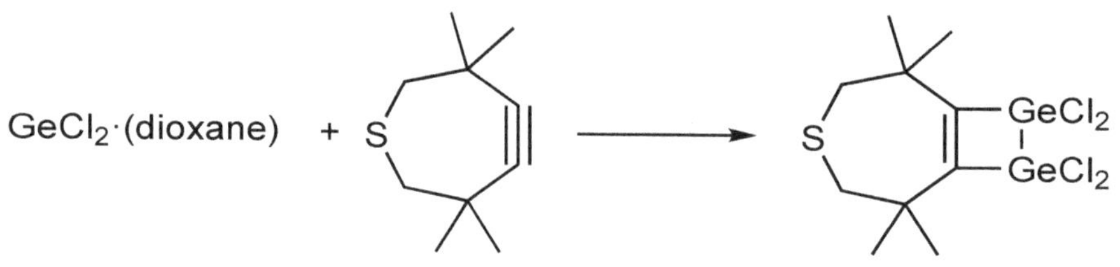

Scheme 4. Reported reaction of $GeCl_2$·(dioxane) with the highly strained alkyne to give the first example of chlorinated 1,2-digerma-3-cyclobutadiene derivative **7** [30].

3. Materials and Methods

3.1. General Information

All manipulations were carried out under an argon atmosphere using either a Schlenk line techniques or glove boxes. Solvents were purified by the Ultimate Solvent System, Glass Contour Company (Laguna Beach, CA, USA) [31]. ^{1}H, ^{13}C, and ^{29}Si NMR spectra were measured on a JEOL AL-300 spectrometer (^{1}H: 300 MHz, ^{13}C: 75 MHz, ^{29}Si: 59 MHz). Signals arising from residual C_6D_5H (7.15 ppm) in the C_6D_6 were used as an internal standard for the ^{1}H NMR spectra, and that of C_6D_6 (128.0 ppm) for the ^{13}C NMR spectra, and external $SiMe_4$ 0.0 ppm for the ^{29}Si NMR spectra. High-resolution mass spectra (HRMS) were measured on a Bruker micrOTOF focus-Kci mass spectrometer (on ESI-positive mode). All melting points were determined on a Büchi Melting Point Apparatus M-565 and are uncorrected. 1,2-digermacyclobutadiene **1** was prepared according to literature procedure [24].

3.2. *Experimental Details*

3.2.1. Reaction of 1,2-Tbb$_2$-1,2-Digermacyclobutadiene **1** with an Excess of SiCl$_4$

A solution of 1,2-Tbb$_2$-1,2-digermacyclobutadiene **1** (56.0 mg, 0.046 mmol) in SiCl$_4$ (1.0 mL, 8.8 mmol, excess) was treated at 55 °C for 48 h, and the color of the dark red solution disappeared. After removal of residual SiCl$_4$ under the reduced pressure, the residue was recrystallized from THF at room temperature to give compound **3** as colorless crystals in quantitative yield (64.2 mg, 0.046 mmol).

Data for **3**: colorless crystals, m.p. = 68.7–69.7 °C (dec.); ^{1}H NMR (300 MHz, C$_6$D$_6$, r.t.): δ 0.11 (s, 36H, SiMe$_3$), 0.32 (s, 36 H, SiMe$_3$), 1.24 (s, 18H, *t*-Bu), 2.51 (bs, 4H, CH), 6.70–6.76 (m, 2H, ArH), 6.84–6.90 (m, 8H, ArH), 7.22 (d, 4H, *J* = 7.2 Hz, ArH); ^{13}C NMR (75 MHz, C$_6$D$_6$, 298 K): δ 1.80 (SiMe$_3$), 2.17 (SiMe$_3$), 30.96 (CMe$_3$), 30.99 (CH), 34.32 (CMe$_3$), 124.20 (ArH), 126.87 (ArH), 127.81 (ArH), 130.97 (ArH), 132.28 (Ar), 140.70 (Ar), 150.13 (Ar), 151.14 (Ar), 162.66 (CAr); ^{29}Si NMR (59 MHz, C$_6$D$_6$, 298 K): δ 3.61 (SiMe$_3$), 3.84 (SiMe$_3$), 28.83 (GeSiGe); MS (DART-TOF, positive mode): *m*/*z* calcd. for C$_{62}$H$_{109}$^{35}Cl$_4$^{74}Ge$_2$Si$_9$ 1393.3630 ([M + H]$^+$), found 1393.3681 ([M + H]$^+$).

3.2.2. Reaction of 1,2-Tbb$_2$-1,2-Digermacyclobutadiene **1** with an Excess of GeCl$_4$

A C$_6$D$_6$ solution of 1,2-Tbb$_2$-1,2-digermacyclobutadiene **1** (38.3 mg, 0.0313 mmol) was treated with an excess amount of GeCl$_4$ (0.3 mL, 2.6 mmol) at room temperature. After stirring of the reaction mixture for 10 min, the solvent and GeCl$_4$ were removed under reduced pressure. The residue was recrystallized from benzene at room temperature to give compound **5** as main product in 61% yield (24.6 mg, 0.0190 mmol).

Data for **5**: colorless crystals, m.p. 90.4–91.4 °C; ^{1}H NMR (300 MHz, C$_6$D$_6$, r.t.): δ 0.13 (s, 36H, SiMe$_3$), 0.27 (s, 36 H, SiMe$_3$), 1.26 (s, 18H, *t*-Bu), 2.58 (s, 4H, CH), 6.88–6.93 (m, 6H, ArH), 7.00 (t, 4H, *J* = 7.2 Hz, ArH), 7.39 (d, 4H, *J* = 7.2 Hz, ArH); ^{13}C NMR (75 MHz, C$_6$D$_6$, 298 K): δ 1.83 (SiMe$_3$), 1.86 (SiMe$_3$), 30.26 (CMe$_3$), 31.03 (CH), 34.39 (CMe$_3$), 124.16 (ArH), 128.00 (ArH), 128.67 (ArH), 129.79 (ArH), 133.71 (Ar), 139.21 (Ar), 150.45 (Ar), 151.29 (Ar), 167.29 (CAr); MS (DART-TOF, positive mode): *m*/*z* calcd. for C$_{62}$H$_{109}$^{35}Cl$_2$^{74}Ge$_2$Si$_8$ 1295.4484 ([M + H]$^+$), found 1295.4492 ([M + H]$^+$).

3.2.3. Reaction of 1,2-Tbb$_2$-1,2-Digermacyclobutadiene **1** with an Excess of CCl$_4$

A C$_6$D$_6$ solution of 1,2-Tbb$_2$-1,2-digermacyclobutadiene **1** (32.9 mg, 0.0269 mmol) was treated with an excess amount of CCl$_4$ (0.2 mL, 2.1 mmol) at room temperature. After stirring of the reaction mixture for 10 min, the solvent and CCl$_4$ were removed under reduced pressure. The residue was recrystallized from benzene at room temperature to give compound **3** as main product in 55% yield (22.8 mg, 0.0175 mmol).

3.3. *Computational Methods*

The level of theory and the basis sets used for the structural optimization are contained within the main text. Frequency calculations confirmed minimum energies for all optimized structures. All calculations were carried out using the *Gaussian 09* program package [28]. Computational time was generously provided by the Supercomputer Laboratory in the Institute for Chemical Research of Kyoto University.

3.4. *X-ray Crystallographic Analysis*

Single crystals of [**3**·(thf)] and [**5**·2(benzene)] were obtained from recrystallization from THF and benzene, respectively. Intensity data were collected on a RIGAKU Saturn70 CCD system with VariMax Mo Optics using Mo Kα radiation (λ = 0.71075 Å). The structures were solved by a direct method (SIR2004 [32]) and refined by a full-matrix least square method on F^2 for all reflections (*SHELXL*-97 [33]). All hydrogen atoms were placed using AFIX instructions, while all other atoms were refined anisotropically. Supplementary crystallographic data were deposited at the Cambridge Crystallographic Data Centre (CCDC; under reference numbers: CCDC-1578241 and 1578242 for [**3**·(thf)] and [**5**·2(benzene)], respectively) and can be obtained free of charge via https://www.ccdc.cam.ac.uk/structures/. X-ray crystallographic data for [**3**·(thf)] and [**5**·2(benzene)]. Data for [**3**·(thf)] ($C_{66}H_{116}Cl_4Ge_2OSi_9$): M = 1465.37, triclinic, P–1 (no.2), a = 12.6367(7) Å, b = 16.9170(6) Å, c = 20.4887(10) Å, α= 91.3815(14)°, β = 105.252(2)°, γ = 109.642(3)°, V = 3949.2(3) Å^3, Z = 2, $D_{calc.}$ = 1.232 g·cm^{-3}, μ = 1.070 mm^{-1}, $2\theta_{max}$ = 51.0°, measd./unique refls. = 83580/14641 ($R_{int.}$ = 0.1095), param = 767, GOF = 1.117, R_1 = 0.0683/0.1122 [I>2σ(I)/all data], wR_2 = 0.1188/0.1359 [I>2σ(I)/all data], largest diff. peak and hole 1.681 and −0.592 e.Å^{-3} (CCDC-1578241). Data for [**5**·2(benzene)] ($C_{74}H_{120}Cl_2Ge_2Si_8$): M = 1450.49, triclinic, P–1 (no.2), a = 11.6792(2) Å, b = 15.7581(3) Å, c = 24.7906(5) Å, α = 76.2640(10)°, β = 88.0800(10)°, γ = 70.2510(10)°, V = 4165.89(14) Å^3, Z = 2, $D_{calc.}$ = 1.156 g·cm^{-3}, μ = 0.937 mm^{-1}, $2\theta_{max}$ = 50.0°, measd./unique refls. = 64887/14555 ($R_{int.}$ = 0.0810), param = 805, GOF = 1.289, R_1 = 0.0637/0.0804 [I>2σ(I)/all data], wR_2 = 0.1247/0.1311 [I>2σ(I)/all data], largest diff. peak and hole 0.983 and −0.689 e.Å^{22123} (CCDC-1578242).

4. Conclusions

It was demonstrated that 1,2-digermacyclobutadiene **1** could work as a bis-germylene building block (>Ge: + :Ge<) towards double Si–Cl activation in the reaction of **1** with $SiCl_4$ to give the halogenated oligometallane, bis(chlorogermyl)dichlorosilane **3**. Conversely, $GeCl_4$ and CCl_4 were found to work as double-chlorinating reagents towards **1** giving dichlorodigermacyclobutene **5**. Thus, **1** would be an interesting building block for oligohalo-oligometallanes.

Acknowledgments: This work was partially supported by the following grants: Grant-in-Aid for Scientific Research (B) (no. 15H03777), Grant-in-Aid for Challenging Exploratory Research (no. 15K13640), Scientific Research on Innovative Areas, "New Polymeric Materials Based on Element-Blocks" (#2401) (no. 25102519), "Stimuli-Responsive Chemical Species for the Creation of Functional Molecules" (#2408) (no. 24109013), and the project of Integrated Research on Chemical Synthesis from the Japanese Ministry of Education, Culture, Sports, Science, and Technology (MEXT), as well as by the "Molecular Systems Research" project of the RIKEN Advanced Science Institute and the Collaborative Research Program of the Institute for Chemical Research, Kyoto University. We would like to thank Toshiaki Noda and Hideko Natsume at Nagoya University for the expert manufacturing of custom-tailored glasswares. T. Sugahara would like to thank the Japan Society for the Promotion of Science (JSPS) for a fellowship (no. JP16J05501).

Author Contributions: Takahiro Sasamori and Norihiro Tokitoh conceived and designed the experiments; Tomohiro Sugahara performed the experiments; Takahiro Sasamori and Tomohiro Sugahara performed the XRD analysis and wrote the paper; and Takahiro Sasamori performed theoretical calculations.

References

1. Miller, R.D.; Michl, J. Polysilane High Polymers. *Chem. Rev.* **1989**, *89*, 1359–1410. [CrossRef]
2. Amadoruge, M.L.; Weinert, C.S. Singly Bonded Catenated Germanes: Eighty Years of Progress. *Chem. Rev.* **2008**, *108*, 4253–4294. [CrossRef] [PubMed]
3. Mochida, K.; Chiba, H. Synthesis, absorption characteristics and some reactions of polygermanes. *J. Organomet. Chem.* **1994**, *473*, 45–54. [CrossRef]
4. Lickiss, P.D.; Smith, C.M. Silicon derivatives of the metals of groups 1 and 2. *Coord. Chem. Rev.* **1995**, *145*, 75–124. [CrossRef]
5. Sekiguchi, A.; Lee, V.Y.; Nanjo, M. Lithiosilanes and their application to the synthesis of polysilane dendrimers. *Coord. Chem. Rev.* **2000**, *210*, 11–45. [CrossRef]
6. Kyushin, S.; Matsumoto, H. Ladder Polysilanes. *Adv. Organomet. Chem.* **2003**, *49*, 133–166. [CrossRef]
7. Weinert, C.S. Syntheses, structures and properties of linear and branched oligogermanes. *Dalton Trans.* **2009**, 1691–1699. [CrossRef] [PubMed]
8. Drost, C.; Hitchcock, P.B.; Lappert, M.F. Unprecedented Oxidative Chlorosilylation Addition Reactions to a Diarylgermylene and -stannylene. *Organometallics* **2002**, *21*, 2095–2100. [CrossRef]
9. Iwamoto, T.; Masuda, H.; Kabuto, C.; Kira, M. Trigermaallene and 1,3-Digermasilaallene. *Organometallics* **2005**, *24*, 197–199. [CrossRef]

10. Kira, M.; Iwamoto, T.; Ishida, S.; Masuda, H.; Abe, T.; Kabuto, C. Unusual Bonding in Trisilaallene and Related Heavy Allenes. *J. Am. Chem. Soc.* **2009**, *131*, 17135–17144. [CrossRef] [PubMed]
11. Sugahara, T.; Sasamori, T.; Tokitoh, N. Highly Bent 1,3-Digerma-2-silaallene. *Angew. Chem. Int. Ed.* **2017**, *56*, 9920–9923. [CrossRef] [PubMed]
12. Mallela, S.P.; Geanangel, R.A. New Cyclic and Acyclic Silicon–Germanium and Silicon–Germanium–Tin Derivatives. *Inorg. Chem.* **1994**, *33*, 1115–1120. [CrossRef]
13. Mallela, S.P.; Geanangel, R.A. Preparation and structural characterization of new derivatives of digermane bearing tris(trimethylsilyl)silyl substituents. *Inorg. Chem.* **1991**, *30*, 1480–1482. [CrossRef]
14. Ichinohe, M.; Sekiyama, H.; Fukaya, N.; Sekiguchi, A. On the Role of *cis,trans*-(t-$Bu_3SiGeCl)_3$ in the Reaction of $GeCl_2$·Dioxane with Tri-*tert*-butylsilylsodium: Evidence for Existence of Digermanylsodium t-$Bu_3SiGe(Cl)_2Ge(Cl)(Na)Si$$t$-$Bu_3$ and Digermene t-$Bu_3Si(Cl)Ge{=}Ge(Cl)Si$$t$-$Bu_3$. *J. Am. Chem. Soc.* **2000**, *122*, 6781–6782. [CrossRef]
15. Fukaya, N.; Sekiyama, H.; Ichinohe, M.; Sekiguchi, A. Photochemical Generation of Chlorine-substituted Digermenes and Their Rearrangement to Germylgermylenes. *Chem. Lett.* **2002**, 802–803. [CrossRef]
16 Sekiguchi, A.; Ishida, Y.; Fukaya, N.; Ichinohe, M. The First Halogen-Substituted Cyclotrigermenes: A Unique Halogen Walk over the Three-Membered Ring Skeleton and Facial Stereoselectivity in the Diels-Alder Reaction. *J. Am. Chem. Soc.* **2002**, *124*, 1158–1159. [CrossRef] [PubMed]
17. Lee, V.Y.; Yasuda, H.; Ichinohe, M.; Sekiguchi, A. $SiGe_2$ and Ge_3: Cyclic Digermenes that Undergo Unexpected Ring-Expansion Reactions. *Angew. Chem. Int. Ed.* **2005**, *44*, 6378–6381. [CrossRef] [PubMed]
18. Lee, V.Y.; Yasuda, H.; Ichinohe, M.; Sekiguchi, A. Heavy cyclopropene analogues R_4SiGe_2 and R_4Ge_3 (R = SiMe*t*Bu_2)—New members of the cyclic digermenes family. *J. Organomet. Chem.* **2007**, *692*, 10–19. [CrossRef]
19. Wagler, J.; Brendler, E.; Langer, T.; Pöttgen, R.; Heine, T.; Zhechkov, L. Ylenes in the $M^{II}\rightarrow Si^{IV}$ (M=Si, Ge, Sn) Coordination Mode. *Chem. Eur. J.* **2010**, *16*, 13429–13434. [CrossRef] [PubMed]
20. Al-Rafia, S.M.; Malcolm, A.C.; McDonald, R.; Ferguson, M.J.; Rivard, E. Trapping the Parent Inorganic Ethylenes H_2SiGeH_2 and H_2SiSnH_2 in the Form of Stable Adducts at Ambient Temperature. *Angew. Chem. Int. Ed.* **2011**, *50*, 8354–8357. [CrossRef] [PubMed]
21. Katir, N.; Matioszek, D.; Ladeira, S.; Escudié, J.; Castel, A. Stable *N*-Heterocyclic Carbene Complexes of Hypermetallyl Germanium(II) and Tin(II) Compounds. *Angew. Chem. Int. Ed.* **2011**, *50*, 5352–5355. [CrossRef] [PubMed]
22. Lee, V.Y.; Ito, Y.; Yasuda, H.; Takanashi, K.; Sekiguchi, A. From Tetragermacyclobutene to Tetragermacyclobutadiene Dianion to Tetragermacyclobutadiene Transition Metal Complexes. *J. Am. Chem. Soc.* **2011**, *133*, 5103–5108. [CrossRef] [PubMed]
23. Cui, C.; Olmstead, M.M.; Power, P.P. Reactivity of Ar′GeGeAr′ (Ar′ = C_6H_3-2,6-$Dipp_2$, Dipp = C_6H_3-2,6-$i$$Pr_2$) toward Alkynes: Isolation of a Stable Digermacyclobutadiene. *J. Am. Chem. Soc.* **2004**, *126*, 5062–5063. [CrossRef] [PubMed]
24. Sugahara, T.; Guo, J.-D.; Sasamori, T.; Karatsu, Y.; Furukawa, Y.; Ferao, A.E.; Nagase, S.; Tokitoh, N. Reaction of a Stable Digermyne with Acetylenes: Synthesis of a 1,2-Digermabenzene and a 1,4-Digermabarrelene. *Bull. Chem. Soc. Jpn.* **2016**, *89*, 1375–1384. [CrossRef]
25. Tashkandi, N.Y.; Pavelka, L.C.; Caputo, C.A.; Boyle, P.D.; Power, P.P.; Baines, K.M. Addition of alkynes to digermynes: Experimental insight into the reaction pathway. *Dalton Trans.* **2016**, *45*, 7226–7230. [CrossRef] [PubMed]
26. Zhao, L.; Jones, C.; Frenking, G. Reaction Mechanism of the Symmetry-Forbidden [2+2] Addition of Ethylene and Acetylene to Amido-Substituted Digermynes and Distannynes Ph_2N-EE-NPh_2, (E = Ge, Sn): A Theoretical Study. *Chem. Eur. J.* **2015**, *21*, 12405–12413. [CrossRef] [PubMed]
27. Tillmann, J.; Meyer, L.; Schweizer, J.I.; Bolte, M.; Lerner, H.-W.; Wagner, M.; Holthausen, M.C. Chloride-Induced Aufbau of Perchlorinated Cyclohexasilanes from Si_2Cl_6: A Mechanistic Scenario. *Chem. A-Eur. J.* **2014**, *20*, 9234–9239. [CrossRef] [PubMed]
28. *Gaussian 09 Program*; Gaussian, Inc.: Wallingford, CT, USA, 2009.
29. Ohtaki, T.; Ando, W. Dichlorodigermacyclobutanes and Digermabicyclo[2.2.0]hexanes from the Reactions of [Tris(trimethylsilyl)methyl]chlorogermylene with Olefins. *Organometallics* **1996**, *15*, 3103–3105. [CrossRef]

30. Espenbetov, A.A.; Struchkov Yu, T.; Kolesnikov, S.P.; Nefedov, O.M. Crystal and Molecular Structure of $\Delta^{1,7}$2,2,6,6,-Tetramethyl-4-thia-8,8,9,9-tetrachloro-8,9-digermabicyclo[5.2.0]nonene; The First Representative of 1,2-Digermacyclobutenes. *J. Organomet. Chem.* **1984**, *275*, 33–37. [CrossRef]
31. Pangborn, A.B.; Giardello, M.A.; Grubbs, R.H.; Rosen, R.K.; Timmers, F.J. Safe and Convenient Procedure for Solvent Purification. *Organometallics* **2004**, *15*, 1518–1520. [CrossRef]
32. Burla, M.C.; Caliandro, R.; Camalli, M.; Carrozzini, B.; Cascarano, G.L.; De Caro, L.; Giacovazzo, C.; Polidori, G.; Spagna, R. *SIR2004*: An improved tool for crystal structure determination and refinement. *J. Appl. Cryst.* **2005**, *38*, 381–388. [CrossRef]
33. Sheldrick, G.M. A short history of *SHELX*. *Acta Crystallogr. Sect. A* **2008**, *64*, 112–122. [CrossRef] [PubMed]

12

One-Pot Synthesis of Heavier Group 14 *N*-Heterocyclic Carbene using Organosilicon Reductant

Ravindra K. Raut, Sheikh Farhan Amin, Padmini Sahoo, Vikas Kumar and Moumita Majumdar *

Indian Institute of Science Education and Research Pune, Dr. Homi Bhabha Road, Pashan, Pune-411008, India; ravindra.raut@students.iiserpune.ac.in (R.K.R.); sheikh.farhan@students.iiserpune.ac.in (S.F.A.); padmini.sahoo@students.iiserpune.ac.in (P.S.); kumar.vikas@students.iiserpune.ac.in (V.K.)

* Correspondence: moumitam@iiserpune.ac.in

Abstract: Syntheses of heavier Group 14 analogues of "Arduengo-type" *N*-heterocyclic carbene majorly involved the use of conventional alkali metal-based reducing agents under harsh reaction conditions. The accompanied reductant-derived metal salts and chances of over-reduced impurities often led to isolation difficulties in this multi-step process. In order to overcome these shortcomings, we have used 1,4-bis-(trimethylsilyl)-1,4-diaza-2,5-cyclohexadiene as a milder reducing agent for the preparation of *N*-heterocyclic germylenes (NHGe) and stannylenes (NHSn). The reaction occurs in a single step with moderate yields from the mixture of *N*-substituted 1,4-diaza-1,3-butadiene, E(II) (E(II) = $GeCl_2$·dioxane, $SnCl_2$) and the organosilicon reductant. The volatile byproducts trimethylsilyl chloride and pyrazine could be removed readily under vacuum. No significant over reduction was observed in this process. However, *N*-heterocyclic silylene (NHSi) could not be synthesized using an even stronger organosilicon reductant under thermal and photochemical conditions.

Keywords: organosilicon; reductant; *N*-Heterocyclic tetrylene; salt-free

1. Introduction

Carbene chemistry kick-started with the ground-breaking discovery of the bottle-able *N*-heterocyclic carbene (NHC) by Arduengo in 1991 [1]. Since this seminal work, the versatile NHCs have replaced the classical phosphine-based ligands for transition-metal catalysts [2]. Subsequently, the isolation of heavier Group 14 analogues of 'Arduengo type' carbene NHE (E = Si, Ge, and Sn) have been the subject of intense study, both for fundamental interests and potential applications in transition metal catalysis, similar to the NHCs. To date, a handful of NHEs have been synthesized and structurally characterized [3,4]. Obviously, the heavier analogues possess distinct electronic features compared to the NHCs, and hence exhibit different reactivities [3,4]. While *N*-heterocyclic silylenes (NHSi) have been engaged as ancillary ligands in numerous homogeneous catalysis [5–7], *N*-heterocyclic germylenes (NHGe) serve as a precursor for polymerization chemistry [8–10] and also for chemical vapour depositions [11].

Typically, the synthesis of the five-membered *N*-heterocyclic tetrylenes involves the reaction between *N*-substituted 1,4-diaza-1,3-butadiene, Group 14 halides, and the harsh alkali metal based reducing agents (Scheme 1) [3,4]. In the case of NHGe and NHSn, the initial step involves the reduction of *N*-substituted 1,4-diaza-1,3-butadiene by lithium metal, followed by cyclization of the dianion with the corresponding Group 14 E(II) halides [12–15]. While in the case of NHSi, the precursor cyclic diaminodichlorosilane was obtained by the cyclization of the dilithiated diazabutadiene with $SiCl_4$ [16–18]. West et al. reported the first synthesis of NHSi by the reduction of this cyclic diaminodichlorosilane with potassium metal in tetrahydrofuran [18]. In these approaches, the choice of reductant play crucial role in the reductive dehalogenation step [12–18]. Moreover, the syntheses suffer from involvement of multiple steps, associated metal salts as byproduct and hence product isolation difficulties, and sometimes cases of over reduction. Certainly, these shortcomings urge a careful revisit into the synthetic methodology involved and finding an alternate milder reducing agent.

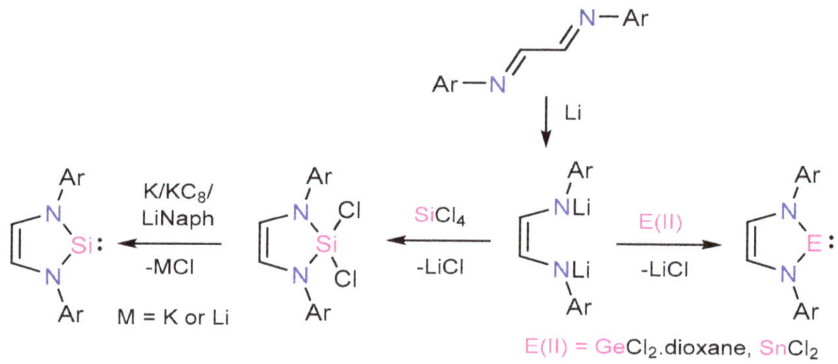

Scheme 1. Typical synthetic route for heavier Group 14 analogue of "Arduengo-type" carbene.

The milder organosilicon reductants have been well-known to efficiently reduce early transition metals without the formation of reductant-derived metal salts and over-reduced impurities [19,20]. Very recently, 2,3,5,6-tetramethyl-1,4-bis(trimethylsilyl)-1,4-diaza-2,5-cyclohexadiene have been employed for metal salt-free reduction of dibromobismuthine and dibromostibine to dibismuthine and distibine, respectively [21]. Worth mentioning, there are few discrete examples where a NHE (E = Si, Ge, and Sn) has been synthesized under metal-free conditions: Dehydrochlorination of cyclic diaminohydrochlorosilane using bulky NHC [22], dehydrogenation of dihydrogermane by frustrated Lewis pair [23], and transamination of $Sn\{N(SiMe_3)_2\}_2$ with α-amino-aldimines [24,25], respectively. In this study, we have developed a simple one-pot synthetic route under milder conditions to synthesize NHGe and NHSn, respectively, free from reductant-derived metal salts using 1,4-bis-(trimethylsilyl)-1,4-diaza-2,5-cyclohexadiene [19,20] as the reductant.

2. Results and Discussion

The *N*-heterocyclic germylene (NHGe) **1** (Scheme 2) has been synthesized in one step by the low temperature addition of $GeCl_2$·dioxane to a mixture of N^1,N^2-dimesitylethane-1,2-diimine and 1,4-bis-(trimethylsilyl)-1,4-diaza-2,5-cyclohexadiene in tetrahydrofuran. The volatile byproducts trimethylsilyl chloride and pyrazine, were easily removable under vacuum (Figures S1–S3). Since small amounts of insoluble solids appeared upon hexane addition, compound **1** was isolated as yellow solid from its hexane extract in an acceptable yield of 80%. Compound **1** was characterized by NMR study (Figures S4 and S5). Single crystals of **1** were grown from hexane at −40 °C, and its structure was determined using single crystal X-ray crystallography (Figure S23 and Table S1). This method is also applicable for other aromatic substituents, such as 2,6-diisopropylphenyl (Figures S6 and S7).

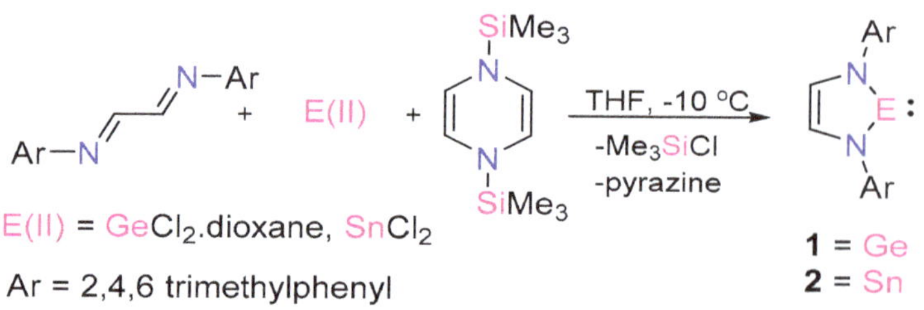

Scheme 2. One-pot synthesis of NHE (E = Ge, Sn) using a salt-free reduction method.

A similar method has been employed for the synthesis of *N*-heterocyclic stannylene (NHSn) **2** (Scheme 2). Formation of metallic tin was observed in the reaction mixture, which was removed by filtration. This arises due to the thermolabile nature of NHSn [24,25], leading to difficulties in acquiring NMR data (Figures S8 and S9). The presence of additional peaks in the ^{119}Sn NMR (Figure S10) of the reaction mixture may be attributed to the formation of Sn(IV) compounds [26,27]. Single crystals of compound **2** were obtained from hexane extract at −40 °C in a yield of 35% (Figures S11–S13 and S24, Table S2). Notably, Gudat et al. reported the first synthesis of NHSn by transamination of $Sn\{N(SiMe_3)_2\}_2$ with α-amino-aldimines [25].

Despite several trials, we have been unsuccessful in synthesizing *N*-heterocyclic silylene (NHSi). Low temperature one-pot reaction (Scheme 3(i)) of N^1,N^2-dimesitylethane-1,2-diimine, $SiCl_4$, and two equivalents of stronger reducing agent 1,1′-bis(trimethylsilyl)-1,1′-dihydro-4,4′-bipyridine (R2) led to a mixture of cyclic diaminodichlorosilane (A) and the disilylated product (B), along with a large amount of insoluble solid (Figures S14 and S15). Notably, the organosilicon reducing agents do not reduce diazabutadiene (Figure S22), while they react with Group 14 halides. Subsequently, we attempted to reduce the synthesized cyclic diaminodichlorosilane [16] using both 1,4-bis-(trimethylsilyl)-1,4-diaza-2,5-cyclohexadiene (R1) and also the stronger reductant 1,1′-bis(trimethylsilyl)-1,1′-dihydro-4,4′-bipyridine (R2) (Scheme 3(ii)) under thermal conditions. The NMR data always reflected the presence of unreacted cyclic diaminodichlorosilane, along with other unidentifiable products (Figures S16–S19). On a related context, 1,4-bis(trimethylsilyl)-substituted 1,4-dihydropyrazine has been reported to be capable of organosilyl group exchange reactions [28]. We have also tried reductions under UV irradiation conditions, anticipating enhanced Si–Cl bond cleavage [29]. However, each time NMR led to a mixture of unidentifiable products, in addition to the unreacted cyclic diaminodichlorosilane (Figures S20 and S21). Probably, the relatively lower reduction potentials of these organosilicon reductants do not allow the reduction of diaminodichlorosilanes.

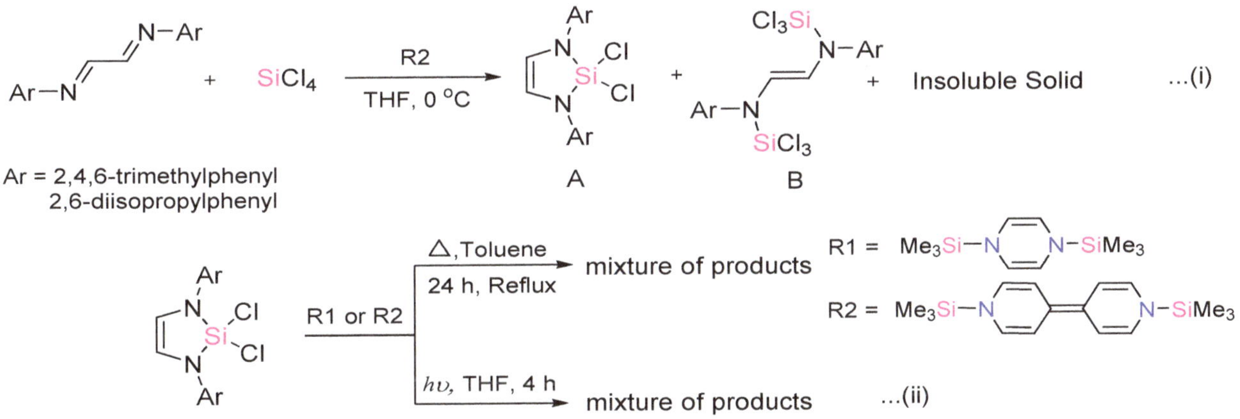

Scheme 3. Attempts to synthesize *N*-heterocyclic silylene (NHSi) using organosilicon reductants.

3. Materials and Methods

3.1. General Information

All manipulations were carried out under a protective atmosphere of argon, applying standard Schlenk techniques or in a dry box. Tetrahydrofuran, toluene, and hexane were refluxed over sodium/benzophenone. All solvents were distilled and stored under argon and degassed prior to use. C_6D_6 was purchased from Sigma Aldrich (Sigma Aldrich Co., St. Louis, MO, USA) and dried over potassium. All chemicals were used as purchased. Diazabutadiene [30] ligand, and the reducing agents 1,4-bis-(trimethylsilyl)-1,4-diaza-2,5-cyclohexadiene, 1,1′-bis(trimethylsilyl)-1, 1′-dihydro-4,4′-bipyridine [19,20] were synthesized according to reported literature procedure. Photochemical reactions were performed in Peschl Photoreactorsystem (Peschl Ultraviolet GmbH,

Mainz, Germany). ^{1}H, $^{13}C\{1H\}$, and $^{29}Si\{^{1}H\}$ NMR spectra were referenced to external $SiMe_4$ using the residual signals of the deuterated solvent (^{1}H) or the solvent itself (^{13}C). ^{119}Sn NMR was referenced to $SnCl_4$ as the external standard. Melting points were determined under argon in closed NMR tubes and are uncorrected. Elemental analyses were performed on Elementar vario EL analyzer (Elementar Analysensysteme GmbH, Langenselbold, Germany). Single crystal data were collected on a Bruker SMART APEX four-circle diffractometer equipped with a CMOS photon 100 detector (Bruker Systems Inc., Fällanden, Switzerland) with a Cu Kα radiation (1.5418 Å). Data were integrated using Bruker SAINT software and absorption correction using SADABS. Structures were solved by Intrinsic Phasing module of the direct methods (*SHELXS*) [31] and refined using the *SHELXL* 2014 [32] software suite. All hydrogen atoms were assigned using AFIX instructions, while all other atoms were refined anisotropically.

3.2. Experimental Detail

Synthesis of compound **1**: $GeCl_2$ Dioxane (0.087 g. 0.38 mmol) dissolved in THF was added drop-wise to a schlenk flask containing N^1,N^2-dimesitylethane-1,2-diimine (0.11 g, 0.38 mmol) and 1,4-bis(trimethylsilyl)-1,4-dihydropyrazine (0.084 g, 0.38 mmol) in THF kept at −10 °C. The reaction mixture was slowly warmed to room temperature and stirred overnight. The volatiles were removed under vacuum and extracted in hexane. The solvent was removed completely to give compound **1** as a yellow solid ((0.110 g, % yield = 80), which was further bulk crystallized from hexane at −40 °C with a crystallization yield of (0.097 g, % yield = 71) (Decomp. 115–118 °C). ^{1}H NMR (400 MHz, C_6D_6, TMS) δ = 6.87 (s, 4H, Ar*H*); 6.57 (s, 2H, NC*H*); 2.23 (s, 12H, *o*-C*H*$_3$); 2.19 (s, 6H, *p*-C*H*$_3$) ppm; $^{13}C\{^{1}H\}$ NMR (101 MHz, C_6D_6, TMS) δ = 142.41 (*i*-ArC); 134.87 (*o*-ArC); 133.61 (*p*-ArC); 128.88 (*m*-ArC); 125.16 (NCH); 20.60 (CH_3); 18.06 (CH_3) ppm. Elemental Analysis: Calcd. For $C_{20}H_{24}GeN_2$: C, 65.80; H, 6.63; N, 7.67. Found: C, 65.82; H, 6.68; N, 7.62.

Synthesis of compound **2**: $SnCl_2$ (0.324 g, 1.71 mmol) dissolved in THF was added drop-wise to a schlenk flask containing N^1,N^2-dimesitylethane-1,2-diimine (0.5 g, 1.71 mmol) and 1,4-bis(trimethylsilyl)-1,4-dihydropyrazine (0.386 g, 1.71 mmol) in THF kept at −10 °C. The reaction mixture was stirred for 10 min, maintaining the same temperature. Subsequently, the volatiles were removed under vacuum. Hexane was added to the red-brown residue and the product was extracted in hexane. The red filtrate of hexane was concentrated and kept at −40 °C to give red crystals of **2** (0.247 g, crystallization yield = 35%). (Decomp. 145–147 °C). ^{1}H NMR (400 MHz, Toluene-D_8, 248 K) δ = 6.92 (s, 4H, *m*-C*H*); 6.91 (s, 2H, NC*H*); 2.33 (s, 12H, C*H*$_3$); 2.29 (s, 6H, C*H*$_3$) ppm. ^{13}C NMR (101 MHz, Toluene-D_8, 248 K) δ = 145.91 (*i*-ArC); 133.97 (*o*-ArC); 133.01 (*p*-ArC); 129.01 (*m*-ArC); 128.32 (NCH); 20.82 (CH_3); 18.48 (CH_3) ppm. ^{119}Sn NMR (149.74 MHz, Toluene-D_8, 248 K) δ = 250 ppm. Elemental Analysis: Calcd. For $C_{20}H_{24}SnN_2$: C, 58.43; H, 5.88; N, 6.81. Found: C, 58.46; H, 5.95; N, 6.87.

4. Conclusions

We have established a salt-free reductive route for the synthesis of *N*-heterocyclic germylene and stannylene in one step. The easily removable volatile byproducts of the reaction leads to easy isolation of *N*-heterocyclic tetrylenes in acceptable yields. However, we have not been able to synthesize *N*-heterocyclic silylene using organosilicon reductants under thermal or photochemical conditions. The potential of this benign salt-free reduction method in the synthesis of other interesting low-valent main-group compounds is a much coveted area to explore.

Author Contributions: M.M. conceptualized the work, designed the experiments and wrote the paper; R.K.R. performed most of the experiments; S.F.A. and P.S. performed the photochemical reactions; V.K. carried out the NMR studies.

References

1. Arduengo, A.J., III; Harlow, R.L.; Kline, M. A stable crystalline carbene. *J. Am. Chem. Soc.* **1991**, *113*, 361–363. [CrossRef]
2. Peris, E. Smart *N*-heterocyclic carbene ligands in catalysis. *Chem. Rev.* **2017**. [CrossRef] [PubMed]
3. Asay, M.; Jones, C.; Driess, M. *N*-Heterocyclic carbene analogues with low-valent group 13 and group 14 elements: Syntheses, structures, and reactivities of a new generation of multitalented ligands. *Chem. Rev.* **2011**, *111*, 354–396. [CrossRef] [PubMed]
4. Mizuhata, Y.; Sasamori, T.; Tokitoh, N. Stable Heavier Carbene Analogues. *Chem. Rev.* **2009**, *109*, 3479–3511. [CrossRef] [PubMed]
5. Raoufmoghaddam, S.; Zhou, Y.-P.; Wang, Y.; Driess, M. *N*-Heterocyclic Silylenes as powerful steering ligands in catalysis. *J. Organomet. Chem.* **2017**, *829*, 2–10. [CrossRef]
6. Blom, B.; Gallego, D.; Driess, M. *N*-Heterocyclic Silylene Complexes in Catalysis: New frontiers in an emerging field. *Inorg. Chem. Front.* **2014**, *1*, 134–148. [CrossRef]
7. Blom, B.; Stoelzel, M.; Driess, M. New Vistas in *N*-Heterocyclic Silylene (NHSi) Transition-Metal Coordination Chemistry: Syntheses, Structures and Reactivity towards Activation of Small Molecules. *Chem. Eur. J.* **2013**, *19*, 40–62. [CrossRef] [PubMed]
8. Shoda, S.-I.; Iwata, S.; Yajima, K.; Yagi, K.; Ohnishi, Y.; Kobayashi, S. Synthesis of germanium enolate polymers from germylene monomers. *Tetrahedron* **1997**, *53*, 15281–15295. [CrossRef]
9. Shoda, S.-I.; Iwata, S.; Kim, H.J.; Hiraishi, M.; Kobayashi, S. Poly(germanium thiolate): A new class of organometallic polymers having a germanium-sulfur bond in the main chain. *Macromol. Chem. Phys.* **1996**, *197*, 2437–2445. [CrossRef]
10. Kobayashi, S.; Iwata, S.; Hiraishi, M. Novel 2:1 periodic copolymers from cyclic germylenes and p-benzoquinone derivatives. *J. Am. Chem. Soc.* **1994**, *116*, 6047–6048. [CrossRef]
11. Veprek, S.; Prokop, J.; Glatz, F.; Merica, R.; Klingan, F.R.; Herrmann, W.A. Organometallic chemical vapour deposition of germanium from a cyclic germylene, 1,3-Di-*tert*-butyl-1,3,2-diazagermolidin-2-ylidene. *Chem. Mater.* **1996**, *8*, 825–831. [CrossRef]
12. Hermann, W.A.; Denk, M.; Behm, J.; Scherer, W.; Klingan, F.; Bock, H.; Solouki, B.; Wagner, M. Stable cyclic germanediyls ("cyclogermylenes"): Synthesis, structure, metal complexes and thermolysis. *Angew. Chem. Int. Ed.* **1992**, *31*, 1485–1488. [CrossRef]
13. Baker, R.J.; Jones, C.; Mills, D.P.; Pierce, G.A.; Waugh, M. Investigation into the preparation of groups 13-15 *N*-heterocyclic carbene analogues. *Inorg. Chim. Acta* **2008**, *361*, 427–435. [CrossRef]
14. Piskunov, A.V.; Aivaz'yan, I.A.; Cherkasov, V.K.; Abakumov, G.A. New paramagnetic *N*-heterocyclic stannylenes: An EPR study. *J. Organomet. Chem.* **2006**, *691*, 1531–1534. [CrossRef]
15. Veith, M. Cyclic nitrogen derivatives of tetra- and divalent tin. *Angew. Chem. Int. Ed.* **1975**, *14*, 263–264. [CrossRef]
16. Park, P.; Schäfer, A.; Mitra, A.; Haase, D.; Saak, W.; West, R.; Müller, T. Synthesis and reactivity of *N*-aryl substituted *N*-heterocyclic silylenes. *J. Organomet. Chem.* **2010**, *695*, 398–408. [CrossRef]
17. Kong, L.; Zhang, J.; Song, H.; Cui, C. *N*-Aryl substituted heterocyclic silylenes. *Dalton Trans.* **2009**, 5444–5446. [CrossRef] [PubMed]
18. Denk, M.; Lennon, R.; Hayashi, R.; West, R.; Belyakov, A.V.; Verne, H.P.; Haaland, A.; Wagner, M.; Metzler, N. Synthesis and structure of a stable silylene. *J. Am. Chem. Soc.* **1994**, *116*, 2691–2692. [CrossRef]
19. Saito, T.; Nishiyama, H.; Tanahashi, H.; Kawakita, K.; Tsuragi, H.; Mashima, K. 1,4-Bis(trimethylsilyl)-1,4-diaza-2,5-cyclohexadienes as strong salt-free reductants for generating low-valent early transition metals with electron-donating ligands. *J. Am. Chem. Soc.* **2014**, *136*, 5161–5170. [CrossRef] [PubMed]
20. Kaim, W. Effects of cyclic 8π-electron conjugation in reductively silylated nitrogen heterocycles. *J. Am. Chem. Soc.* **1983**, *105*, 707–713. [CrossRef]
21. Majhi, P.K.; Ikeda, H.; Sasamori, T.; Tsuragi, H.; Mashima, K.; Tokitoh, N. Inorganic-salt-free reduction in main-group chemistry: Synthesis of a dibismuthene and a distibene. *Organometallics* **2017**, *36*, 1224–1226. [CrossRef]
22. Cui, H.; Shao, Y.; Li, X.; Kong, L.; Cui, C. Dehydrochlorination to silylenes by *N*-heterocyclic carbenes. *Organometallics* **2009**, *28*, 5191–5195. [CrossRef]

23. Jana, A.; Tavčar, G.; Roesky, H.; Schulzke, C. Facile synthesis of dichlorosilane by metathesis reaction and dehydrogenation. *Dalton Trans.* **2010**, *39*, 6217–6220. [CrossRef] [PubMed]
24. Gans-Eichler, T.; Gudat, D.; Nättinen, K.; Nieger, M. The transfer of tin and germanium atoms from *N*-heterocyclic stannylenes and germylenes to diazadienes. *Chem. Eur. J.* **2006**, *12*, 1162–1173. [CrossRef] [PubMed]
25. Gans-Eichler, T.; Gudat, D.; Nieger, M. Tin analogues of "arduengo carbenes": Synthesis of 1,3,2λ^2-diazastannoles and transfer of Sn atoms between a 1,3,2λ^2-diazastannole and a diazadiene. *Angew. Chem. Int. Ed.* **2002**, *41*, 1888–1891. [CrossRef]
26. Mansell, S.M.; Russel, C.A.; Wass, D.F. Synthesis of chelating diamido Sn(IV)compounds from oxidation of Sn(II) and directly from Sn(IV) precursors. *Dalton Trans.* **2015**, *44*, 9756–9765. [CrossRef] [PubMed]
27. Sarazin, Y.; Coles, S.J.; Hughes, D.L.; Hursthouse, M.B.; Bochmann, M. Cationic brønsted acids for the preparation of Sn^{IV} salts: Synthesis and characterization of $[Ph_3Sn(OEt_2)][H_2N\{B(C_6F_5)_3\}_2]$, $[Sn(NMe_2)_3(HNMe_2)_2][B(C_6F_5)_4]$ and $[Me_3Sn(HNMe_2)_2][B(C_6F_5)_4]$. *Eur. J. Inorg. Chem.* **2006**, 3211–3220. [CrossRef]
28. Lichtblau, A.; Eblend, A.; Hausen, H.-D.; Kaim, W. *N,N'*-disilylated 1,4-dihydropyrazines: Organosilyl substitution reactions, structural effects of steric hindrance, and electron exchange with C60. *Chem. Ber.* **1995**, *128*, 745–750. [CrossRef]
29. Rej, S.; Pramanik, S.; Tsurugi, H.; Mashima, K. Dehalogenation of vicinal dihalo compounds by 1,1′-bis(trimethylsilyl)-1*H*, 1′*H*-4,4′-bipyridinylidene for giving alkenes and alkynes in a salt-free manner. *Chem. Commun.* **2017**, *53*, 13157–13160. [CrossRef] [PubMed]
30. Koten, G.V.; Vrieze, K. 1,4-Diaza-1,3-butadiene (α-diimine) ligands: Their coordination modes and the reactivity of their metal complexes. *Adv. Organomet. Chem.* **1982**, *21*, 151–239. [CrossRef]
31. Sheldrick, G.M. A short history of *SHELX*. *Acta Cryst.* **2008**, *64*, 112–122. [CrossRef] [PubMed]
32. Sheldrick, G.M. Crystal Structure Refinement with *SHELXL*. *Acta Cryst.* **2015**, *C71*, 3–8. [CrossRef]

13

Modification of TiO_2 Surface by Disilanylene Polymers and Application to Dye-Sensitized Solar Cells

Yohei Adachi, Daiki Tanaka, Yousuke Ooyama and Joji Ohshita *

Department of Applied Chemistry, Graduate School of Engineering, Hiroshima University, Higashi-Hiroshima 739-8527, Japan; yadachi@hiroshima-u.ac.jp (Y.A.); tanashi0920@gmail.com (D.T.); yooyama@hiroshima-u.ac.jp (Y.O.)

* Correspondence: jo@hiroshima-u.ac.jp

Abstract: The surface modification of inorganic materials with organic units is an important process in device preparation. For the modification of TiO_2, organocarboxylic acids (RCO_2H) are usually used. Carboxylic acids form ester linkages (RCO_2Ti) with hydroxyl groups on the TiO_2 surface to attach the organic groups on the surface. However, the esterification liberates water as a byproduct, which may contaminate the surface by affecting TiO_2 electronic states. In addition, the ester linkages are usually unstable towards hydrolysis, which causes dye detachment and shortens device lifetime. In this review, we summarize our recent studies of the use of polymers composed of disilanylene and π-conjugated units as new modifiers of the TiO_2 surface. The TiO_2 electrodes modified by those polymers were applied to dye-sensitized solar cells.

Keywords: dye-sensitized solar cell; disilanylene polymer; photoreaction; surface modification; TiO_2

---

1. Introduction

Dye-sensitized solar cells (DSSCs) are of current interest because of the advantages they offer, including low fabrication cost and possible color tuning of the cells. The cells possess dye-attached TiO_2 as photoactive electrodes [1]. This system involves electron injection from photoexcited dyes to the conduction band of TiO_2 as the key step of the photocurrent generation. The resulting oxidized dyes are reduced by accepting an electron from the redox system, such as I_2/I^- in acetonitrile to recover the neutral state. Subsequently, electron-flow takes place from TiO_2 to the redox system through electrodes generating the photo-current of the device. Conventionally, the modification of the TiO_2 surface by organic dyes is performed by the formation of ester linkages between the Ti–OH bonds of the surface and the carboxylic acid groups of the dyes, as shown in Scheme 1 (1). However, the esterification produces water as a byproduct, which may contaminate the surface and thus change the electronic properties of TiO_2. Furthermore, the ester linkages on the TiO_2 surface are usually unstable towards hydrolysis and react with moisture to detach the dyes, shortening cell lifetime.

On the other hand, polymers having backbones composed of alternating organosilicon units and π-conjugated systems have been investigated as functional materials, such as carrier transporting and emissive materials [2–4]. Photoactive properties are also an important characteristic of Si–π polymers. In particular, those with Si–Si bonds are photoactive and UV irradiation of the polymer solutions leads to the cleavage of the Si–Si bonds. When the polymer films are irradiated in air, siloxane (Si–O–Si) and silanol (Si–OH) bonds arising from the reactions of the photoexcited Si–Si bonds with oxygen and moisture are formed. The formation of these relatively polar units increases the solubility of the polymers in alcohols, making it possible to utilize the polymer films as positive photoresists [2,5,6]. In fact, irradiation of the polymer films through a photomask followed by the development of the irradiated films by washing with alcohols provides sub–micron–order fine patterning. Utilizing the photoactivities, disilanylene–π alternating polymers are photochemically attached to the TiO_2 surface through the formation of Si–O–Ti bonds, as presented in Scheme 1 (2). In addition, disilanobithiophene is also investigated as a binding unit to TiO_2, and those resulting in polymer-attached TiO_2 materials are applied to DSSCs. Hanaya and coworkers reported a similar modification of TiO_2 electrodes by silanol and alkoxysilane dyes via the formation of Si–O–Ti linkages [7–11]. They demonstrated that the resulting electrodes show high performance as DSSC electrodes with high robustness towards hydrolysis in particular, as compared with electrodes with conventional ester linkages. For example, DSSCs based on a dye with a trimethoxysilyl anchor (**1**), shown in Chart 1, exhibited high performance with a power conversion efficiency (PCE) over 12%. This was higher than the DSSCs with a similar dye that had a carboxylic acid unit as the anchor (**2**) [9,11], clearly indicating the high potential of the Si–O–Ti bond as an efficient anchoring linkage. Dye **1**-attached TiO_2 showed higher stability towards hydrolysis and nearly no detach of the dye was observed after soaking for 2 h at 85 °C, while TiO_2 with **2** underwent the liberation of approximately 70% of the dye under the same conditions. It was also demonstrated that an aminoazobenzene dye with a triethoxysilyl anchor (**3**) showed higher sensitizing ability than a similar one bearing a carboxylic acid unit (**4**) (Chart 1), because of an improved open-circuit voltage (V_{oc}) arising from suppressed charge recombination [10].

In general, the anchors of the sensitizing dyes should have electron deficiency for the smooth electron injection from the photo-excited dye to the TiO_2 conduction band. Silicon units are generally recognized as electron-rich units, because of the low electronegativity of silicon. However, it is also known that silicon substituents work as electron-accepting units when attached to π–electron systems. In this review, we summarize our recent studies of the use of the disilanylene polymers as new modifiers of the TiO_2 surface. Applications of the modified TiO_2 electrodes to DSSCs are also described.

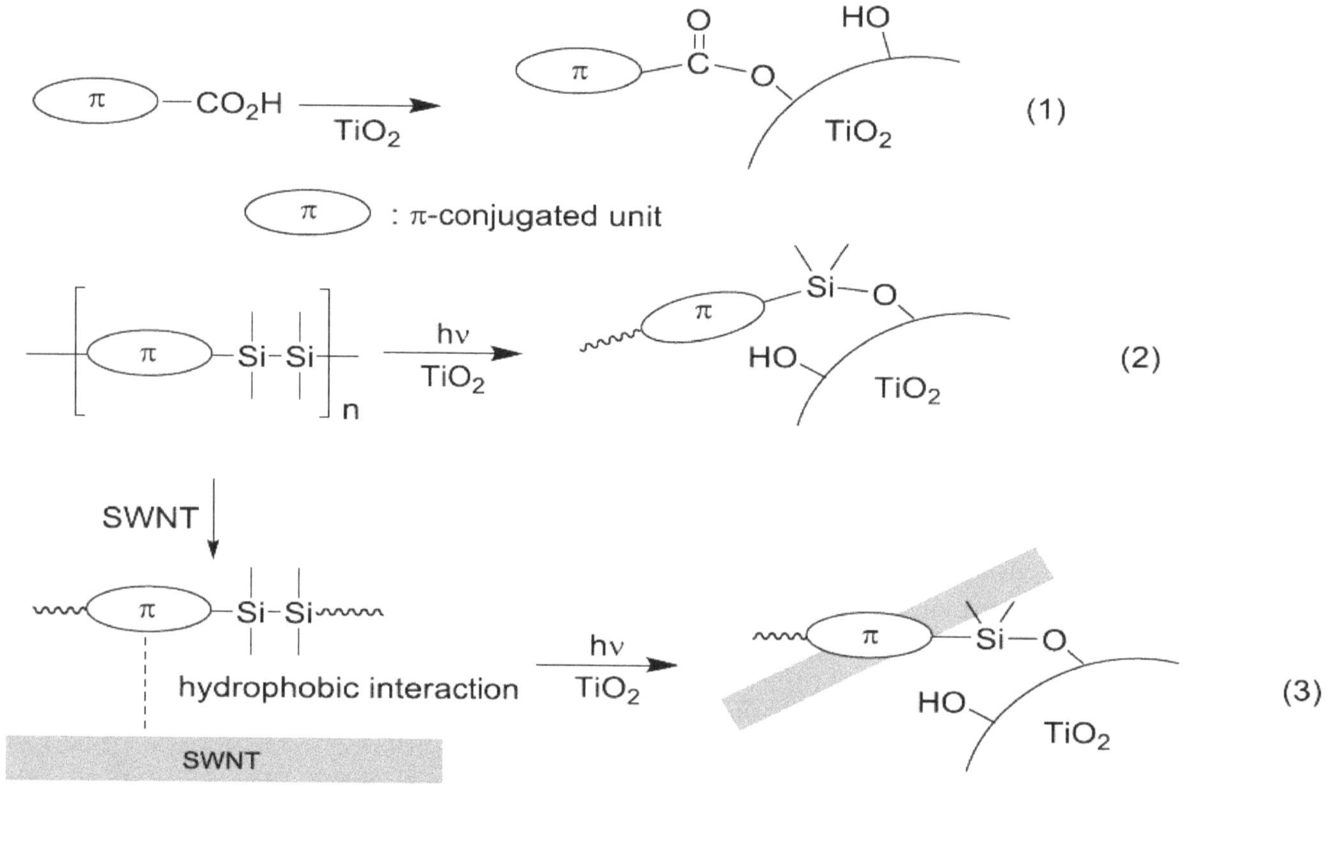

Scheme 1. Modification of TiO_2 surface with (1) organocarboxylic acid; (2) Si–Si–π polymer; and (3) s hybrid of Si–Si–π polymer and SWNT (single-walled carbon nanotube).

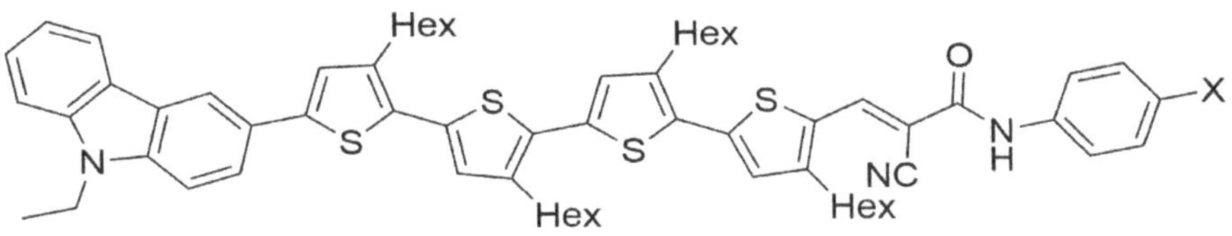

1 X = $Si(OMe)_3$ 12.5% (12.8% with co-sensitizer)
2 X = CO_2H 9.32%

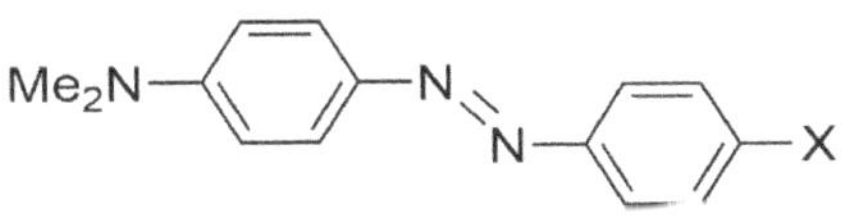

3 X = $Si(OEt)_3$ 2.6%
4 X = CO_2H 2.4%

Chart 1. Structures of sensitizing dyes with a trialkoxysilyl or carboxylic acid anchor reported by Hanaya et al. and PCEs of the dye-sensitized solar cells (DSSCs) utilizing the dyes [9–11].

2. Results and Discussion

2.1. Photochemical Attachment of Si–Si–π Polymers to TiO_2 Surface

When a TiO_2 electrode was irradiated (>400 nm) in a chloroform solution of poly[(disilanylene)quinquethienylene] (**DS5T** in Chart 2) with a Xe lamp bearing a cut filter, the colorless electrode turned yellowish brown [12]. In this process, light longer than 400 nm was used to avoid the activation of TiO_2.

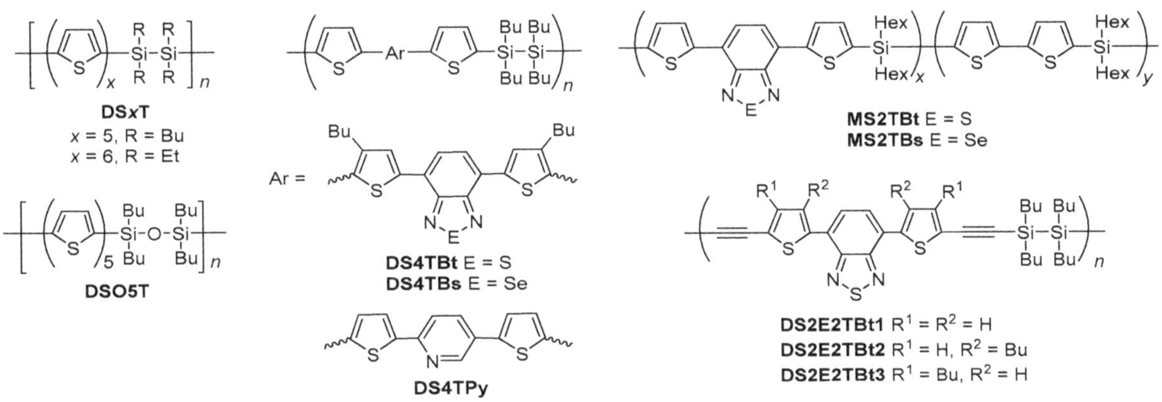

Chart 2. Si–Si–π and Si–π alternating polymers for a DSSC.

The photoreactions of compounds and polymers with Si–Si–π units have been studied in detail, and three types of reactions have been suggested as the major photodegradation pathways, as illustrated for poly(disilanylenephenylene) in Scheme 2 [2,5,6]: (1) 1,3-silyl shift from the disilanylene unit to the π–electron system forming silenes; (2) homolytic cleavage of the Si–Si bonds; and (3) direct reactions of the Si–Si bonds of the photoexcited molecules with alcohols. The alcoholysis of disilane units (route 3) proceeds dominantly over routes 1 and 2 when a large excess of alcohol is present in the reaction media [13,14]. The homolytic cleavage of Si–C bonds (route 4) is occasionally involved as a minor pathway, and routes 2 and 4 are preferred to route 1 in the polymeric systems [15,16]. However, the expansion of the π-conjugation usually suppresses the photoreactivities of the Si–Si–π compounds and polymers [17,18] and indeed, **DS5T** is basically not photoactive in an inert atmosphere [18]. Given these considerations, it seems most likely that the photochemical modification of the TiO_2 surface with **DS5T** occurs via direct reactions of the photoexcited polymer with TiOH groups on the surface. The reactions of **DS5T** with water adsorbed to the TiO_2 surface to form silanols and the subsequent condensation of the silanols with TiOH groups may also take place to form Si–O–Ti bonds.

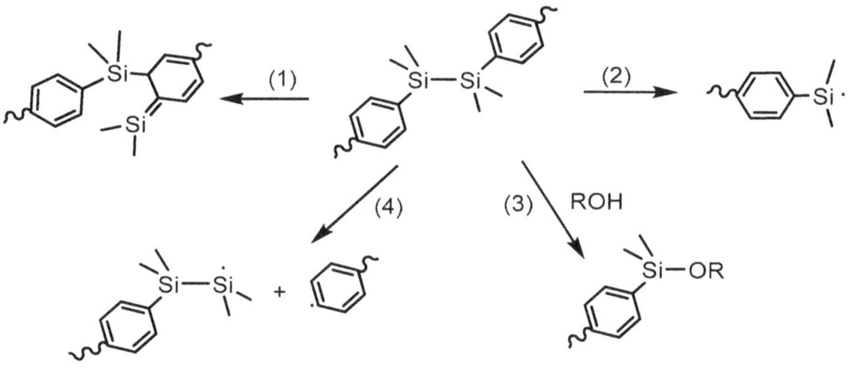

Scheme 2. Photodegradation of poly(disilanylenephenylene): (1) 1,3-silyl shift; (2) homolytic cleavage and (3) alcoholysis of Si–Si bond; and (4) homolytic cleavage of Si–C bond.

A similar treatment of TiO_2 with **DS6T** (Chart 2) also provided a polymer-attached TiO_2. This modified TiO_2 was examined as a photoelectrode of DSSCs. As presented in Table 1, the DSSCs showed photocurrent conversion, although the activities were not very high and the PCEs were approximately 0.1%. A colored TiO_2 electrode was also obtained by dipping the electrode into a solution of the corresponding siloxane polymer **DSO5T** (Chart 2), presumably owing to the interaction between the polymer chain and the TiO_2 surface, such as the coordination of the siloxane oxygen to the Lewis-acidic Ti site (Si_2O–Ti) and hydrogen bonding to TiOH (Si_2O–HOTi). However, the device with **DSO5T**-attached TiO_2 showed much less efficient photocurrent generation (PCE = 0.05%), indicating that a chemically bound polymer on TiO_2 is necessary to improve the activity. The attachment of poly[(ethoxysilanylene)quinquethiophene] (**MS5T**) on the TiO_2 surface was also examined, as shown in Scheme 3 [19]. A DSSC with TiO_2 modified by **MS5T** provided a PCE of 0.13% with J_{sc} (short-circuit current density) = 0.44 mA/cm^2, V_{oc} = 338 mV, and FF (fill factor) = 0.48.

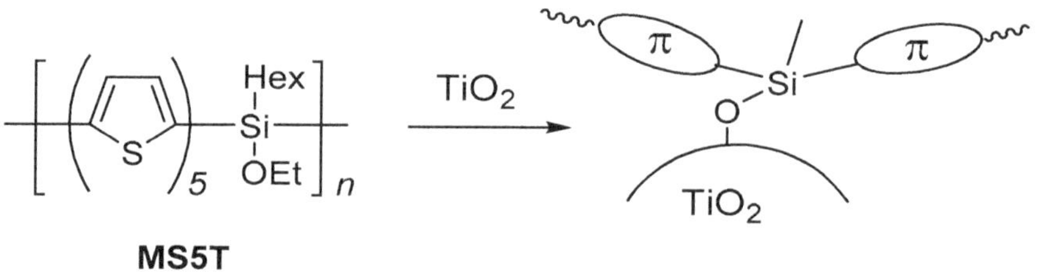

Scheme 3. Reaction of **MS5T** with TiO_2.

Table 1. Polymer absorption maximum and DSSC performance.

Polymer	λ_{max} [a]/nm	DSSC [b] J_{sc}/mA cm^{-2}	V_{oc}/mV	FF	PCE/%
DS5T	436	0.76	292	0.52	0.11
DS6T	418	0.86	296	0.48	0.12
DSO5T [c]	426	0.57	234	0.39	0.05
DS5T/SWNT	-	1.84	340	0.62	0.39
DS2E2TBt1	487	1.30	308	0.61	0.25
DS2E2TBt2	430	0.29	228	0.46	0.03
DS2E2TBt3	504	1.08	392	0.60	0.26
DS4TBt	508	1.30	324	0.63	0.28
DS4TBs	546	0.61	324	0.57	0.11
DS4TPy	416	2.15	296	0.63	0.40
MS2TBt [c]	451	0.42	358	0.40	0.06
MS2TBs [c]	482	0.54	336	0.49	0.09

[a] UV–Vis absorption maximum in solution; [b] FTO/TiO_2–polymer/$I_2 \cdot I^-$/Pt; [c] dye-attached TiO_2 electrode was prepared by dipping the electrode into the polymer solution without irradiation.

The three-component hybridization of the polymer, TiO_2, and carbon nanotube was also possible, as shown in Scheme 1 (3) [20]. Mixing **DS5T** with single-walled carbon nanotubes (SWNTs) by ball milling provided the hybrid material **DS5T**/SWNT, which was soluble in organic solvents. Irradiation of the TiO_2 electrode in a solution of the hybrid gave a TiO_2/SWNT/**DS5T** hybrid electrode. Application of the electrode to a DSSC led to improved performance with a PCE of 0.39%, which was 3.5 times higher than that based on the **DS5T**-modified TiO_2 (Table 1). Improvement of the device performance was likely ascribed primarily to the enhanced carrier transporting properties by hybridization with SWNTs.

The rather low DSSC performance based on the **DS5T**- and **DS6T**-modified TiO_2 was presumably due to the narrow absorption windows of the polymers, and thus donor-acceptor type π-conjugated systems were introduced to the Si–Si–π polymers in order to obtain more red-shifted absorption bands (Chart 2, Table 1) [21,22]. Photochemical treatment of TiO_2 electrodes in the polymer solutions, similarly to that for **DS5T** and **DS6T** mentioned above, provided polymer-modified TiO_2 that showed improved DSSC performance, as expected (Table 1). Among them, the best performance was obtained using the pyridine-containing polymer (**DS4TPy**). The pyridine unit would participate in a secondary coordinative interaction with the Lewis-acidic Ti site of the TiO_2 surface to facilitate the electron injection from the photoexcited polymer to TiO_2, as shown in Scheme 4. Enhanced electron injection through pyridine–Ti coordination has been reported [23]. Monosilane polymers with D–A type π-conjugated units (**MS2TBt** and **MS2TBs** in Chart 2) were also examined as DSSC dyes, which may attach to the TiO_2 surface via coordination of the benzothiadiazole or selenadiazole units to the Lewis-acidic Ti sites [24]. Although the DSSCs based on these polymers showed photocurrent conversion, the performance was low with PCE < 0.1%, again indicating that chemical bonding to polymer is important to improve DSSC performance.

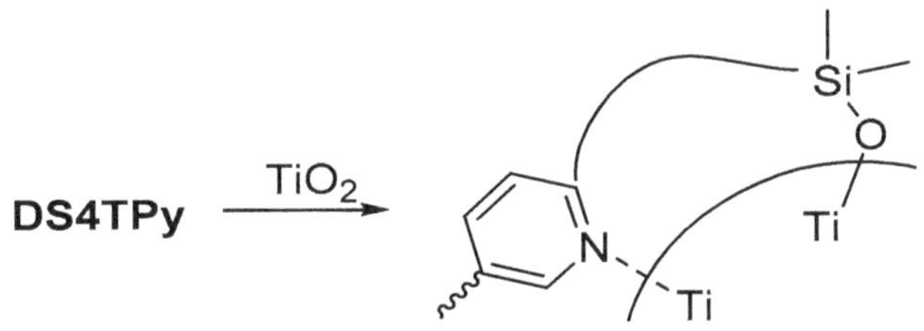

Scheme 4. Modification of TiO_2 surface with **DS4TPy**.

The thermal attachment of **DS2E2TBt1** and **DS4TPy** to the TiO_2 surface was also examined. However, DSSCs using the thermally modified TiO_2 electrodes showed lower PCEs (0.17% and 0.23% for **DS2E2TBt1** and **DS4TPy**, respectively) than those with the corresponding photochemically modified electrodes. This is due to the smaller amount of dye adsorbed to the surface, which was estimated to be approximately half of those of the photochemically modified electrodes.

2.2. Dithienosilole- and Disilanobithiophene-Containing π-Conjugated Polymers as Modifiers of the TiO_2 Surface

In spite of our efforts to develop new and efficient sensitizing dyes for DSSCs based on Si–Si–π polymers, DSSC performance was rather low, with a maximal PCE of 0.40%. The absorption windows seemed to be still narrow even though the D–A type π-conjugated units were introduced to the polymers. This was presumably because the polymer π-conjugation was interrupted by disilanylene units, although there might be some interaction between the Si–Si σ-orbital and the π-electron systems, namely, σ–π conjugation [2]. In addition, the Si–Si bonds might be cleaved on photolysis to produce silyl radicals that compete with the TiO_2 surface modification. The silyl radicals might add the π-electron systems to decompose the conjugated structures. We therefore prepared dithienosilole–pyridine fully conjugated polymers **DTSPy** and **DTS2TPy** (Chart 3), expecting that the polymers would interact with the TiO_2 surface via pyridine–Ti coordination [25]. The polymers could be attached to TiO_2 electrodes by dipping the electrodes into the polymer solutions without UV irradiation, and PCEs of 0.55% and 0.54% were obtained from the DSSCs based on **DTSPy** and **DTS2TPy**, respectively.

On the basis of these results, we designed and synthesized disilanobithiophene (DSBT)–pyridine and –pyrazine alternating polymers (Chart 3). We recently demonstrated that DSBT is an efficient donor unit of D–A π-polymers that are potentially useful as active materials of bulk hetero-junction polymer solar cells [26–28]. These DSBT–pyridine and –pyrazine polymers have fully conjugated systems in their backbones and show red-shifted absorption bands around 500 nm, as illustrated in Figure 1 [29]. They are able to attach to the TiO_2 surface by both Si–O–Ti bonding and pyridine– or pyrazine–Ti coordination (Scheme 5). Interestingly, DSBT showed high reactivity arising from the ring strain and reacted with the TiO_2 surface even in the dark. Indeed, homopolymer **pDSBT** that has no Lewis-base site could be attached to TiO_2 by dipping a TiO_2 electrode into the chloroform solution in the dark. As presented in Figure 2, the electrode thermally modified by **pDSBT** shows a darker color than that modified photochemically. This is most likely because the degradation of π-systems occurred to some extent under photochemical conditions, competing with the photo-derived modification of TiO_2. In some cases, however, the thermally modified TiO_2 electrode showed inferior performance as the photo-electrode of DSSCs to that modified under photochemical conditions, because smaller amounts of polymers could attach to TiO_2 in the dark.

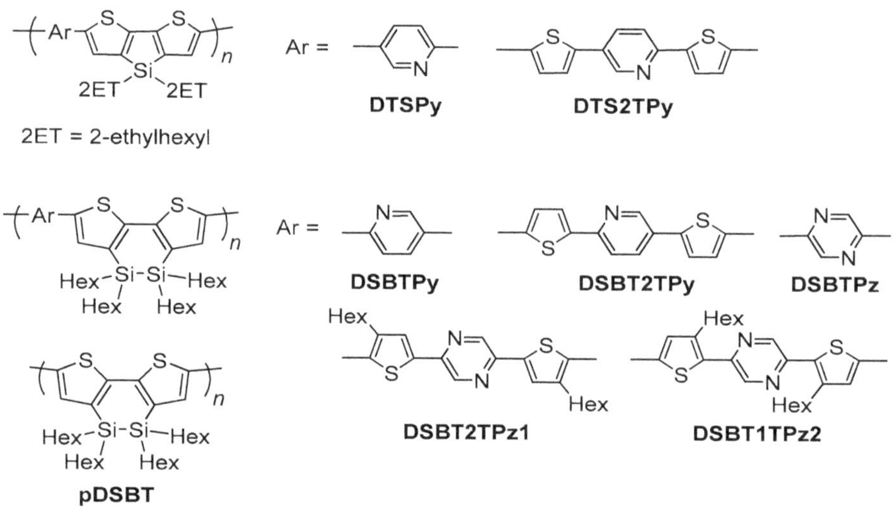

Chart 3. DTS–pyridine and DSBT–pyridine polymers.

DSSCs using TiO_2 electrodes modified by the DSBT–pyridine and –pyrazine polymers exhibited good performance with a maximal PCE of 0.89%, as presented in Table 2 and Figure 3, using a TiO_2 electrode thermally modified with **DSBTPz**. Presumably, thermal modification in the dark led to the introduction of smaller amounts of polymers on the surface. However, in the photochemical modification, it is speculated that silyl and aryl radicals would be formed from the photo-induced homolysis of the Si–Si and Si–C bonds to some extent, as illustrated in Scheme 2, routes 2 and 4, as the minor photodegradation pathways for Si–Si–π polymers. The radicals add to the π-conjugated systems to suppress the conjugation, thus leading to the decreased efficiencies. In fact, DSSCs using photochemically modified TiO_2 usually show IPCE (incident photon to current conversion efficiency) maxima at higher energies than DSSCs with thermally modified TiO_2 [29]. As can be seen in Table 2, the performance changed depending on the conditions of attaching the dyes on TiO_2. Some polymers showed higher performance when attached to TiO_2 photochemically, but some others gave rise to better results under thermal conditions. Establishing a balance between the amount of polymer loaded and the degree of photodegradation seems important to further improve DSSC performance. This may be achieved by optimizing the polymer structure.

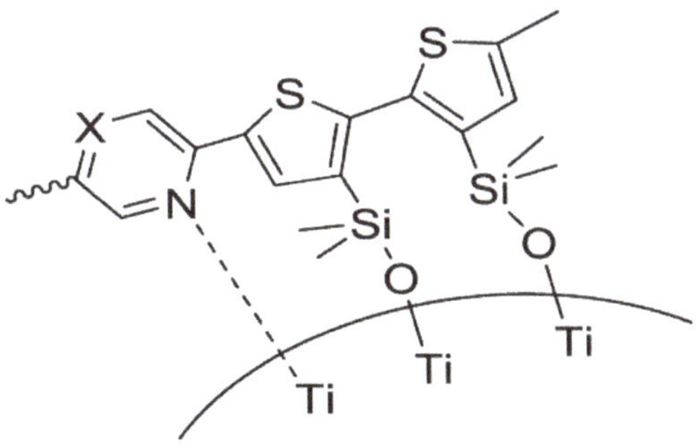

Scheme 5. Modification of TiO_2 surface with DSBT–pyridine and –pyrazine polymers.

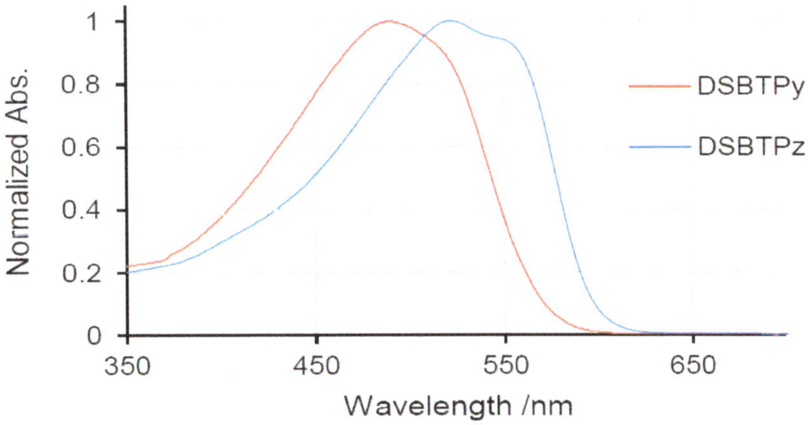

Figure 1. Absorption spectra of **DSBTPy** and **DSBTPz** in *o*-dichlorobenzene. Reproduced from Reference [29]—Published by the Royal Society of Chemistry.

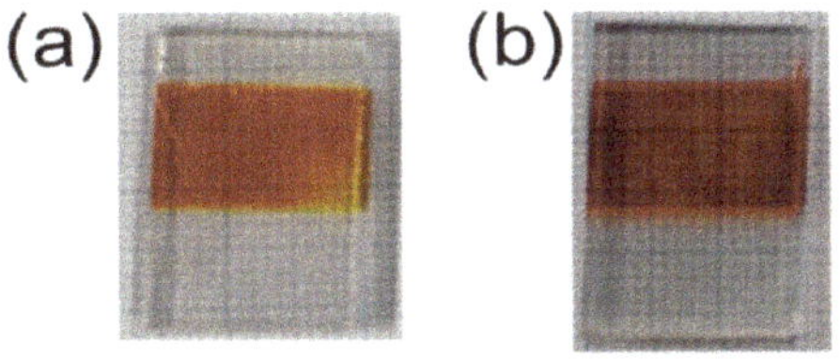

Figure 2. Photographs of (**a**) photochemically and (**b**) thermally modified TiO_2 electrode by **pDSBT**.

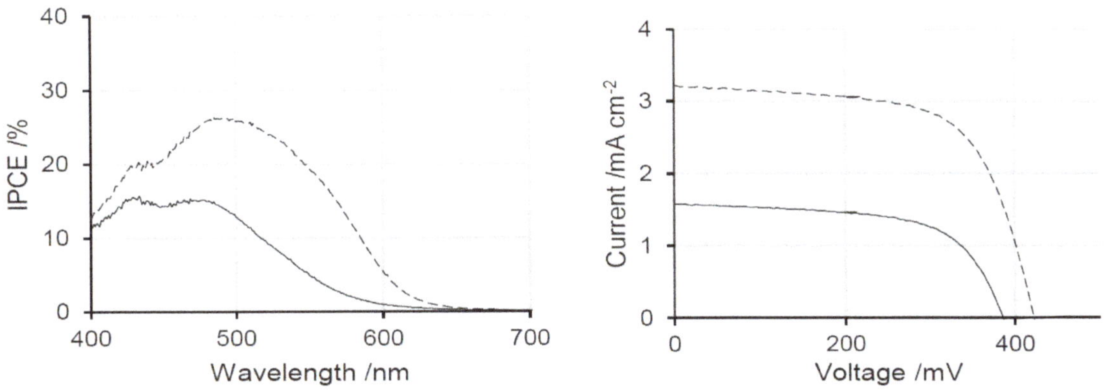

Figure 3. IPCE spectra and *J*–*V* curves of DSSCs based on photochemically (solid line) and thermally (dashed line) modified TiO_2 electrodes by **DSBTPz**. Reproduced from Reference [29]—Published by the Royal Society of Chemistry.

Table 2. Polymer absorption maximum and DSSC performance.

Polymer	λ_{max} [a]/nm	DSSC [b] J_{sc}/mA cm^{-2}	V_{oc}/mV	FF	PCE/%
DTSPy [c]	-	2.17	400	0.63	0.54
DTS2TPy [c]	-	2.03	390	0.69	0.55
pDSBT	451	2.10	308	0.61	0.39
pDSBT [c]	520	0.69	356	0.67	0.16
DSBTPy	439	1.91	344	0.63	0.41
DSBTPy [c]	484	1.67	396	0.63	0.42
DSBT2Tpy	468	3.11	380	0.63	0.74
DSBT2TPy [c]	475	1.34	392	0.66	0.35
DSBTPz	468	1.58	384	0.62	0.38
DSBTPz [c]	496	3.22	424	0.65	0.89
DSBT2TPz1	482	2.21	396	0.64	0.56
DSBT2TPz1 [c]	489	2.28	432	0.68	0.67
DSBT2TPz2	490	2.70	384	0.59	0.61
DSBT2TPz2 [c]	503	1.58	420	0.66	0.44

[a] UV–Vis absorption maximum of polymer-attached TiO_2 electrode; [b] FTO/TiO_2–polymer/$I_2 \cdot I^-$/Pt; [c] Dye-attached TiO_2 electrode was prepared by dipping the electrode into the polymer solution without irradiation.

3. Conclusions

We have demonstrated that the reactions of Si–Si bonds with hydroxyl groups on the TiO_2 surface provide an efficient route to modify the surface. These reactions proceeded cleanly without forming byproducts that might affect the properties of the TiO_2. The Si–O–Ti bonds were known to be stable towards hydrolysis and seem to be useful for DSSCs with long lifetime. This process may be also applied to modify inorganic oxide surfaces other than TiO_2, providing a hydrophobic surface with functional dye structures, thereby useful to control the surface and interface fine structures of organic optoelectronic devices such as organic thin film transistors and sensors. It has been also demonstrated that the attachment of azine-containing disilanylene polymers by both Si–O–Ti bonding and azine–Ti coordination improves the DSSC performance. This is likely ascribed to enhanced electron-injection through the azine–Ti coordination site. A similar function-separated dual site attachment of dyes on TiO_2 electrodes by the simultaneous formation of an anchoring unit and an electron-injecting unit has been recently applied to DSSCs [30–32]. The present system with disilane and azine units as the anchoring and electron-injecting units, respectively, seems to provide a new molecular design for robust sensitizing dyes.

Acknowledgments: This work was partly supported by JSPS KAKENHI Grant Nos. JP26288094 and JP17H03105.

References

1. Ooyama, Y.; Harima, Y. Molecular designs and syntheses of organic dyes for dye-sensitized solar cells. *Eur. J. Org. Chem.* **2009**, 2903–2934. [CrossRef]
2. Ohshita, J.; Kunai, A. Polymers with alternating organosilicon and π–conjugated units. *Acta Polym.* **1998**, *49*, 379–403. [CrossRef]
3. Uhlig, W. Synthesis, functionalization, and cross-linking reactions of organosilicon polymers using silyl triflate intermediates. *Prog. Polym. Sci.* **2002**, *27*, 255–305. [CrossRef]
4. Ponomarenko, S.A.; Kirchmeyer, S. Conjugated Organosilicon Materials for Organic Electronics and Photonics. *Adv. Polym. Sci.* **2011**, *235*, 33–110.
5. Ishikawa, M.; Nate, K. Photochemical behavior of organosilicon polymers bearing phenyldisilanyl units. *ACS Symp. Ser.* **1988**, *360*, 209–223.
6. Nate, K.; Ishikawa, M.; Ni, H.; Watanabe, H.; Saheki, Y. Photolysis of polymeric organosilicon systems. 4. photochemical behavior of poly[*p*-(disilanylene)phenylene]. *Organometallics* **1987**, *6*, 1673–1679. [CrossRef]

7. Kakiage, K.; Yamamura, M.; Fujimura, E.; Kyomen, T.; Unno, M.; Hanaya, M. High performance of Si–O–Ti bonds for anchoring sensitizing dyes on TiO_2 electrodes in dye-sensitized solar cells evidenced by using alkoxysilylazobenzenes. *Chem. Lett.* **2010**, *39*, 260–262. [CrossRef]
8. Unno, M.; Kakiage, K.; Yamamura, M.; Kogure, T.; Kyomen, T.; Hanaya, M. Silanol dyes for solar cells: Higher efficiency and significant durability. *Appl. Organometal. Chem.* **2010**, *24*, 247–250. [CrossRef]
9. Kakiage, K.; Aoyama, Y.; Yano, T.; Otsuka, T.; Kyomen, T.; Unno, M.; Hanaya, M. An achievement of over 12 percent efficiency in an organic dye-sensitized solar cell. *Chem. Commun.* **2014**, *50*, 6379–6381. [CrossRef] [PubMed]
10. Matta, S.K.; Kakiage, K.; Makuta, S.; Veamatshau, A.; Aoyama, Y.; Yano, T.; Hanaya, M.; Tachibana, Y. Dye-anchoring functional groups on the performance of dye-sensitized solar cells: Comparison between alkoxysilyl and carboxyl groups. *J. Phys. Chem. C* **2014**, *118*, 28425–28434. [CrossRef]
11. Kakiage, K.; Aoyama, Y.; Yano, T.; Oya, K.; Kyomen, T.; Hanaya, M. Fabrication of a high-performance dye-sensitized solar cell with 12.8% conversion efficiency using organic silyl-anchor dyes. *Chem. Commun.* **2015**, *51*, 6315–6317. [CrossRef] [PubMed]
12. Ohshita, J.; Matsukawa, J.; Hara, M.; Kunai, A.; Kajiwara, S.; Ooyama, Y.; Harima, Y.; Kakimoto, M. Attachment of disilanylene–oligothienylene polymers on TiO_2 surface by photochemical cleavage of the Si–Si bonds. *Chem. Lett.* **2008**, *37*, 316–317. [CrossRef]
13. Kira, M.; Miyazawa, T.; Sugiyama, H.; Yamaguchi, M.; Sakurai, H. $\sigma\pi^*$ Orthogonal intramolecular charge-transfer (OICT) excited states and photoreaction mechanism of trifluoromethyl-substituted phenyldisilanes. *J. Am. Chem. Soc.* **1993**, *115*, 3116–3124. [CrossRef]
14. Ohshita, J.; Ohsaki, H.; Ishikawa, M.; Minato, A. Silicon–carbon unsaturated compounds. 26. Photochemical behavior of 1,4- and 1,5-bis(pentamethyldisilanyl)naphthalene. *Organometallics* **1991**, *10*, 880–887. [CrossRef]
15. Ishikawa, M.; Watanabe, K.; Sakamoto, H.; Kunai, A. Silicon carbon unsaturated-compounds. 40. Photolysis of 1,4-bis(2-phenyltetramethyldisilanyl)benzene. *J. Organomet. Chem.* **1992**, *435*, 249–256. [CrossRef]
16. Ishikawa, M.; Watanabe, K.; Sakamoto, H.; Kunai, A. Silicon-carbon unsaturated-compounds. 44. Photochemical behavior of permethylated *p*-(disilanylene)phenylene oligomers. *J. Organomet. Chem.* **1993**, *455*, 61–68. [CrossRef]
17. Ohshita, J.; Watanabe, T.; Kanaya, D.; Ohsaki, H.; Ishikawa, M.; Ago, H.; Tanaka, K.; Yamabe, T. Polymeric organosilicon systems. 22. Synthesis and photochemical properties of poly[(disilanylene) oligophenylylenes] and poly[(silylene)biphenylylenes]. *Organometallics* **1994**, *13*, 5002–5012. [CrossRef]
18. Kunai, A.; Ueda, T.; Horata, K.; Toyoda, E.; Nagamoto, I.; Ohshita, J.; Ishikawa, M. Polymeric organosilicon systems. 26. Synthesis and photochemical and conducting properties of poly[(tetraethyldisilanylene) oligo(2,5-thienylenes)]. *Organometallics* **1996**, *15*, 2000–2008. [CrossRef]
19. Ohshita, J.; Matsukawa, J.; Iwawaki, T.; Matsui, S.; Ooyama, Y.; Harima, Y. Attachment of poly[(ethoxyhexylsilylene)oligothienylene]s to inorganic oxide surface. *Synth. Met.* **2009**, *159*, 817–820. [CrossRef]
20. Ohshita, J.; Tanaka, D.; Matsukawa, J.; Mizumo, T.; Yoshida, H.; Ooyama, Y.; Harima, Y. Hybridization of carbon nanotubes with Si–π polymers and attachment of resulting hybrids to TiO_2 surface. *Chem. Lett.* **2011**, *40*, 87–89. [CrossRef]
21. Tanaka, D.; Ohshita, J.; Ooyama, Y.; Mizumo, T.; Harima, Y. Synthesis of disilanylene polymers with donor–acceptor-type π–conjugated units and applications to dye-sensitized solar cells. *J. Organomet. Chem.* **2012**, *719*, 30–35. [CrossRef]
22. Tanaka, D.; Ohshita, J.; Mizumo, T.; Ooyama, Y.; Harima, Y. Synthesis of donor–acceptor type new organosilicon polymers and their applications to dye-sensitized solar cells. *J. Organomet. Chem.* **2013**, *741–742*, 97–101. [CrossRef]
23. Ooyama, Y.; Inoue, S.; Nagano, T.; Kushimoto, K.; Ohshita, J.; Imae, I.; Komaguchi, K.; Harima, Y. Dye-sensitized solar cells based on donor–acceptor π–conjugated fluorescent dyes with a pyridine ring as an electron-withdrawing anchoring group. *Angew. Chem. Int. Ed.* **2011**, *50*, 7429–7433. [CrossRef] [PubMed]
24. Ohshita, J.; Kangai, S.; Yoshida, H.; Kunai, A.; Kajiwara, S.; Ooyama, Y.; Harima, Y. Synthesis of organosilicon polymers containing donor–acceptor type π–conjugated units and their applications to dye-sensitized solar cells. *J. Organomet. Chem.* **2007**, *692*, 801–805. [CrossRef]

25. Tanaka, D.; Ohshita, J.; Ooyama, Y.; Morihara, Y. Synthesis and optical and photovoltaic properties of dithienosilole–dithienylpyridine and dithienosilole–pyridine alternate polymers and polymer–B(C_6F_5) complexes. *Polym. J.* **2013**, *45*, 1153–1158. [CrossRef]
26. Ohshita, J.; Nakashima, M.; Tanaka, D.; Morihara, Y.; Fueno, H.; Tanaka, K. Preparation of a D–A polymer with disilanobithiophene as a new donor component and application to high-voltage bulk heterojunction polymer solar cells. *Polym. Chem.* **2014**, *5*, 346–349. [CrossRef]
27. Nakashima, M.; Otsura, T.; Naito, H.; Ohshita, J. Synthesis of new D–A polymers containing disilanobithiophene donor and application to bulk heterojunction polymer solar cells. *Polym. J.* **2015**, *47*, 733–738. [CrossRef]
28. Nakashima, M.; Ooyama, Y.; Sugiyama, T.; Naito, H.; Ohshita, J. Synthesis of a Conjugated D–A Polymer with Bi(disilanobithiophene) as a New Donor Component. *Molecules* **2016**, *21*, 789. [CrossRef] [PubMed]
29. Ohshita, J.; Adachi, Y.; Tanaka, D.; Nakashima, M.; Ooyama, Y. Synthesis of D–A polymers with a disilanobithiophene donor and a pyridine or pyrazine acceptor and their applications to dye-sensitized solar cells. *RSC Adv.* **2015**, *5*, 36673–36679. [CrossRef]
30. Ooyama, Y.; Hagiwara, Y.; Oda, Y.; Mizumo, T.; Harima, Y.; Ohshita, J. Dye-sensitized solar cells based on a functionally separated D–π–A fluorescent dye with an aldehyde as an electron-accepting group. *New J. Chem.* **2013**, *37*, 2336–2340. [CrossRef]
31. Ooyama, Y.; Uenaka, K.; Ohshita, J. Development of a functionally separated D–π–A fluorescent dye with a pyrazyl group as an electron-accepting group for dye-sensitized solar cells. *Org. Chem. Front.* **2015**, *2*, 552–559. [CrossRef]
32. Zhang, L.; Cole, J.M. Anchoring groups for dye-sensitized solar cells. *ACS Appl. Mater. Interfaces* **2015**, *7*, 3427–3455. [CrossRef] [PubMed]

14

Stepwise Introduction of Different Substituents to α-Chloro-ω-hydrooligosilanes: Convenient Synthesis of Unsymmetrically Substituted Oligosilanes

Ken-ichiro Kanno *, Yumi Aikawa, Yuka Niwayama, Misaki Ino, Kento Kawamura and Soichiro Kyushin *

Division of Molecular Science, Graduate School of Science and Technology, Gunma University, Kiryu, Gunma 376-8515, Japan; yumi.a174.4@gmail.com (Y.A.); yu.niwa06@gmail.com (Y.N.); misaki.ino512@gmail.com (M.I.); t12301059@gunma-u.ac.jp (K.K.)

* Correspondence: kkanno@gunma-u.ac.jp (K.-i.K.); kyushin@gunma-u.ac.jp (S.K.)

Abstract: A series of unsymmetrically substituted oligosilanes were synthesized via stepwise introduction of different substituents to α-chloro-ω-hydrooligosilanes. The reactions of α-chloro-ω-hydrooligosilanes with organolithium or Grignard reagents gave hydrooligosilanes having various alkyl, alkenyl, alkynyl and aryl groups. Thus-obtained hydrooligosilanes were converted into alkoxyoligosilanes by ruthenium-catalyzed dehydrogenative alkoxylation with alcohols.

Keywords: α-chloro-ω-hydrooligosilane; titanium; ruthenium; dehydrogenative alkoxylation

---

1. Introduction

Recently, various functionalized oligosilanes have attracted growing interests due to their unique photophysical and electronic properties [1–13]. In the synthesis of such compounds, introduction of functionality to oligosilanes is an important process. Oligosilanes having both chlorosilane and hydrosilane moieties are especially interesting because two different functionalities could be introduced to the oligosilanes. For example, α-chloro-ω-hydrooligosilanes are potential precursors for such modification.

A problem to be overcome is limitation of facile functionalization of the hydrosilane moiety in the oligosilanes. Thus far, halogenation [14–16], Lewis acid-catalyzed reactions [17–19] and radical-promoted reactions [20–24] have been used for the Si–H modification of hydrooligosilanes. Transition metal-catalyzed reactions of hydrosilanes are well known as Si–H conversion methods, including hydrosilylation [25–27], dehydrogenative silylation of OH and NH groups [28–31], silylation of aromatic rings [32–36] and so on. however, most of them are applicable only to hydromonosilanes. When hydrooligosilanes are subjected to these transition metal-catalyzed reactions, major products come from the cleavage of Si–Si bonds. Although the transition metal-catalyzed reactions are utilized as valuable organic synthesis methodologies [37–40], it is not the case for the transformation of hydrooligosilanes. As a rare example, Yamanoi and Nishihara have reported the palladium-catalyzed arylation of hydrooligosilanes with aryl iodides to afford the corresponding arylated oligosilanes with preservation of the Si–Si bonds [41,42].

Previously, we reported titanium-catalyzed synthesis of the hydrogen-terminated oligosilanes from α,ω-dichlorooligosilanes [43]. This method enabled us to synthesize α-chloro-ω-hydrooligosilanes with high selectivity. The α-chloro-ω-hydrooligosilanes are good precursors of variously substituted hydrooligosilanes via nucleophilic substitution of the chlorosilane moiety. Furthermore, we have recently found ruthenium-catalyzed dehydrogenative alkoxylation of a hydrodisilane with alcohols [44]. Interestingly, the reactions proceed with preserving the Si–Si bond despite its susceptible nature to transition metals [37–40]. Combination of these two methods leads to convenient synthesis of unsymmetrically substituted oligosilanes. In this paper, we report the synthesis of various substituted hydrooligosilanes from α-chloro-ω-hydrooligosilanes and transformation of thus-obtained hydrooligosilanes to the corresponding alkoxy oligosilanes.

2. Results and Discussion

2.1. Synthesis of 1-Hydrooligosilanes

As reported previously, α-chloro-ω-hydrooligosilanes **1–3** are synthesized by the selective monoreduction of α,ω-dichlorooligosilanes with alkylmagnesium chlorides in the presence of a catalytic amount of $TiCl_4$ (Scheme 1) [43]. The remaining chlorosilane moiety is possible functionality for introduction of various substituents.

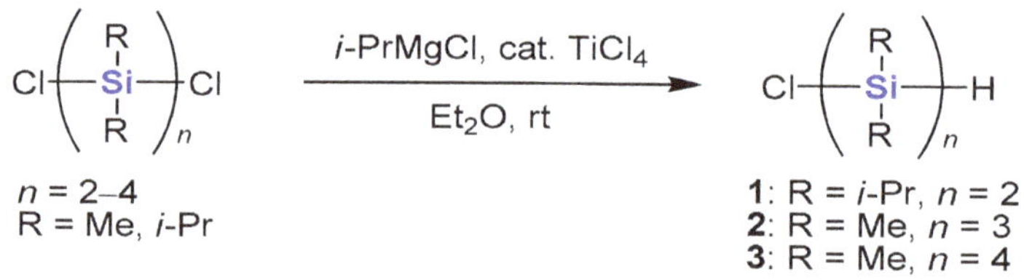

Scheme 1. Monoreduction of α,ω-dichlorooligosilanes with *i*-PrMgCl in the presence of a catalytic amount of $TiCl_4$.

As shown in Scheme 2, the chlorosilane moiety of **1** was smoothly substituted by organolithium reagents. When **1** was treated with 2-thienyllithium, 2-thienyldisilane **4** was obtained in high yield. 4-Methoxyphenyl group was also introduced to afford 4-methoxyphenyldisilane **5**.

Scheme 2. Synthesis of substituted hydrodisilanes from 1-chloro-2-hydrodisilane.

In the cases of polymethylated oligosilanes **2** and **3**, Grignard reagents are more suitable for the substitution of the chlorosilane moiety. The Grignard reagents were prepared in the presence of lithium chloride and DIBAL-H to activate magnesium metal [45]. As shown in Scheme 3, various alkyl, alkenyl, alkynyl and aryl groups were introduced, and substituted hydrooligosilanes were successfully synthesized. As alkyl group installation, 5-hexenyl- and 2-phenylethylmagnesium reagents were used to afford hydrotrisilanes **6** and **7**. Mono- and disubstituted alkenyl groups such as styryl and 2-buten-2-yl ones were also introduced to the trisilane skeleton successfully. The stereochemistry of the alkene moieties of trisilanes **8** and **9** was determined by ^{1}H NMR spectroscopy. The *E* geometry of the styryl group in **8** was confirmed by the *J* value of the vinylic coupling (19 Hz). The major *Z* isomer of **9** was confirmed by NOE experiment. Three types of alkynyl Grignard reagents with aliphatic, aryl and silyl substituents were subjected to the reaction to afford alkynyltrisilanes **10–12**. Various aryl groups were also introduced to **2** to afford phenyl-, 4-methoxyphenyl-, 4-dimethylaminophenyl-, and 2-thienyltrisilanes **13–16**. The similar arylation was also applicable to **3**, and 2-thienyltetrasilane **17** was obtained. All of these reactions proceed without loss of the hydrosilane moieties, which can be used for further modification of the hydrooligosilanes. The reason for low yields in some cases is attributed to loss during the isolation procedure by column chromatography over silica gel.

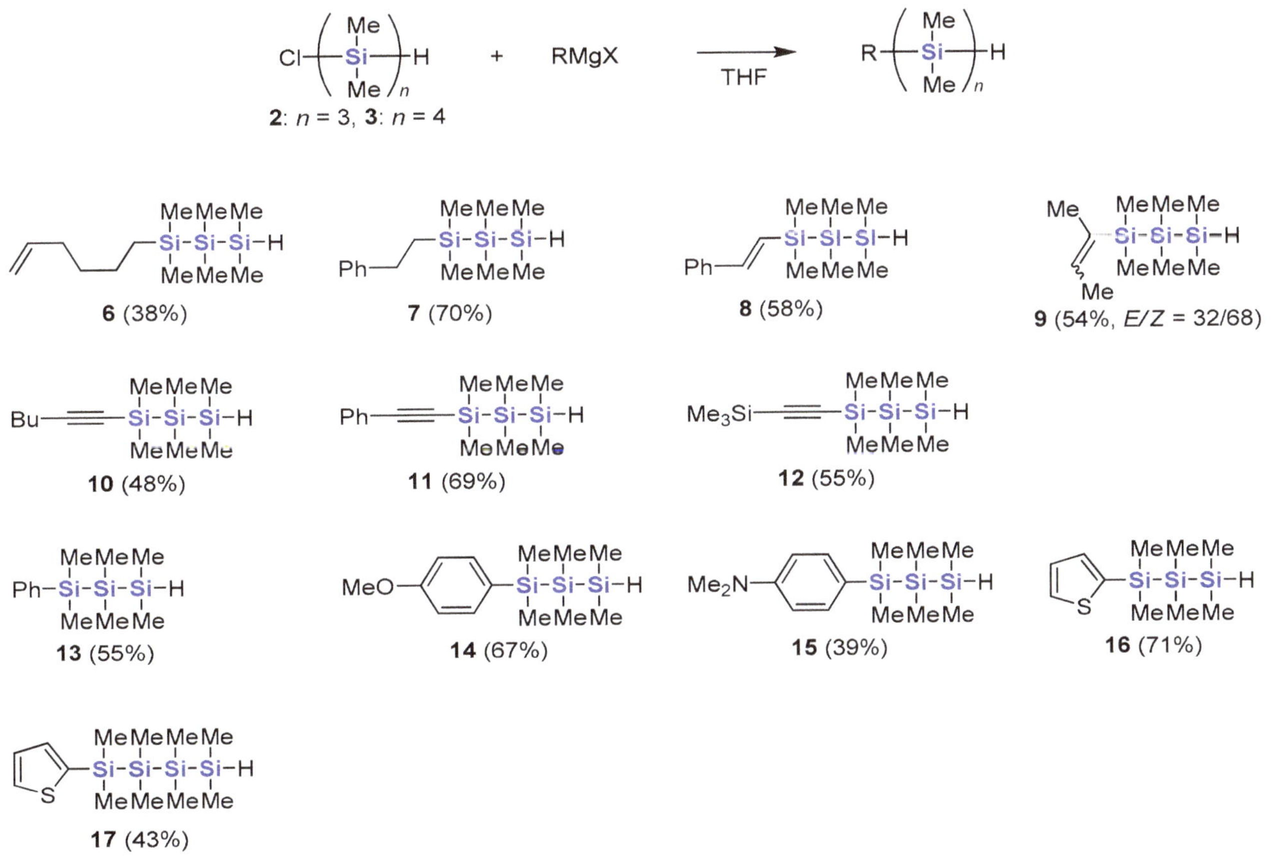

Scheme 3. Synthesis of substituted hydrooligosilanes from α-chloro-ω-hydrooligosilanes.

2.2. Dehydrogenative Alkoxylation of hydrooligosilanes with Alcohols

As mentioned above, we have reported ruthenium-catalyzed dehydrogenative alkoxylation of a hydrodisilane with alcohols without Si–Si bond cleavage [44]. The reactions are also applicable to various hydrotrisilanes. As shown in Scheme 4, some of the hydrotrisilanes synthesized above were subjected to the ruthenium-catalyzed dehydrogenative alkoxylation with methanol. All reactions proceeded smoothly in toluene at room temperature in the presence of 2.5 mol % of $[RuCl_2(p\text{-cymene})]_2$ to afford methoxytrisilanes **18–21** in good yields. It is worth noting that the alkenyl and alkynyl moieties tolerate the reactions. Neither hydrosilylation nor bis-silylation occurred under these reaction conditions.

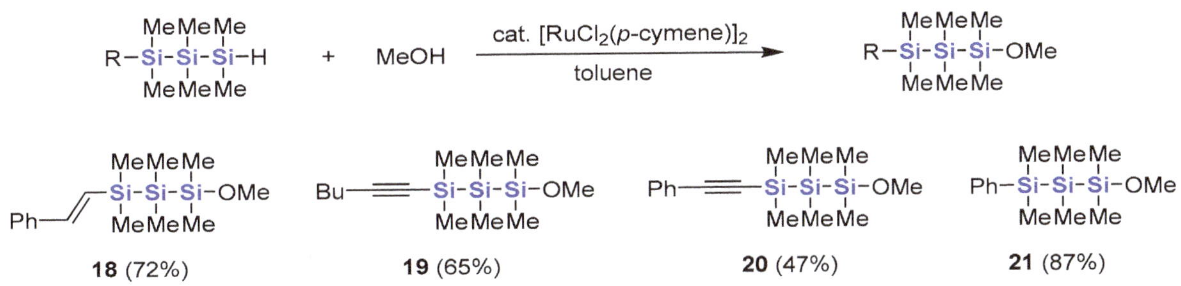

Scheme 4. Synthesis of 1-methoxytrisilanes from 1-hydrotrisilanes.

To gain further insight into the substituent effects in the ruthenium-catalyzed alkoxylation, ethyl-substituted hydrodisilane **22** and 2-hydrotrisilane **26** were used for the reactions with methanol. The results are summarized in Tables 1 and 2. Even though an excess amount of methanol (10 equiv) was used in the reaction of **22**, the reaction rate is much slower than that of $PhMe_2SiSiMe_2H$, which finished within 2 h under the same reaction conditions [44]. When bulkier alcohols are used, the reaction rate becomes slower. The reaction of **22** with ethanol needed heating at 50 °C to be completed within one day. The reaction with a large excess of 2-propanol is much more sluggish. Even though the reaction was carried out on heating, more than 40 h were needed for complete consumption of **22**.

Table 1. Reactions of **22** with various alcohols in the presence of $[RuCl_2(p\text{-cymene})]_2$.

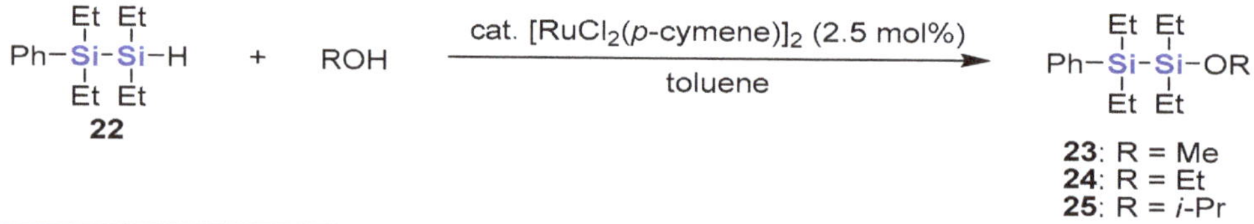

Entry	Alcohol (equiv)	Temperature	Reaction Time/h	Product	Isolated Yield/%
1	MeOH (10)	0 °C–rt	20	**23**	83
2	EtOH (10)	50 °C	23	**24**	45
3	*i*-PrOH (100)	0–50 °C	41	**25**	58

Table 2. Reactions of **26** with methanol in the presence of various ruthenium catalysts.

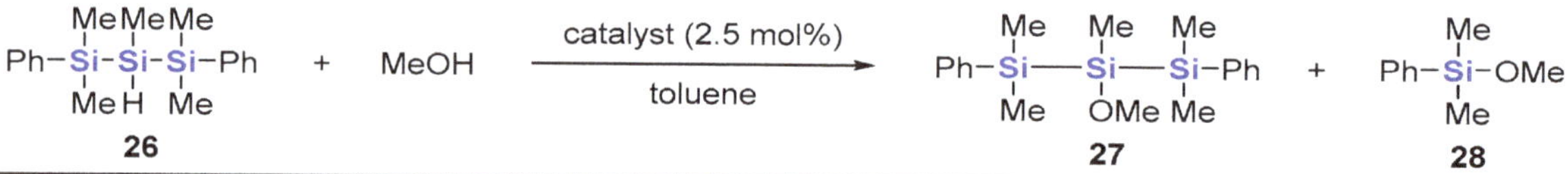

Entry	Catalyst	Methanol (equiv)	Temperature/°C	Reaction Time/h	GC Yield/% [1]		
					26	27	28
1	$[RuCl_2(p\text{-cymene})]_2$	10	0	23	19	16	47
2	$[RuCl_2(\text{mesitylene})]_2$	10	0–50	51	12	22	11
3	$[RuCl_2(\text{benzene})]_2$	10	0–50	100	3	59 (29)	5
4	$RuHCl(CO)(PPh_3)_3$ [2]	10	0	19	41	0	2

[1] Isolated yield was given in parentheses. [2] 5 mol %.

For the alkoxylation of **26**, optimization of the ruthenium catalyst was necessary, as shown in Table 2. The reaction of **26** with methanol in the presence of the (*p*-cymene)ruthenium catalyst gave the desired 2-methoxytrisilane **27** in low yield along with monosilane **28**, which was produced by Si–Si bond cleavage (Entry 1). Changing the aromatic ligand to mesitylene slightly improved the formation of **27**, but the Si–Si bond cleavage still occurred significantly (Entry 2). In contrast, the (benzene)ruthenium catalyst gave **27** more selectively and suppressed the formation of **28** (Entry 3). Compared with the (arene)ruthenium complexes, $RuHCl(CO)(PPh_3)_3$ showed little catalytic activity, and no alkoxylation product was detected in the reaction mixture (Entry 4).

The superior performance of the (benzene)ruthenium catalyst over the (*p*-cymene)ruthenium catalyst might be attributed to the less steric hindrance around the coordinated arenes. The intermediate of the reaction might be the hydrosilane-bound ruthenium complex via σ-coordination or oxidative addition of the Si–H bond to the ruthenium atom. Nucleophilic attack of methanol to the silicon atom having the Si–H bond produces **27**. At this step, more crowded *p*-cymene prevents methanol from attacking the central silicon atom. As a result, methanol attacks the terminal silicon atom of **26** to afford **28** via Si–Si bond cleavage.

3. Materials and Methods

All reactions were carried out under an argon atmosphere using standard Schlenk techniques unless otherwise noted. THF and diethyl ether were distilled from sodium benzophenone ketyl under a nitrogen atmosphere. Toluene was distilled from sodium under a nitrogen atmosphere. $Cl(i\text{-}Pr)_2SiSi(i\text{-}Pr)_2H$ (**1**) [43], $Cl(SiMe_2)_3H$ (**2**) [43], $Cl(SiMe_2)_4H$ (**3**) [43], $PhMe_2SiSi(H)MeSiMe_2Ph$ (**26**) [46] and $Et_3SiSiEt_3$ [47] were prepared according to the reported procedures. Silica gel for column chromatography (Kanto Chemical, silica gel 60N, spherical, neutral, particle size 100–210 μm) was purchased and used as received. The other chemicals were purchased (Kanto Chemical, Tokyo, Japan; Kishida Chemical, Osaka, Japan; Sigma-Aldrich Japan, Tokyo, Japan; Tokyo Chemical Industry, Tokyo, Japan; Wako Pure Chemical Industries, Osaka, Japan) and used without further purification.

GC analysis was performed on a Shimadzu GC-8A gas chromatograph equipped with packed columns containing 10% silicone SE-30 on Uniport B (GL Sciences, Tokyo, Japan) and a Shimadzu (Kyoto, Japan) C-R8A Chromatopack integrator. ^{1}H, ^{13}C and ^{29}Si NMR spectra were measured with JEOL (Akishima, Japan) JNM-ECA600, JNM-ECS400 and JNM-ECS300 spectrometers. Dodecane, tricosane or mesitylene were used as internal standards for NMR yield estimation. IR spectra were recorded on Shimadzu FTIR-8700, JASCO (Hachioji, Japan) FT/IR-4600 and hitachi (Tokyo, Japan) FTIR 270-50 spectrophotometers. Mass spectra and high-resolution mass spectra were recorded on Shimadzu GCMS-QP2010 SE and JEOL JMS-T100GCV mass spectrometers. Spectral data of all new compounds are showed in Figure S1–S113 in supplementary materials.

3.1. *Synthesis of 1,1,2,2-Tetraisopropyl-1-(2′-thienyl)disilane* (4)

A 1.64 M solution of butyllithium in hexane (1.5 mL, 2.5 mmol) was added dropwise to a solution of thiophene (212 mg, 2.52 mmol) in THF (8 mL) at 0 °C, and the mixture was stirred at 0 °C for 1 h. Compound **1** (536 mg, 2.02 mmol) was added, and the mixture was stirred for 2 days at room temperature. The reaction was quenched by adding 1.0 M hydrochloric acid. The reaction mixture was extracted with hexane. The organic layer was washed with water and brine, and dried over anhydrous sodium sulfate. After evaporation of the solvents, the residue was distilled with a Kugelrohr distillation apparatus (100–130 °C/1 mmHg) to give **4** (537 mg, 85%) as a yellow oil.

4. ^{1}H NMR (301 MHz, $CDCl_3$): δ 1.07 (d, 6H, J = 6.9 Hz), 1.10 (d, 6H, J = 6.9 Hz), 1.12 (d, 6H, J = 7.2 Hz), 1.14 (d, 6H, J = 8.1 Hz), 1.16–1.31 (m, 2H), 1.40 (sept, 2H, J = 7.3 Hz), 3.73 (t, 1H, J = 3.2 Hz), 7.20 (dd, 1H, J = 4.7, 3.5 Hz), 7.31 (dd, 1H, J = 3.5, 1.1 Hz), 7.61 (dd, 1H, J = 4.7, 1.1 Hz). ^{13}C NMR (76 MHz, $CDCl_3$): δ 11.7, 13.7, 19.3, 19.5, 20.8, 21.6, 128.0, 130.6, 134.8, 136.2. ^{29}Si NMR (60 MHz, $CDCl_3$): δ −14.7, −9.0. IR (NaCl): 2940, 2860, 2070, 1460, 1210, 1000, 880, 750, 700 cm^{-1}. MS (EI, 70 eV): m/z 312 (M^+, 16), 269 (100), 227 (37), 197 (16), 185 (25), 155 (17), 141 (15), 127 (20). hRMS (EI): found 312.1774, calcd for $C_{16}H_{32}SSi_2$ 312.1763.

3.2. *Synthesis of 1,1,2,2-Tetraisopropyl-1-(4′-methoxyphenyl)disilane* (5)

A 1.53 M solution of *tert*-butyllithium in pentane (3.7 mL, 5.7 mmol) was added dropwise to a solution of 4-iodoanisole (260 mg, 1.11 mmol) in THF (5 mL) at −78 °C, and the mixture was stirred at −78 °C for 1 h. After warming to room temperature, **1** (244 mg, 0.921 mmol) was added, and the mixture was stirred at room temperature for 24 h. The reaction was quenched by adding 1.0 M hydrochloric acid. The reaction mixture was extracted with hexane. The organic layer was washed with water and brine, and dried over anhydrous sodium sulfate. After evaporation of the solvents, the residue was distilled with a Kugelrohr distillation apparatus (190–220 °C/1 mmHg) to give **5** (141 mg, 45%) as a yellow oil.

5. ^{1}H NMR (301 MHz, $CDCl_3$): δ 1.03 (d, 6H, J = 7.2 Hz), 1.05 (d, 6H, J = 7.2 Hz), 1.08 (d, 6H, J = 7.2 Hz), 1.09 (d, 6H, J = 7.2 Hz), 1.14–1.27 (m, 2H), 1.40 (sept, 2H, J = 7.2 Hz), 3.73 (t, 1H, J = 3.3 Hz), 3.81 (s, 3H), 6.89 (d, 2H, J = 8.7 Hz), 7.45 (d, 2H, J = 8.7 Hz). ^{13}C NMR (76 MHz, $CDCl_3$): δ 11.7, 12.7, 19.2, 19.4, 21.0, 21.7, 55.0, 113.4, 126.5, 137.2, 160.1. ^{29}Si NMR (99 MHz, $CDCl_3$): δ −15.0, −6.1. IR (NaCl): 2940, 2860,

2070, 1590, 1500, 1460, 1280, 1250, 1180, 1100 cm^{-1}. MS (EI, 70 eV): *m*/*z* 336 (M^+, 43), 293 (100), 251 (48), 221 (77), 209 (38), 179 (43), 167 (19), 165 (16), 151 (27), 137 (14). hRMS (EI): found 336.2305, calcd for $C_{19}H_{36}OSi_2$ 336.2305.

3.3. Representative Procedure: Synthesis of 1,1,2,2,3,3-Hexamethyl-1-(2′-thienyl)trisilane (**16**)

A 1.64 M solution of butyllithium in hexane (4.8 mL, 7.9 mmol) was added dropwise to a solution of chlorotrimethylsilane (847 mg, 7.80 mmol) in THF (15 mL) at −78 °C, and the mixture was stirred at −78 °C for 10 min. After warming to room temperature, magnesium turnings (392 mg, 16.1 mmol) and a 1.0 M solution of DIBAL-H in toluene (0.1 mL, 0.1 mmol) were added. After stirring for 5 min, the mixture was cooled to −10 °C, 2-bromothiophene (974 mg, 5.97 mmol) was added. After stirring for 1 h, **2** (876 mg, 4.15 mmol) was added, and the mixture was stirred at room temperature overnight. The reaction was quenched with 1.0 M hydrochloric acid. The reaction mixture was extracted with hexane. The organic layer was washed with water and brine, and dried over anhydrous sodium sulfate. After evaporation of the solvents, the residue was separated by column chromatography over silica gel with hexane to give **16** (758 mg, 71%) as a colorless oil.

16. ^1H NMR (600 MHz, $CDCl_3$): δ 0.09 (d, 6H, *J* = 4.5 Hz), 0.15 (s, 6H), 0.41 (s, 6H), 3.73 (sept, 1H, *J* = 4.5 Hz), 7.19 (dd, 1H, *J* = 4.5, 3.2 Hz), 7.21 (dd, 1H, *J* = 3.2, 0.8 Hz), 7.59 (dd, 1H, *J* = 4.5, 0.8 Hz). ^{13}C NMR (151 MHz, $CDCl_3$): δ −6.6, −5.9, −1.8, 128.3, 130.6, 134.2, 139.1. ^{29}Si NMR (119 MHz, $CDCl_3$): δ −47.2, −36.3, −20.5. IR (NaCl): 2950, 2090, 1400, 1250, 1030, 880, 830, 770 cm^{-1}. MS (EI, 70 eV): *m*/*z* 258 (M^+, 5), 243 (59), 199 (21), 173 (26), 170 (23), 141 (82), 116 (52), 73 (100). hRMS (EI): found 257.0669, calcd for $C_{10}H_{21}SSi_3$ (M^+ − h) 257.0672.

3.4. Synthesis of 1-(5′-Hexenyl)-1,1,2,2,3,3-hexamethyltrisilane (**6**)

Synthesis of **6** was carried out by the same procedure as **16** using THF (4 mL), chlorotrimethylsilane (178 mg, 1.64 mmol), a 1.64 M solution of butyllithium in hexane (1.14 mL, 1.87 mmol), magnesium turnings (95 mg, 3.9 mmol), a 1.0 M solution of DIBAL-H in toluene (0.024 mL, 0.024 mmol), 6-bromo-1-hexene (234 mg, 1.44 mmol) and **2** (178 mg, 0.844 mmol). Separation by column chromatography over silica gel with hexane gave **6** (84 mg, 38%) as a colorless oil.

6. ^1H NMR (600 MHz, $CDCl_3$): δ 0.06 (s, 6H), 0.12 (s, 6H), 0.15 (d, 6H, *J* = 4.5 Hz), 0.59–0.64 (m, 2H), 1.30–1.37 (m, 2H), 1.40–1.44 (m, 2H), 2.03–2.07 (m, 2H), 3.72 (sept, 1H, *J* = 4.5 Hz), 4.92–4.94 (m, 1H), 4.98–5.01 (m, 1H), 5.77–5.84 (m, 1H). ^{13}C NMR (151 MHz, $CDCl_3$): δ −6.4, −5.8, −3.4, 15.5, 24.3, 33.1, 33.7, 114.3, 139.2. ^{29}Si NMR (119 MHz, $CDCl_3$): δ −48.0, −36.1, −13.5. IR (NaCl): 3080, 2950, 2920, 2090, 1640, 1411, 1250, 910, 880, 830, 790 cm^{-1}. MS (EI, 70 eV): *m*/*z* 258 (M^+, 9), 215 (20), 141 (64), 127 (18), 117 (20), 116 (19), 73 (100), 59 (51). hRMS (EI): found 258.1649, calcd for $C_{12}H_{30}Si_3$ 258.1655.

3.5. Synthesis of 1,1,2,2,3,3-Hexamethyl-1-(2′-phenylethyl)trisilane (**7**)

Synthesis of **7** was carried out by the same procedure as **16** using THF (20 mL), chlorotrimethylsilane (1.01 g, 9.32 mmol), a 1.60 M solution of butyllithium in hexane (6.0 mL, 9.6 mmol), magnesium turnings (472 mg, 19.4 mmol), a 1.0 M solution of DIBAL-H in toluene (0.12 mL, 0.12 mmol), (2-bromoethyl)benzene (1.30 g, 7.04 mmol) and **2** (1.01 g, 4.79 mmol). Separation by column chromatography over silica gel with hexane containing 1% triethylamine gave **7** (939 mg, 70%) as a colorless oil.

7. ^1H NMR (600 MHz, $CDCl_3$): δ 0.16 (s, 6H), 0.19 (s, 6H), 0.21 (d, 6H, *J* = 4.5 Hz), 1.01–1.05 (m, 2H), 2.65–2.70 (m, 2H), 3.81 (sept, 1H, *J* = 4.5 Hz), 7.17–7.26 (m, 3H), 7.29–7.33 (m, 2H). ^{13}C NMR (151 MHz, $CDCl_3$): δ −6.4, −5.8, −3.5, 17.9, 30.9, 125.7, 127.9, 128.5, 145.5. ^{29}Si NMR (119 MHz, $CDCl_3$): δ −47.9, −36.2, −13.3. IR (NaCl): 3030, 2950, 2890, 2090, 1600, 1490, 1450, 1410, 1250, 880, 830, 790 cm^{-1}. MS (EI, 70 eV): *m*/*z* 265 ($M^+ - CH_3$, 11), 221 (76), 163 (71), 135 (100), 117 (84), 116 (38), 73 (78), 59 (84). hRMS (EI): found 265.1263, calcd for $C_{13}H_{25}Si_3$ ($M^+ - CH_3$) 265.1264.

3.6. *Synthesis of 1,1,2,2,3,3-Hexamethyl-1-(E)-styryltrisilane* (**8**)

Synthesis of **8** was carried out by the same procedure as **16** using THF (8 mL), chlorotrimethylsilane (407 mg, 3.75 mmol), a 1.64 M solution of butyllithium in hexane (2.3 mL, 3.8 mmol), magnesium turnings (188 mg, 7.73 mmol), a 1.0 M solution of DIBAL-H in toluene (0.064 mL, 0.064 mmol), β-bromostyrene (527 mg, 2.88 mmol) and **2** (430 mg, 2.04 mmol). Separation by column chromatography over silica gel with hexane gave **8** (332 mg, 58%) as a colorless oil.

8. ^{1}H NMR (600 MHz, $CDCl_3$): δ 0.09 (s, 6H), 0.10 (d, 6H, *J* = 4.5 Hz), 0.20 (s, 6H), 3.70 (sept, 1H, *J* = 4.5 Hz), 6.45 (d, 1H, *J* = 19 Hz), 6.78 (d, 1H, *J* = 19 Hz), 7.16–7.19 (m, 1H), 7.24–7.28 (m, 2H), 7.33–7.38 (m, 2H). ^{13}C NMR (151 MHz, $CDCl_3$): δ −6.5, −5.8, −3.3, 126.4, 128.0, 128.7 (2 peaks are overlapped.), 138.6, 143.8. ^{29}Si NMR (119 MHz, $CDCl_3$): δ −47.2, −36.1, −20.2. IR (NaCl): 3020, 2950, 2890, 2090, 1710, 1600, 1570, 1490, 1450, 1400, 1250, 990, 880, 840, 790, 730 cm^{-1}. MS (EI, 70 eV): *m/z* 278 (M^+, 4), 219 (18), 204 (22), 161 (26), 145 (79), 135 (32), 117 (25), 116 (91), 102 (17), 73 (100), 59 (68). hRMS (EI): found 278.1334, calcd for $C_{14}H_{26}Si_3$ 278.1342.

3.7. *Synthesis of 1-(2-Buten-2-yl)-1,1,2,2,3,3-hexamethyltrisilane* (**9**)

Synthesis of **9** was carried out by the same procedure as **16** using THF (8 mL), chlorotrimethylsilane (407 mg, 3.75 mmol), a 1.64 M solution of butyllithium in hexane (2.3 mL, 3.8 mmol), magnesium turnings (188 mg, 7.73 mmol), a 1.0 M solution of DIBAL-H in toluene (0.048 mL, 0.048 mmol), 2-bromo-2-butene (388 mg, 2.87 mmol) and **2** (420 mg, 1.99 mmol). Separation by column chromatography over silica gel with hexane gave **9** (248 mg, 54%, *E*/*Z* = 32/68) as a colorless oil.

(*Z*)-**9**. ^{1}H NMR (600 MHz, $CDCl_3$): δ 0.15 (s, 6H), 0.15 (d, 6H, *J* = 4.8 Hz), 0.25 (s, 6H), 1.67–1.71 (m, 3H), 1.74–1.76 (m, 3H), 3.74 (sept, 1H, *J* = 4.2 Hz), 6.07–6.11 (m, 1H). ^{13}C NMR (151 MHz, $CDCl_3$): δ −5.9, −5.8, −1.9, 18.4, 25.3, 134.8, 136.4. ^{29}Si NMR (119 MHz, $CDCl_3$): δ −46.6, −35.7, −22.3.

(*E*)-**9**. ^{1}H NMR (600 MHz, $CDCl_3$): δ 0.13 (d, 6H, *J* = 4.2 Hz), 0.15 (s, 6H), 0.25 (s, 6H), 1.68–1.71 (m, 3H), 1.74–1.75 (m, 3H), 3.71 (sept, 1H, *J* = 4.8 Hz), 5.77–5.80 (m, 1H). ^{13}C NMR (151 MHz, $CDCl_3$): δ −6.1, −5.9, −3.7, 14.4, 15.4, 133.9, 136.2. ^{29}Si NMR (119 MHz, $CDCl_3$): δ −47.8, −35.7, −17.3.

Mixture of (*Z*)-**9** and (*E*)-**9**. IR (NaCl): 2950, 2900, 2090, 1250, 880, 830, 790 cm^{-1}. MS (EI, 70 eV): *m/z* 230 (M^+, 7), 156 (14), 141 (16), 131 (19), 117 (35), 116 (82), 97 (19), 73 (100), 59 (41). hRMS (EI): found 230.1335, calcd for $C_{10}H_{26}Si_3$ 230.1342.

3.8. *Synthesis of 1-(1′-Hexynyl)-1,1,2,2,3,3-hexamethyltrisilane* (**10**)

A 0.90 M solution of isopropylmagnesium chloride in THF (1.11 mL, 1.0 mmol) was added to a solution of 1-hexyne (83 mg, 1.0 mmol) in THF (5 mL), and the mixture was stirred at 40 °C for 1 h. After cooling to room temperature, **2** (316 mg, 1.50 mmol) was added, and the mixture was stirred at room temperature for 1 h. The reaction was quenched with 1.0 M hydrochloric acid. The reaction mixture was extracted with hexane. The organic layer was washed with water, saturated aqueous sodium hydrogencarbonate and brine, and dried over anhydrous sodium sulfate. After evaporation of the solvents, the residue was separated by column chromatography over silica gel with hexane to give **10** (123 mg, 48%) as a colorless oil.

10. ^{1}H NMR (600 MHz, $CDCl_3$): δ 0.16 (s, 6H), 0.18 (d, 6H, *J* = 4.5 Hz), 0.20 (s, 6H), 0.90 (t, 3H, *J* = 7.2 Hz), 1.37–1.44 (m, 2H), 1.46–1.51 (m, 2H), 2.24 (t, 2H, *J* = 6.9 Hz), 3.75 (sept, 1H, *J* = 4.5 Hz). ^{13}C NMR (151 MHz, $CDCl_3$): δ −6.7, −5.8, −1.8, 13.7, 19.9, 22.0, 31.0, 82.9, 110.3. ^{29}Si NMR (119 MHz, $CDCl_3$): δ −46.8, −36.2, −34.9. IR (NaCl): 2960, 2170, 2090, 1710, 1260, 800 cm^{-1}. MS (EI, 70 eV): *m/z* 256 (M^+, 3), 241 (14), 197 (16), 141 (38), 139 (22), 117 (37), 116 (100), 101 (14), 83 (22), 73 (99), 59 (36). hRMS (EI): found 256.1491, calcd for $C_{12}H_{28}Si_3$ 256.1499.

3.9. *Synthesis of 1,1,2,2,3,3-Hexamethyl-1-(phenylethynyl)trisilane* (**11**)

Synthesis of **11** was carried out by the same procedure as **10** using THF (25 mL), phenylacetylene (811 mg, 7.94 mmol), a 0.81 M solution of isopropylmagnesium chloride in THF (9.5 mL, 7.7 mmol) and **2** (1.55 g, 7.36 mmol). Separation by column chromatography over silica gel with hexane containing 1% triethylamine gave **11** (1.41 g, 69%) as a colorless oil.

11. ^{1}H NMR (600 MHz, $CDCl_3$): δ 0.22 (d, 6H, *J* = 4.5 Hz), 0.23 (s, 6H), 0.31 (s, 6H), 3.81 (sept, 1H, *J* = 4.5 Hz), 7.27–7.30 (m, 3H), 7.44–7.46 (m, 2H). ^{13}C NMR (151 MHz, $CDCl_3$): δ −6.6, −5.8, −2.1, 93.3, 107.8, 123.6, 128.3, 128.5, 132.0. ^{29}Si NMR (119 MHz, $CDCl_3$): δ −46.3, −36.2, −33.7. IR (NaCl): 2950, 2360, 2150, 1490, 1250 cm^{-1}. MS (EI, 70 eV): *m*/*z* 276 (M^+, 14), 261 (15), 217 (63), 203 (27), 159 (64), 135 (26), 116 (71), 73 (100). hRMS (EI): found 276.1178, calcd for $C_{14}H_{24}Si_3$ 276.1186.

3.10. *Synthesis of 1,1,2,2,3,3-Hexamethyl-1-(trimethylsilylethynyl)trisilane* (**12**)

Synthesis of **12** was carried out by the same procedure as **10** using THF (5 mL), trimethylsilylacetylene (98 mg, 1.0 mmol), a 0.90 M solution of isopropylmagnesium chloride in THF (1.11 mL, 1.0 mmol) and **2** (316 mg, 1.50 mmol). Separation by column chromatography over silica gel with hexane gave **12** (149 mg, 55%) as a colorless oil.

12. ^{1}H NMR (600 MHz, $CDCl_3$): δ 0.16 (s, 9H), 0.17 (s, 6H), 0.19 (d, 6H, *J* = 4.5 Hz), 0.22 (s, 6H), 3.77 (sept, 1H, *J* = 4.5 Hz). ^{13}C NMR (151 MHz, $CDCl_3$): δ −6.7, −5.8, −2.2, 0.1, 113.0, 116.9. ^{29}Si NMR (119 MHz, $CDCl_3$): δ −46.6, −36.1, −34.7, −18.9. IR (NaCl): 2960, 2090, 1260, 840, 800 cm^{-1}. MS (EI, 70 eV): *m*/*z* 272 (M^+, 4), 257 (12), 213 (41), 155 (18), 117 (16), 116 (100), 73 (80). hRMS (EI): found 272.1267, calcd for $C_{11}H_{28}Si_4$ 272.1268.

3.11. *Synthesis of 1,1,2,2,3,3-Hexamethyl-1-phenyltrisilane* (**13**)

Synthesis of **13** was carried out by the same procedure as **16** using THF (10 mL), chlorotrimethylsilane (0.7 mL, 6 mmol), a 1.60 M solution of butyllithium in hexane (3.4 mL, 5.4 mmol), magnesium turnings (290 mg, 11.9 mmol), a 1.0 M solution of DIBAL-H in toluene (0.073 mL, 0.073 mmol), bromobenzene (0.44 mL, 4.2 mmol) and **2** (609 mg, 2.89 mmol). Separation by column chromatography over silica gel with hexane containing 1% triethylamine gave **13** (404 mg, 55%) as a colorless oil. The ^{1}H NMR spectrum is identical to the reported data [48].

13. ^{1}H NMR (600 MHz, $CDCl_3$): δ 0.08 (d, 6H, *J* = 4.5 Hz), 0.13 (s, 6H), 0.40 (s, 6H), 3.73 (sept, 1H, *J* = 4.5 Hz), 7.33–7.36 (m, 3H), 7.46–7.48 (m, 2H). ^{13}C NMR (151 MHz, $CDCl_3$): δ −6.5, −5.9, −3.2, 127.9, 128.5, 133.9, 139.8. ^{29}Si NMR (119 MHz, $CDCl_3$): δ −47.4, −36.1, −18.0.

3.12. *Synthesis of 1-(4′-Methoxyphenyl)-1,1,2,2,3,3-hexamethyltrisilane* (**14**)

Synthesis of **14** was carried out by the same procedure as **16** using THF (5 mL), chlorotrimethylsilane (301 mg, 2.77 mmol), a 1.64 M solution of butyllithium in hexane (1.7 mL, 2.8 mmol), magnesium turnings (127 mg, 5.22 mmol), a 1.0 M solution of DIBAL-H in toluene (0.06 mL, 0.06 mmol), 4-bromoanisole (380 mg, 2.03 mmol) and **2** (301 mg, 1.43 mmol). Separation by column chromatography over silica gel with hexane gave **14** (270 mg, 67%) as a colorless oil.

14. ^{1}H NMR (600 MHz, $CDCl_3$): δ 0.08 (d, 6H, *J* = 4.5 Hz), 0.11 (s, 6H), 0.35 (s, 6H), 3.71 (sept, 1H, *J* = 4.5 Hz), 3.82 (s, 3H), 6.90 (d, 2H, *J* = 8.7 Hz), 7.37 (d, 2H, *J* = 8.7 Hz). ^{13}C NMR (151 MHz, $CDCl_3$): δ −6.4, −5.9, −2.9, 55.1, 113.7, 130.3, 135.2, 160.2. ^{29}Si NMR (119 MHz, $CDCl_3$): δ −47.5, −36.1, −18.4. IR (NaCl): 2950, 2090, 1590, 1500, 1280, 1250, 1180, 1110, 880, 830, 780 cm^{-1}. MS (EI, 70 eV): *m*/*z* 282 (M^+, 8), 281 (20), 193 (36), 165 (100), 135 (16), 116 (54), 73 (29). hRMS (EI): found 281.1213, calcd for $C_{13}H_{25}OSi_3$ (M^+ − h) 281.1213.

3.13. Synthesis of 1-(4′-Dimethylaminophenyl)-1,1,2,2,3,3-hexamethyltrisilane (**15**)

Synthesis of **15** was carried out by the same procedure as **16** using THF (5 mL), chlorotrimethylsilane (290 mg, 2.67 mmol), a 1.64 M solution of butyllithium in hexane (1.6 mL, 2.6 mmol), magnesium turnings (122 mg, 5.02 mmol), a 1.0 M solution of DIBAL-H in toluene (0.06 mL, 0.06 mmol), 4-bromo-*N*,*N*-dimethylaniline (400 mg, 2.00 mmol) and **2** (277 mg, 1.31 mmol). The crude product was separated by column chromatography over silica gel with hexane–ethyl acetate (5:1). Compound **15** (150 mg, 39%) was obtained as a colorless oil.

15. ^{1}H NMR (600 MHz, $CDCl_3$): δ 0.09 (d, 6H, *J* = 4.5 Hz), 0.11 (s, 6H), 0.33 (s, 6H), 2.95 (s, 6H), 3.71 (sept, 1H, *J* = 4.5 Hz), 6.73 (d, 2H, *J* = 8.4 Hz), 7.32 (d, 2H, *J* = 8.4 Hz). ^{13}C NMR (151 MHz, $CDCl_3$): δ –6.4, –5.8, –2.8, 40.4, 112.3, 124.5, 135.0, 150.8. ^{29}Si NMR (119 MHz, $CDCl_3$): δ −47.5, −36.0, −19.0. IR (NaCl): 2950, 2890, 2080, 1600, 1510, 1350, 1240, 1110, 880, 830, 790 cm^{-1}. MS (EI, 70 eV): *m*/*z* 295 (M^+, 15), 178 (100), 162 (9), 134 (12), 116 (9), 102 (12), 73 (9). hRMS (EI): found 294.1528, calcd for $C_{14}H_{28}NSi_3$ (M^+ − h) 294.1529.

3.14. Synthesis of 1,1,2,2,3,3,4,4-Octamethyl-1-(2′-thienyl)tetrasilane (**17**)

Synthesis of **17** was carried out by the same procedure as **16** using THF (5 mL), chlorotrimethylsilane (282 mg, 2.60 mmol), a 1.64 M solution of butyllithium in hexane (1.6 mL, 2.6 mmol), magnesium turnings (127 mg, 5.22 mmol), a 1.0 M solution of DIBAL-H in toluene (0.08 mL, 0.08 mmol), 2-bromothiophene (324 mg, 1.99 mmol) and **3** (381 mg, 1.42 mmol). Separation by column chromatography over silica gel with hexane gave **17** (191 mg, 43%) as a colorless oil.

17. ^{1}H NMR (600 MHz, $CDCl_3$): δ 0.09 (s, 6H), 0.12 (d, 6H, *J* = 4.5 Hz), 0.15 (s, 6H), 0.42 (s, 6H), 3.72 (sept, 1H, *J* = 4.5 Hz), 7.19 (dd, 1H, *J* = 4.5, 3.3 Hz), 7.20 (dd, 1H, *J* = 3.0, 0.6 Hz), 7.59 (dd, 1H, *J* = 4.2, 1.2 Hz). ^{13}C NMR (151 MHz, $CDCl_3$): δ −5.9, −5.81, −5.79, −1.4, 128.3, 130.5, 134.1, 139.3. ^{29}Si NMR (119 MHz, $CDCl_3$): δ −44.1, −43.9, −35.9, −19.9. IR (NaCl): 2950, 2890, 2090, 1400, 1250, 1210, 990, 880, 830, 780, 700 cm^{-1}. MS (EI, 70 eV): *m*/*z* 316 (M^+, 1), 257 (100), 173 (17), 167 (34), 159 (19), 141 (39), 116 (18), 73 (93), 59 (15). hRMS (EI): found 301.0753, calcd for $C_{11}H_{25}SSi_4$ (M^+ − CH_3) 301.0754.

3.15. Synthesis of 1-Methoxy-1,1,2,2,3,3-hexamethyl-3-(E)-styryltrisilane (**18**)

[$RuCl_2$(*p*-cymene)]$_2$ (15 mg, 0.024 mmol) was added to a solution of **8** (271 mg, 0.973 mmol) and methanol (65 μL, 1.6 mmol) in toluene (4 mL) at 0 °C. The mixture was stirred with gradual warming to room temperature overnight. After evaporation of the solvent, the residue was distilled with a Kugelrohr distillation apparatus (146–152 °C/0.90–1.4 mmHg) to give **18** (216 mg, 72%) as a colorless oil.

18. ^{1}H NMR (600 MHz, $CDCl_3$): δ 0.19 (s, 6H), 0.25 (s, 6H), 0.28 (s, 6H), 3.43 (s, 3H), 6.54 (d, 1H, *J* = 18.9 Hz), 6.86 (d, 1H, *J* = 18.9 Hz), 7.23–7.25 (m, 1H), 7.31–7.35 (m, 2H), 7.42–7.45 (m, 2H). ^{13}C NMR (151 MHz, $CDCl_3$): δ −6.4, −3.4, −0.1, 51.4, 126.4, 127.9, 128.58, 128.64, 138.6, 143.8. ^{29}Si NMR (119 MHz, $CDCl_3$): δ −50.2, −20.7, 21.3. IR (NaCl): 2950, 1590, 1570, 1490, 1450, 1400, 1250, 1080, 990, 830, 780, 730, 690 cm^{-1}. MS (EI, 70 eV): *m*/*z* 308 (M^+, 10), 293 (44), 219 (48), 173 (52), 145 (72), 135 (65), 133 (37), 117 (41), 116 (57), 89 (39), 73 (100), 59 (71). hRMS (FD): found 308.1449, calcd for $C_{15}H_{28}OSi_3$ 308.1448.

3.16. Synthesis of 1-(1′-Hexynyl)-3-methoxy-1,1,2,2,3,3-hexamethyltrisilane (**19**)

Synthesis of **19** was carried out by the same procedure as **18** using toluene (4 mL), **10** (264 mg, 1.03 mmol), methanol (65 μL, 1.6 mmol) and [$RuCl_2$(*p*-cymene)]$_2$ (15 mg, 0.024 mmol). The crude product was distilled with a Kugelrohr distillation apparatus (112–116 °C/1.2–1.3 mmHg) to give **19** (191 mg, 65%) as a colorless oil.

19. ^{1}H NMR (600 MHz, $CDCl_3$): δ 0.17 (s, 6H), 0.20 (s, 6H), 0.25 (s, 6H), 0.89 (t, 3H, *J* = 7.2 Hz), 1.35–1.42 (m, 2H), 1.45–1.50 (m, 2H), 2.22 (t, 2H, *J* = 7.2 Hz), 3.43 (s, 3H). ^{13}C NMR (151 MHz, $CDCl_3$): δ −6.6,

−1.9, −0.1, 13.7, 19.9, 22.0, 30.9, 51.4, 82.9, 110.2. ^{29}Si NMR (119 MHz, $CDCl_3$): δ −49.9, −35.1, 21.2. IR (NaCl): 2960, 2170, 1250, 1090, 1040, 830, 780 cm^{-1}. MS (EI, 70 eV): m/z 286 (M^+, 1), 257 (19), 229 (16), 133 (20), 117 (35), 116 (54), 89 (41), 73 (100), 59 (61). hRMS (FD): found 286.1612, calcd for $C_{13}H_{30}OSi_3$ 286.1604.

3.17. *Synthesis of 1-Methoxy-1,1,2,2,3,3-hexamethyl-3-(phenylethynyl)trisilane* (**20**)

Synthesis of **20** was carried out by the same procedure as **18** using toluene (4 mL), **11** (280 mg, 1.01 mmol), methanol (65 μL, 1.6 mmol) and $[RuCl_2(p\text{-cymene})]_2$ (15 mg, 0.024 mmol). The crude product was distilled with a Kugelrohr distillation apparatus (134–135 °C/0.90–0.93 mmHg) to give **20** (147 mg, 47%) as a colorless oil.

20. ^{1}H NMR (600 MHz, $CDCl_3$): δ 0.25 (s, 6H), 0.30 (s, 6H), 0.33 (s, 6H), 3.46 (s, 3H), 7.29–7.30 (m, 3H), 7.43–7.45 (m, 2H). ^{13}C NMR (151 MHz, $CDCl_3$): δ −6.5, −2.1, 0.0, 51.5, 93.3, 107.8, 123.5, 128.3, 128.5, 132.0. ^{29}Si NMR (119 MHz, $CDCl_3$): δ −49.5, −33.9, 21.0. IR (NaCl): 2970, 2940, 2900, 2150, 1490, 1250, 1080, 840, 780, 690 cm^{-1}. MS (EI, 70 eV): m/z 306 (M^+, 7), 305 (23), 193 (41), 159 (37), 116 (34), 89 (30), 73 (100), 59 (64). hRMS (FD): found 306.1293, calcd for $C_{15}H_{26}OSi_3$ 306.1291.

3.18. *Synthesis of 1-Methoxy-1,1,2,2,3,3-hexamethyl-3-phenyltrisilane* (**21**)

Synthesis of **21** was carried out by the same procedure as **18** using toluene (4 mL), **13** (258 mg, 1.02 mmol), methanol (60 μL, 1.5 mmol) and $[RuCl_2(p\text{-cymene})]_2$ (15 mg, 0.024 mmol). The crude product was distilled with a Kugelrohr distillation apparatus (153–157 °C/11 mmHg) to give **21** (251 mg, 87%) as a colorless oil. The NMR data are identical to the reported data [48].

21. ^{1}H NMR (600 MHz, $CDCl_3$): δ 0.156 (s, 6H), 0.160 (s, 6H), 0.42 (s, 6H), 3.35 (s, 3H), 7.32–7.37 (m, 3H), 7.46–7.49 (m, 2H). ^{13}C NMR (151 MHz, $CDCl_3$): δ −6.3, −3.2, −0.2, 51.3, 127.9, 128.5, 133.9, 139.6. ^{29}Si NMR (119 MHz, $CDCl_3$): δ −50.2, −18.4, 21.3. IR (NaCl): 2950, 2890, 1250, 1080, 830, 780, 730, 700 cm^{-1}. MS (EI, 70 eV): m/z 282 (M^+, 2), 267 (38), 193 (14), 135 (54), 116 (100), 89 (15), 73 (23), 59 (17). hRMS (FD): found 282.1300, calcd for $C_{13}H_{26}OSi_3$ 282.1291.

3.19. *Synthesis of 1,2-Dichloro-1,1,2,2-tetraethyldisilane*

Freshly distilled acetyl chloride (14.0 mL, 197 mmol) was added dropwise to the mixture of aluminum chloride (26.04 g, 195 mmol) and hexaethyldisilane (21.8 mL, 79.5 mmol). The flask was immersed in a water bath to prevent raising the reaction temperature too much. After stirring for 3 h, the reaction mixture was distilled under reduced pressure (bp 100–131 °C/9 mmHg) to give 1,2-dichloro-1,1,2,2-tetraethyldisilane (16.47 g, 85%) as a colorless oil. The ^{1}H NMR spectrum is identical to the reported data [14].

3.20. *Synthesis of 1,1,2,2-Tetraethyl-1-phenyldisilane* (**22**)

Phenylmagnesium bromide was prepared from magnesium turnings (2.17 g, 89.4 mmol), bromobenzene (11.72 g, 74.6 mmol) and diethyl ether (42 mL). The Grignard reagent was added dropwise to a solution of 1,2-dichloro-1,1,2,2-tetraethyldisilane (16.47 g, 67.7 mmol) in diethyl ether (30 mL) at 0 °C. The mixture was stirred at room temperature for 12 h. After filtration of the reaction mixture, the filtrate was evaporated to give a crude product of 1-chloro-1,1,2,2-tetraethyl-2-phenyldisilane (19.29 g).

A solution of 1-chloro-1,1,2,2-tetraethyl-2-phenyldisilane in diethyl ether (60 mL) was added dropwise to a mixture of lithium aluminum hydride (1.21 g, 32.0 mmol) and diethyl ether (100 mL) at 0 °C. The mixture was stirred with gradual warming to room temperature overnight. The reaction was quenched with 1.0 M hydrochloric acid. The reaction mixture was extracted with hexane. The organic layer was washed with brine and dried over anhydrous sodium sulfate. After evaporation of the solvents, the residue was distilled under reduced pressure according to a normal distillation procedure (bp 129–131 °C/5 mmHg) to give **22** (12.30 g, 73% in two steps) as a colorless oil.

22. ^{1}H NMR (600 MHz, $CDCl_3$): δ 0.69–0.76 (m, 4H), 0.95–1.02 (m, 10H), 1.03–1.06 (m, 6H), 3.67 (quin, 1H, J = 3.8 Hz), 7.33–7.36 (m, 3H), 7.48–7.50 (m, 2H). ^{13}C NMR (151 MHz, $CDCl_3$): δ 2.2, 4.4, 8.4, 10.2, 127.9, 128.6, 134.6, 137.6. ^{29}Si NMR (119 MHz, $CDCl_3$): δ −26.4, −13.1. IR (NaCl): 3070, 2950, 2910, 2870, 2080, 1460, 1430, 1230, 1100, 1010, 970, 800, 770, 700 cm^{-1}. MS (EI, 70 eV): m/z 250 (M^+, 13), 163 (61), 135 (100), 107 (55). hRMS (FD): found 250.1574, calcd for $C_{14}H_{26}Si_2$ 250.1573.

3.21. *Synthesis of 1,1,2,2-Tetraethyl-1-methoxy-2-phenyldisilane* (**23**)

$[RuCl_2(p\text{-cymene})]_2$ (16 mg, 0.026 mmol) was added to a solution of **22** (260 mg, 1.04 mmol) and methanol (338 mg, 10.5 mmol) in toluene (4 mL) at 0 °C. The mixture was stirred with gradual warming to room temperature for 1 day. After evaporation of the solvent, the residue was distilled with a Kugelrohr distillation apparatus (81–117 °C/0.5 mmHg) to give **23** (241 mg, 83%) as a colorless oil.

23. ^{1}H NMR (600 MHz, $CDCl_3$): δ 0.70–0.81 (m, 4H), 0.95–1.02 (m, 10H), 1.06–1.10 (m, 6H), 3.41 (s, 3H), 7.31–7.36 (m, 3H), 7.51–7.53 (m, 2H). ^{13}C NMR (151 MHz, $CDCl_3$): δ 4.4, 7.0, 7.3, 8.3, 51.7, 127.9, 128.5, 134.8, 137.7. ^{29}Si NMR (119 MHz, $CDCl_3$): δ −17.3, 19.2. IR (NaCl): 3070, 2960, 2880, 1460, 1430, 1090, 1010, 700 cm^{-1}. MS (EI, 70 eV): m/z 280 (M^+, 40), 279 (100), 265 (21), 251 (20), 237 (40), 223 (27), 209 (28), 195 (23), 135 (39), 117 (25), 107 (61), 89 (35), 59 (21). hRMS (FD): found 280.1670, calcd for $C_{15}H_{28}OSi_2$ 280.1678.

3.22. *Synthesis of 1-Ethoxy-1,1,2,2-tetraethyl-2-phenyldisilane* (**24**)

Synthesis of **24** was carried out by the almost same procedure as **23** using toluene (4 mL), **22** (253 mg, 1.01 mmol), ethanol (476 mg, 10.3 mmol) and $[RuCl_2(p\text{-cymene})]_2$ (15 mg, 0.024 mmol). The mixture was stirred at 50 °C for 1 day. The solvent was evaporated, and the residue was distilled with a Kugelrohr distillation apparatus (95–110 °C/0.5 mmHg) to give **24** (135 mg, 45%) as a colorless oil.

24. ^{1}H NMR (600 MHz, $CDCl_3$): δ 0.69–0.80 (m, 4H), 0.94–1.01 (m, 10H), 1.05–1.08 (m, 6H), 1.14 (t, 3H, J = 7.0 Hz), 3.61 (q, 2H, J = 7.0 Hz), 7.32–7.35 (m, 3H), 7.49–7.53 (m, 2H). ^{13}C NMR (151 MHz, $CDCl_3$): δ 4.4, 7.1, 7.6, 8.3, 18.8, 59.5, 127.9, 128.5, 134.8, 137.9. ^{29}Si NMR (119 MHz, $CDCl_3$): δ −17.4, 16.5. IR (NaCl): 2960, 2880, 1460, 1430, 1100, 1080, 1000, 700 cm^{-1}. MS (EI, 70 eV): m/z 294 (M^+, 2), 293 (4), 265 (21), 237 (37), 209 (33), 135 (36), 131 (47), 107 (100), 105 (39), 103 (63), 87 (19), 75 (35), 59 (23). hRMS (FD): found 294.1849, calcd for $C_{16}H_{30}OSi_2$ 294.1835.

3.23. *Synthesis of 1,1,2,2-Tetraethyl-1-isopropoxy-2-phenyldisilane* (**25**)

$[RuCl_2(p\text{-cymene})]_2$ (17 mg, 0.028 mmol) was added to a solution of **22** (265 mg, 1.06 mmol) and 2-propanol (675 mg, 11.2 mmol) in toluene (4 mL) at 0 °C. The mixture was stirred with gradual warming to room temperature for 12 h. Additional 2-propanol (5.84 g, 97.1 mmol) was added, and the mixture was heated at 50 °C for 1 day. After cooling to room temperature, the solvent and excess 2-propanol were evaporated, and the residue was distilled with a Kugelrohr distillation apparatus (91–111 °C/0.5 mmHg) to give **25** (188 mg, 58%) as a colorless oil.

25. ^{1}H NMR (600 MHz, $CDCl_3$): δ 0.71–0.75 (m, 4H), 0.93–1.00 (m, 10H), 1.04–1.07 (m, 6H), 1.09 (d, 6H, J = 6.0 Hz), 3.90 (sept, 1H, J = 6.0 Hz), 7.31–7.34 (m, 3H), 7.51–7.53 (m, 2H). ^{13}C NMR (151 MHz, $CDCl_3$): δ 4.4, 7.3, 8.2, 8.4, 26.1, 66.0, 127.8, 128.4, 134.8, 138.0. ^{29}Si NMR (119 MHz, $CDCl_3$): δ −17.5, 14.1. IR (NaCl): 2960, 2880, 1460, 1430, 1380, 1120, 1020, 700 cm^{-1}. MS (EI, 70 eV): m/z 308 (M^+, 0.2), 265 (73), 237 (84), 221 (17), 209 (54), 193 (25), 181 (14), 165 (15), 145 (41), 135 (41), 107 (100), 105 (35), 103 (79), 75 (75). hRMS (FD): found 308.1997, calcd for $C_{17}H_{32}OSi_2$ 308.1991.

3.24. *Synthesis of 2-Methoxy-1,1,2,3,3-pentamethyl-1,3-diphenyltrisilane* (**27**)

$[RuCl_2(benzene)]_2$ (13 mg, 0.026 mmol) was added to a solution of **26** (317 mg, 1.01 mmol) and methanol (328 mg, 10.2 mmol) in toluene (4 mL) at 0 °C. The mixture was stirred with gradual warming to room temperature for 1 day and at 50 °C for 5 days. After cooling to room temperature, the solvent was evaporated, and the residue was distilled with a Kugelrohr distillation apparatus (128–149 °C/0.5 mmHg) to give **27** (100 mg, 29%) as a colorless oil.

27. ^{1}H NMR (600 MHz, $CDCl_3$): δ 0.32 (s, 6H), 0.34 (s, 6H), 0.37 (s, 3H), 3.32 (s, 3H), 7.32–7.33 (m, 6H), 7.42–7.43 (m, 4H). ^{13}C NMR (151 MHz, $CDCl_3$): δ −3.8, −2.90, −2.87, 53.2, 127.9, 128.7, 134.1, 139.1. ^{29}Si NMR (119 MHz, $CDCl_3$): δ −21.8, 12.6. IR (NaCl): 3070, 2950, 1430, 1250, 1070, 770, 730, 700 cm^{-1}. MS (EI, 70 eV): m/z 344 (M^+, 5), 329 (25), 251 (14), 209 (17), 197 (10), 193 (13), 179 (33), 178 (54), 163 (24), 135 (100), 122 (11), 117 (11), 107 (11), 105 (12), 59 (17). hRMS (FD): found 344.1460, calcd for $C_{18}H_{28}OSi_3$ 344.1447.

4. Conclusions

We found that α-chloro-ω-hydrooligosilanes, synthesized by the titanium-catalyzed monoreduction of α,ω-dichlorooligosilanes, are good precursors for the synthesis of unsymmetrically substituted oligosilanes. Each functional group, the chlorosilane and hydrosilane moieties, is independently substituted by the reactions with organolithium or Grignard reagents and the ruthenium-catalyzed alkoxylations. In both steps, little or no cleavage of the Si–Si bond occurred. Further studies on these synthetic reactions are now in progress.

Author Contributions: Conceptualization, K.-i.K.; Experiments, Y.A., Y.N., M.I. and K.K.; Project Administration, S.K.; Preparation of the manuscript, K.-i.K. and S.K.

Acknowledgments: Some of the spectral data of the new compounds were measured in the Instrumental Analysis Division, Equipment Management Center, Creative Research Institution, hokkaido University, Japan.

References

1. Miller, R.D.; Michl, J. Polysilane high polymers. *Chem. Rev.* **1989**, *89*, 1359–1410. [CrossRef]
2. Beckmann, J. Oligosilanes. In *Comprehensive Organometallic Chemistry III*; Crabtree, R.H., Mingos, D.M.P., Eds.; Elsevier: Oxford, UK, 2007; Volume 3, pp. 409–512.
3. Marschner, C. Oligosilanes. In *Functional Molecular Silicon Compounds I: Regular Oxidation States*; Scheschkewitz, D., Ed.; Springer: Cham, Switzerland, 2014; pp. 163–228.
4. Mignani, G.; Krämer, A.; Puccetti, G.; Ledoux, I.; Soula, G.; Zyss, J.; Meyrueix, R. A new class of silicon compounds with interesting nonlinear optical effects. *Organometallics* **1990**, *9*, 2640–2643. [CrossRef]
5. Mignani, G.; Krämer, A.; Puccetti, G.; Ledoux, I.; Zyss, J.; Soula, G. Effect of a weak donor on the intramolecular charge transfer of molecules containing two neighboring silicon atoms. *Organometallics* **1991**, *10*, 3656–3659. [CrossRef]
6. Mignani, G.; Barzoukas, M.; Zyss, J.; Soula, G.; Balegroune, F.; Grandjean, D.; Josse, D. Improved transparency–efficiency trade-off in a new class of nonlinear organosilicon compounds. *Organometallics* **1991**, *10*, 3660–3668. [CrossRef]
7. Tsuji, H.; Sasaki, M.; Shibano, Y.; Toganoh, M.; Kataoka, T.; Araki, Y.; Tamao, K.; Ito, O. Photoinduced electron transfer of dialkynyldisilane-linked zinc porphyrin–[60]fullerene dyad. *Bull. Chem. Soc. Jpn.* **2006**, *79*, 1338–1346. [CrossRef]
8. Shibano, Y.; Sasaki, M.; Tsuji, H.; Araki, Y.; Ito, O.; Tamao, K. Conformation effect of oligosilane linker on photoinduced electron transfer of tetrasilane-linked zinc porphyrin–[60]fullerene dyads. *J. Organomet. Chem.* **2007**, *692*, 356–367. [CrossRef]
9. Sasaki, M.; Shibano, Y.; Tsuji, H.; Araki, Y.; Tamao, K.; Ito, O. Oligosilane chain-length dependence of electron transfer of zinc porphyrin–oligosilane–fullerene molecules. *J. Phys. Chem. A* **2007**, *111*, 2973–2979. [CrossRef] [PubMed]
10. Hiratsuka, H.; horiuchi, H.; Takanoha, Y.; Matsumoto, H.; Yoshihara, T.; Okutsu, T.; Negishi, K.; Kyushin, S.; Matsumoto, H. Excited-state property of 1-(4-cyanophenyl)-2-(4-methoxyphenyl)-1,1,2,2-tetramethyl-disilane. *Chem. Lett.* **2007**, *36*, 1168–1169. [CrossRef]

11. Iwamoto, T.; Tsushima, D.; Kwon, E.; Ishida, S.; Isobe, H. Persilastaffanes: Design, synthesis, structure, and conjugation between silicon cages. *Angew. Chem., Int. Ed.* **2012**, *51*, 2340–2344. [CrossRef] [PubMed]
12. Surampudi, S.; Yeh, M.-L.; Siegler, M.A.; hardigree, J.F.M.; Kasl, T.A.; Katz, H.E.; Klausen, R.S. Increased carrier mobility in end-functionalized oligosilanes. *Chem. Sci.* **2015**, *6*, 1905–1909. [CrossRef] [PubMed]
13. Shimada, M.; Yamanoi, Y.; Matsushita, T.; Kondo, T.; Nishibori, E.; hatakeyama, A.; Sugimoto, K.; Nishihara, H. Optical properties of disilane-bridged donor–acceptor architectures: Strong effect of substituents on fluorescence and nonlinear optical properties. *J. Am. Chem. Soc.* **2015**, *137*, 1024–1027. [CrossRef] [PubMed]
14. Kunai, A.; Kawakami, T.; Toyoda, E.; Ishikawa, M. highly selective synthesis of chlorosilanes from hydrosilanes. *Organometallics* **1992**, *11*, 2708–2711. [CrossRef]
15. Kunai, A.; Ochi, T.; Iwata, A.; Ohshita, J. Synthesis of bromohydrosilanes: Reactions of hydrosilanes with $CuBr_2$ in the presence of CuI. *Chem. Lett.* **2001**, 1228–1229. [CrossRef]
16. Kunai, A.; Ohshita, J. Selective synthesis of halosilanes from hydrosilanes and utilization for organic synthesis. *J. Organomet. Chem.* **2003**, *686*, 3–15. [CrossRef]
17. Harrison, D.J.; McDonald, R.; Rosenberg, L. Borane-catalyzed hydrosilylation of thiobenzophenone: A new route to silicon–sulfur bond formation. *Organometallics* **2005**, *24*, 1398–1400. [CrossRef]
18. Kato, N.; Tamura, Y.; Kashiwabara, T.; Sanji, T.; Tanaka, M. $AlCl_3$-catalyzed hydrosilylation of alkynes with hydropolysilanes. *Organometallics* **2010**, *29*, 5274–5282. [CrossRef]
19. Oestreich, M.; hermeke, J.; Mohr, J. A unified survey of Si–H and h–H bond activation catalysed by electron-deficient boranes. *Chem. Soc. Rev.* **2015**, *44*, 2202–2220. [CrossRef] [PubMed]
20. Chatgilialoglu, C. *Organosilanes in Radical Chemistry*; Wiley: Chichester, UK, 2004.
21. Chatgilialoglu, C. $(Me_3Si)_3SiH$: Twenty years after its discovery as a radical-based reducing agent. *Chem. Eur. J.* **2008**, *14*, 2310–2320. [CrossRef] [PubMed]
22. Urenovitch, J.V.; West, R. Pentamethyldisilane and 1,1,2,2-tetramethyldisilane and their addition to olefins. *J. Organomet. Chem.* **1965**, *3*, 138–145. [CrossRef]
23. Sakurai, H.; Kishida, T.; hosomi, A.; Kumada, M. Decomposition of some free radical initiators in hexamethyldisilane. *J. Organomet. Chem.* **1967**, *8*, 65–68. [CrossRef]
24. Hsiao, Y.-L.; Waymouth, R.M. Free-radical hydrosilylation of poly(phenylsilane): Synthesis of functional polysilanes. *J. Am. Chem. Soc.* **1994**, *116*, 9779–9780. [CrossRef]
25. Ojima, I. The hydrosilylation reaction. In *The Chemistry of Organic Silicon Compounds*; Patai, S., Rappoport, Z., Eds.; Wiley: Chichester, UK, 1989; pp. 1479–1526.
26. Ojima, I.; Li, Z.; Zhu, J. Recent advances in the hydrosilylation and related reactions. In *The Chemistry of Organic Silicon Compounds*; Rappoport, Z., Apeloig, Y., Eds.; Wiley: Chichester, UK, 1998; Volume 2, pp. 1687–1792.
27. Marciniec, B.; Maciejewski, H.; Pietraszuk, C.; Pawluć, P. *Hydrosilylation: A Comprehensive Review on Recent Advances*; Marciniec, B., Ed.; Springer: Berlin, Germany, 2009.
28. Lukevics, E.; Dzintara, M. The alcoholysis of hydrosilanes. *J. Organomet. Chem.* **1985**, *295*, 265–315. [CrossRef]
29. Corey, J.Y. Dehydrogenative coupling reactions of hydrosilanes. In *Advances in Silicon Chemistry*; Larson, G.L., Ed.; JAI Press: Greenwich, CT, USA, 1991; Volume 1, pp. 327–387.
30. Gauvin, F.; harrod, J.F.; Woo, H.G. Catalytic dehydrocoupling: A general strategy for the formation of element–element bonds. *Adv. Organomet. Chem.* **1998**, *42*, 363–405.
31. Reichl, J.A.; Berry, D.H. Recent progress in transition metal-catalyzed reactions of silicon, germanium, and tin. *Adv. Organomet. Chem.* **1999**, *43*, 197–265.
32. Murata, M.; Suzuki, K.; Watanabe, S.; Masuda, Y. Synthesis of arylsilanes via palladium(0)-catalyzed silylation of aryl halides with hydrosilane. *J. Org. Chem.* **1997**, *62*, 8569–8571. [CrossRef] [PubMed]
33. Tsukada, N.; hartwig, J.F. Intermolecular and intramolecular, platinum-catalyzed, acceptorless dehydrogenative coupling of hydrosilanes with aryl and aliphatic methyl C–H bonds. *J. Am. Chem. Soc.* **2005**, *127*, 5022–5023. [CrossRef] [PubMed]
34. Yamanoi, Y. Palladium-catalyzed silylations of hydrosilanes with aryl halides using bulky alkyl phosphine. *J. Org. Chem.* **2005**, *70*, 9607–9609. [CrossRef] [PubMed]
35. Murata, M.; Yamasaki, H.; Ueta, T.; Nagata, M.; Ishikura, M.; Watanabe, S.; Masuda, Y. Synthesis of aryltriethoxysilanes via rhodium(I)-catalyzed cross-coupling of aryl electrophiles with triethoxysilane. *Tetrahedron* **2007**, *63*, 4087–4094. [CrossRef]

36. Yamanoi, Y.; Nishihara, H. Direct and selective arylation of tertiary silanes with rhodium catalyst. *J. Org. Chem.* **2008**, *73*, 6671–6678. [CrossRef] [PubMed]
37. Horn, K.A. Regio- and stereochemical aspects of the palladium-catalyzed reactions of silanes. *Chem. Rev.* **1995**, *95*, 1317–1350. [CrossRef]
38. Sharma, H.K.; Pannell, K.H. Activation of the Si–Si bond by transition metal complexes. *Chem. Rev.* **1995**, *95*, 1351–1374. [CrossRef]
39. Suginome, M.; Ito, Y. Transition-metal-catalyzed additions of silicon–silicon and silicon–heteroatom bonds to unsaturated organic molecules. *Chem. Rev.* **2000**, *100*, 3221–3256. [CrossRef] [PubMed]
40. Beletskaya, I.; Moberg, C. Element–element additions to unsaturated carbon–carbon bonds catalyzed by transition metal complexes. *Chem. Rev.* **2006**, *106*, 2320–2354. [CrossRef] [PubMed]
41. Lesbani, A.; Kondo, H.; Sato, J.; Yamanoi, Y.; Nishihara, H. Facile synthesis of hypersilylated aromatic compounds by palladium-mediated arylation reaction. *Chem. Commun.* **2010**, *46*, 7784–7786. [CrossRef] [PubMed]
42. Inubushi, H.; hattori, Y.; Yamanoi, Y.; Nishihara, H. Structures and optical properties of *tris*(trimethylsilyl)silylated oligothiophene derivatives. *J. Org. Chem.* **2014**, *79*, 2974–2979. [CrossRef] [PubMed]
43. Kanno, K.; Niwayama, Y.; Kyushin, S. Selective catalytic monoreduction of dichlorooligosilanes with Grignard reagents. *Tetrahedron Lett.* **2013**, *54*, 6940–6943. [CrossRef]
44. Kanno, K.; Aikawa, Y.; Kyushin, S. Ruthenium-catalyzed alkoxylation of a hydrodisilane without Si–Si bond cleavage. *Tetrahedron Lett.* **2017**, *58*, 9–12. [CrossRef]
45. Piller, F.M.; Metzger, A.; Schade, M.A.; haag, B.A.; Gavryushin, A.; Knochel, P. Preparation of polyfunctional arylmagnesium, arylzinc, and benzylic zinc reagents by using magnesium in the presence of LiCl. *Chem. Eur. J.* **2009**, *15*, 7192–7202. [CrossRef] [PubMed]
46. Russell, A.G.; Guveli, T.; Kariuki, B.M.; Snaith, J.S. Synthesis and characterisation of two new binaphthyl trisilanes. *J. Organomet. Chem.* **2009**, *694*, 137–141. [CrossRef]
47. Ahmed, M.A.K.; Wragg, D.S.; Nilsen, O.; Fjellvåg, H. Synthesis and properties of ethyl, propyl, and butyl hexa-alkyldisilanes and tetrakis(tri-alkylsilyl)silanes. *Z. Anorg. Allg. Chem.* **2014**, *640*, 2956–2961. [CrossRef]
48. Hoffmann, F.; Wagler, J.; Roewer, G. Selective synthesis of functional alkynylmono- and -trisilanes. *Eur. J. Inorg. Chem.* **2010**, 1133–1142. [CrossRef]

15

Construction of a Planar Tetrapalladium Cluster by the Reaction of Palladium(0) Bis(isocyanide) with Cyclic Tetrasilane

Yusuke Sunada [1,2,*], Nobuhiro Taniyama [1], Kento Shimamoto [2], Soichiro Kyushin [3] and Hideo Nagashima [4]

1 Institute of Industrial Science, The University of Tokyo, 4-6-1 Komaba Meguro-ku, Tokyo 153-8580, Japan; taniyama@iis.u-tokyo.ac.jp
2 Department of Applied Chemistry, School of Engineering, The University of Tokyo, 4-6-1 Komaba Meguro-ku, Tokyo 153-8580, Japan; shimamo@iis.u-tokyo.ac.jp
3 Division of Molecular Science, Graduate School of Science and Technology, Gunma University, Kiryu, Gunma 376-8515, Japan; kyushin@gunma-u.ac.jp
4 Institute for Materials Chemistry and Engineering, Kyushu University, 6-1 Kasugakoen Kasuga, Fukuoka 816-8580, Japan; nagasima@cm.kyushu-u.ac.jp
* Correspondence: sunada@iis.u-tokyo.ac.jp

Abstract: The planar tetrapalladium cluster $Pd_4\{Si(^iPr)_2\}_3(CN^tBu)_4$ (**4**) was synthesised in 86% isolated yield by the reaction of palladium(0) bis(isocyanide) $Pd(CN^tBu)_2$ with octaisopropylcyclotetrasilane (**3**). In the course of this reaction, the palladium atoms are clustered via insertion into the Si–Si bonds of **3**, followed by extrusion of one Si^iPr_2 moiety and reorganisation to afford **4** with a 54-electron configuration. The CN^tBu ligand in **4** was found to be easily replaced by *N*-heterocyclic carbene ($^iPr_2IM^{Me}$) to afford the more coordinatively unsaturated cluster $Pd_4\{Si(^iPr)_2\}_3(^iPr_2IM^{Me})_3$ (**5**) having the planar Pd_4Si_3 core. On the other hand, the replacement of CN^tBu with a sterically compact ligand trimethylolpropane phosphite $\{P(OCH_2)_3CEt\}$ led to a planar tripalladium cluster $Pd_3\{Si(^iPr)_2\}_3\{P(OCH_2)_3CEt\}_3$ (**6**) and $Pd\{P(OCH_2)_3CEt\}_4$ in 1:1 molar ratio as products.

Keywords: palladium; cluster; cyclic organopolysilane; template; bridging silylene ligand; isocyanide

---

1. Introduction

Transition metal clusters have attracted much attention because of their unique chemical properties. These clusters have been extensively studied in homogeneous catalysis, in which the substrate can be cooperatively activated by dual metal components in the cluster [1–3]. In many cases, the clusters are also used as structural models of the active sites in heterogeneous catalysts. Detailed spectroscopic analysis of the substrates coordinated to the cluster framework has provided unique insight into the function of the active surface in, for example, the chemisorption process [4,5]. Moreover, some of the clusters provide structural and/or functional mimics of the active sites in enzymes [6–9]. A typical way to prepare the clusters is the self-assembly of metals and bridging ligands, by simply mixing the appropriate metal precursors in proper molar ratios under suitable reaction conditions. However, the self-assembly process generally forms a mixture of clusters of different nuclearities. Thus, a separation step is required to isolate the desired clusters, thereby lowering the yield of the product. In addition, the molecular structure of the formed clusters is not very predictable; in other words, this strategy is unsuitable for custom-designed clusters.

To overcome these drawbacks, a new methodology called template synthesis has recently been introduced [10–28]. In this method, the products are synthesized by clustering multiple metal atoms on the template. The structure of the clusters obtained in this method is highly controllable by using template molecules with appropriate structures; the yield of the product is generally high. Using this synthetic strategy, many transition metal clusters having 1D chain-like structures have been prepared by employing polyenes [10–13], multidentate phosphines [14–17], and nitrogen-containing compounds as templates [18–22]. Similarly, the use of (poly)cyclic aromatic hydrocarbons as a template led to the formation of metal clusters with 2D sheet-like structures [23–28]. However, only a limited number of 2D sheet-like clusters have been prepared, due to the lack of suitable template molecules.

We are interested in producing more 2D sheet-like clusters based on the template synthesis approach, focusing on the use of cyclic organopolysilanes as the template molecules. In our previous study, we found that a ladder polysilane, decaisopropylbicyclo[2.2.0]hexasilane (**1**), acts as a good template molecule for the preparation of large palladium clusters [29]. It is known that some transition metal species can be inserted into the Si-Si bonds of organosilanes [30,31], and "$Pd(CNR)_2$ (CNR = isocyanide)" is one of the representative [32,33].

In the reaction of $Pd(CN^tBu)_2$ with **1**, the Pd_{11} cluster $Pd_{11}\{Si(^iPr)\}_2\{Si(^iPr)_2\}_4(CN^tBu)_{10}$ (**2**), having a "folding" nanosheet structure, was confirmed to form in high isolated yield. As shown in Scheme 1, **1** behaves as a template to fix seven palladium species between seven different Si–Si bonds. Four additional palladium species participated to form the Pd_{11} cluster **2**. In addition, the ligand exchange of CN^tBu in **2** to $CN(2,4,6\text{-}Me_3\text{-}C_6H_2)$ triggered the skeletal rearrangement to produce another Pd_{11} cluster to have a "folding" nanosheet structure but with a wider dihedral angle. This indicates that the structure of the cluster can be tuned by changing the auxiliary ligand on the metal. Theoretical calculations were also used to elucidate the electronic structure and bonding nature of these clusters [34].

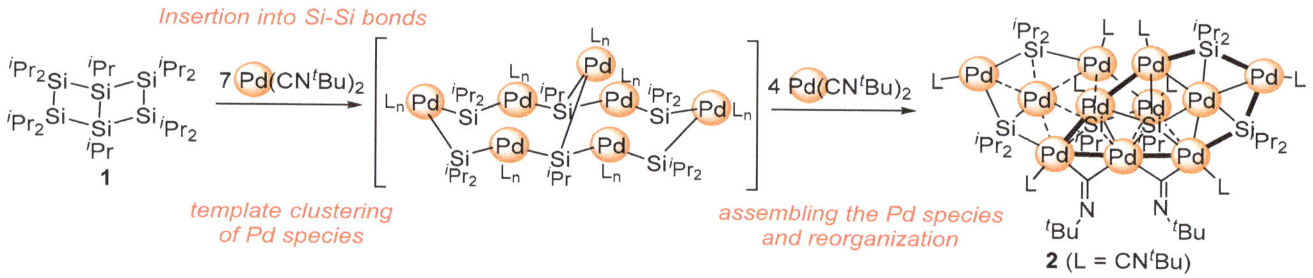

Scheme 1. Reaction of palladium(0) bis(isocyanide) $Pd(CN^tBu)2$ with ladder polysilane **1** to afford the Pd_{11} cluster **2**.

Along the same line, in this study we intend to use octaisopropylcyclotetrasilane (**3**) (Chart 1) having four Si–Si bonds as a template for the clustering of palladium species. As we expected, the reaction of **3** with $Pd(CN^tBu)_2$ effectively formed a planar tetrapalladium (Pd_4) cluster framework. Meanwhile, an unexpected outcome was that the reaction was accompanied by extrusion of one Si^iPr_2 moiety, leading to the formation of a coordinatively unsaturated cluster $Pd_4\{Si(^iPr)_2\}_3(CN^tBu)_4$ (**4**) with 54-electron configuration as a single product. Ligand exchange of CN^tBu in **4** by an *N*-heterocyclic carbene ($^iPr_2IM^{Me}$) proceeded with preserved Pd_4Si_3 core structure to afford a planar Pd_4 cluster $Pd_4\{Si(^iPr)_2\}_3(^iPr_2IM^{Me})_3$ (**5**) quantitatively, which has 52 electrons and coordinatively more unsaturated than **4**. On the other hand, ligand exchange reaction of **4** with trimethylolpropane phosphite decreased the cluster nuclearity to afford the planar tripalladium cluster $Pd_3\{Si(^iPr)_2\}_3\{P(OCH_2)_3CEt\}_3$ (**6**) and the mononuclear $Pd\{P(OCH_2)_3CEt\}_4$ concomitantly.

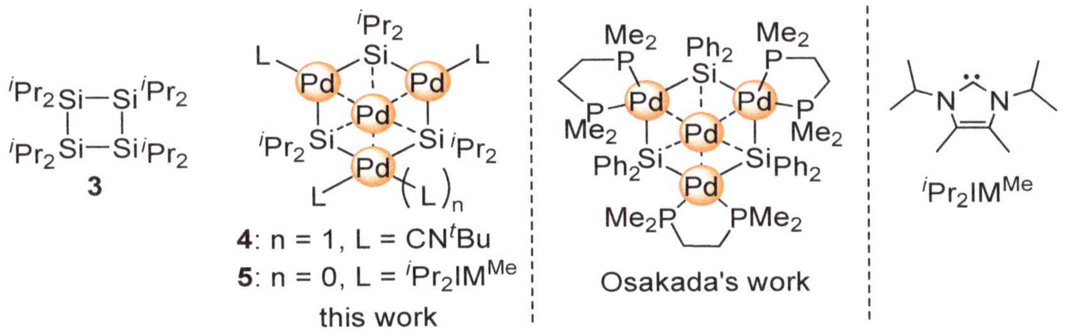

Chart 1. Structures of octaisopropylcyclotetrasilane **3**, palladium clusters having bridging silylene ligands, and *N*-heterocyclic carbene ligand (^{i}Pr$_2$IM^{Me}).

2. Results and Discussion

2.1. Synthesis of Planar Pd$_4$ Cluster 4 by Reaction of Pd(CN^{t}Bu)$_2$ with Cyclotetrasilane 3

As mentioned in the introduction, Pd(CN^{t}Bu)$_2$ is known to show high reactivity toward insertion into the Si–Si bonds of various organosilanes. Indeed, the reaction of Pd(CN^{t}Bu)$_2$ with **3** proceeded smoothly in toluene at 65 °C to afford the planar Pd$_4$ cluster **4** in 86% isolated yield (Scheme 2). By monitoring this reaction by ^{1}H NMR spectroscopy, we found that **4** was formed as the sole product, since no other silicon-containing by-products were observed. This indicates that all the silylene (Si^{i}Pr$_2$) moieties in **3** were incorporated into **4**. In addition, no intermediary palladium species were found in this reaction. It should be noted that Osakada et al. synthesized a structurally similar planar Pd$_4$ cluster having three bridging silylene ligands (Chart 1) by the reaction of dinuclear {Pd(PCy$_3$)}$_2$(μ-η^2-SiHPh$_2$)$_2$ and 1,2-bis(dimethylphosphino)ethane (dppm) at 80 °C [35]. In that cluster, dppm was used as the auxiliary ligand for the palladium centre on the edge. In contrast, two of the three palladium atoms on the edge of **4** (Pd(1) and Pd(2) in Figure 1) bear only one isocyanide ligand each, giving a more coordinatively unsaturated cluster with 54 cluster valence electrons, compared with that prepared by Osakada et al.

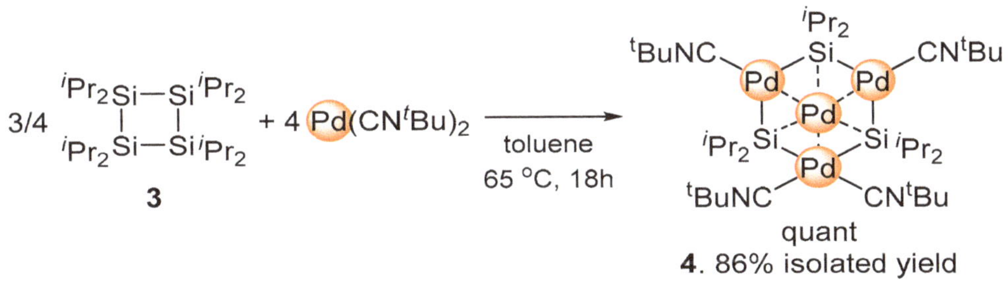

Scheme 2. Reaction of octaisopropylcyclotetrasilane **3** with Pd(CN^{t}Bu)$_2$ to afford **4**.

The molecular structure of **4** was determined by X-ray diffraction analysis. The ORTEP drawing of **4** is given in Figure 1a, and the side view of its core Pd$_4$Si$_3$ fragment is shown in Figure 1b. The selected bond distances are summarized in Table 1. Three palladium atoms and three silicon atoms derived from the bridging silylene moieties form an almost planar six-membered ring, and the fourth palladium atom (Pd(4)) is located at the centre of this ring. Therefore, the four palladium atoms and three silicon atoms lie on a plane, and deviations of all seven atoms from this plane are within the range of 0.028–0.178 Å. The Pd(4)–Pd(1)–C(CN^{t}Bu) and Pd(4)–Pd(2)–C(CN^{t}Bu) axes slightly deviate from linearity (162.37(14) and 163.14(14)°, respectively), which may originate from the strong trans-influence of the central Pd(4) atom whose formal oxidation state is Pd(0). The sum of the two Pd(cent)–Si–C$_{ipso}$ angles and the C$_{ipso}$–Si–C$_{ipso}$ angle is close to 360° (355.5°–360.0°). This suggests that the coordination geometry around the Si atoms can be regarded as pseudo-trigonal bipyramidal, and the two Pd(edge) atoms for each Si atom (Pd(1) and Pd(2) for Si(1), Pd(2) and Pd(3) for Si(2), and Pd(1) and Pd(3) for Si(3)) are located on the axial position of the Si atom. The silylene ligand including Si(1) bridges over two Pd cores (Pd(1) and Pd(2)). The bond distances of Pd(1)–Si(1) and Pd(2)–Si(1) are 2.4252(12) and 2.4339(12) Å, respectively, which are considerably shorter than those observed in other Pd(edge)–Si moieties (2.5094(11)–2.6401(12) Å). In contrast, the Pd(edge)–Si(silylene) bond distances in Osakada's Pd_4Si_3 cluster are reported to be 2.505–2.546 Å with no significant deviation [35–39]. This difference may be derived from the formal electron configurations of Pd(1) and Pd(2). As suggested by our previous theoretical calculations, the metal-to-silylene charge transfer, from the occupied d-orbital of the metal to the empty p-orbital of the silylene moieties, plays a crucial role in the bonding interaction between Pd(edge) and the bridging silylene moieties. Because Pd(3) has two isocyanide ligands, π-back donation from Pd to the two isocyanide ligands causes the Pd(3) centre to be more electron-deficient. This leads to longer Pd(3)–silylene bond distances compared with those of Pd(1)–Si(1) and Pd(2)–Si(1). The Pd(3)–Pd(4) bond length (2.7523(5) Å) is slightly longer than Pd(1)–Pd(4) (2.6812(6) Å) and Pd(2)–Pd(4) (2.6778(6) Å); however, they are within the range of metal-metal bonding interaction reported in the literature [10].

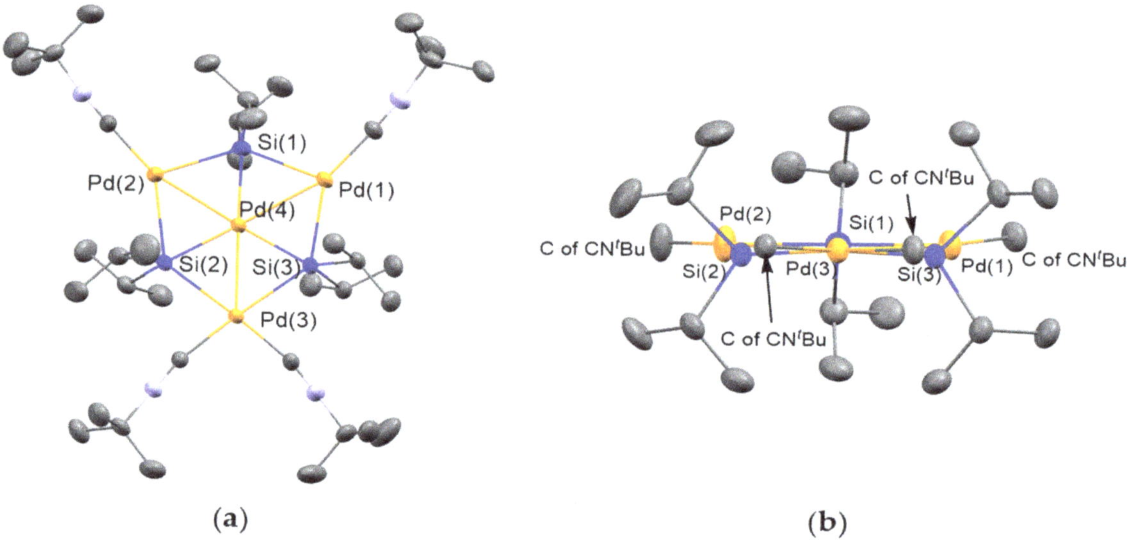

Figure 1. (**a**) Top view of the molecular structure of **4** with 50% probability ellipsoids. Hydrogen atoms were omitted for clarity; (**b**) Side view of **4**. All atoms derived from the isocyanide ligands except for the coordinated carbon atoms, and all hydrogen atoms were omitted for clarity.

Although there are three inequivalent palladium centres in the solid-state structure of **4**, only one singlet ^{1}H NMR peak derived from the ^{t}Bu moiety of the isocyanide ligand was observed at 0.99 ppm in C_6D_6 at room temperature. In the variable temperature ^{1}H NMR spectrum of **4** in toluene-d_8, this ^{t}Bu signal (appeared at 1.02 ppm at r.t.) started to broaden at around 0 °C, almost coalesced at −70 °C, then two signals appeared at 0.90 and 1.05 ppm at −90 °C. Similarly, methyl peaks corresponding to the ^{i}Pr moieties on the Si atoms also began to broaden at around 0 °C, then split into three slightly broad signals at −90 °C in an intensity ratio of ca. 1:1:1. This suggests the presence of two inequivalent isocyanide ligands, as well as three inequivalent methyl groups of the Si^{i}Pr$_2$ moieties at lower temperatures. These spectral features suggest the presence of fluxional behaviour, due to the facile site exchange of the isocyanide ligands. The fluxional behaviour of the coordinated isocyanide ligands has also been observed in the previously reported transition metal clusters [40,41]. The IR spectrum of the crystals of **4** displays two absorption bands at 2103 and 2065 cm^{-1} along with a shoulder band at 2125 cm^{-1}. This spectral feature is consistent with the solid-state structure determined by X-ray diffraction analysis. Unfortunately, no signal appeared in the ^{29}Si{^{1}H} NMR spectrum of **4** with longer (5 sec) or shorter (0.2 sec) relaxation time, even when using very concentrated sample with many scans.

*2.2. Ligand Exchange of **4** with N-Heterocyclic Carbene*

In our previous paper, all 10 CN^tBu ligands in **2** were found to be easily and quantitatively replaced by another isocyanide ligand $CN(2,4,6\text{-}Me_3\text{-}C_6H_2)$ to form a new "folding" Pd_{11} cluster at −35 °C [29]. Thus, the ligand exchange reaction of **4** with $CN(2,4,6\text{-}Me_3\text{-}C_6H_2)$ was attempted. However, no reaction took place even at higher temperatures. Instead, **4** underwent the facile ligand exchange with 3 equiv. of *N*-heterocyclic carbene, ${}^iPr_2IM^{Me}$, to give the new planar Pd_4Si_3 cluster $Pd_4\{Si({}^iPr)_2\}_3({}^iPr_2IM^{Me})_3$ (**5**) at room temperature (Scheme 3). ^1H NMR monitoring of this reaction indicated that **5** was formed in quantitative yield concomitant with the formation of free isocyanide, and it was isolated in 63% yield after purification. During this reaction, the four isocyanide ligands in **4** were replaced by three ${}^iPr_2IM^{Me}$ ligands. Consequently, a more coordinatively unsaturated Pd_4Si_3 cluster with 52 valence electrons was formed. An possible alternative synthetic route for cluster **5** may be the reaction of cyclotetrasilane **3** with "$Pd({}^iPr_2IM^{Me})_2$". To check this possibility, we performed the reaction of **3** with "$Pd({}^iPr_2IM^{Me})_2$" which was generated in situ from the reaction of $CpPd(\eta^3\text{-allyl})$ with 2 equiv. of ${}^iPr_2IM^{Me}$. We confirmed that no reaction took place even at higher temperatures such as 65 °C in C_6D_6, suggesting that cluster **5** is available only in the reaction shown in Scheme 3.

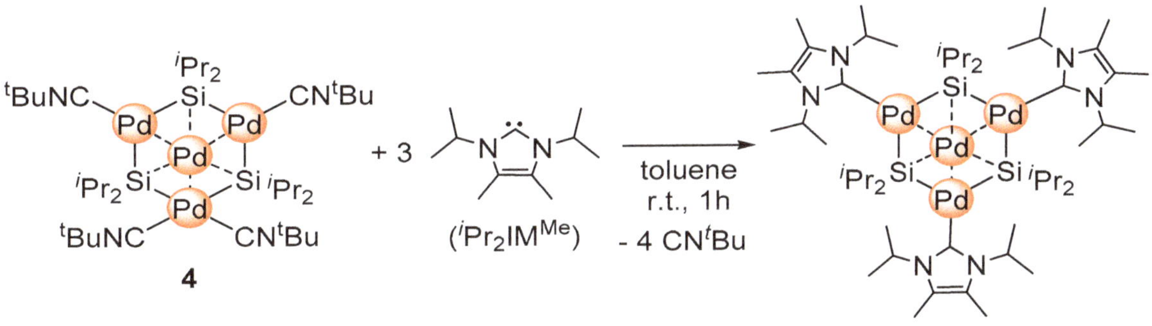

Scheme 3. Synthesis of **5** via ligand exchange reaction.

The molecular structure of **5** is shown in Figure 2, and selected bond distances are summarized in Table 1. Cluster **5** has a pseudo-C_3 symmetric structure, and each Pd(edge) atom bears only one ${}^iPr_2IM^{Me}$ ligand. Four Pd atoms and three Si(silylene) atoms are located on a plane, and deviations of these atoms from the plane are within the range of 0.031–0.178 Å. As seen in cluster **4**, all three Pd(4)–Pd(edge)–C(${}^iPrIM^{Me}$) axes deviate slightly from linearity (Pd(4)–Pd(1)–C(${}^iPrIM^{Me}$): 163.61(13)°, Pd(4)–Pd(2)–C(${}^iPrIM^{Me}$): 170.18(12)°, Pd(4)–Pd(3)–C(${}^iPrIM^{Me}$): 169.15(9)°). The dihedral angle between the Pd_4Si_3 plane and the five-membered ring of the ${}^iPr_2IM^{Me}$ ligands are 87.5°–104.3°, suggesting the almost perpendicular orientation of the ${}^iPr_2IM^{Me}$ moieties. Because cluster **5** has a pseudo-C_3 symmetric structure, the three Pd(edge)–Pd(4) distances are essentially the same (2.6991(8)–2.7030(7) Å). Each silylene ligand bridges over two Pd(edge) atoms in an asymmetric manner. For instance, the bond distance of Pd(1)–Si(1) (2.3668(13) Å) is significantly shorter compared with that in Pd(2)–Si(1) (2.6359(14) Å). This may originate from the biased charge transfer from Pd to silylene moieties. Further studies including theoretical calculations to elucidate the details of the bonding interaction in the planar Pd_4Si_3 clusters are now underway.

The ^1H and $^{13}C\{^1H\}$ NMR spectra of **5** are consistent with the expectation from its pseudo-C_3 symmetric structure. One singlet derived from the carbene carbon atom of the ${}^iPr_2IM^{Me}$ moiety was observed at 194.0 ppm in the $^{13}C\{^1H\}$ NMR spectrum. The $^{29}Si\{^1H\}$ NMR spectrum of **4** showed a singlet at 223.1 ppm, which is comparable to those derived from the silylene moieties of **2** (191.85 and

226.63 ppm). This signal appeared at slightly lower field compared with Osakada's Pd_4Si_3 cluster (195 ppm) [35], but it is within the range for bridging silylene ligands reported in the literature [42–44].

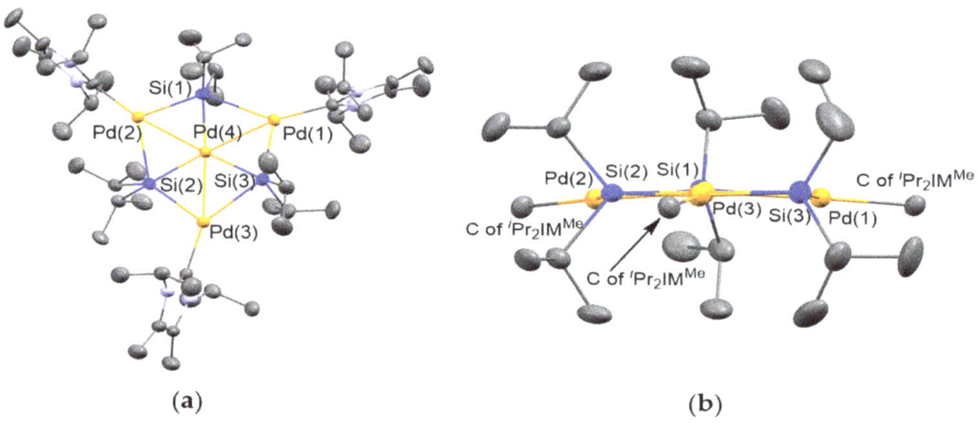

Figure 2. (**a**) Top view of the molecular structure of **5** with 50% probability ellipsoids. Hydrogen atoms were omitted for clarity; (**b**) Side view of **5**. All atoms derived from the $^{i}Pr_2$IMMe ligands except for the coordinated carbon atoms, and all hydrogen atoms were omitted for clarity.

*2.3. Ligand Exchange of **4** with Trimethylolpropane Phosphite to Afford the Planar Pd_3 Cluster **6***

The facile ligand exchange of **4** with $^{i}Pr_2IM^{Me}$ prompted us to examine the reaction of **4** with trimethylolpropane phosphite. This reaction was first monitored by $^{31}P\{^1H\}$ NMR. Treating **4** with 7 equiv. of trimethylolpropane phosphite provides two signals at 118.9 and 143.9 ppm. The former was assignable to the mononuclear $Pd\{P(OCH_2)_3CEt\}_4$ complex by comparison with the independently prepared $Pd\{P(OCH_2)_3CEt\}_4$ from the reaction of CpPd(η^3-allyl) with 4 equiv. of $P(OCH_2)_3CEt$. With a lower amount of $P(OCH_2)_3CEt$, no intermediary species were visible in the reaction of **4** to form **6** and $Pd\{P(OCH_2)_3CEt\}_4$. For example, treating **4** with 4 equiv. of $P(OCH_2)_3CEt$ resulted in the exclusive formation of **6** and $Pd\{P(OCH_2)_3CEt\}_4$ with recovery of **4**. Because $Pd\{P(OCH_2)_3CEt\}_4$ shows low solubility in diethyl ether or pentane, it can be easily removed from the crude product. Subsequent recrystallisation from pentane gave the planar tripalladium cluster **6** in a pure form. During this reaction, the sterically compact phosphite ligand may first attack the central palladium atom in **4**, followed by the fragmentation to afford **6** and $Pd\{P(OCH_2)_3CEt\}_4$ as products in 1:1 molar ratio (Scheme 4). It should be noted that the synthesis of structurally similar triplatinum clusters bearing bridging silylene ligands has been reported [45–48]. However, to the best of our knowledge, cluster **6** is the first example of the palladium analogue of these clusters.

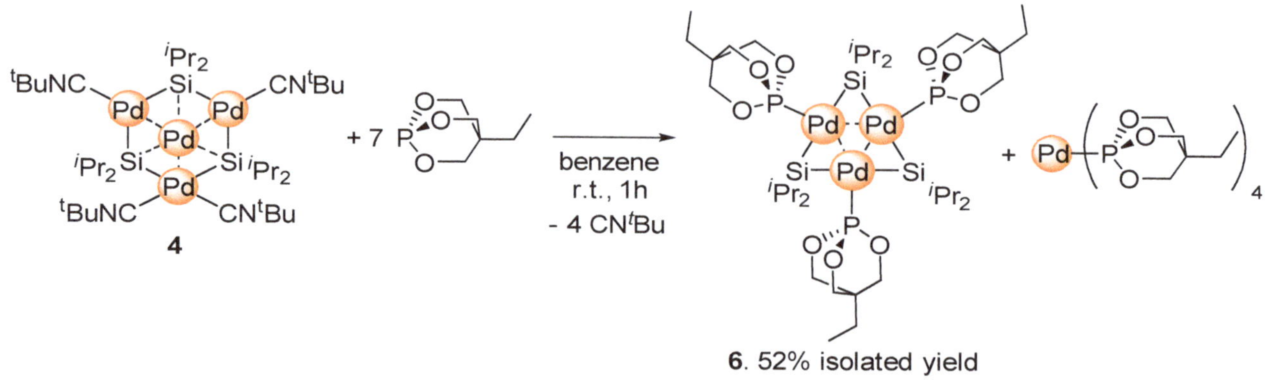

Scheme 4. Ligand exchange of **4** with trimethylolpropanephosphite to give the planar tripalladium cluster **6**.

The molecular structure of **6** was confirmed by X-ray diffraction analysis. The ORTEP drawing is shown in Figure 3, and the selected bond distances are depicted in Table 1. Cluster **6** can be regarded formally as a 42-electron cluster with three Pd–Pd bonds similar to those found in the previously reported trinuclear Pt_3Si_3 clusters [45–48]. As shown in Figure 3b, there is a planar Pd_3Si_3 unit, and the deviations of all atoms from the plane are within the range of 0.000–0.172 Å. One of the three phosphorus atoms (P(3) in Figure 3b) is located below the plane by 0.504 Å. The Pd–Pd bond distances are 2.7041(6)–2.7117(5) Å, which are almost comparable to those found in **5** and reported polynuclear palladium clusters [10]. In contrast to clusters **4** and **5**, the silylene ligands symmetrically bridge over two palladium atoms, and the Pd–Si bond lengths are 2.3557(9)–2.3754(11) Å, which are in the normal range for those found in palladium complexes having bridging silylene ligands.

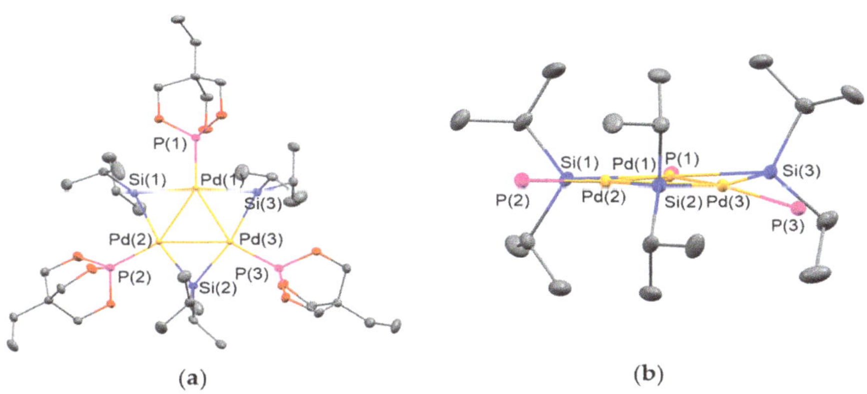

Figure 3. (**a**) Top view of the molecular structure of **6** with 50% probability ellipsoids. Hydrogen atoms were omitted for clarity; (**b**) Side view of **6**. All atoms derived from the phosphite ligands except for the coordinated phosphorus atoms, and all hydrogen atoms were omitted for clarity.

Table 1. Selected bond distances for **4**, **5** and **6**.

	4	5	6
Pd(1)–Pd(4)	2.6812(6)	2.7029(6)	-
Pd(2)–Pd(4)	2.6778(6)	2.6991(8)	-
Pd(3)–Pd(4)	2.7523(5)	2.7030(7)	-
Pd(1)–Pd(2)	-	-	2.7117(5)
Pd(1)–Pd(3)	-	-	2.7041(6)
Pd(2)–Pd(3)	-	-	2.7050(4)
Pd(1)–Si(1)	2.4252(12)	2.3668(13)	2.3557(9)
Pd(1)–Si(3)	2.5827(12)	2.6359(14)	2.3602(9)
Pd(2)–Si(1)	2.4339(12)	2.6518(12)	2.3707(12)
Pd(2)–Si(2)	2.6401(12)	2.3769(15)	2.3754(11)
Pd(3)–Si(2)	2.5094(11)	2.6659(14)	2.3658(11)
Pd(3)–Si(3)	2.5341(12)	2.3693(12)	2.3691(11)
Pd(4)–Si(1)	2.3106(12)	2.2638(14)	-
Pd(4)–Si(2)	2.2621(10)	2.2558(11)	-
Pd(4)–Si(3)	2.2692(12)	2.2603(13)	-

The NMR spectra of **6** are consistent with those expected from the molecular structure determined by X-ray diffraction analysis. For instance, the methyl and methine groups of the ^{i}Pr groups appeared at 1.86 and 1.93 ppm as doublet and septet, respectively. The $^{31}P\{^{1}H\}$ NMR spectrum of **6** exhibited an intense singlet at 143.9 ppm.

3. Materials and Methods

Manipulation of air and moisture sensitive compounds was carried out under a dry nitrogen atmosphere, using standard Schlenk tube techniques associated with a high-vacuum line, or in the glove box filled with dry nitrogen. All solvents were distilled over appropriate drying reagents prior to use (toluene, benzene, ether, pentane, hexamethyldisiloxane (HMDSO); Ph_2CO/Na). ^1H, $^{13}C\{^1H\}$, $^{29}Si\{^1H\}$ and $^{31}P\{^1H\}$ NMR spectra were recorded on a Lambda 400 spectrometer at ambient temperature unless otherwise noted. ^1H, $^{13}C\{^1H\}$, $^{29}Si\{^1H\}$ and $^{31}P\{^1H\}$ NMR chemical shifts (δ values) were given in ppm relative to the solvent signal or standard resonances ($^{29}Si\{^1H\}$: external tetramethylsilane, $^{31}P\{^1H\}$: external H_3PO_4). Elemental analyses were performed at the A Rabbit Science Co., Ltd. (Ayako Sato, 5-4-21 Nishihashimoto, Midori, Sagamihara, Kanagawa 252–0131, Japan). IR spectra were recorded on a PerkinElmer Spectrum Two spectrometer. Starting materials, octaisopropylcyclotetrasilane (**3**) [49], $Pd(CN^tBu)_2$ [31], $CpPd(\eta^3\text{-allyl})$ and $^iPr_2IM^{Me}$ [50] were synthesized by the method reported in the literature.

*3.1. Synthesis of $Pd_4(Si(^iPr)_2)_3(CN^tBu)_4$ (**4**)*

In a 50-mL Schlenk tube, $Pd(CN^tBu)_2$ (273 mg, 1.00 mmol) was dissolved in toluene (20 mL), then octaisopropylcyclotetrasilane (**3**) (86 mg, 0.19 mmol) was added to this solution at room temperature. The solution was stirred at 65°C for 18 h, then the solvent was removed in vacuo. The remaining crude product was dissolved in pentane (40 mL), and centrifuged to remove the insoluble materials. The supernatant was collected, concentrated to ca. 5 mL, and cooled at −35°C to give **4** as yellow crystals (237 mg, 0.22 mmol, 86%).^1H NMR (400 MHz, r.t., C_6D_6): δ = 0.99 (s, 36H, ^tBu), 1.85 (d, J = 7.3 Hz, 36H, $CH(CH_3)_2$), 2.42 (sept, J = 7.3 Hz, 6H, $CH(CH_3)_2$). $^{13}C\{^1H\}$ NMR (100 MHz, r.t., C_6D_6): δ = 23.8 (s, $Si\{CH(CH_3)_2\}_2$), 26.3 (s, $C(CH_3)_3$), 30.1 (s, $Si\{CH(CH_3)_2\}_2$), 53.1 (s, $C(CH_3)_3$) (one peak due to the $\underline{C}N^tBu$ moiety was not detectable). $^{29}Si\{^1H\}$ NMR (119 MHz, r.t., C_6D_6): no signal appeared. IR (ATR): ν_{CN} = 2125, 2103, 2065 cm^{-1}. Anal calcd for $C_{38}H_{78}N_4Pd_4Si_3$; C 41.45, H 7.14, N 5.09; found: C 41.17, H 6.84, N 5.20.

*3.2. Synthesis of $Pd_4(Si(^iPr)_2)_3(^iPr_2IM^{Me})_4$ (**5**)*

In a 20-mL Schlenk tube, **4** (67 mg, 0.061 mmol) was dissolved in toluene (20 mL), then $^iPr_2IM^{Me}$ (33 mg, 0.183 mmol) was added to this solution at room temperature. The solution was stirred at room temperature for 1 h, then the solvent was removed in vacuo. The remaining solid was again dissolved in toluene (20 mL), stirred at room temperature for 1 h, then the solvent was removed in vacuo. The remaining crude product was extracted with pentane (10 mL), and the mother liquid was centrifuged to remove the small amount of insoluble materials. The supernatant was collected, then the solvent was removed in vacuo. The remaining powder was washed with HMDSO (5 mL × 2) to afford **5** as a brown powder (50 mg, 0.038 mmol, 63%). Crystals suitable for X-ray diffraction analysis were obtained by cooling the saturated pentane solution at −35 °C. ^1H NMR (400 MHz, r.t., C_6D_6): δ = 1.44 (d, J = 7.3 Hz, 36H, $CH(CH_3)_2$), 1.69 (s, 18H, C=CMe), 1.85 (d, J = 7.3 Hz, 36H, $CH(CH_3)_2$), 2.10 (sept, J = 7.3 Hz, 6H, $Si\{CH(CH_3)_2\}_2$), 6.24 (sept, J = 7.3 Hz, 6H, $N\{CH(CH_3)_2\}$). $^{13}C\{^1H\}$ NMR (100 MHz, r.t., C_6D_6): δ = 10.2 (s, $Si\{CH(CH_3)_2\}_2$), 22.5 (s, $Si\{CH(CH_3)_2\}_2$), 23.2 (s, $N\{CH(CH_3)_2\}$), 23.6 (s, C=C*Me*), 54.0 (s, $N\{CH(CH_3)_2\}$), 122.6 (s, C=CMe), 194.0 (s, Pd–C). $^{29}Si\{^1H\}$ NMR (119 MHz, r.t., C_6D_6): δ = 243.3 (s, *Si*Me_2). Anal calcd for $C_{51}H_{102}N_6Pd_4Si_3$; C 46.78, H 7.85, N 6.42; found: C 46.75, H 7.74, N 6.25.

*3.3. Synthesis of $Pd_3\{Si(^iPr)_2\}_3\{P(OCH_2)_3CEt\}_3$ (**6**)*

In a 20-mL Schlenk tube, **4** (30 mg, 0.027 mmol) was dissolved in benzene (20 mL), then trimethylolpropane phosphite (31 mg, 0.19 mmol) was added to this solution at room temperature. The solution was stirred at room temperature for 1 h. After freeze-drying, the remaining solid was dissolved in ether (20 mL), and centrifuged to remove the insoluble materials. The supernatant was collected, then the solvent was removed in vacuo. The remaining powder was washed with HMDSO (5 mL) and cold pentane (3 mL) to afford **6** as a brown powder (16 mg, 0.014 mmol, 52%). Crystals suitable for X-ray diffraction analysis were obtained by cooling the saturated ether solution at −35 °C. ^1H NMR (400 MHz, r.t., C_6D_6): δ = 0.06 (t, J = 7.6 Hz, 9H, CH_2CH_3), 0.19 (q, J = 7.6 Hz, 6H, CH_2CH_3), 1.86 (d, J = 7.6 Hz, 36H, $CH(CH_3)_2$), 1.93 (sept, J = 7.6 Hz, 6H, $Si\{CH(CH_3)_2\}_2$), 3.56 (s, 18H, OCH_2–).

$^{13}C\{^1H\}$ NMR (100 MHz, r.t., C_6D_6): δ = 6.7 (s, CH_2CH_3), 21.3 (s, $Si\{CH(CH_3)_2\}_2$), 23.6 (s, CH_2CH_3), 29.6 (s, $Si\{CH(CH_3)_2\}_2$), 34.1 (dd, J = 11.5, 22.0 Hz, CCH_2CH_3), 72.4 (s, OCH_2). $^{29}Si\{^1H\}$ NMR (119 MHz, r.t., C_6D_6); no signal appeared. ^{31}P NMR (162 MHz, r.t., C_6D_6): δ = 143.9. Even though a number of attempts have been made, we have been unable to obtain satisfactory elemental analysis values for **6**, presumably due to the high instability toward both air and moisture. Anal calcd for $C_{36}H_{75}O_9P_3Pd_3Si_3$; C 37.65, H 6.58; found: C 38.40, H 5.83.

3.4. In Situ *Preparation of* $Pd\{P(OCH_2)_3CEt\}_4$

In a J. Young NMR tube, CpPd(η^3-allyl) (20 mg, 0.094 mmol) was dissolved in C_6D_6 (0.5 mL), then trimethylolpropane phosphite (61 mg, 0.38 mmol) was added to this solution at room temperature. The solution was allowed to stand at room temperature for 45 min. ^1H and $^{31}P\{^1H\}$ NMR spectra indicates quantitative formation of $Pd\{P(OCH_2)_3CEt\}_4$. ^1H NMR (400 MHz, r.t., C_6D_6): δ = 0.03 (t, J = 7.6 Hz, 12H, CH_2CH_3), 0.14 (q, 6H, J = 7.6 Hz, 8H, CH_2CH_3), 3.65 (s, 24H, OCH_2–). $^{31}P\{^1H\}$ NMR (162 MHz, r.t., C_6D_6): δ = 118.9.

3.5. *X-Ray Data Collection and Reduction*

X-ray crystallography for complex **4** and **5** was performed on a Rigaku Saturn CCD area detector with graphite monochromated Mo Kα radiation (λ = 0.71075 Å), and single crystals of **6** suitable for X-ray crystallography were analyzed by synchrotron radiation at beam line BL02B1 (λ = 0.71075 Å) of Spring-8 (Hyogo, Japan) using Rigaku Mercury II detector. The data obtained were processed using Crystal-Clear (Rigaku) on a Pentium computer, and were corrected for Lorentz and polarization effects. The structures were solved by direct methods [51], and expanded using Fourier techniques [52]. Hydrogen atoms were refined using the riding model. The final cycle of full-matrix least-squares refinement on F^2 was based on 12,489 observed reflections and 470 variable parameters for **4**; 14,053 observed reflections and 577 variable parameters for **5**; 11,457 observed reflections and 507 variable parameters for **6**. Neutral atom scattering factors were taken from Cromer and Waber [53]. All calculations were performed using the Crystal Structure [54] crystallographic software package except for refinement, which was performed using *SHELXL*-97 [55]. Details of final refinement, as well as the bond lengths and angle, are summarized in Tables S1–S3, and the numbering scheme employed is also shown in Figures S7–S9, which were drawn with ORTEP at 50% probability ellipsoids. CCDC numbers 1582225 (**4**), 1582226 (**5**) and 1582227 (**6**) contain the supplementary crystallographic data for this paper. These data can be obtained free of charge from the Cambridge Crystallographic Data Centre via www.ccdc.cam.ac.uk/data_request/cif.

4. Conclusions

In the present study, we found that the planar tetrapalladium cluster can be easily obtained by the reaction of cyclic tetrasilane with $Pd(CN^tBu)_2$. The ligand exchange of the cluster led to the formation of new clusters, with or without maintaining the core structure. These results indicate that clustering metal atoms through insertion into the Si–Si bonds of cyclic organopolysilanes is an effective way to synthesise transition metal clusters with bridging organosilylene ligands. Efforts are underway to synthesise a series of new metal clusters by the reaction of appropriate transition metal precursors with cyclic organopolysilanes. Application of these cluster molecules as functional materials, such as catalysts, will also be investigated in the near future.

Acknowledgments: This work was supported by the Core Research Evolutional Science and Technology (CREST) Program of Japan Science and Technology Agency (JST) Japan, the Cooperative Research Program of "Network Joint Research Center for Materials and Devices.", Grant in Aid for Scientific Research (B) (No. 16H04120) and Challenging Exploratory Research (No. 26620047) from the Ministry of Education, Culture, Sports, Science and Technology, Japan, the Iwatani Naoji Foundation, and the JFE 21st Century Foundation. This work was also supported by the special fund of Institute of Industrial Science, The University of Tokyo. The synchrotron radiation experiments were performed at the BL02B1 of SPring-8 with the approval of the Japan Synchrotron Radiation Research Institute (JASRI) (Proposal No. 2016A1118, 2016B1392, and 2017A1400).

Author Contributions: Yusuke Sunada, Soichiro Kyushin and Hideo Nagashima conceived and designed the experiments; Yusuke Sunada, Nobuhiro Taniyama and Kento Shimamoto performed the experiments; Yusuke Sunada analyzed the data; Yusuke Sunada and Hideo Nagashima wrote the paper.

References

1. Markó, L.; Vizi-Orosz, A. *Metal Clusters in Catalysis*; Gates, B.C., Guczi, L., Knözinger, H., Eds.; Elsevier: Amsterdam, The Netherlands, 1986.
2. Süss-Fink, G.; Jahncke, M. *Catalysis by Di- and Polynuclear Metal Cluster Complexes*; Adams, R.D., Cotton, F.A., Eds.; Wiley-VCH: Toronto, ON, Canada, 1998.
3. Puddephatt, R.J. *Metal Clusters in Chemistry*; Braunstein, P., Oro, L.A., Raithby, P.R., Eds.; Wiley-VCH: Toronto, ON, Canada, 1999.
4. Braga, D.; Dyson, F.; Grepioni, F.; Johnson, B.F.G. Arene clusters. *Chem. Rev.* **1994**, *94*, 1585–1620. [CrossRef]
5. Inagaki, A.; Takaya, Y.; Takemori, T.; Suzuki, H.; Tanaka, M.; Haga, M. Trinuclear ruthenium complex with a face-capping benzene ligand. Hapticity change induced by two-electron redox reaction. *J. Am. Chem. Soc.* **1997**, *119*, 625–626. [CrossRef]
6. Holm, R.H.; Lo, W. Structural Conversions of synthetic and protein-bound iron–sulfur clusters. *Chem. Rev.* **2016**, *116*, 13685–13713. [CrossRef] [PubMed]
7. Groysman, S.; Holm, R.H. Biomimetic chemistry of iron, nickel, molybdenum, and tungsten in sulfur-ligated protein sites. *Biochemistry* **2009**, *48*, 2310–2320. [CrossRef] [PubMed]
8. Lee, S.C.; Lo, W.; Holm, R.H. Developments in the biomimetic chemistry of cubane-type and higher nuclearity iron–sulfur cluster. *Chem. Rev.* **2014**, *114*, 3579–3600. [CrossRef] [PubMed]
9. Ohki, Y.; Tatsumi, K. New synthetic routes to metal-sulfur clusters relevant to the nitrogenase metallo-clusters. *Z. Anorg. Allg. Chem.* **2013**, *639*, 1340–1349. [CrossRef]
10. Murahashi, T.; Kurosawa, H. Organopalladium complexes containing palladium–palladium bonds. *Coord. Chem. Rev.* **2002**, *231*, 207–228. [CrossRef]
11. Horiuchi, S.; Tachibana, Y.; Yamashita, M.; Yamamoto, K.; Masai, K.; Takase, K.; Matsutani, T.; Kawamata, S.; Kurashige, Y.; Yanai, T.; et al. Multinuclear metal-binding ability of a carotene. *Nat. Commun.* **2015**, *6*, 7742/1–7742/8. [CrossRef] [PubMed]
12. Murahashi, T.; Mochizuki, E.; Kai, Y.; Kurosawa, H. Organometallic sandwich chains made of conjugated polyenes and metal-metal chains. *J. Am. Chem. Soc.* **1999**, *121*, 10660–10661. [CrossRef]
13. Tatsumi, Y.; Nagai, T.; Nakashima, H.; Murahashi, T.; Kurosawa, H. Stepwise growth of polypalladium chains in 1,4-diphenyl-1,3-butadiene sandwich complexes. *Chem. Commun.* **2004**, 1430–1431. [CrossRef] [PubMed]
14. Nakamae, K.; Takeemura, Y.; Kure, B.; Nakajima, T.; Kitagawa, Y.; Tanase, T. Self-alignment of low-valent octanuclear palladium atoms. *Angew. Chem. Int. Ed.* **2015**, *54*, 1016–1023. [CrossRef] [PubMed]
15. Goto, E.; Begum, R.A.; Zhan, S.; Tanase, T.; Tanigaki, K.; Sakai, K. Linear, redox-active Pt_6 and $Pt_2Pd_2Pt_2$ clusters. *Angew. Chem. Int. Ed.* **2004**, *43*, 5029–5032. [CrossRef] [PubMed]
16. Goto, E.; Begum, R.A.; Ueno, C.; Hosokawa, A.; Yamamoto, C.; Nakamae, K.; Liure, B.; Nakajima, T.; Kajiwara, T.; Tanase, T. Electron-deficient $Pt_2M_2Pt_2$ hexanuclear metal strings (M = Pt, Pd) supported by triphosphine ligands. *Organometallics* **2014**, *33*, 1893–1904. [CrossRef]
17. Takemura, Y.; Takenaka, H.; Nakajima, T.; Tanase, T. Hexa- and octagold chains from flexible tetragold molecular units supported by linear tetraphosphine ligands. *Angew. Chem. Int. Ed.* **2009**, *48*, 2157–2161. [CrossRef] [PubMed]
18. Hua, S.-A.; Cheng, M.-C.; Chen, C.-H.; Peng, S.-M. From homonuclear metal string complexes to heteronuclear metal string complexes. *Eur. J. Inorg. Chem.* **2015**, *2015*, 2510–2523. [CrossRef]
19. Hurley, T.J.; Robinson, M.A. Nickel(II)-2,2′-dipyridylamine system. I. Synthesis and stereochemistry of the complexes. *Inorg. Chem.* **1968**, *7*, 33–38. [CrossRef]

20. Wu, L.P.; Field, P.; Morrissey, T.; Murphy, C.; Nagle, P.; Hathaway, B.; Simmons, C.; Thornton, P. Crystal structure and electronic properties of dibromo- and dichloro-tetrakis[μ_3-bis(2-pyridyl)amido]tricopper(II) hydrate. *J. Chem. Soc. Dalton Trans.* **1990**, 3835–3840. [CrossRef]
21. Liu, I.P.-O.; Bénard, M.; Hasanov, H.; Chen, I.-W.P.; Tseng, W.-H.; Fu, M.-D.; Rohmer, M.-M.; Chen, C.-H.; Lee, G.-H.; Peng, S.-M. A new generation of metal string complexes: Structure, magnetism, spectroscopy, theoretical analysis, and single molecular conductance of an unusual mixed-valence linear $[Ni_5]^{8+}$ complex. *Chem. Eur. J.* **2007**, *13*, 8667–8677. [CrossRef] [PubMed]
22. Ismayilov, R.-H.; Wang, W.-Z.; Lee, G.-H.; Yeh, C.-Y.; Hua, S.-A.; Song, Y.; Rohmer, M.-M.; Bénard, M.; Peng, S.-M. Two linear undecanickel mixed-valence complexes: Increasing the size and the scope of the electronic properties of nickel metal strings. *Angew. Chem. Int. Ed.* **2011**, *50*, 2045–2048. [CrossRef] [PubMed]
23. Murahashi, T.; Fujimoto, M.; Oka, M.; Hashimoto, Y.; Uemura, T.; Tatsumi, Y.; Nakao, Y.; Ikeda, A.; Sakaki, S.; Kurosawa, H. Discrete sandwich compounds of monolayer palladium sheets. *Science* **2006**, *313*, 1104–1107. [CrossRef] [PubMed]
24. Murahashi, T.; Inoue, R.; Usui, K.; Ogoshi, S. Square tetrapalladium sheet sandwich complexes: Cyclononatetraenyl as a versatile face-capping ligand. *J. Am. Chem. Soc.* **2009**, *131*, 9888–9889. [CrossRef] [PubMed]
25. Murahashi, T.; Usui, K.; Inoue, R.; Ogoshi, S.; Kurosawa, H. Metallocenoids of platinum: Syntheses and structures of triangular triplatinum sandwich complexes of cycloheptatrienyl. *Chem. Sci.* **2011**, *2*, 117–122. [CrossRef]
26. Murahashi, T.; Kato, N.; Uemura, T.; Kurosawa, H. Rearrangement of a Pd_4 skeleton from a 1D Chain to a 2D sheet on the face of a perylene or fluoranthene ligand caused by exchange of the binder molecule. *Angew. Chem. Int. Ed.* **2007**, *46*, 3509–3512. [CrossRef] [PubMed]
27. Ishikawa, Y.; Kimura, S.; Yamamoto, K.; Murahashi, T. Bridging coordination of vinylarenes to Pd_3- or Pd_4 cluster sites. *Chem. Eur. J.* **2017**, *23*, 14149–14152. [CrossRef] [PubMed]
28. Murahashi, T.; Uemura, T.; Kurosawa, H. Perylene—Tetrapalladium sandwich complexes. *J. Am. Chem. Soc.* **2003**, *125*, 8436–8437. [CrossRef] [PubMed]
29. Sunada, Y.; Haige, R.; Otsuka, K.; Kyushin, S.; Nagashima, H. A ladder polysilane as a template for folding palladium nanosheets. *Nat. Commun.* **2013**, *4*, 3014/1–3014/7. [CrossRef] [PubMed]
30. Suginome, M.; Kato, Y.; Takeda, N.; Oike, H.; Ito, Y. Reactions of a spiro trisilane with palladium complexes: synthesis and structure of Tris(organosilyl)$CpPd^{IV}$ and Bis(organosilyl)(μ-organosilylene)Pd^{II}_2 complexes. *Organometallics* **1998**, *17*, 495–497. [CrossRef]
31. Ansell, M.B.; Navarro, O.; Spencer, J. Transition metal catalyzed element–element' additions to alkynes. *Coord. Chem. Rev.* **2017**, *336*, 54–77. [CrossRef]
32. Ansell, M.B.; Roberts, D.E.; Cloke, F.G.N.; Navarro, O.; Spencer, J. Synthesis of an $[(NHC)_2Pd(SiMe_3)_2]$ complex and catalytic *cis*-bis(silyl)ations of alkynes with unactivated disilanes. *Angew. Chem. Int. Ed.* **2015**, *54*, 5578–5582. [CrossRef] [PubMed]
33. Suginome, M.; Oike, H.; Park, S.-S.; Ito, Y. Reactions of Si–Si σ-bonds with Bis(*t*-alkyl isocyanide)palladium(0) complexes, synthesis and reactions of cyclic Bis(organosilyl)palladium complexes. *Bull. Chem. Soc. Jpn.* **1996**, *69*, 289–299. [CrossRef]
34. Chen, Y.; Sunada, Y.; Nagashima, H.; Sakaki, S. Theoretical study of $Pd_{11}Si_6$ nanosheet compounds including seven-coordinated Si species and its Ge analogues. *Chem. Eur. J.* **2016**, *22*, 1076–1087. [CrossRef] [PubMed]
35. Yamada, T.; Mawatari, A.; Tanabe, M.; Osakada, K.; Tanase, T. Planar tetranuclear and dumbbell-shaped octanuclear palladium complexes with bridging silylene ligands. *Angew. Chem. Int. Ed.* **2009**, *48*, 568–571. [CrossRef] [PubMed]
36. Tanabe, M.; Yumoto, R.; Yamada, T.; Fukuta, T.; Hoshino, T.; Osakada, K.; Tanase, T. Planar $PtPd_3$ complexes stabilized by three bridging silylene ligands. *Chem. Eur. J.* **2017**, *23*, 1386–1392. [CrossRef] [PubMed]
37. Shimada, S.; Li, Y.-H.; Choe, Y.-K.; Tanaka, M.; Bao, M.; Uchimaru, T. Multinuclear palladium compounds containing palladium centers ligated by five silicon atoms. *Proc. Nat. Acad. Sci. USA* **2007**, *104*, 7758–7763. [CrossRef] [PubMed]
38. Tanabe, M.; Ishikawa, N.; Chiba, M.; Ide, T.; Osakada, K.; Tanase, T. Tetrapalladium complex with bridging germylene ligands. Structural change of the planar Pd_4Ge_3 core. *J. Am. Chem. Soc.* **2011**, *133*, 18598–18601. [CrossRef] [PubMed]

39. Beck, R.; Johnson, S.A. Structural similarities in dinuclear, tetranuclear, and pentanuclear nickel silyl and silylene complexes obtained via Si–H and Si–C activation. *Organometallics* **2012**, *31*, 3599–3609. [CrossRef]
40. Bradford, A.M.; Kristof, E.; Rashidi, M.; Yang, D.-S.; Payne, N.C.; Puddephatt, R.J. Isocyanide and diisocyanide complexes of a triplatinum cluster: Fluxionality, isomerism, structure, and bonding. *Inorg. Chem.* **1994**, *33*, 2355–2363. [CrossRef]
41. Mann, B.E. Mechanism of the low-energy fluxional process in $[Fe_3(CO)_{12-n}L_n]$ (n = 0–2): A perspective. *J. Chem. Soc. Dalton Trans.* **1997**, 1457–1471. [CrossRef]
42. Corey, J.Y. Reactions of hydrosilanes with transition metal complexes. *Chem. Rev.* **2016**, *116*, 11291–11435. [CrossRef] [PubMed]
43. Corey, J.Y. Reactions of hydrosilanes with transition metal complexes and characterization of the products. *Chem. Rev.* **2011**, *111*, 863–1071. [CrossRef] [PubMed]
44. Ogino, H.; Tobita, H. Bridged silylene and germylene complexes. *Adv. Organomet. Chem.* **1998**, *42*, 223–290. [CrossRef]
45. Tanaka, K.; Kamono, M.; Tanabe, M.; Osakada, K. Ring expansion of cyclic triplatinum(0) silylene complexes induced by insertion of alkyne into a Si–Pt bond. *Organometallics* **2015**, *34*, 2985–2990. [CrossRef]
46. Braddock-Wilking, J.; Corey, J.Y.; Dill, K.; Rath, N.P. Formation and X-ray crystal structure determination of the novel triplatinum cluster $[(Ph_3P)Pt(\mu\text{-}SiC_{12}H_8)]_3$ from reaction of silafluorene with $(Ph_3P)_2Pt(\eta^2\text{-}C_2H_4)$. *Organometallics* **2002**, *21*, 5467–5469. [CrossRef]
47. Osakada, K.; Tanabe, M.; Tanase, T. A triangular triplatinum complex with electron-releasing $SiPh_2$ and PMe_3 ligands: $[\{Pt(\mu\text{-}SiPh_2)(PMe_3)\}_3]$. *Angew. Chem. Int. Ed.* **2000**, *39*, 4053–4055. [CrossRef]
48. Braddock-Wilking, J.; Corey, J.Y.; French, L.M.; Choi, E.; Speedie, V.J.; Rutheeford, M.F.; Yao, S.; Xu, H.; Rath, N.P. Si–H bond activation by $(Ph_3P)_2Pt(\eta^2\text{-}C_2H_4)$ in dihydrosilicon tricycles that also contain O and N heteroatoms. *Organometallics* **2006**, *25*, 3974–3988. [CrossRef]
49. Watanebe, H.; Muraoka, T.; Kageyama, M.; Yoshizumi, K.; Nagai, Y. Synthesis and some spectral properties of peralkylcyclopolysilanes, $[R^1R^2Si]_n$ (n = 4–7). *Organometallics* **1984**, *3*, 141–147. [CrossRef]
50. Ryan, S.J.; Schimler, S.D.; Bland, D.C.; Sanford, M.S. Acyl azolium fluorides for room temperature nucleophilic aromatic fluorination of chloro- and nitroarenes. *Org. Lett.* **2015**, *17*, 1866–1869. [CrossRef] [PubMed]
51. Burla, M.C.; Caliandro, R.; Camalli, M.; Carrozzini, B.; Cascarano, G.L.; Giacovazzo, C.; Mallamo, M.; Mazzone, A.; Polidori, G.; Spagna, R. SIR2011: A new package for crystal structure determination and refinement. *J. Appl. Cryst.* **2012**, *45*, 357–361. [CrossRef]
52. Beurskens, P.T.; Admiraal, G.; Beurskens, G.; Bosman, W.P.; de Gelder, R.; Israel, R.; Smits, J.M.M. *The DIRDIF-99 program system*; Technical Report of the Crystallography Laboratory; University of Nijmegen: Nijmegen, The Netherlands, 1999.
53. Cromer, D.T.; Waber, J.T. *International Tables for X-ray Crystallography*; Kynoch Press: Birmingham, UK, 1974.
54. *CrystalStructure 4.0*, Crystal Structure Analysis Package; Rigaku Corporation: Tokyo, Japan, 2000–2010.
55. Sheldrick, G.M. A short history of *SHELX*. *Acta Cryst.* **2008**, *A64*, 112–122. [CrossRef] [PubMed]

16

Synthesis and Functionalization of a 1,2-Bis(trimethylsilyl)-1,2-disilacyclohexene that can Serve as a Unit of *cis*-1,2-Dialkyldisilene

Naohiko Akasaka, Kaho Tanaka, Shintaro Ishida and Takeaki Iwamoto *

Department of Chemistry, Graduate School of Science, Tohoku University, Aoba-ku, Sendai 980-8578, Japan; nao.akasaka1216@gmail.com (N.A.); kaho.tanaka.q2@dc.tohoku.ac.jp (K.T.); sishida@m.tohoku.ac.jp (S.I.)

* Correspondence: iwamoto@m.tohoku.ac.jp

Abstract: π-Electron compounds that include multiple bonds between silicon atoms have received much attention as novel functional silicon compounds. In the present paper, 1,2-bis(trimethylsilyl)-1,2-disilacyclohexene **1** was successfully synthesized as thermally stable yellow crystals. Disilene **1** was easily converted to the corresponding potassium disilenide **4**, which furnished novel functionalized disilenes after the subsequent addition of an electrophile. Interestingly, two trimethylsilyl groups in **1** can be stepwise converted to anthryl groups. The novel disilenes derived from **1** were characterized by a combination of nuclear magnetic resonance (NMR) spectroscopy, mass spectrometry (MS), elemental analyses, and X-ray single crystal diffraction analysis. The present study demonstrates that disilene **1** can serve as a unit of *cis*-1,2-dialkyldisilene.

Keywords: disilene; functionalization; π-electron systems

---

1. Introduction

Stable silicon compounds that include double bonds between silicon atoms (disilenes, $R_2Si{=}SiR_2$) have received much attention over the last three decades [1–3]. Because such silicon π-electron systems have an intrinsically higher π-orbital level and a narrower HOMO (highest occupied molecular orbital)–LUMO (lowest unoccupied molecular orbital) gap compared to those of the corresponding organic π-electron systems, extended π-electron systems that include the Si=Si double bond(s) should be anticipated to be unprecedented functional π-electron materials. In this context, the reactions of disilenide (a disilicon analogue of vinyl anion $[R_2Si{=}SiR]^-$) with electrophiles is one of the promising routes to introduce a functional group into the silicon π-electron systems [4–9]. Disilenides have been synthesized by reductive dehalogenation of the corresponding trihalodisilane [10] or reductive cleavage of R–Si(sp^2) bond on the Si=Si double bond in a stable disilene [11–15].

Very recently, we found a novel route to a disilenide from a stable disilene under milder conditions [16]. In this route, the disilenide was generated from the reaction of a trimethylsilyl-substituted disilene and potassium *t*-butoxide [17] via selective cleavage of the Si(sp^2)–Si(sp^3) bond on the Si=Si double bond. In this reaction, neither an undesired reduction of the Si=Si double bond nor addition of *t*-BuOK across the Si=Si double bond occurred. During the course of our study, we designed a novel cyclic disilene, 1,2-bis(trimethylsilyl)-substituted 1,2-disilacyclohexene **1**. Compound **1** has two trimethylsilyl groups that can be converted stepwise to the corresponding potassium derivatives after treatment of *t*-BuOK, as well as alkyl groups that cause at least perturbation to the electronic structure of the Si=Si double bond and should be suitable for the investigation of interactions between the Si=Si double bond and the functional group. Herein, we report successful synthesis, molecular structure and functionalization of **1** and its derivatives. Although **1** has six trimethylsilyl groups on the six-membered ring, one trimethylsilyl group on the Si=Si double bond is selectively eliminated to provide the corresponding disilenide after treatment with *t*-BuOK. Noticeably, two trimethylsilyl groups in **1** were converted stepwise to anthryl groups to furnish the corresponding 1,2-dianthryldisilene in good yield.

2. Results and Discussion

2.1. Synthesis and Molecular Structure of Disilacyclohexene 1

1,2-Bis(trimethylsilyl)-1,2-disilacyclohexene **1** was synthesized as shown in Scheme 1 similar to that of 1,2-diphenyl-3,3,6,6-tetrakis(trimethylsilyl)-1,2-disilacyclohexene **A** [18] (Figure 1). The reaction of Me_3SiSiH_2Cl, which was generated by the reaction of Me_3SiSiH_2Ph with triflic acid followed by addition of $NEt_3{\cdot}HCl$ salt, with 1,4-dilithio-1,1,4,4-tetrakis(trimethylsilyl)butane, which was also generated from 1,1-bis(trimethylsilyl)ethene and lithium in tetrahydrofuran (THF), afforded 1,4-bis(2,2,2-trimethyldisilanyl)butane **2** in 89% yield. Bromination of **2** with Br_2 provided tetrabromo derivative **3** in 70% yield. Finally, treatment of **3** with potassium graphite (KC_8) in THF furnish **1** in 99% yield as yellow crystals. The structure of **1** was determined by a combination of multinuclear nuclear magnetic resonance (NMR) spectroscopy, mass spectrometry (MS) and X-ray diffraction (XRD) analysis (Figure 2, Table 1 (vide infra)). The six-membered ring of **1** adopts a half-chair conformation similar to the corresponding 1,2-diphenyl derivative **A**.

The Si1–Si2 bond length in **1** (2.1762(5) Å) is slightly longer than that in **A** (2.1595(5) Å). The geometry around the Si=Si double bond of **1** is trans-bent and twisted; bent angles β of **1** (Si1: 12.9°; Si2: 6.5° (Si2)) are smaller than those of **A** (Si1: 13.7°; Si2: 19.2° (Si2)), while the twist angle τ of **1** (17.7°) is larger than that of **A** (4.8°). These substantial differences in the geometry around the Si=Si bond would be due to a combination of the electronic effects of silyl substituents that favor a planar geometry [19–21] and the steric effects of the silyl substituents, as the potential energy surface of the bending of the Si=Si double bond has been predicted to be shallow [20,22]. In ^{1}H, ^{13}C and ^{29}Si NMR spectra, three singlet signals of trimethylsilyl groups (one is on the unsaturated silicon atom and the other two are on the alkyl substituent), which indicates that the ring inversion occurs slowly on the NMR time scale in solution. The ^{29}Si resonance of unsaturated silicon nuclei of **3** (131.4 ppm) are slightly downfield-shifted compared to that of **A** (100.9 ppm), which should be due to the substituent effects of silyl group on the Si=Si double bond [23]. In the UV–Vis absorption spectrum of **1** in hexane, a distinct absorption band assignable to a π(Si=Si)$\rightarrow\pi^*$(Si=Si) transition (see, Figure S46) appeared at 420 nm (ε 1.57 × 10^4), which is slightly hypsochromically shifted compared to that of **A** (427 nm (ε 8400) in hexane).

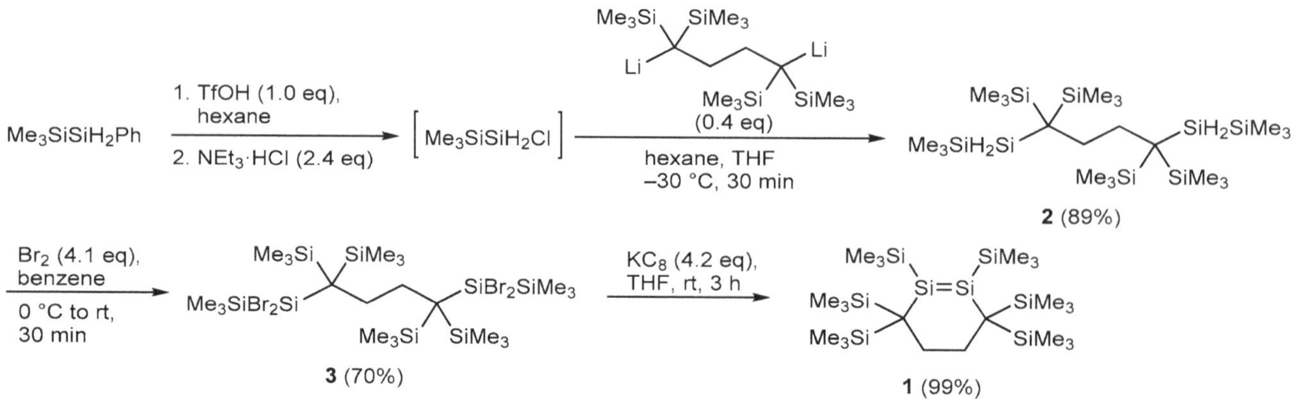

Scheme 1. Synthesis of disilene **1**.

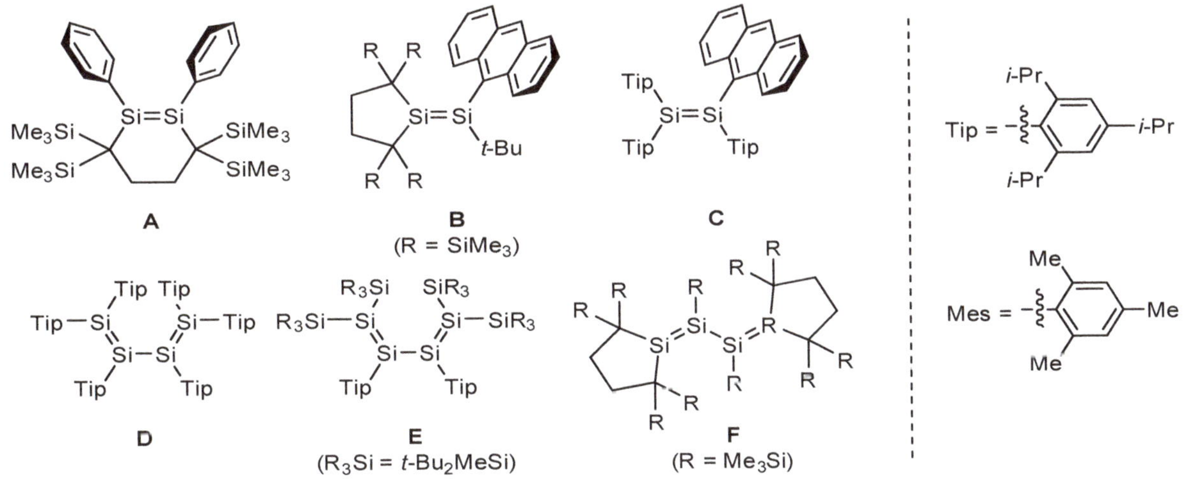

Figure 1. Related disilenes **A** [18], **B** [10], and **C** [24], as well as tetrasila-1,3-dienes **D** [25], **E** [12], and **F** [26].

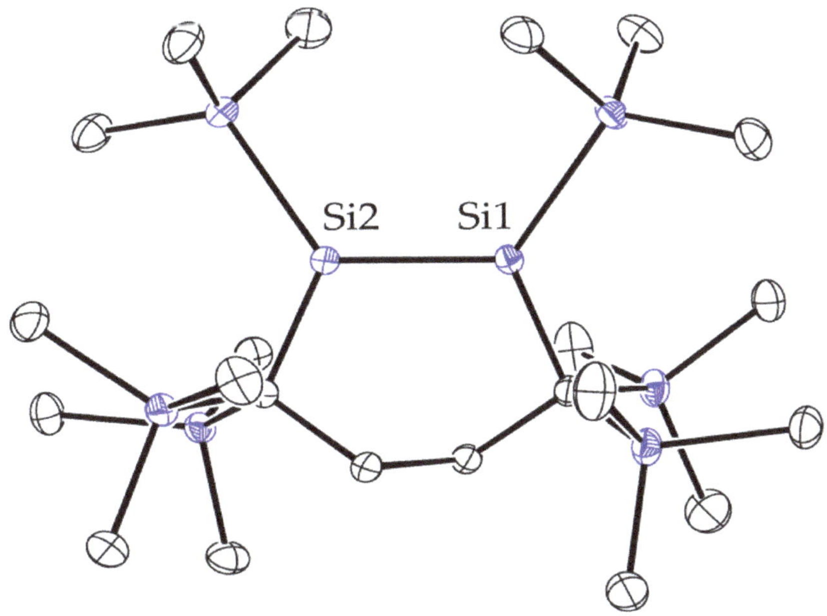

Figure 2. ORTEP (Oak Ridge Thermal Ellipsoid Plot) drawing of **1** with thermal ellipsoids set at 50% probability and hydrogen atoms omitted for clarity.

2.2. *Conversion of* **1** *to Disilenide* **4** *and Functionalized Disilenes*

Disilene **1** can be effectively converted to the corresponding disilenide through a selective cleavage of the Si(sp^2)–Si(sp^3) bond (Scheme 2) [16]. When **1** was treated with one equivalent of *t*-BuOK in THF, the color of solution gradually turned from yellow to orange. The formation of potassium disilenide **4** as a sole product in the resulting reaction mixture was confirmed by a combination of the multinuclear NMR spectra in C_6D_6 and the results of the reaction with electrophiles (vide infra). In this reaction, 1,2-addition of *t*-BuOK across the Si=Si double bond of **1** as well as elimination of a trimethylsilyl group on the carbon atom in the disilacyclohexene ring was not observed at all. The ^{1}H, ^{13}C, and ^{29}Si NMR spectra of **4**(thf) in C_6D_6 exhibits the presence of five trimethylsilyl groups. The ^{29}Si NMR spectra exhibited two signals due to the unsaturated silicon nuclei at 146.9 ppm (=*Si*$SiMe_3$) and 219.3 ppm (=Si^-). Similar large differences between the chemical shifts of the unsaturated silicon nuclei have been found in those of structurally related disilenides (Me_3Si)TipSi=SiTip[K(thf)$_n$] (101.4 (=*Si*($SiMe_3$)Tip), 186.6 (=Si^-) in THF-d_8; Tip = 2,4,6-triisopropylphenyl) [16] and reported disilenides [27].

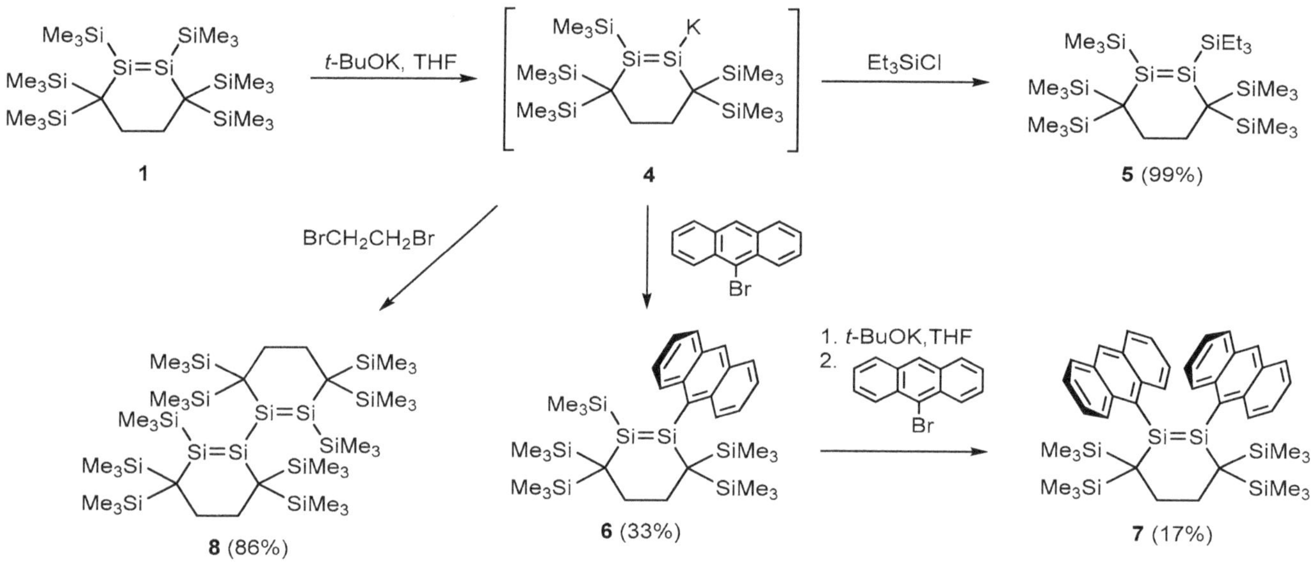

Scheme 2. Synthesis of functionalized disilenes through desilylation of **1**.

After removal of the volatiles, recrystallization from toluene provided suitable single crystals of (**4**(toluene))$_2$. In the single crystals, disilenide **4** forms a dimer with a crystallographic inversion center (Figure 3a). Interestingly, the potassium cations are coordinated by one anionic silicon atom in an η^1-fashion, one toluene molecule in an η^6-fashion, and the Si=Si double bond in the other disilenide moiety in an η^2-fashion. The distance of Si1–K1 is 3.4655(5) Å, which falls in typical range of Si1–K1 distance in the reported potassium disilenides (3.33–3.52 Å) [10,13,28], while the Si1···K1′ and Si2···K1′ distances are 3.5066(5) and 3.7645(5) Å. Although the dimeric structure of metal disilenides in the solid state have been reported [10,28,29], to the best of our knowledge, such η^2-coordination of the Si=Si double bond in a disilenide to the metal cation in the solid state is unprecedented. The Si=Si distance in (**4**(toluene))$_2$ (2.2035(5) Å) is substantially elongated compared that of neutral disilene **1** (2.1762(5) Å) and the geometry around the Si=Si double bond is slightly *cis*-bent (bent angle β: 3.7° (*Si1*–K1), 1.5° (*Si2*–SiMe$_3$); twist angle τ: 1.5°), which may result from the η^2-coordination of the Si=Si double bond to the K cation.

Treatment of **4**, which was generated from **1** and *t*-BuOK in THF, with an electrophile provide various functionalized disilenes (Scheme 2). For instance, reaction of **4** with triethylchlorosilane gave Et$_3$Si-substituted disilene **5** as yellow crystals in 99% yield. The reaction of **4** with one equivalent of 9-bromoanthracene furnished the corresponding anthryldisilene **6** (33% yield). In a similar manner, (10-bromo-9-anthryl)disilene **6^{Br}** was obtained as a major product from the reaction of **4** and one equivalent of 9,10-dibromoanthracene, although it was not obtained in a pure form due to the inseparable byproducts such as anthracene and 9-bromoanthracene. Although the reactions of **4** with less bulky bromoarenes such as bromobenzene or bromomesitylene (2,4,6-trimethylbromobenzene) afforded a mixture which may contain the desired phenyl or mesityl-substituted disilenes, isolation of these disilenes was unsuccessful probably due to the instability of the resulting less bulky aryldisilenes under these reaction conditions. Noticeably, disilene **6** underwent a further desilylation reaction followed by addition of 9-bromoanthracene to furnish 1,2-dianthryldisilene **7** as black purple crystals in 17% yield. The reaction of **4** with 0.5 equivalent of 1,2-dibromoethane afforded the corresponding tetrasiladiene **8** in 86% yield as red-orange crystals similar to the reaction of Tip$_2$Si=SiTipLi and mesityl bromide leading to the corresponding hexaryltetrasila-1,3-diene **D** (Figure 1) [25].

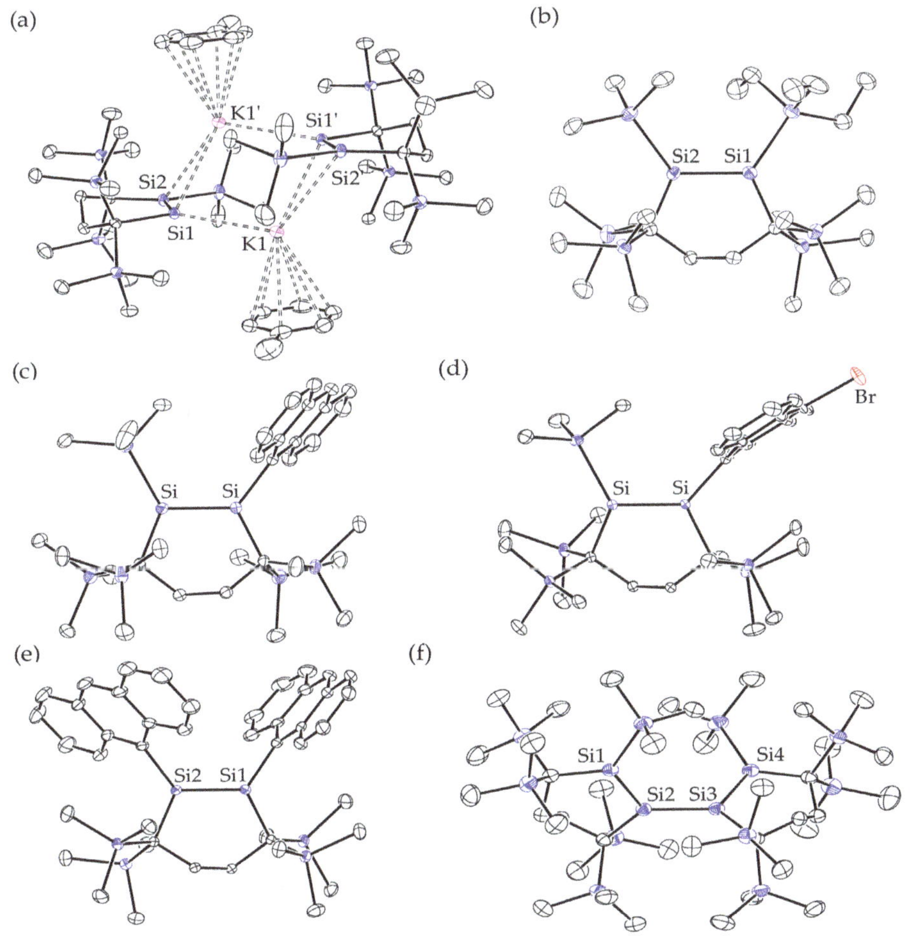

Figure 3. ORTEP drawings of (**a**) (**4**(toluene))$_2$, (**b**) **5**, (**c**) **6**, (**d**) **6^{Br}**, (**e**) **7**, and (**f**) **8** with thermal ellipsoids set at 50% probability and hydrogen atoms omitted for clarity.

Table 1. Selected structural parameters and spectral data of disilenes.

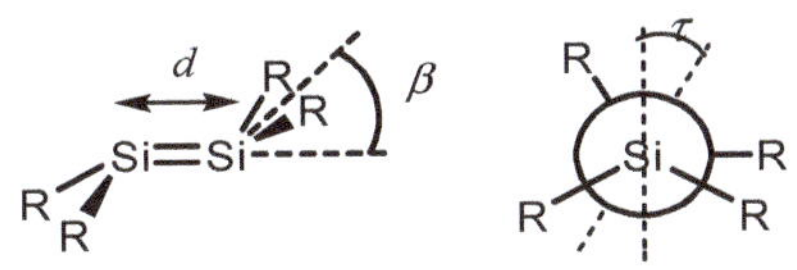

Cpd	d/Å	Angle Sum at Si/°	β/°	τ/°	δ(^{29}Si) [1]	λ_{max}/nm (ε) [2]
1	2.1762(5)	358.52(3) (*Si1*–SiMe$_3$), 359.61(3) (*Si2*–SiMe$_3$)	12.9 (*Si1*–SiMe$_3$), 6.5 (*Si2*–SiMe$_3$)	17.7	131.4 (=*Si*–SiMe$_3$)	420 (1.57 × 10^4), 343 (4.63 × 10^3)
4	2.2035(5)	358.04(3) (=*Si*···K), 359.94(3) (=*Si*–SiMe$_3$)	3.7 (=*Si*···K) [3], 1.5 (=*Si*–SiMe$_3$) [3]	2.5	146.9 (=*Si*–SiMe$_3$) 219.3 (=*Si*$^-$)	–
5	2.1860(19)	358.8(1) (=*Si*–SiEt$_3$), 358.6(1) (=*Si*–SiMe$_3$)	11.2 (=*Si*–SiEt$_3$), 12.1 (=*Si*–SiMe$_3$)	17.6	123.5 (=*Si*–SiEt$_3$), 134.2 (=*Si*–SiMe$_3$)	–
6	2.1598(6)	360.00(5) (=*Si*–Ant), 355.98(3) (=*Si*–SiMe$_3$)	0.1 (=*Si*–Ant), 22.5 (=*Si*–SiMe$_3$)	0.1	130.3 (=*Si*–Ant), 74.9 (=*Si*–SiMe$_3$)	535 (9.25 × 10^2), 399 (2.64 × 10^4)
6^{Br}	2.1711(7)	359.76(6) (=*Si*–Ant^{Br}), 356.04(5) (=*Si*–SiMe$_3$)	4.7 (=*Si*–Ant^{Br}), 22.4 (=*Si*–SiMe$_3$)	8.5	75.1 (=*Si*–SiMe$_3$), 129.1 (=*Si*–Ant^{Br})	[578 (9.67 × 10^2), 410 (3.08 × 104)] [5]
7	2.1525(7)	357.10(6) (*Si1*–Ant), 355.71(6) (*Si2*–Ant)	17.4 (=*Si1*–Ant), 21.0 (=*Si2*–Ant)	3.3	88.2 (=*Si*–Ant)	543 (1.68 × 10^3) 399 (1.85 × 10^4)
8 [4]	2.1850(4) (*Si1*=*Si2*) 2.1915(5) (*Si3*=*Si4*)	359.99(4) (*Si1*), 359.99(3) (*Si2*) 359.98(3) (*Si3*) 359.94(4) (*Si4*)	0.3 (*Si1*) 1.2 (*Si2*) 1.0 (*Si3*) 2.4 (*Si4*)	0.7 (*Si1*=*Si2*) 1.8 (*Si3*=*Si4*)	106.4 (*Si*=SiSiMe$_3$), 143.0 (Si=*Si*SiMe$_3$)	~500 (sh, 3.57 × 10^3), 437 (3.42 × 10^4)

[1] Measured in benzene-d_6. [2] Measured in hexane. [3] *cis*-bent. [4] Dihedral angle Si1–Si2–Si3–Si4 = −88.36(2)°. [5] The sample contains **6** (~10%).

2.3. Molecular Structures of Disilenes 5–8

Similar to **1**, the Si=Si double bond in triethylsilyl-substituted disilene **5** adopts a *trans*-bent and twist geometry (Figure 3b, Table 1): the Si=Si double bond distance (2.1860(19) Å) is slightly longer than that in **1** (2.1762(5) Å) possibly due to the increased steric bulkiness of the silyl group. The ^{29}Si chemical shifts of double bonded Si nuclei in **5** (123.5 (Si–$SiEt_3$) and 134.2 ppm (Si–$SiMe_3$)) are close to that of **1** (131.4 ppm).

Monoanthryldisilene **6** has a slightly shorter Si=Si double bond distance (2.1598(6) Å) compared to that of **1** (Figure 3c). The geometry around the Si atom bonded to the anthryl group is almost planar (the angle sum: 360.00(5)°), while that bonded to the $SiMe_3$ group is significantly pyramidalized (355.98 (3)°). Such larger pyramidalization at the silyl-substituted silicon center compared to that at the phenyl-substituted center have been predicted theoretically in an unsymmetrical substituted disilene $Ph_2Si{=}Si(SiH_3)_2$ [30]. The disilene π (πSi) and aromatic π (πC) systems are almost perpendicular to each other (dihedral angle δ: 76.0°). A similar structural feature was found in 10-bromoanthryl-substituted disilene $\mathbf{6^{Br}}$ (the angle sum: 359.76(6)° (Si–Ant^{Br}), 356.04(5)° (Si–$SiMe_3$); dihedral angle δ: 81.9°) except for the longer Si=Si double bond distance (2.1711(7) Å) (Figure 3d). Although the structural characteristics of **6** and $\mathbf{6^{Br}}$ were essentially consistent with the optimized structures of **6** and $\mathbf{6^{Br}}$ ($\mathbf{6_{opt}}$ and $\mathbf{6^{Br}{}_{opt}}$) at the B3PW91-D3/6-31G(d) level of theory (Table S6), the Si=Si distance of $\mathbf{6^{Br}}$ (2.1711(7) Å) observed in the single crystals is considerably longer than that of the calculated value of $\mathbf{6^{Br}{}_{opt}}$ (2.1574 Å). The observed longer Si=Si distance in $\mathbf{6^{Br}}$ may be due to the significant intermolecular C–H···Br–C interaction in the crystals (H···Br distance: 3.08 Å) [31–34] (Figure 4). The substantial upfield-shifted ^{29}Si resonances of silyl-substituted double bond Si nuclei compared to that of aryl-substituted Si nuclei in **6** and $\mathbf{6^{Br}}$ (**6**: 74.9 (Si–$SiMe_3$) and 130.3 ppm (Si–anthryl); $\mathbf{6^{Br}}$: 75.1 (Si–$SiMe_3$) and 129.1 ppm (Si–anthryl)) have often been observed in 1-aryl-2-silyldisilenes [11,15,35].

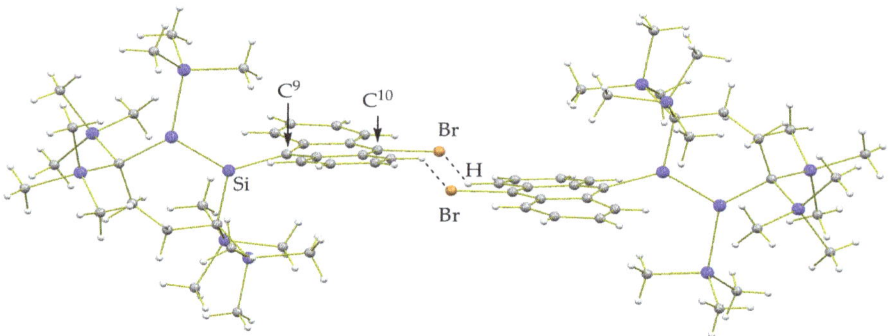

Figure 4. Crystal structure of $\mathbf{6^{Br}}$. The nearest Br···H distance is 3.08 Å and the Si–C^9–C^{10} angle is 164.7°.

1,2-Dianthryldisilene **7** also adopts a moderately *trans*-bent and slightly twisted geometry around the Si=Si double bond (bent angle β: 17.4° (Si1) and 21.0° (Si2); twist angle τ: 3.3°) (Figure 3e). The Si=Si distance of **7** (2.1525(7) Å) is shorter than those of disilenes **1**, **5**, **6** and $\mathbf{6^{Br}}$ probably due to the absence of the $SiMe_3$ group. The disilene π (πSi) and aromatic π (πC) systems in **7** are considerably twisted with respect to each other (dihedral angle δ: 79.7° (Si1) and 63.5° (Si2)). The upfield-shifted ^{29}Si resonance of this double bonded silicon nuclei in **7** (88.2 ppm) compared to that in **1** (131.4 ppm) is often observed in symmetrically-substituted silyl-substituted disilenes and aryl-substituted disilenes [1–3].

Similar to the reported anthryl-substituted disilenes [10,24,36], **6**, $\mathbf{6^{Br}}$, and **7** exhibit a weak and broad absorption band I assignable to π(Si=Si)→π*(anthryl) transition in the visible region as well as a structured intense band II that involves π(anthryl)→π*(anthryl) and π(Si=Si)→π*(Si=Si) transitions (300–400 nm). The absorption maximum of band I of **6** in hexane (535 nm (ε 9.3 × 10^2)) is slightly bathochromically shifted compared to that of trialkylanthryldisilene **B** (525 nm (ε 4.2 × 10^2)) (Figure 1) [10], which would be due to the presence of electron-donating silyl-substituents, while it is moderately hypsochromically shifted compared to triarylanthryldisilene **C** (550 nm (ε 3800)) [24]

(Figure 1) probably due to the absence of extra aryl groups. The maximum of band I of $\mathbf{6^{Br}}$ (578 nm) is considerably bathochromically shifted compared to that of **6**. The lower-lying π^* orbitals in 10-bromo-9-anthryl group compared to 9-anthryl group may be responsible for the bathochromic shift of band I, which was qualitatively reproduced by density functional theory (DFT) calculations of **6** and $\mathbf{6^{Br}}$ (see, Figures S47 and S48). The absorption maximum of band I of 1,2-dianthryldisilene **7** (543 nm (1.68×10^3)) (Figure S49) is close to that of **6** and almost twice as large as that of **6**, which is consistent with the presence of two anthryl groups on the Si=Si double bond.

Two Si=Si double bonds in **8** are almost perpendicular to each other with the dihedral angle Si1–Si2–Si3–Si4 of 88.35(3)° (Figure 3f), which would be mainly due to the severe steric congestions around the central Si2–Si3 bond similar to the hitherto known tetrasila-1,3-dienes, such as Tip_2Si=SiTip–SiTip=$SiTip_2$ (**D**, dihedral angle: 51°) [25], ((t-$Bu_2MeSi)_2Si$=SiMes–SiMes=Si(Sit-$Bu_2Me)_2$) (**E**, 72(2)°) [12]), and R_2Si=Si($SiMe_3$)–Si($SiMe_3$)=SiR_2 (R_2 = 1,1,4,4-tetrakis(trimethylsilyl)butan-1,4-diyl) (**F**, 122.56(7)°) [26] (Figure 1). Each Si=Si double bond in **8** adopts a planar structure with the bent angles of 0.3–2.4°. While the Si2–Si3 distance (2.3392(5) Å) falls in the typical range of the Si–Si single bond, the Si=Si bond lengths of Si1–Si2 and Si3–Si4 in **8** (2.1850(5) and 2.1914(5) Å, respectively) are slightly longer than that of **1** (2.1762(5) Å). The ultraviolet-visible (UV-vis) absorption spectrum of **8** exhibits an intense and broad absorption band (437 nm (ε 3.42×10^4)) (Figure 5) accompanied by a shoulder peak at around 500 nm (ε 3.57×10^3). These bands are substantially bathochromically shifted compared to that of monodisilene **1** [420 (ε 1.57 × 104)], which suggests significant interactions between two Si=Si double bonds in **8**. DFT calculation provided further information on the interactions between two Si=Si double bonds in **8**. The optimized structure of **8** ($\mathbf{8_{opt}}$) at the B3PW91-D3/6-31G(d) level of theory is roughly consistent with the structure obtained from XRD analysis: the dihedral angle of two Si=Si bond of $\mathbf{8_{opt}}$ is 71.4° and the Si=Si distances are 2.1904 and 2.1907 Å. Even though the dihedral angle between the two Si=Si double bonds is large, the frontier orbitals involve two split π orbitals delocalized over two Si=Si double bonds (−4.61 and −4.79 eV at the B3LYP/6-311G(d)//B3PW91-D3/6-31G(d) level of theory), while LUMO and LUMO+1 correspond to π^* orbitals (−1.75 and −1.46 eV). Time-dependent (TD) DFT calculation at the same level predicted four $\pi \rightarrow \pi^*$ transitions, which is qualitatively consistent with the observed absorption bands of **8** (Figure S50). These results are consistent with a recent theoretical study that two Si=Si in tetrasila-1,3-diene can interact each other even at the dihedral angle of 90° due to the intrinsic non-planar geometry around the Si=Si double bond [16]. The longest wavelength absorption bands found in **8** (~500 nm) are hypsochromically-shifted compared those of the reported aryl-substituted tetrasila-1,3-dienes **D** (518 nm) [25], **E** (531 nm) [12] probably due to the absence of aryl-substituents, but close to that of structurally similar tetrasila-1,3-diene **F** (510 nm) [26] (Figure 1). The ^{29}Si resonances of **8** (106.4 (–Si=$SiSiMe_3$) and 143.0 ppm (–Si=$Si$$SiMe_3$)), which are upfield- and downfield-shifted compared to that of **1** (131.4 ppm), may also be consistent with the substantial interactions between two Si=Si double bonds.

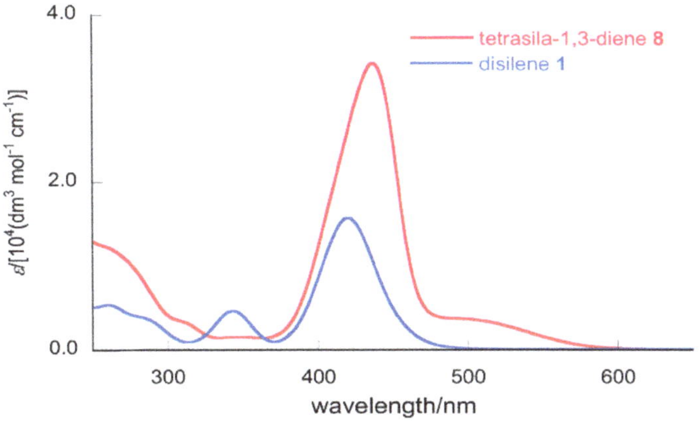

Figure 5. Ultraviolet–Visible absorption spectra of **1** and **8** in hexane.

3. Materials and Methods

3.1. General Procedure

All reactions involving air-sensitive compounds were performed under a nitrogen atmosphere using a high-vacuum line and standard Schlenk techniques, or a glove box, as well as dry and oxygen-free solvents. NMR spectra were recorded on a Bruker Avance 500 FT NMR spectrometer (Bruker Japan, Yokohama, Japan). The ^{1}H and ^{13}C NMR chemical shifts were referenced to residual ^{1}H and ^{13}C signals of the solvents: benzene-d_6 (^{1}H: δ 7.16 and ^{13}C: δ 128.0). The ^{29}Si NMR chemical shifts were relative to Me_4Si (δ 0.00). Sampling of air-sensitive compounds was carried out using a VAC NEXUS 100027 type glove box (Vacuum Atmospheres Co., Hawthorne, CA, USA). UV–Vis spectra were recorded on a JASCO V-660 spectrometer (JASCO, Tokyo, Japan). Melting points were measured on a SRS OptiMelt MPA100 (SRS, Sunnyvale, CA, USA) without correction.

3.2. Materials

Hexane, toluene and THF were dried using a VAC solvent purifier 103991 (Vacuum Atmospheres Co., Hawthorne, CA, USA). Benzene and 1,2-dimethoxyethane (DME) were dried over $LiAlH_4$, and then distilled under reduced pressure prior to use via a vacuum line. Benzene-d_6 and 1,2-dibromoethane were degassed and dried over 4A molecular sieves (activated). $CDCl_3$ was dried over 4A molecular sieves (activated). Potassium graphite (KC_8) [37] and Me_3SiSiH_2Ph [38] were prepared according to the procedure described in the literature. Bromine, 9-bromoanthracene, 9,10-dibromoanthracene, chlorotriethylsilane, lithium, magnesium sulfate (anhydrous), potassium *t*-butoxide (*t*-BuOK), $Et_3N{\cdot}HCl$ salt and triflic acid (TfOH) were purchased from commercial sources and used without further purification.

3.3. Preparation of 1,4-Bis(2,2,2-trimethyldisilanyl)-1,1,4,4-tetrakis(trimethylsilyl)butane 2

In a Schlenk flask (200 mL) equipped with a magnetic stir bar, Me_3SiSiH_2Ph (15.5 g, 0.0859 mol) and hexane (30 mL) were placed. To the solution, TfOH (12.9 g, 0.0895 mol) was added at 0 °C. After stirring for 30 min, the reaction mixture was transferred to an another Schlenk flask (200 mL) equipped with a magnetic stir bar and $Et_3N{\cdot}HCl$ salt (28.3 g, 0.206 mol). After stirring for 30 min at room temperature, the resulting insoluble materials were removed by filtration. To the solution, a THF solution (100 mL) of $(Li(Me_3Si)_2CCH_2)_2{\cdot}(thf)_5$ (24.7 g, 0.0343 mmol), which was prepared from 1,1-bis(trimethylsilyl)ethene (11.9 g, 0.0690 mmol) and lithium (722 mg, 0.104 mmol) in THF [39], was added at −30 °C. After the mixture was stirred for 30 min, the resulting solution was hydrolyzed and extracted with hexane. The organic layer was separated, washed with a brine, dried over anhydrous magnesium sulfate, and finally concentrated under reduced pressure. Kugelrohr distillation (115 °C, 0.1 Pa) of the residue furnished 1,4-bis(2,2,2-trimethydisilanyl)-1,1,4,4-tetrakis(trimethylsilyl)butane (**2**) as a colorless oil (16.8 g, 0.305 mmol) in 89% yield.

2: a colorless oil; b.p. 115 °C/0.1 Pa (Kugelrohr); ^{1}H NMR ($CDCl_3$, 500 MHz, 300 K) δ 0.14 (s, 36H, $Si(CH_3)_3$), 0.22 (s, 18H, $Si(CH_3)_3$), 1.94 (s, 4H, CH_2), 3.51 (s, 4H, SiH_2); ^{13}C NMR ($CDCl_3$, 125 MHz, 300 K) δ 0.9 ($Si(CH_3)_3$), 1.8 (($Si(CH_3)_3$), 3.8 (C), 32.5 (CH_2); ^{29}Si NMR ($CDCl_3$, 99 MHz, 301 K) δ −54.5 (*Si*H_2), −15.6 (*Si*Me_3), 3.2 (*Si*Me_3); MS (EI, 70 eV) m/z (%) 535.0 (15, M^+ − Me), 477.0 (41, M^+ − $SiMe_3$), 402.0 (76, M^+ − 2H − $2SiMe_3$); Anal. Calcd. for $C_{22}H_{62}Si_8$: C, 47.92; H, 11.33; found: C, 47.84; H, 11.33%.

3.4. Preparation of 1,4-Bis(1,1-dibromo-2,2,2-trimethyldisilanyl)-1,1,4,4-tetrakis(trimethylsilyl)butane 3

In a two-necked flask (100 mL), **2** (1.00 g, 1.81 mmol) and benzene (20 mL) were placed. To the mixture, bromine (1.19 g, 7.44 mmol) was added dropwise at 0 °C and then the mixture was stirred for 30 min at room temperature. The volatiles were removed in vacuo, and washing the residue with hexane afforded pure **3** as colorless crystals (1.09 g, 1.26 mmol) in 70% yield.

3: colorless crystals; m.p. 212 °C; ^{1}H NMR (C_6D_6, 500 MHz, 296 K) δ 0.36 (s, 18H, Si(CH_3)$_3$), 0.44 (s, 36H, Si(CH_3)$_3$), 2.74 (s, 4H, CH_2); ^{13}C NMR (C_6D_6, 125 MHz, 297 K) δ 0.4 (Si*Me*$_3$), 4.5 (Si*Me*$_3$), 19.2 (C), 32.6 (CH_2); ^{29}Si NMR (C_6D_6, 99 MHz, 296 K) δ −4.3 (*Si*Me$_3$), 2.5 (*Si*Me$_3$), 28.6 (*Si*Br$_2$); MS (EI, 70 eV) m/z (%) 851.0 (31, M$^+$ − Me), 641.0 (100, M$^+$ − 2SiMe$_3$ − Br), 560.0 (23, M$^+$ − 2SiMe$_3$ − 2Br), 487.0 (77, M$^+$ − 3SiMe$_3$ − 2Br); Anal. Calcd. for $C_{22}H_{58}Br_4Si_8$: C, 30.48; H, 6.74%. Found: C, 30.61; H, 6.95%.

3.5. *Synthesis of 1,2,3,3,6,6-Hexakis(trimethylsilyl)-1,2-disilacyclohexene* **1**

All operations were carried out in a glove box. In a Schlenk tube (50 mL), **3** (1.26 g, 1.45 mmol), KC_8 (825 mg, 6.10 mmol) and THF (70 mL) were placed. After stirring the mixture at room temperature for three hours, the volatiles were removed under reduced pressure. Then dry hexane was added to the residue and the resulting salt was filtered off. Removal of hexane under reduced pressure gave disilene **1** as yellow crystals (789.0 mg, 1.44 mmol) in 99% yield.

1: yellow crystals; m.p. 120 °C; ^{1}H NMR (C_6D_6, 500 MHz, 296 K) δ 0.25 (s, 18H, Si(CH_3)$_3$), 0.34 (s, 18H, Si(CH_3)$_3$), 0.47 (s, 18H, Si(CH_3)$_3$), 1.90–2.00 (m, 2H, CH_2), 2.45–2.57 (m, 2H, CH_2); ^{13}C NMR (C_6D_6, 125 MHz, 297 K) δ 1.1 (Si(CH_3)$_3$), 4.32 (Si(CH_3)$_3$), 4.34 (Si(CH_3)$_3$), 20.7 (C), 34.5 (CH_2); ^{29}Si NMR (C_6D_6, 99 MHz, 296 K) δ −11.9 (*Si*Me$_3$), 0.8 (*Si*Me$_3$), 1.6 (*Si*Me$_3$), 131.4 (*Si*=*Si*); UV–Vis (hexane, room temperature) λ_{max}/nm (ε) 420 (1.57 × 10^4), 343 (4.63 × 10^3); HRMS (APCI) m/z [M]$^+$ Calcd. for $C_{22}H_{58}Si_8$: 546.2687. Found. 546.2687; Anal. Calcd. for $C_{22}H_{58}Si_8$: C, 48.27; H, 10.68%. Found: C, 48.21; H, 10.55%.

3.6. *Synthesis of Potassium 2,3,3,6,6-Pentakis(trimethylsilyl)-1,2-disilacyclohexen-1-ide* **4**

In a Schlenk tube (50 mL) equipped with a magnetic stir bar, disilene **1** (81.4 mg, 0.149 mmol) and *t*-BuOK (17.6 mg, 0.157 mmol) and THF (9.0 mL) were placed. After stirring for one hour at room temperature, disilenide **4** was formed as a sole product, which was confirmed by NMR spectroscopy. The volatiles including THF and the resulting *t*-BuOSiMe$_3$ were removed under reduced pressure and then the resulting residue was washed with hexane to afford **4**(thf) as an orange powder (43.0 mg, 7.35 × 10^{-2} mmol) in 49% yield.

4(thf): an orange powder; m.p. 157 °C; ^{1}H NMR (C_6D_6, 500 MHz, 323 K) δ 0.33 (s, 27H, 3 × Si(CH_3)$_3$), 0.47 (s, 9H, Si(CH_3)$_3$), 0.49 (s, 9H, Si(CH_3)$_3$), 1.43–1.45 (m, 4H, 2 × CH_2 of THF), 2.15–2.24 (m, 2H, CH_2), 2.69–2.83 (m, 2H, CH_2), 3.54–3.56 (m, 4H, 2 × CH_2 of THF); ^{13}C NMR (C_6D_6, 125 MHz, 323 K) δ 1.9 (Si(CH_3)$_3$), 2.1 (Si(CH_3)$_3$), 4.6 (Si(CH_3)$_3$), 5.1 (Si(CH_3)$_3$), 6.4 (Si(CH_3)$_3$), 20.8 (C), 24.2 (C), 33.6 (CH_2), 35.2 (CH_2), 25.9 (CH_2 in THF), 68.0 (CH_2 in THF); ^{29}Si NMR (C_6D_6, 99 MHz, 323 K) δ −15.9 (*Si*Me$_3$), −3.1 (*Si*Me$_3$), −2.9 (*Si*Me$_3$), −2.3 (*Si*Me$_3$), −1.4 (*Si*Me$_3$), 146.9 (=*Si*SiMe$_3$), 219.3 (=*Si*$^-$). The elemental analysis was unsuccessful as **4**(thf) is very sensitive to air and moisture.

Similar to the reaction in THF, **4**(dme) (63.2 mg, 9.11 × 10^{-2} mmol, 49% yield) was obtained from **1** (102 mg, 1.86 × 10^{-1} mmol), *t*-BuOK (22.5 mg, 2.0 × 10^{-1} mmol) and DME (9.0 mL).

4(dme): an orange powder; m.p. 135 °C; ^{1}H NMR (C_6D_6, 500 MHz, 323 K) δ 0.37 (s, 9H, 2 × Si(CH_3)$_3$), 0.39 (s, 9H, Si(CH_3)$_3$), 0.49 (s, 9H, Si(CH_3)$_3$), 0.54 (s, 9H, Si(CH_3)$_3$), 2.24–2.30 (m, 2H, CH_2), 2.70–2.87 (m, 2H, CH_2), 3.05 (s, 6H, 2 × CH_3O of DME), 3.07 (s, 2H, OCH_2CH_2O), 3.08 (s, 2H, OCH_2CH_2O of DME); ^{13}C NMR (C_6D_6, 125 MHz, 323 K) δ 1.9 (Si(CH_3)$_3$), 2.1 (Si(CH_3)$_3$), 4.6 (Si(CH_3)$_3$), 5.0 (Si(CH_3)$_3$), 6.3 (Si(CH_3)$_3$), 33.6 (CH_2), 35.1 (CH_2), 58.9 (CH_3O of DME), 71.6 (OCH_2CH_2O in DME), 71.7 (OCH_2CH_2O of DME); ^{29}Si NMR (C_6D_6, 99 MHz, 323 K) δ −15.9 (*Si*Me$_3$), −3.2 (*Si*Me$_3$), −2.3 (*Si*Me$_3$), −1.5 (*Si*Me$_3$), 142.9, 143.1 (=*Si*SiMe$_3$), 222.2 (=*Si*$^-$). The reasons for the observation of very slightly split signals that can be assigned to one of the double bonded Si nuclei remain unclear at this point.

3.7. *Synthesis of 1-Triethylsilyl-2,3,3,6,6-pentakis(trimethylsilyl)-1,2-disilacyclohexene* **5**

In a Schlenk tube (50 mL) equipped with a magnetic stir bar, disilene **1** (450.0 mg, 0.822 mmol), *t*-BuOK (92.2 mg, 0.822 mmol) and THF (8 mL) were placed. After stirring for one hour at room temperature, disilenide **4** was formed quantitatively, which was confirmed by NMR spectroscopy.

Then, THF was removed in vacuo. To the residue toluene (10 mL) and chlorotriethylsilane (150.0 mg, 0.995 mmol) were added. After stirring for 15 min at room temperature, the volatiles were removed in vacuo. To the residue dry hexane was added and the resulting salt was filtered off. Removal of hexane from the filtrate gave Et_3Si-substituted disilene **5** as yellow crystals (481 mg, 0.816 mmol) in 99% yield.

5: yellow crystals; m.p. 180 °C; ^{1}H NMR (C_6D_6, 500 MHz) δ 0.270 (s, 9H, Si(CH_3)$_3$), 0.273 (s, 9H, Si(CH_3)$_3$), 0.35 (s, 9H, Si(CH_3)$_3$), 0.37 (s, 9H, Si(CH_3)$_3$), 0.48 (s, 9H, Si(CH_3)$_3$), 0.94–1.09 (m, 6H, Si(CH_2CH$_3$)$_3$), 1.17 (t, J = 7.8 Hz, 9H, Si(CH$_2$$CH_3$)$_3$), 1.88–2.02 (m, 2H, CH_2), 2.49–2.62 (m, 2H, CH_2); ^{13}C NMR (C_6D_6, 125 MHz, 300 K) δ 1.01 (Si(CH$_3$)$_3$), 1.03 (Si(CH$_3$)$_3$), 4.1 (Si(CH$_3$)$_3$), 4.3 (Si(CH$_3$)$_3$), 4.4 (Si(CH$_3$)$_3$), 7.8 (Si(CH$_2$CH$_3$)$_3$), 9.2 (Si(CH$_2$CH$_3$)$_3$), 20.7 (CH$_2$), 21.3 (CH$_2$), 34.5 (C), 34.9 (C); ^{29}Si NMR (C_6D_6, 99 MHz, 300 K) δ −12.5 (*Si*Me$_3$), 0.4 (SiMe$_3$), 0.5 (SiMe$_3$), 1.0 (*Si*Et$_3$), 1.3 (*Si*Me$_3$), 1.8 (*Si*Me$_3$), 123.5 (Et$_3$Si*Si*=), 134.2 (Me$_3$Si*Si*=); HRMS (APCI) m/z [M]$^+$ Calcd. for $C_{25}H_{64}Si_8$: 588.3157, Found. 588.3156; Anal. Calcd. for $C_{25}H_{64}Si_8$: C, 50.94; H, 10.94%. Found: C, 50.58; H, 10.70.

3.8. Synthesis of 1-(9-Anthryl)-2,3,3,6,6-pentakis(trimethylsilyl)-1,2-disilacyclohexene **6**

In a Schlenk tube (50 mL) equipped with a magnetic stir bar, disilene **1** (181 mg, 3.30×10^{-1} mmol) and *t*-BuOK (38.0 mg, 3.39×10^{-1} mmol) and DME (15.0 mL) were placed. After stirring for one hour at room temperature, disilenide **4** was formed as a sole product, which was confirmed by NMR spectroscopy. Then DME was removed in vacuo, and the resulting residue was dissolved in benzene (11.0 mL). To the solution, a benzene solution (7.5 mL) of 9-bromoanthracene (86.2 mg, 3.35×10^{-1} mmol) was added and the mixture was stirred for 1 min at room temperature. The resulting salt was removed by filtration and washed with benzene and the solvent was removed from the filtrate in vacuo. Formation of the anthryl-substituted disilene **6** as a major product was observed by ^{1}H NMR spectrum. Recrystallization from hexane and DME at −35 °C gave **6** as reddish purple crystals (83.6 mg, 1.28×10^{-1} mmol) in 33% yield.

6: reddish purple crystals; m.p. 178 °C; ^{1}H NMR (C_6D_6, 500 MHz, 298 K) δ −0.21 (s, 9H, Si(CH_3)$_3$), −0.01 (s, 9H, Si(CH_3)$_3$), 0.32 (s, 9H, Si(CH_3)$_3$), 0.40 (s, 9H, Si(CH_3)$_3$), 0.48 (s, 9H, Si(CH_3)$_3$), 2.34–2.45 (m, 1H, CH$_2$CH*H*), 2.52–2.59 (m, 3H, CH$_2$CHH), 7.21–7.24 (m, 2H, 2 × *CH*), 7.37–7.42 (m, 2H, 2 × *CH*), 7.72–7.76 (m, 2H, 2 × *CH*), 8.19 (s, 1H, *CH*), 9.45 (d, J = 8.7 Hz, 1H, *CH*), 9.75 (d, J = 8.8 Hz, 1H, *CH*); ^{13}C NMR (C_6D_6, 125 MHz, 298 K) δ 1.68 (Si(CH$_3$)$_3$), 1.73 (Si(CH$_3$)$_3$), 3.1 (Si(CH$_3$)$_3$), 3.6 (Si(CH$_3$)$_3$), 4.5 (Si(CH$_3$)$_3$), 15.0 (C), 23.6 (C), 32.3 (CH$_2$), 32.5 (CH$_2$), 125.2 (2 × CH), 125.5 (CH), 125.6 (CH), 129.46 (CH), 129.49 (CH), 131.0 (CH), 131.9 (C), 132.4 (C), 132.8 (CH), 133.7 (CH), 137.7 (C), 138.1 (C), 138.5 (C); ^{29}Si NMR (C_6D_6, 99 MHz, 297 K) δ −10.0 (*Si*Me$_3$), 1.6 (*Si*Me$_3$), 1.7 (*Si*Me$_3$), 2.2 (*Si*Me$_3$), 2.3 (*Si*Me$_3$), 74.9 (=*Si*(SiMe$_3$)), 130.3 (Ant*Si*=); UV–Vis (hexane) λ_{max}/nm (ε) 399 (26400), 535 (925); MS (EI, 70 eV) m/z (%) (682.0 (53, M$^+$ + O$_2$)), 650.0 (19, M$^+$), 577.0 (100, M$^+$ − SiMe$_3$); Anal. Calcd. for $C_{33}H_{58}Si_7$: C, 60.85; H, 8.970. Found: C, 60.69; H, 9.042%.

3.9. Synthesis of 1-(10-Bromo-9-anthryl)-2,3,3,6,6-pentakis(trimethylsilyl)-1,2-disilacyclohexene **6^{Br}**

In a Schlenk tube (50 mL) equipped with a magnetic stir bar, were placed disilene **1** (169 mg, 3.08×10^{-1} mmol) and *t*-BuOK (36.0 mg, 3.21×10^{-1} mmol) and DME (6.0 mL). After stirring for one hour at room temperature, disilenide **4** was formed as a sole product, which was confirmed by NMR spectroscopy. Then DME was removed in vacuo and benzene (7.5 mL) was added. The resulting benzene solution of **4** was added to a benzene (17 mL) solution of 9,10-dibromoanthracene (103 mg, 3.06×10^{-1} mmol) in another Schlenk tube (50 mL) equipped with a magnetic stir bar. After the mixture was stirred for 1 min at room temperature, the resulting salt was removed by filtration and washed with benzene and the volatiles were removed from the filtrate in vacuo. Formation of 9-bromoanthryl-substituted disilene **6^{Br}** as a major product was observed by ^{1}H NMR spectrum. Recrystallization from hexane at –35 °C gave **6^{Br}** (32.0 mg, 4.38×10^{-2} mmol) as blue crystals in 14% yield including anthryldisilene **6** as a major contaminant. Further separation of **6** and **6^{Br}** was unsuccessful.

6^{Br} (including **6** (~10%)): blue crystals; ^{1}H NMR (C_6D_6, 500 MHz, 298 K) δ −0.26 (s, 9H, Si(CH_3)$_3$), −0.04 (s, 9H, Si(CH_3)$_3$), 0.27 (s, 9H, Si(CH_3)$_3$), 0.38 (s, 9H, Si(CH_3)$_3$), 0.45 (s, 9H, Si(CH_3)$_3$), 2.35–2.40

(m, 1H, CHHCH$_2$), 2.49–2.56 (m, 3H, CHHCH$_2$), 7.23–7.27 (m, 2H, 2 × CH), 7.30–7.35 (m, 2H, 2 × CH), 8.62 (d, *J* = 8.6, 1H, CH), 8.65 (dd, *J* = 8.7 Hz, 14.6 Hz, 1H, CH), 9.53 (d, *J* = 8.5 Hz, 1H, CH), 9.85 (d, *J* = 8.7 Hz, 1H, CH); ^{13}C NMR (C$_6$D$_6$, 125 MHz, 299 K) δ 1.65 (Si(CH$_3$)$_3$), 1.70 (Si(CH$_3$)$_3$), 3.07 (Si(CH$_3$)$_3$), 3.60 (Si(CH$_3$)$_3$), 4.44 (Si(CH$_3$)$_3$), 15.1 (C), 23.6 (C), 32.3 (CH$_2$) 32.4 (CH$_2$), 125.36 (CH), 125.38 (CH), 127.5 (CH), 127.6 (CH), 128.9 (CH), 129.0 (CH), 131.0 (C), 131.5 (C), 133.2 (CH), 134.2 (CH), 138.2 (C), 138.6 (C), 140.3 (C); ^{29}Si NMR (C$_6$D$_6$, 99 MHz, 298 K) δ −9.9 (*Si*Me$_3$), 1.7 (*Si*Me$_3$), 1.8 (*Si*Me$_3$), 2.29 (*Si*Me$_3$), 2.34 (*Si*Me$_3$), 75.1 (=*Si*(SiMe$_3$)), 129.1 [(BrAnt)*Si*=]; UV–Vis (hexane) λ_{max}/nm (ε) 391 (24800), 410 (30800), 578 (967); HRMS (APCI) *m/z* [M]$^+$ Calcd. for C$_{33}$H$_{57}$BrSi$_7$: 728.2029. Found. 728.2026; [M$^+$ + H$^+$]. for C$_{33}$H$_{58}$BrSi$_7$: 729.2107. Found: 729.2104; [M$^+$ − H$^+$] Calcd. for C$_{33}$H$_{56}$BrSi$_7$: 727.1950. Found: 727.1947. A quaternary carbon signal was not observed in the ^{13}C NMR spectrum.

3.10. *Synthesis of 1,2-Di(9-anthryl)-3,3,6,6-tetrakis(trimethylsilyl)-1,2-disilacyclohexene* **7**

In a Schlenk tube (50 mL) equipped with a magnetic stir bar, were placed **6** (38.7 mg, 7.07 ×10^{-2} mmol), *t*-BuOK (8.70 mg, 7.75 × 10^{-2} mmol) and DME (7.5 mL). After stirring for one hour at room temperature, the corresponding disilenide was formed as a sole product, as confirmed by NMR spectroscopy. Then the volatiles were removed in vacuo and benzene was added. To the solution, a benzene solution of 9-bromoanthracene (17.1 mg, 6.65 × 10^{-2} mmol) was added. The ^{1}H NMR spectrum of the reaction mixture indicated formation of the desired bisanthryl-substituted disilene. The resulting salt was removed by filtration and washed with benzene and the volatiles were removed from the filtrate in vacuo. Recrystallization from toluene at −35 °C gave disilene **7** as black purple crystals (7.50 mg, 9.93 × 10^{-3} mmol) in 17% yield.

7: black purple crystals; m.p. 220 °C (decomp.); ^{1}H NMR (C$_6$D$_6$, 500 MHz, 298 K) δ −0.11 (s, 18H, 2 × Si(CH$_3$)$_3$), 0.58 (s, 18H, 2 × Si(CH$_3$)$_3$), 2.81 (s, 4H, (CH$_2$)$_2$), 6.84–6.87 (m, 2H, 2 × CH), 7.10–7.14 (m, 2H, 2 × CH), 7.23–7.30 (m, 4H, 4 × CH), 7.55–7.58 (m, 4H, 4 × CH), 7.78 (s, 2H, 2 × CH), 9.57–9.59 (m, 2H, 2 × CH), 9.82–9.84 (m, 2H, 2 × CH); ^{13}C NMR (C$_6$D$_6$, 125 MHz, 299 K) δ 1.5 (Si(CH$_3$)$_3$), 4.6 (Si(CH$_3$)$_3$), 21.3 (CH$_2$), 32.6 (C), 125.06 (CH), 125.09 (CH), 125.14 (CH), 129.1 (CH), 129.7 (CH), 130.8 (CH), 131.7 (C), 132.3 (C), 132.6 (CH), 132.9 (CH), 136.2 (C), 137.6 (C), 137.7 (C); ^{29}Si NMR (C$_6$D$_6$, 99 MHz, 298 K) δ 2.6 (*Si*Me$_3$), 3.8 (*Si*Me$_3$), 88.2 (*Si*=*Si*); UV–Vis (hexane) λ_{max}/nm (ε) 340 (5250), 357 (8640), 375 (14100), 543 (1680); MS (EI, 70 eV) *m/z* (%) 755 (11, M$^+$), 681 (2.9, M$^+$ − SiMe$_3$), 607 (1.5, M$^+$ − 2SiMe$_3$); Anal. Calcd. for C$_{44}$H$_{58}$Si$_6$: C, 69.96; H, 7.74. Found: C, 68.54; H, 7.774%. Although the reason why the observed relative ratio of carbon atoms in the elemental analysis was less than the calculated value remains unclear at this point, flame resistant materials such as silicon carbide may be formed during elemental analysis.

3.11. *Synthesis of 1,4-Bis(trimethylsilyl)tetrasila-1,3-diene* **8**

In a Schlenk tube (50 mL) equipped with a magnetic stir bar, were placed disilene **1** (139.9 mg, 0.256 mmol), *t*-BuOK (29.0 mg, 0.258 mmol) and THF (5 mL). After the mixture was stirred for one hour at room temperature, ^{1}H NMR spectrum of the mixture showed that disilenide **4** was formed quantitatively. Then, to the mixture, 1,2-dibromoethane (24.0 mg, 0.128 mmol) was added at 0 °C. After additional stirring of the mixture for 15 min at 0 °C, the volatiles were removed under reduced pressure. To the residue dry hexane was added and the resulting insoluble materials were filtered off. Removal of hexane from the filtrate and subsequent recrystallization from diethyl ether at −35 °C gave tetrasila-1,3-diene **8** as red crystals (104.0 mg, 0.110 mmol) in 86% yield.

8: red crystals; m.p. 220 °C (decomp.); ^{1}H NMR (C$_6$D$_6$, 500 MHz, 296 K) δ 0.30 (s, 18H, Si(CH$_3$)$_3$), 0.37 (s, 18H, Si(CH$_3$)$_3$), 0.468 (s, 18H, Si(CH$_3$)$_3$), 0.470 (s, 18H, Si(CH$_3$)$_3$), 0.68 (s, 18H, Si(CH$_3$)$_3$), 2.06–2.15 (m, 4H, CH$_2$), 2.45–2.64 (m, 4H, CH$_2$); ^{13}C NMR (C$_6$D$_6$, 125 MHz, 296 K) δ 1.8 (Si(CH$_3$)$_3$), 3.2 (Si(CH$_3$)$_3$), 4.8 (Si(CH$_3$)$_3$), 5.66 (Si(CH$_3$)$_3$), 5.74 (Si(CH$_3$)$_3$), 23.5 (C), 26.9 (C), 33.5 (CH$_2$), 36.7 (CH$_2$); ^{29}Si NMR (C$_6$D$_6$, 99 MHz, 296 K) δ −10.9 (*Si*Me$_3$), 1.4 (*Si*Me$_3$), 2.8 (*Si*Me$_3$), 3.3 (*Si*Me$_3$), 3.6 (*Si*Me$_3$), 106.4 (*Si*=SiSiMe$_3$), 143.0 (Si=*Si*SiMe$_3$); UV–Vis (hexane, room temperature) λ_{max}/nm (ε) ~500 (sh,

3.57×10^3), 437 (3.42×10^4); HRMS (APCI) m/z $[M]^+$ Calcd. for $C_{38}H_{98}Si_{14}$: 946.4433. Found. 946.4430; Anal. Calcd. for $C_{38}H_{98}Si_{14}$: C, 48.13; H, 10.42%. Found: C, 48.48; H, 10.58%.

3.12. Single Crystal X-ray Diffraction Analyses

Single crystals for data collection were obtained by recrystallization under the following conditions: from hexane at −35 °C (**1**, **5**, **6**, **6**^{Br}), from toluene at −35 °C ((**4**(toluene))$_2$, **7**) or from diethyl ether at −35 °C (**8**). Each single crystal coated by Apiezon® grease was mounted on the glass fiber and transferred to the cold nitrogen gas stream of the diffractometer. X-ray data were collected on a BrukerAXS APEXII diffractometer with graphite monochromated Mo-Kα radiation (λ = 0.71073 Å). The data were corrected for Lorentz and polarization effects. An empirical absorption correction based on the multiple measurement of equivalent reflections was applied using the program SADABS [40]. The structures were solved by direct methods and refined by full-matrix least squares against F^2 using all data (SHELXL-2014/7) [41]. CCDC-1813641–1813647 contain the supplementary crystallographic data for this paper. These data can be obtained free of charge via http://www.ccdc.cam.ac.uk/conts/retrieving.html (or from the CCDC, 12 Union Road, Cambridge CB2 1EZ, UK.

1: CCDC-1813641; 100 K; $C_{22}H_{58}Si_8$; Fw 547.40; orthorhombic; space group P_{bca} (#61); a = 17.3380(16) Å, b = 18.0108(16) Å, c = 22.317(2) Å, V = 6969.0(11) Å^3, Z = 8, D_{calcd} = 1.043 Mg/m^3, $R1$ = 0.0218 ($I > 2\sigma(I)$), $wR2$ = 0.0619 (all data), GOF = 1.051.

4: CCDC-1813642; 100 K; $C_{26}H_{57}KSi_7$; Fw 605.45; monoclinic; space group $P2_1/n$ (#14), a = 15.7323(17) Å, b = 11.7003(13) Å, c = 21.031(2) Å, β = 107.1880(10)°, V = 3698.4(7) Å^3, Z = 4, D_{calcd} = 1.087 Mg/m^3, $R1$ = 0.0234 ($I > 2\sigma(I)$), $wR2$ = 0.0642 (all data), GOF = 1.049.

5: CCDC-1813643; 100 K; $C_{25}H_{64}Si_8$; Fw 589.48; monoclinic; space group $P2_1/n$ (#14), a = 9.0608(19) Å, b = 37.640(8) Å, c = 11.263(2) Å, β = 103.230(2)°, V = 3739.3(14) Å^3, Z = 4, D_{calcd} = 1.047 Mg/m^3, $R1$ = 0.0729 ($I > 2\sigma(I)$), $wR2$ = 0.1806 (all data), GOF = 1.335.

6: CCDC-1813644; 100 K; $C_{33}H_{58}Si_7$; Fw 651.42; monoclinic; space group $P2_1/c$ (#14), a = 17.7493(8) Å, b = 9.0323(4) Å, c = 23.7624(10) Å, β = 91.2830(10)°, V = 3808.6(3) Å^3, Z = 4, D_{calcd} = 1.136 Mg/m^3, $R1$ = 0.0320 ($I > 2\sigma(I)$), $wR2$ = 0.0847 (all data), GOF = 1.038.

6^{Br}: CCDC-1813645; 100 K; $C_{33}H_{57}BrSi_7$; Fw 730.32; triclinic; space group P-1 (#2), a = 9.0529(3) Å, b = 11.9109(4) Å, c = 19.1793(7) Å, α = 93.9480(10)°, β = 90.1810(10)°, γ = 109.1330(10)°, V = 1947.47(12) Å^3, Z = 2, D_{calcd} = 1.245 Mg/m^3, $R1$ = 0.0263 ($I > 2\sigma(I)$), $wR2$ = 0.0701 (all data), GOF = 1.043.

7: CCDC-1813646; 100 K; $C_{44}H_{58}Si_6 \cdot C_7H_8$; Fw 847.57; triclinic; space group P-1 (#2), a = 11.0302(9) Å, b = 12.8081(10) Å, c = 15.7049(13) Å, α = 95.628(2)°, β = 98.481(2)°, γ = 90.755(2)°, V = 2361.1(3) Å^3, Z = 2, D_{calcd} = 1.192 Mg/m^3, $R1$ = 0.0350 ($I > 2\sigma(I)$), $wR2$ = 0.0954 (all data), GOF = 1.025.

8: CCDC-1813647; 173 K; $C_{38}H_{98}Si_{14}$; Fw 948.42; triclinic; space group P-1 (#2), a = 11.4093(11) Å, b = 15.6821(15) Å, c = 19.5623(19) Å, α = 84.8020(10)°, β = 75.9110(10)°, γ = 71.9750(10)°, V = 3227.8(5) Å^3, Z = 2, D_{calcd} = 0.976 Mg/m^3, $R1$ = 0.0266 ($I > 2\sigma(I)$), $wR2$ = 0.0759 (all data), GOF = 1.039.

4. Conclusions

We successfully synthesized 1,2-bis(trimethylsilyl)-1,2-disilacyclhexene **1** and converted the silyl group(s) to the functional group(s). Noticeably, two silyl groups in **1** can be replaced stepwise by anthryl groups, which suggests that **1** can serve as a unit of *cis*-1,2-dialkyldisilene for novel extended π-electron systems. As the introduction of heteroatoms into sp^2-carbon-based nanocarbon materials such as polycyclic aromatic hydrocarbons, fullerenes and graphenes has received substantial attention, nanocarbon materials in which the Si=Si double bond was introduced into the peripheral position by using functionalizable disilenes such as **1** may provide novel, fascinating nanosilicon–carbon hybrid materials.

Acknowledgments: This work was supported by JSPS KAKENHI grant JP24655024, JPK1513634 (Takeaki Iwamoto), and Grant-in-Aid for JSPS Fellows (Naohiko Akasaka).

Author Contributions: Naohiko Akasaka, Kaho Tanaka and Takeaki Iwamoto conceived and designed the experiments; Naohiko Akasaka and Kaho Tanaka performed the experiments; Shintaro Ishida, Naohiko Akasaka and Takeaki Iwamoto performed the X-ray single crystal analyses; Naohiko Akasaka, Kaho Tanaka and Takeaki Iwamoto analyzed the NMR data; Takeaki Iwamoto performed theoretical calculations; Naohiko Akasaka, Kaho Tanaka and Takeaki Iwamoto wrote the manuscript.

References

1. Okazaki, R.; West, R. Chemistry of stable disilenes. *Adv. Organomet. Chem.* **1996**, *39*, 231–273.
2. Kira, M.; Iwamoto, T. Progress in the chemistry of stable disilenes. *Adv. Organomet. Chem.* **2006**, *54*, 73–148.
3. Lee, V.Y.; Sekiguchi, A. Heavy analogs of alkenes, 1,3-dienes, allenes and alkynes: Multiply bonded derivatives of Si, Ge, Sn and Pb. In *Organometallic Compounds of Low-Coordinate Si, Ge, Sn and Pb: From Phantom Species to Stable Compounds*; Wiley-VCH: Chichester, Germany, 2010; pp. 199–334.
4. Scheschkewitz, D. Anionic reagents with silicon-containing double bonds. *Chem. Eur. J.* **2009**, *15*, 2476–2485. [CrossRef] [PubMed]
5. Abersfelder, K.; Scheschkewitz, D. Synthesis of homo- and heterocyclic silanes via intermediates with Si=Si bonds. *Pure Appl. Chem.* **2010**, *82*, 595–602. [CrossRef]
6. Scheschkewitz, D. The versatile chemistry of disilenides: Disila analogues of vinyl anions as synthons in low-valent silicon chemistry. *Chem. Lett.* **2011**, *40*, 2–11. [CrossRef]
7. Präsang, C.; Scheschkewitz, D. Silyl anions. *Struct. Bonding* **2013**, *156*, 1–47.
8. Präsang, C.; Scheschkewitz, D. Reactivity in the periphery of functionalised multiple bonds of heavier group 14 elements. *Chem. Soc. Rev.* **2016**, *45*, 900–921. [CrossRef] [PubMed]
9. Rammo, A.; Scheschkewitz, D. Functional disilenes in synthesis. *Chem. Eur. J.* **2017**. [CrossRef] [PubMed]
10. Iwamoto, T.; Kobayashi, M.; Uchiyama, K.; Sasaki, S.; Nagendran, S.; Isobe, H.; Kira, M. Anthryl-substituted trialkyldisilene showing distinct intramolecular charge-transfer transition. *J. Am. Chem. Soc.* **2009**, *131*, 3156–3157. [CrossRef] [PubMed]
11. Scheschkewitz, D. A silicon analogue of vinyllithium: Structural characterization of a disilenide. *Angew. Chem. Int. Ed.* **2004**, *43*, 2965–2967. [CrossRef] [PubMed]
12. Ichinohe, M.; Sanuki, K.; Inoue, S.; Sekiguchi, A. Disilenyllithium from tetrasila-1,3-butadiene: A silicon analogue of a vinyllithium. *Organometallics* **2004**, *23*, 3088–3090. [CrossRef]
13. Ichinohe, M.; Sanuki, K.; Inoue, S.; Sekiguchi, A. Tetrasila-1,3-butadiene and its transformation to disilenyl anions. *Silicon Chem.* **2006**, *3*, 111–116. [CrossRef]
14. Meltzer, A.; Majumdar, M.; White, A.J.P.; Huch, V.; Scheschkewitz, D. Potential protecting group strategy for disila analogues of vinyllithiums: Synthesis and reactivity of a 2,4,6-trimethoxyphenyl-substituted disilene. *Organometallics* **2013**, *32*, 6844–6850. [CrossRef]
15. Abersfelder, K.; Zhao, H.; White, A.J.P.; Präsang, C.; Scheschkewitz, D. Synthesis of the first homoleptic trisilaallyl chloride: 3-Chloro-1,1,2,3,3-pentakis(2′,4′,6′-triisopropylphenyl)trisil-1-ene. *Z. Anorg. Allg. Chem.* **2015**, *641*, 2051–2055. [CrossRef]
16. Akasaka, N.; Fujieda, K.; Garoni, E.; Kamada, K.; Matsui, H.; Nakano, M.; Iwamoto, T. Synthesis and functionalization of a 1,4-bis(trimethylsilyl)tetrasila-1,3-diene through the selective cleavage of Si(sp^2)-Si(sp^3) bonds under mild reaction conditions. *Organometallics* **2018**, *37*, 172–175. [CrossRef]
17. Marschner, C. Silicon-centered anions. In *Organosilicon Compounds Theory and Experiment (Synthesis)*; Lee, V.Y., Ed.; Academic Press: Oxford, UK, 2017; pp. 295–360.
18. Abe, T.; Iwamoto, T.; Kira, M. A stable 1,2-disilacyclohexene and its 14-electron palladium(0) complex. *J. Am. Chem. Soc.* **2010**, *132*, 5008–5009. [CrossRef] [PubMed]
19. Liang, C.; Allen, L.C. Group IV double bonds: Shape deformation and substituent effects. *J. Am. Chem. Soc.* **1990**, *112*, 1039–1041. [CrossRef]
20. Karni, M.; Apeloig, Y. Substituent effects on the geometries and energies of the Si=Si double bond. *J. Am. Chem. Soc.* **1990**, *112*, 8589–8590. [CrossRef]
21. Apeloig, Y.; Müller, T. Do silylenes always dimerize to disilenes? Novel silylene dimers with unusual structures. *J. Am. Chem. Soc.* **1995**, *117*, 5363–5364. [CrossRef]

22. Goldberg, D.E.; Hitchcock, P.B.; Lappert, M.F.; Thomas, K.M.; Thorne, A.J.; Fjeldberg, T.; Haaland, A.; Schilling, B.E.R. Subvalent Group 4B metal alkyls and amides. Part 9. Germanium and tin alkene analogues, the dimetallenes M_2R_4[M = Ge or Sn, R = $CH(SiMe_3)_2$]: X-ray structures, molecular orbital calculations for M_2H_4, and trends in the series $M_2R'_4$[M = C, Si, Ge, or Sn; R′ = R, Ph, $C_6H_2Me_3$-2,4,6, or $C_6H_3Et_2$-2,6]. *J. Chem. Soc. Dalton Trans.* **1986**, 2387–2394.
23. West, R.; Cavalieri, J.D.; Buffy, J.J.; Fry, C.; Zilm, K.W.; Duchamp, J.C.; Kira, M.; Iwamoto, T.; Müller, T.; Apeloig, Y. A solid-state NMR and theoretical study of the chemical bonding in disilenes. *J. Am. Chem. Soc.* **1997**, *119*, 4972–4976. [CrossRef]
24. Obeid, N.M.; Klemmer, L.; Maus, D.; Zimmer, M.; Jeck, J.; Bejan, I.; White, A.J.P.; Huch, V.; Jung, G.; Scheschkewitz, D. (Oligo)Aromatic Species With One or Two Conjugated Si=Si Bonds: Near-IR Emission of Anthracenyl-Bridged Tetrasiladiene. *Dalton Trans.* **2017**, *46*, 8839–8848. [CrossRef] [PubMed]
25. Weidenbruch, M.; Willms, S.; Saak, W.; Henkel, G. Hexaaryltetrasilabuta-1,3-diene: A molecule with conjugated Si=Si double bonds. *Angew. Chem. Int. Ed. Engl.* **1997**, *36*, 2503–2504. [CrossRef]
26. Uchiyama, K.; Nagendran, S.; Ishida, S.; Iwamoto, T.; Kira, M. Thermal and photochemical cleavage of Si=Si double bond in tetrasila-1,3-diene. *J. Am. Chem. Soc.* **2007**, *129*, 10638–10639. [CrossRef] [PubMed]
27. Iwamoto, T.; Ishida, S. Multiple bonds with silicon: Recent advances in synthesis, structure, and functions of stable disilenes. In *Structure and Bonding: Functional Molecular Silicon Compounds II*; Scheschkewitz, D., Ed.; Springer: Cham, Switzerland, 2017; Volume 156, pp. 125–202.
28. Kosai, T.; Iwamoto, T. Stable push–pull disilene: Substantial donor–acceptor interactions through the Si=Si double bond. *J. Am. Chem. Soc.* **2017**, *139*, 18146–18149. [CrossRef] [PubMed]
29. Cowley, M.J.; Abersfelder, K.; White, A.J.; Majumdar, M.; Scheschkewitz, D. Transmetallation reactions of a lithium disilenide. *Chem. Commun.* **2012**, *48*, 6595–6597. [CrossRef] [PubMed]
30. Auer, D.; Strohmann, C.; Arbuznikov, A.V.; Kaupp, M. Understanding substituent effects on ^{29}Si chemical shifts and bonding in disilenes. A quantum chemical analysis. *Organometallics* **2003**, *22*, 2442–2449. [CrossRef]
31. Mazik, M.; Buthe, A.C.; Jones, P.G. C–H···Br, C–Br···Br, and C–Br···π interactions in the crystal structures of mesitylene- and dimesitylmethane-derived compounds bearing bromomethyl units. *Tetrahedron* **2010**, *66*, 385–389. [CrossRef]
32. Safin, D.A.; Babashkina, M.G.; Robeyns, K.; Garcia, Y. C–H···Br–C vs. C–Br···Br–C vs. C–Br···N bonding in molecular self-assembly of pyridine-containing dyes. *RSC Adv.* **2016**, *6*, 53669–53678. [CrossRef]
33. Cavallo, G.; Metrangolo, P.; Milani, R.; Pilati, T.; Priimagi, A.; Resnati, G.; Terraneo, G. The halogen bond. *Chem. Rev.* **2016**, *116*, 2478–2601. [CrossRef] [PubMed]
34. Pati, N.N.; Kumar, B.S.; Chandra, B.; Panda, P.K. Unsymmetrical bipyrrole-derived β-tetraalkylporphycenes and C–H···Br–C interaction induced 2d arrays of the 2:1 supramolecular sandwich complex of their *cis-/trans*-dibromo isomers. *Eur. J. Org. Chem.* **2017**, 741–745. [CrossRef]
35. Ichinohe, M.; Arai, Y.; Sekiguchi, A.; Takagi, N.; Nagase, S. A new approach to the synthesis of unsymmetrical disilenes and germasilene: Unusual ^{29}Si NMR chemical shifts and regiospecific methanol addition. *Organometallics* **2001**, *20*, 4141–4143. [CrossRef]
36. Kosai, T.; Ishida, S.; Iwamoto, T. Heteroaryldisilenes: heteroaryl groups serve as electron acceptors for Si=Si double bonds in intramolecular charge transfer transitions. *Dalton Trans.* **2017**, *46*, 11271–11281. [CrossRef] [PubMed]
37. Weitz, I.S.; Rabinovitz, M. The application of C_8K for organic synthesis: Reduction of substituted naphthalenes. *J. Chem. Soc. Perkin Trans.* **1993**, *1*, 117–120. [CrossRef]
38. Hengge, E.; Bauer, G.; Marketz, H. Darstellung und Eigenschaften einiger asymmetrisch substituierter 1,1,1-Trimethyldisilan-Verbindungen. *Z. Anorg. Allg. Chem.* **1972**, *394*, 93–100. [CrossRef]
39. Kira, M.; Hino, T.; Kubota, Y.; Matsuyama, N.; Sakurai, H. Preparation and reactions of 1,1,4,4-tetrakis(trimethylsilyl)butane-1,4-diyl dianion. *Tetrahedron Lett.* **1988**, *29*, 6939–6942. [CrossRef]
40. Sheldrick, G.M. *SADABS*; Empirical Absorption Correction Program; University of Göttingen: Göttingen, Germany, 1996.
41. Sheldrick, G.M. *SHELXL-2014/7*; Program for the Refinement of Crystal Structures; University of Göttingen: Göttingen, Germany, 2014.

17

Preparation and Molecular Structure of a Cyclopentyl-Substituted Cage Hexasilsesquioxane T_6 (T = cyclopentyl-$SiO_{1.5}$) Starting from the Corresponding Silanetriol

Jürgen Kahr [1], Ferdinand Belaj [1] and Rudolf Pietschnig [1,2,*]

1 Institut für Chemie, Karl Franzens-Universität, NAWI Graz, Schubertstraße 1, 8010 Graz, Austria; juergen.kahr@gmx.at (J.K.); ferdinand.belaj@uni-graz.at (F.B.)
2 Institut für Chemie und CINSaT, Universität Kassel, Heinrich-Plett-Straße 40, 34132 Kassel, Germany
* Correspondence: pietschnig@uni-kassel.de

Abstract: Cyclopentyl substituted silanetriol can be prepared and isolated. Its condensation yields the corresponding disiloxanetetrol as a primary condensation product. Further condensation leads to the hexameric polyhedral silsesquioxane cage T_6. The latter has been mentioned in the literature before however, lacking structural data. All compounds have been characterized with multinuclear NMR spectroscopy and, in addition, the molecular structures have been determined in the case of the disiloxanetetrol and the hexasilsesquioxane via single crystal X-ray diffraction.

Keywords: silanetriols; disiloxane tetrols; silsesquioxanes; condensation; molecular cage

1. Introduction

Cage silsesquioxanes have attracted much attention in recent years [1,2] owing to their widespread applications for example in catalysis [3], as model systems for silica surfaces [4,5], in the design of superoleophobic surfaces [6], ionic liquids [7], biocompatible materials [8], as well as in polymer chemistry [2]. The synthetic approach towards such octasilsesquioxanes is mainly based on the hydrolytic condensation of trifunctional silanes $RSiX_3$, where R is a stable organic substituent and X a reactive moiety (i.e., X = Cl, OMe etc.) [2,9] and catalysts like tetrabutylammonium fluoride (TBAF) have been shown to improve the yields in the presence of certain organic substituent [10–12]. Recently, it has been demonstrated that silanetriols are suitable starting materials for cage silsesquioxanes, giving access to T_8 cages in a one-pot synthesis which could not be obtained from the corresponding alkoxysilanes via other routes [13,14]. Here we report our investigation to prepare a cyclopentyl substituted silanetriol and its condensation to the T_6 cage via the corresponding tetrahydroxydisiloxane.

2. Results and Discussion

Starting from commercially available cyclopentyltrichlorosilane, the corresponding silanetriol was prepared by careful hydrolysis in ether solution at 0 °C in the presence of three equivalents of aniline in analogy to an established procedure by *Takiguchi* [15]. Silanetriol **1** has been be isolated from the etheral solution as colorless powder in above 80% yield. The ^{29}Si-NMR resonance of the product was observed at −37.7 ppm in D_2O which compares well with the ^{29}Si chemical shifts of related alkylsilanetriols such as *tert*-butylSi$(OH)_3$ (−36.8 ppm, D_2O). While cyclopentyl substituted silanetriol **1** was stable as a solid, it slowly underwent condensation in polar solvents such as THF or DMSO (Scheme 1). The resulting tetrahydroxydisiloxane **2** has been identified as primary condensation product and was characterized by NMR and IR spectroscopy, mass spectrometry and single crystal X-ray diffraction.

The ^{29}Si-NMR chemical shift at −51.8 ppm in THF solution is in good agreement with the known chemical shifts of other alkylsubstituted disiloxane tetrols [14,16], resonating at slightly lower field compared with aryl substituted disiloxane tetrols [17,18].

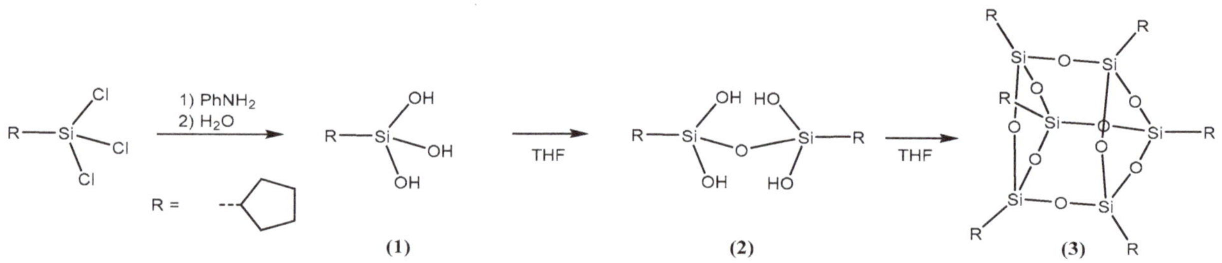

Scheme 1. Formation of **3** via **1** and **2** starting from cyclopentyl trichlorosilane.

Compound **2** could be obtained as single crystals suitable for X-ray diffraction as two polymorphs, one crystallizing in a monoclinic, the other in an orthorhombic crystal system both confirming the constitution of 1,3-di-cyclopentyl-1,1,3,3-tetrahydroxysiloxane. In the monoclinic crystal of **2**, the molecules are lying with O1 on an inversion center resulting in an Si–O–Si angle of 180°. The cyclopentyl rings are disordered over two sites (Figure 1). Each of the four OH groups of the tetrahydroxysiloxanes are involved in one donor and in one acceptor hydrogen bond [O2···O3′ 2.6811(19) Å, O2–H2···O3′ 174.8(9)°; O3···O2″ 2.6718(19) Å, O3–H3···O2″ 177.3(14)°], resp., forming two-dimensional aggregates, in which each molecule is connected to six neighbors showing a two-dimensional closest packing. This planar aggregate is shielded on both sides by the cyclopentyl groups. The molecules show pseudo-mirror planes normal to the c axis; the transformation to orthorhombic symmetry would lead to an angle differing by 0.268(4)° from 90°.

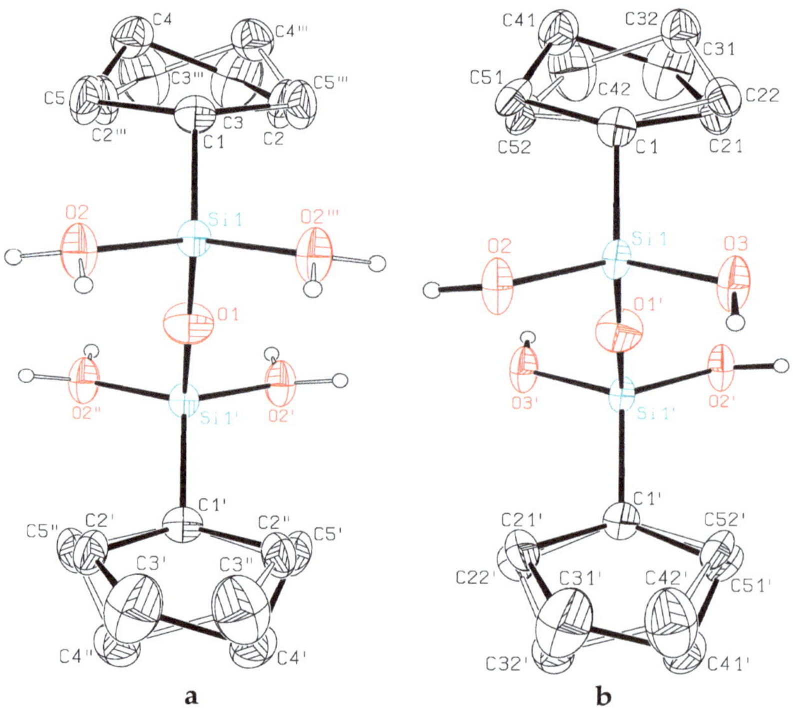

Figure 1. This ORTEP plot of the molecular structure of **2** from the monoclinic (**b**) and the orthorhombic (**a**) polymorph showing the atomic numbering scheme. The probability ellipsoids are drawn at the 50% probability level. The cyclopentyl rings are disordered over two sites. The H atoms bonded to oxygen are drawn with arbitrary radii, the H atoms of the cyclopentyl rings were omitted for the sake of clarity. Red: oxygen; blue: silicon.

In the orthorhombic phase the molecules of **2** adopt C_{2h} (= 2/m) symmetry resulting in a Si–O–Si angle of 180°. Again, the cyclopentyl rings are disordered over two sites (Figure 1). The main difference between the two phases is the fact that the H atoms of the OH groups are ordered in the monoclinic phase but disordered in the orthorhombic phase. Equivalence of the two OH groups bonded to a Si atom and therefore a higher effective symmetry is reached by this disorder in orthorhombic **2**. All four OH groups in orthorhombic **2** are equivalent by symmetry and are involved in two hydrogen bonds [O2···O2′ 2.670(2)Å, O2–H3···O2′ 169(2)°; O2···O2″ 2.678(2)Å, O2–H2···O2″ 172.5(19)°] forming two-dimensional aggregates, in which each molecule is connected to six neighbors, showing a two-dimensional closest packing almost identical to monoclinic **2** (Figure 2). In addition, the supramolecular hydrogen bonding the bond distances and angles of the central $RSi(OH)_2O$– units are in the typical range of such disiloxanetetrols [14,16–20].

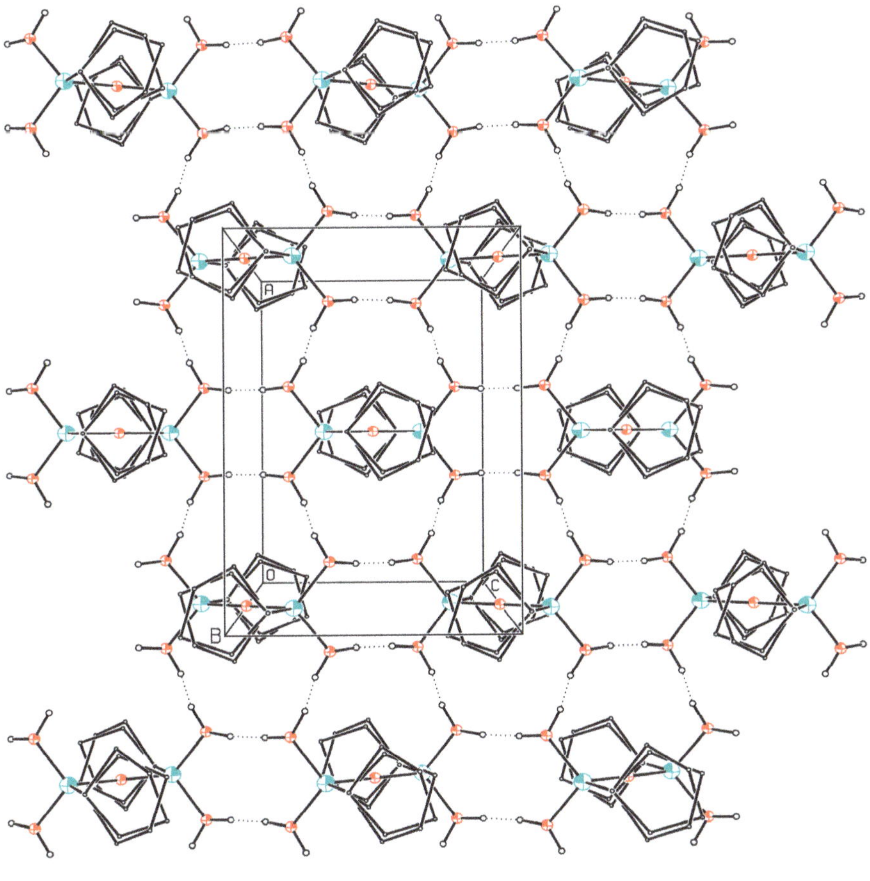

Figure 2. ORTEP plot of the orthorhombic packing of **2**. The atoms are drawn with arbitrary radii, the hydrogen bonds are plotted with dashed lines. Red: oxygen; blue: silicon.

Prolonged condensation of **1** in THF over five months resulted in a mixture containing mainly disiloxane **2** and the corresponding hexasilsesquioxane **3** in a 2:1 ratio. Recrystallization of this mixture in DMSO furnished crystalline **3**, which has been identified via spectroscopic methods and single crystal X-ray diffraction. Compound **3** has been already described in the literature together with its spectroscopic data [21]. The ^{29}Si-chemical shift of T_6 cage **3** at −56.3 ppm in DMSO solution observed by us is very similar to the previously reported one (−54.4, $CDCl_3$) and fits well in the range observed for alkyl substituted T_6 cages [16,21] but resonates at a lower field than the comparable T_8-cages [10,12,14,16,22]. Compound **3** could be obtained as single crystals suitable for X-ray diffraction and crystallizes in the orthorhombic space group *Ccce*. In the crystal structure analysis of **3**, the molecules of hexa(cyclopentylsilsesquioxane) are located with one O atom (O5) on a two-fold rotation axis parallel to the crystallographic a-axis (Figure 3).

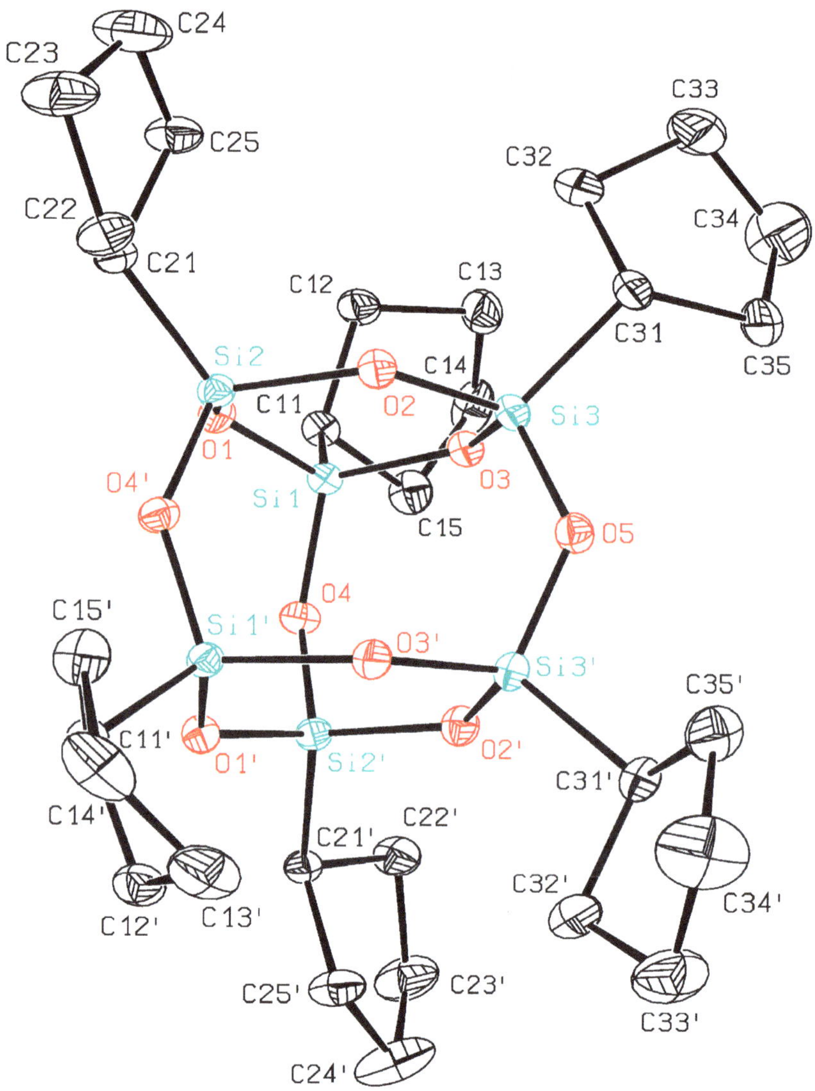

Figure 3. ORTEP plot of **3** showing the atomic numbering scheme. The probability ellipsoids are drawn at the 50% probability level. The hydrogen atoms were omitted for clarity reasons. Selected bond lengths [Å] and angles [°]: Si1–O4 1.6269(12), Si1–O1 1.6390(11), Si1–O3 1.6430(11), Si1–C11 1.8369(15), Si2–O4′ 1.6274(11), Si2–O1 1.6390(11), Si2–O2 1.6417(12), Si2–C21 1.8385(16), Si3–O5 1.6284(7), Si3–O2 1.6367(11), Si3–O3 1.6400(11), Si3–C31 1.8375(16); Si1–O1–Si2 128.69(7), Si3–O2–Si2 131.35(7), Si3–O3–Si1 131.11(7), Si1–O4–Si2′ 139.84(7), Si3–O5–Si3′ 132.34(10).

The molecules are packed in layers normal to the a-axis, leading to mechanically very soft crystals. The central highly-symmetric Si_6O_9 tetracycle shows slight but significant deviations from D_{3h} symmetry (e.g., Si1–O4–Si2′ 139.84(7)° vs Si3–O5–Si3′ 132.34(10)°). The variation of the Si–O bond lengths covers a narrow range between 1.627 Å and 1.643 Å which is smaller than in the other structure reports where variations as much as 0.04 to 0.28 Å are reported. Until now, only a few structure determinations of hexa(alkyl/aryl)silsesquioxanes with this Si_6O_9 cage can be found in the literature including *tert*-butyl [16], cyclohexyl [23], 1,1,2-trimethylpropyl [24], 2,4,6-triisopropylphenyl [25], trimethoxysilyl [26], and isopropyl [27] substituted hexasilsesquioxanes. For these, the mean value of the Si–O–Si angles in the six-membered rings is 130.2(4)°, the mean value of the other Si–O–Si angles is 140.0(8)° (min. 136.5°). Moreover the topic has been reviewed not long ago and also a structure determination of a T_6 cage has been performed in the gas phase [28,29]. The unit cell contains eight equivalent isolated molecules of **3**. Relevant geometric parameters of **3** are listed in the caption of Figure 3 and crystallographic details are summarized in Table 1.

Table 1. Crystal data and structure refinement for **2** and **3**.

Parameter	2m	2o	3
Formula	$C_{10}H_{22}O_5Si_2$	$C_{10}H_{22}O_5Si_2$	$C_{30}H_{54}O_9Si_6$
Formular weight	278.46	278.46	727.27
Temperature [K]	100	100	100
Wavelength [Å]	0.71073	0.71073	0.71073
Crystal system	monoclinic	orthorhombic	orthorhombic
Space group	$P2_1/c$	*Cmce*	*Ccce*
Unit cell dimensions:	-	-	-
a [Å]	11.5359(16)	10.1476(5)	16.2855(8)
b [Å]	6.7079(9)	20.7484(10)	22.3460(11)
c [Å]	10.1493(13)	6.7022(3)	19.6563(9)
α [°]	90	90	90
β [°]	115.830(4)	90	90
γ [°]	90	90	90
Volume [Å^3]	706.90(16)	1411.12(12)	7153.2(6)
Z	2	4	8
Calcd. density [mg/m^3]	1.308	1.311	1.351
μ [mm^{-1}]	0.258	0.258	0.283
Θ-range for data collected [°]	3.62–25.50	3.62–26.00	2.59–30.00
Data/parameters	1303/125	734/66	5219/213
Goodness-of-fit on F^2	1.095	1.159	1.052
R_1 (obsd. data)	0.0326	0.03440.0346	0.0378
wR_2 (all data)	0.0873	0.0903	0.1049
R_{int}	0.0235	0.0264	0.0397
r.e.d. min/max [e Å^{-3}]	−0.247/0.235	−0.291/0.414	−0.346/0.631

r.e.d.: Residual electron density.

3. Experimental Details

All manipulations were carried out under inert argon atmosphere using standard Schlenk technique. All solvents were dried and freshly distilled over Na/K-alloy where applicable. Cyclopentyltrichlorosilane has been purchased and used without further purification. ^{1}H– and ^{13}C–NMR-data have been recorded on a Bruker Avance III (Billerica, MA, USA) 300 MHz spectrometer (operating at 300 MHz, 75.4 MHz) or a Varian MR-400 MHz spectrometer (operating at 400 MHz, 100.5 MHz). ^{29}Si-NMR-data have been recorded on a Bruker Avance III 300 MHz spectrometer (operating at 59.6 MHz). All measurements have been performed at room temperature using TMS as external standard. EI-mass spectra have been recorded on an Agilent Technologies 5975C (Santa Clara, CA, USA) inert XL MSD with SIS Direct Insertion Probe. IR-spectra have been recorded using a Perkin-Elmer 1725X FT/IR (Waltham, MA, USA) spectrometer using KBr plates.

3.1. *Synthesis of cyclopentylsilanetriol* **1**

Cyclopentyltrichlorosilane (3.43 g, 16.9 mmol), dissolved in 13 mL diethylether, was added dropwise to a solution of water (0.91 g, 50.6 mmol) and aniline (4.78 g, 51.4 mmol) in 150 mL of diethylether at 0 °C while stirring. A white precipitate is formed in the reaction mixture and stirring is continued for 2 h upon completed addition while slowly warming to room temperature. The precipitate is filtered off with a fritted funnel and discarded. The solvent of the remaining solution is removed in vacuo. Yield 2.02 g (13.7 mmol, 81%). ^{1}H-NMR (300 MHz, D_2O): δ(ppm): 1.01, 1.48, 1.59, 1.79; ^{29}Si-NMR (59.6 MHz, D_2O): δ(ppm): −37.7.

3.2. *Synthesis of the 1,3-Dicyclopentyldisiloxane-1,1,3,3-tetrol* **2**

Silanetriol **1** (0.5 g) was dissolved in 20 mL THF at room temperature. Slow evaporation of the volatiles took four weeks. The raw material was extracted with pentane which upon evaporation of the solvent yielded compound **2** as colorless crystalline solid (0.3 g, 1.1 mmol, 64%). ^{1}H-NMR (300 MHz, THF-d_8): δ(ppm): 0.83, 1.10, 1.40–1.59, 1.78; ^{13}C-NMR (75.4 MHz, THF-d_8): δ(ppm): 25.4, 27.9, 28.7; ^{29}Si-NMR (59.6 MHz, THF-d_8): δ(ppm): −51.8; IR: 3251 (OH), 1104 (Si-O-Si); MS/EI (70 eV): *m/z* (%) = 209 (100) [M-cyc]$^+$, 141 (72) [M-cyc$_2$]$^+$.

3.3. *Synthesis of Hexa(cyclopentylsilsesquioxane)* **3**

2.5 g (16.9 mmol) of silanetriol **1** are dissolved in THF (100 mL) and are stored at room temperature for five months. The solvent is removed and the resulting solid (2.1 g) contains compounds **2** and **3** in a 2:1 ratio. Extraction with DMSO yields **3** as colorless crystalline material (1.1 g, 1.5 mmol, 53%). ^{1}H-NMR (250 MHz, DMSO-d_6): δ(ppm): 0.80–1.00 (br), 1.42–1.60 (br), 1.70; ^{13}C-NMR (75.4 MHz, DMSO-d_6): δ(ppm): 23.3, 26.38, 27.04; ^{29}Si-NMR (59.6 MHz, DMSO-d_6): δ(ppm): −56.3. MS/EI (70 eV): *m/z* (%) = 726.3 (1) [M]$^+$, 67 (100) [cyc]$^+$.

3.4. *X-ray Crystallography*

X-ray diffraction measurements were performed on a BRUKER-AXS SMART APEX 2 CCD diffractometer using graphite-monochromatized Mo-K_α radiation. Supplementary crystallographic data for this paper can be obtained free of charge quoting CCDC 1575839–1575841 from the Cambridge Crystallographic Data Centre via www.ccdc.cam.ac.uk/data_request/cif.

The structures were solved by direct methods (*SHELXS*-97)[2] and refined by full-matrix least-squares techniques against F^2 (*SHELXL*-97)[2]. The cyclopentyl groups in **2** are disordered over two orientations and were refined with site occupation factors of 0.5. In **2m** the equivalent bonds in these groups were restrained to have the same lengths. In **2o** the same anisotropic displacement parameters were used for atoms C2 and C5. The other non-hydrogen atoms were refined with anisotropic displacement parameters. The positions of the H atoms of the OH groups were taken from a difference Fourier map, the O–H distances were fixed to 0.84 Å, and the H atoms were refined with common isotropic displacement parameters without any constraints to the bond angles. The site occupation factors of the disordered H atoms of the OH groups in **2m** were fixed to 0.5. The H atoms of the tertiary C–H groups were refined with individual isotropic displacement parameters and all X–C–H angles equal at C–H distances of 1.00 Å. The H atoms of the CH_2 groups were refined with common isotropic displacement parameters for the H atoms of the equivalent CH_2 groups and idealized geometries with approximately tetrahedral angles and C–H distances of 0.99 Å.

4. Conclusions

In summary, we have shown that cyclopentyl substituted silanetriol can be prepared and isolated. In polar solvents, spontaneous condensation occurs which yields the corresponding disiloxanetetrol as a primary condensation product. Further condensation leads to the hexameric polyhedral silsesquioxane cage T_6. The latter has been mentioned in the literature before. However, it lacked structural data. All compounds have been characterized with multinuclear NMR spectroscopy and in addition the molecular structures have been determined in the case of the disiloxanetetrol and the hexasilsesquioxane via single crystal X-ray diffraction. Our results show that silanetriols bearing secondary alkyl substituents may be suitable precursors for the synthesis of POSS cages as well.

Acknowledgments: The authors would like to thank the EU-COST network CM1302 "Smart Inorganic Polymers" (SIPs).

Author Contributions: Jürgen Kahr performed the experiments and analyzed the spectroscopic data; Ferdinand Belaj performed the X-ray diffraction and interpretation; Rudolf Pietschnig provided the materials and wrote the paper.

References

1. Hartmann-Thompson, C. *Advances in Silicon Science*; Springer: Heidelberg, Germany, 2011; Volume 3.
2. Cordes, D.B.; Lickiss, P.D.; Rataboul, F. Recent developments in the chemistry of cubic polyhedral oligosilsesquioxanes. *Chem. Rev.* **2010**, *110*, 2081–2173. [CrossRef] [PubMed]
3. Janssen, M.; Wilting, J.; Müller, C.; Vogt, D. Continuous rhodium-catalyzed hydroformylation of 1-octene with polyhedral oligomeric silsesquioxanes (POSS) enlarged triphenylphosphine. *Angew. Chem. (Int. Ed.)* **2010**, *49*, 7738–7741. [CrossRef] [PubMed]
4. Dittmar, U.; Hendan, B.J.; Floerke, U.; Marsmann, H.C. Functionalized octa(propylsilsesquioxanes) (3-xC_3H_6)$_8$(Si_8O_{12})—Model compounds for surface modified silica gel. *J. Organomet. Chem.* **1995**, *489*, 185–194. [CrossRef]
5. Feher, F.J.; Newman, D.A.; Walzer, J.F. Silsesquioxanes as models for silica surfaces. *J. Am. Chem. Soc.* **1989**, *111*, 1741–1748. [CrossRef]
6. Tuteja, A.; Choi, W.; Ma, M.; Mabry, J.M.; Mazzella, S.A.; Rutledge, G.C.; McKinley, G.H.; Cohen, R.E. Designing superoleophobic surfaces. *Science* **2007**, *318*, 1618–1622. [CrossRef] [PubMed]
7. Tanaka, K.; Ishiguro, F.; Chujo, Y. POSS ionic liquid. *J. Am. Chem. Soc.* **2010**, *132*, 17649–17651. [CrossRef] [PubMed]
8. Feher, F.J.; Wyndham, K.D.; Scialdone, M.A. Octafunctionalized polyhedral oligosilsesquioxanes as scaffolds: Synthesis of peptidyl silsesquioxanes. *Chem. Commun.* **1998**, *14*, 1469–1470. [CrossRef]
9. Matisons, J. *Applications of Polyhedral Oligomeric Silsesquioxanes*; Springer: Dordrecht, The Netherlands; Heidelberg, Germany; London, UK; New York, NY, USA, 2011.
10. Bassindale, A.R.; Liu, Z.; MacKinnon, I.A.; Taylor, P.G.; Yang, Y.; Light, M.E.; Horton, P.N.; Hursthouse, M.B. A higher yielding route for T_8 silsesquioxane cages and X-ray crystal structures of some novel spherosilicates. *Dalton Trans.* **2003**, *14*, 2945–2949. [CrossRef]
11. Bassindale, A.R.; Pourny, M.; Taylor, P.G.; Hursthouse, M.B.; Light, M.E. Fluoride-ion encapsulation within a silsesquioxane cage. *Angew. Chem. Int. Ed.* **2003**, *42*, 3488–3490. [CrossRef] [PubMed]
12. Bassindale, A.R.; Chen, H.; Liu, Z.; MacKinnon, I.A.; Parker, D.J.; Taylor, P.G.; Yang, Y.; Light, M.E.; Horton, P.N.; Hursthouse, M.B. A higher yielding route to octasilsesquioxane cages using tetrabutylammonium fluoride, Part 2: Further synthetic advances, mechanistic investigations and X-ray crystal structure studies into the factors that determine cage geometry in the solid state. *J. Organomet. Chem.* **2004**, *689*, 3287–3300. [CrossRef]
13. Pietschnig, R.; Spirk, S. The chemistry of organo silanetriols. *Coord. Chem. Rev.* **2016**, *323*, 87–106. [CrossRef]
14. Hurkes, N.; Bruhn, C.; Belaj, F.; Pietschnig, R. Silanetriols as powerful starting materials for selective condensation to bulky POSS cages. *Organometallics* **2014**, *33*, 7299–7306. [CrossRef] [PubMed]
15. Takiguchi, T. Preparation of some organosilanediols and phenylsilanetriol by direct hydrolysis using aniline as hydrogen chloride acceptor. *J. Am. Chem. Soc.* **1959**, *81*, 2359–2361. [CrossRef]
16. Spirk, S.; Nieger, M.; Belaj, F.; Pietschnig, R. Formation and hydrogen bonding of a novel POSS-trisilanol. *Dalton Trans.* **2009**, 163–167. [CrossRef] [PubMed]
17. Čas, D.; Hurkes, N.; Spirk, S.; Belaj, F.; Bruhn, C.; Rechberger, G.N.; Pietschnig, R. Dimer formation upon deprotonation: Synthesis and structure of a *m*-terphenyl substituted (*R,S*)-dilithium disiloxanolate disilanol. *Dalton Trans.* **2015**, *44*, 12818–12823. [CrossRef] [PubMed]
18. Hurkes, N.; Spirk, S.; Belaj, F.; Pietschnig, R. At the edge of stability—Preparation of methyl-substituted arylsilanetriols and investigation of their condensation behavior. *Z. Anorg. Allg. Chem.* **2013**, *639*, 2631–2636. [CrossRef]
19. Suyama, K.-I.; Nakatsuka, T.; Gunji, T.; Abe, Y. Synthesis and crystal structure of disiloxane-1,3-diols and disiloxane-1,1,3,3-tetraol. *J. Organomet. Chem.* **2007**, *692*, 2028–2035. [CrossRef]
20. Lickiss, P.D.; Litster, S.A.; Redhouse, A.D.; Wisener, C.J. Isolation of a tetrahydroxydisiloxane formed during hydrolysis of an alkyltrichlorosilane: Crystal and molecular structure of $[Bu^t(OH)_2Si]_2O$. *J. Chem. Soc. Chem. Commun.* **1991**, *3*, 173–174. [CrossRef]
21. Bassindale, A.R.; MacKinnon, I.A.; Maesano, M.G.; Taylor, P.G. The preparation of hexasilsesquioxane (T_6) cages by "non aqueous" hydrolysis of trichlorosilanes. *Chem. Commun.* **2003**, *12*, 1382–1383. [CrossRef]
22. Unno, M.; Alias, S.B.; Arai, M.; Takada, K.; Tanaka, R.; Matsumoto, H. Synthesis and characterization of cage and bicyclic silsesquioxanes via dehydration of silanols. *Appl. Organometal. Chem.* **1999**, *13*, 303–310. [CrossRef]

23. Behbehani, T.; Brisdon, H.B.J.; Mahon, M.F.; Molloy, K.C. The structure of hexa(cyclohexylsesquisiloxane), $(C_6H_{11})_6Si_6O_9$. *J. Organomet. Chem.* **1994**, *469*, 19–24. [CrossRef]
24. Unno, M.; Alias, S.B.; Saito, H.; Matsumoto, H. Synthesis of hexasilsesquioxanes bearing bulky substituents: Hexakis((1,1,2-trimethylpropyl)silsesquioxane) and hexakis(tert-butylsilsesquioxane). *Organometallics* **1996**, *15*, 2413–2414. [CrossRef]
25. Unno, M.; Imai, Y.; Matsumoto, H. Hexakis(2,4,6-triisopropylphenylsilsesquioxane). *Silicon Chem.* **2003**, *2*, 175–178. [CrossRef]
26. Hoebbel, D.; Engelhardt, G.; Samoson, A.; Újszászy, K.; Smolin, Y.I. Preparation and constitution of the crystalline silicic acid trimethylsilyl ester $[(CH_3)_3Si]_6Si_6O_{15}$. *Z. Anorg. Allg. Chem.* **1987**, *552*, 236–240. [CrossRef]
27. Unno, M.; Suto, A.; Takada, K.; Matsumoto, H. Synthesis of Ladder and Cage Silsesquioxanes from 1,2,3,4-Tetrahydroxycyclotetrasiloxane. *Bull. Chem. Soc. Jpn.* **2000**, *73*, 215–220. [CrossRef]
28. Lickiss, P.D.; Rataboul, F. Fully condensed polyhedral oligosilsesquioxanes (POSS): From synthesis to application. In *Advances in Organometallic Chemistry*; Academic Press: Oxford, UK, 2008; Volume 57, pp. 1–116.
29. Wann, D.A.; Reilly, A.M.; Rataboul, F.; Lickiss, P.D.; Rankin, D.W.H. The gas-phase structure of the hexasilsesquioxane $Si_6O_9(OSiMe_3)_6$. *Z. Naturforsch. B* **2009**, *64*, 1269–1275. [CrossRef]

18

Transformative Si_8R_8 Siliconoids

Naohiko Akasaka, Shintaro Ishida and Takeaki Iwamoto *

Department of Chemistry, Graduate School of Science, Tohoku University, Sendai 980-8578, Japan; nao.akasaka1216@gmail.com (N.A.); ishida@tohoku.ac.jp (S.I.)

* Correspondence: takeaki.iwamoto@tohoku.ac.jp

Abstract: Molecular silicon clusters with unsubstituted silicon vertices (siliconoids) have received attention as unsaturated silicon clusters and potential intermediates in the gas-phase deposition of elemental silicon. Investigation of behaviors of the siliconoids could contribute to the greater understanding of the transformation of silicon clusters as found in the chemical vapor deposition of elemental silicon. Herein we reported drastic transformation of a Si_8R_8 siliconoid to three novel silicon clusters under mild thermal conditions. Molecular structures of the obtained new clusters were determined by XRD analyses. Two clusters are siliconoids that have unsaturated silicon vertices adopting unusual geometries, and another one is a bis(disilene) which has two silicon–silicon double bonds interacted to each other through the central polyhedral silicon skeleton. The observed drastic transformation of silicon frameworks suggests that unsaturated molecular silicon clusters have a great potential to provide various molecular silicon clusters bearing unprecedented structures and properties.

Keywords: cluster; isomerization; silicon; siliconoid; subvalent compounds

---

1. Introduction

Molecular silicon clusters have attracted much attention as their characteristic electronic properties and reactivity depend on the structures [1–5]. In the chemical vapor deposition (CVD) process of elemental silicon, unsaturated silicon clusters with unsubstituted silicon vertices are considered as potential intermediates [6–8]. Recently, such kinds of unsaturated molecular silicon clusters are synthesized as isolable molecules [9–22] and termed "siliconoids" by Scheschkewitz et al. [10]. The siliconoids show their characteristic distorted structures and large distribution of ^{29}Si nuclear magnetic resonances. As the unsubstituted vertices are considered to be preferred reaction sites for transformation of silicon clusters [23–25], investigation of thermal transformation of the siliconoids would provide fundamental reactions that may contribute to improving our understanding of the mechanism of the CVD process elemental silicon [26–30]. However, thermal transformation of siliconoids is still scarce. Scheschkewitz et al. have reported thermal rearrangement, expansion, and contraction of Si_6Tip_6 (Tip = 2,4,6-triisopropylphenyl) siliconoids [10,13,15–17,22].

Recently, we reported that thermal reaction of Si_5R_6 siliconoid **1** at 40 °C afforded Si_8R_8 siliconoids **4** and **6**, the latter of which undergoes thermal isomerization to **4** upon extra heating (Scheme 1). In this transformation, elimination of a silylene unit (SiR_2, **2**) from **1** generates disilene **3** and silylene **5**, which equilibrate with each other and dimerize to give **4** and **6**, respectively [22]. As **4** has still highly strained structures similar to **1**, we examined further thermal reactions of **4**. Herein, we report thermal transformation of **4** giving new silicon clusters that involve highly distorted silicon atoms, silicon–silicon double bonds, or unsubstituted silicon vertices. The observed drastic transformation suggests that unsaturated molecular silicon clusters have a great potential to provide various molecular silicon clusters bearing unprecedented structures and properties.

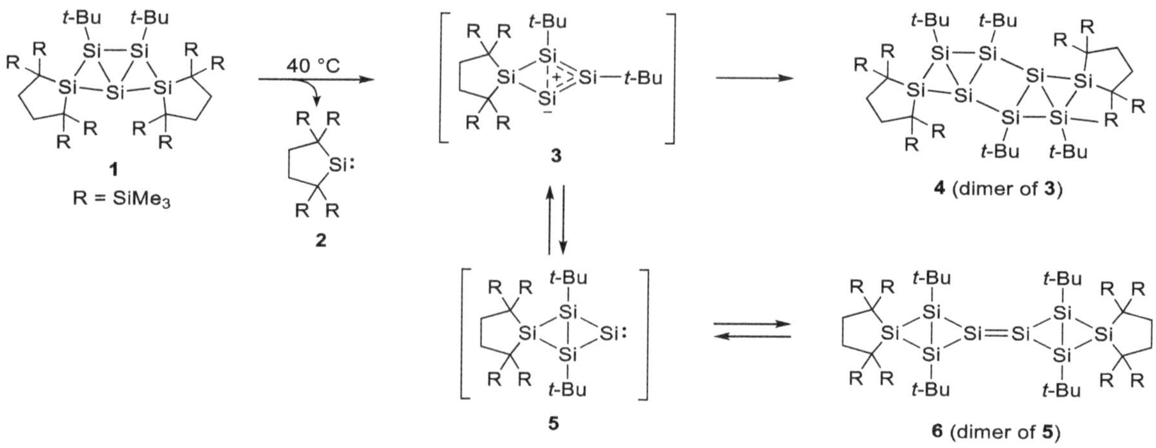

Scheme 1. Thermal transformation of **1**. R = $SiMe_3$.

2. Results

2.1. Thermal Reactions of 4

Upon heating **4** at 75 °C in benzene-d_6, new silicon clusters with unprecedented silicon frameworks **7**, **8**, and **9** were formed together with silylene **2** and its thermal isomerization product, silene **10** (Scheme 2). Compounds **7** and **8** are isomers of **4** and **6** bearing the formula of Si_8R_8, while **9** is a contracted Si_7R_6 cluster. Formation of **9** is consistent with concomitant formations of SiR_2 units (**2** and **10**). The time course of the product yields monitored by ^1H NMR spectrum (see Figure S16 in the Supplementary Materials) indicated that **7** was initially formed and then **8** and **9** were generated in this reaction: after heating for 3 h, **7**, **8**, and **9** were observed in 26%, 3%, and 28% yields, respectively, while 19 h later, **7** disappeared and **8** and **9** were observed in 6% and 59% yields, respectively. Silicon clusters **7**, **8**, and **9** are moisture and air sensitive and isolated by careful recrystallization of the reaction mixture. Details of the structures of **7**–**9** will be discussed in the subsequent sections.

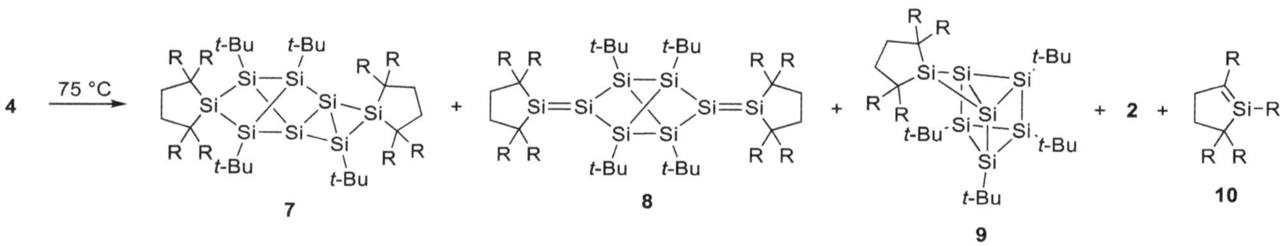

Scheme 2. Thermal transformation of **4**. R = $SiMe_3$.

Interestingly, thermal reaction of pure **7** at 60 °C in benzene-d_6 provided not only **8** and **9** but also **4**. This result indicates that isomerization between **4** and **7** is reversible at 60 °C in solution, though we are not able to distinguish whether **8** and **9** were formed from **4** directly or via **7**. While **8** remains intact after heating even at 80 °C for 10 h in benzene-d_6, **9** was decomposed together with formation of **10** under the same conditions as monitored by ^1H NMR spectroscopy.

2.2. Molecular Structure of 7

XRD analysis exhibits that **7** has unsubstituted vertices at the Si3 and Si4 atoms that were shared by the bicyclo[1.1.0]tetrasilane moiety [Si1, Si2, Si3, and Si4] similar to **4** and the tricyclo[2.2.0.0^{2,5}]hexasilane moiety [Si3, Si4, Si5, Si6, Si7, and Si8] (Figure 1). The Si3 and Si4 atoms adopt a highly distorted umbrella geometry which is very far from the typical tetrahedral configuration:

the bond angles vary from 57.47(2)° [Si2–Si3–Si4] to 149.57(3)° [Si1–Si3–Si6] around the Si3 atom and from 56.02(2)° [Si2–Si4–Si3] to 140.10(3)° [Si2–Si4–Si5] around the Si4 atom. The distance between Si3 and Si4 [2.5477(8) Å] is longer than typical Si–Si single bond length [ca. 2.36 Å] [31] and those of the other Si–Si bonds in the silicon skeleton of **7** [2.3037(8)–2.4618(7) Å] but close to the distance between bridgehead unsubstituted silicon atoms in hexamesitylpentasila[1.1.1]propellane **11** [2.636(1) Å] [18] and siliconoids such as **12** [2.7076(8) Å] [13,14] (Figure 2).

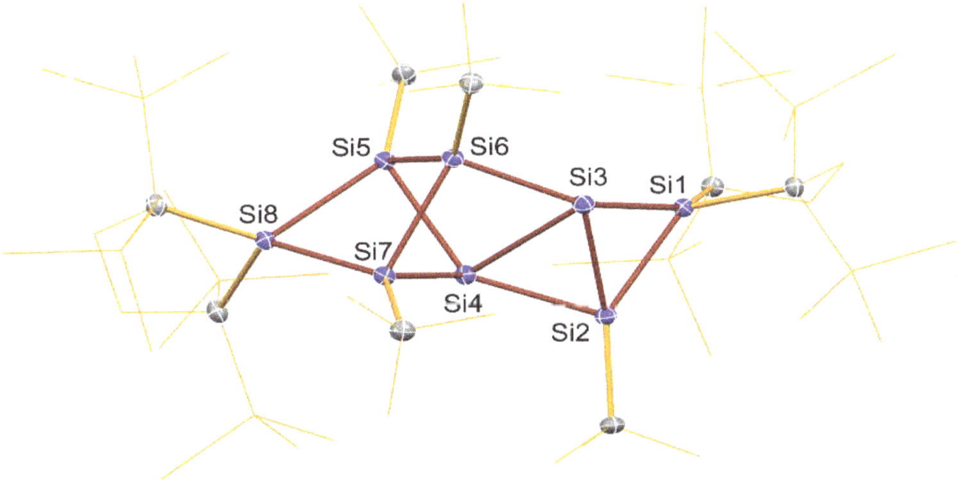

Figure 1. ORTEP drawing of **7** (atomic displacement parameters set at 50% probability; hydrogen atoms and thermal ellipsoids of selected carbon and silicon atoms omitted for clarity). Selected distances [Å] and angles [°]: Si1–Si2 2.3684(7), Si1–Si3 2.3798(8), Si2–Si3 2.3037(8), Si2–Si4 2.3422(8), Si3–Si4 2.5466(8), Si4–Si5 2.4005(8), Si4–Si7 2.4240(7), Si5–Si6 2.3563(8), Si5–Si8 2.4441(7), Si6–Si7 2.3440(8), Si7–Si8 2.4618(7), Si1–Si2–Si3–Si4 122.48(3).

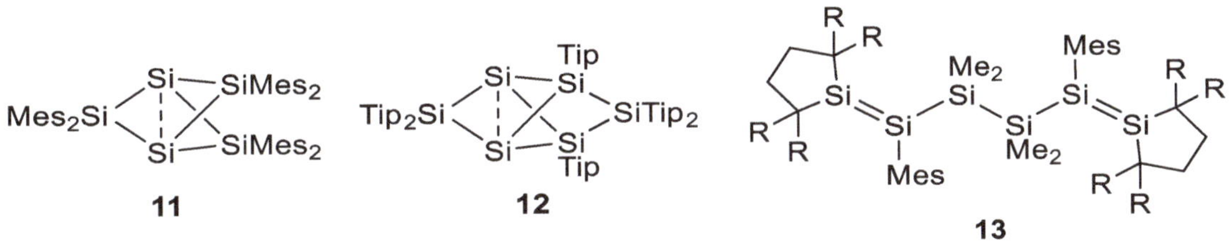

Figure 2. Related silicon clusters and a disilene. Mes = 2,4,6-trimethylphenyl, Tip = 2,4,6-triisopropylphenyl, R = $SiMe_3$.

The ^{1}H NMR spectrum of **7** indicates a facile flipping of the bicyclo[1.1.0]tetrasilane moiety (Si1, Si2, Si3, and Si4) in solution on the NMR time scale: two *t*-Bu groups on Si5 and Si7 atoms were equally observed and four singlet signals for $SiMe_3$ groups on the terminal five-membered rings were observed with the same integral ratio. In the ^{29}Si NMR spectrum of **7**, a large distribution of chemical shifts [δSi = −191.9 (Si4), −66.3 (Si6), −20.0 (Si3), −12.0 (Si5, Si7), 7.5 (Si2), 42.2 (Si1), 64.7 (Si8) ppm] were observed similar to those observed for the reported siliconoids [13,14,18].

The UV–Vis spectrum of **7** in hexane exhibits several absorption bands tailed to ca. 600 nm with several shoulders as found in those of other molecular silicon clusters. The longest wavelength's absorption maximum was observed at 492 nm (ε 1400) (Figure 3). Judging from the results of the TD-DFT calculation of **7** at the B3LYP/6-311G(d)//B3PW91-D3/6-31G(d) level of theory ($\mathbf{7_{opt}}$), the band at 492 nm can be assigned to the transition from HOMO to LUMO. As shown in Figure 3b, HOMO and LUMO are mainly the σ and σ* orbitals of the Si3–Si4 bond with the highly distorted geometry which is the longest Si–Si distance in the silicon framework as mentioned above.

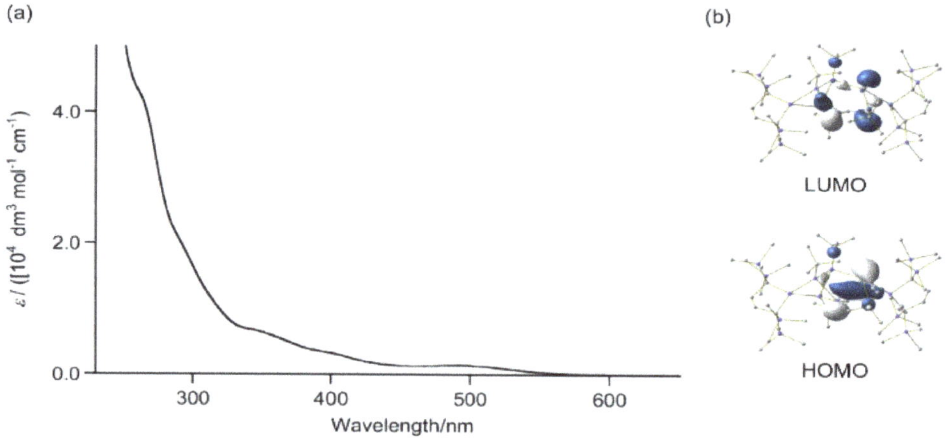

Figure 3. (**a**) UV–Vis spectrum of **7** in hexane at room temperature. (**b**) Frontier Kohn–Sham orbitals of **7** at the B3LYP-D3/6-311G(d)//B3PW91-D3/6-31G(d) level of theory (isosurface value = 0.05 $e^-/Å^3$).

2.3. Molecular Structure of **8**

Although purification of **8** was very difficult because of its very low yield, a few single crystals of **8** were luckily obtained after recrystallization from hexane. Accordingly, **8** was characterized by only XRD analysis and ^{1}H NMR spectroscopy. XRD analysis exhibits that it has two silicon–silicon double bonds at both sides of a tricyclo[2.2.0.0^{2,5}]hexasilane framework (Figure 4). The double-bond silicon atoms Si1, Si2, Si5, and Si6 lie on the crystallographic *C*2 axis. The Si1–Si2 [2.1570(12) Å] and Si5–Si6 [2.1641(12) Å] distances are within the range of the known silicon–silicon double bonds (2.118–2.289 Å) [32,33] and these Si=Si double-bond planes are almost perpendicular to each other with the dihedral angle between the least-square plane of the terminal silacyclopentane rings of 87.3°. The Si–Si single bond distances in the central tricyclo[2.2.0.0^{2,5}]hexasilane skeleton [2.3496(8)–2.3641(9) Å] are similar to that of typical Si–Si single bonds (ca. 2.36 Å) [31] and reported tricyclo[2.2.0.0^{2,5}]hexasilanes [2.33–2.47 Å] [13,34–36].

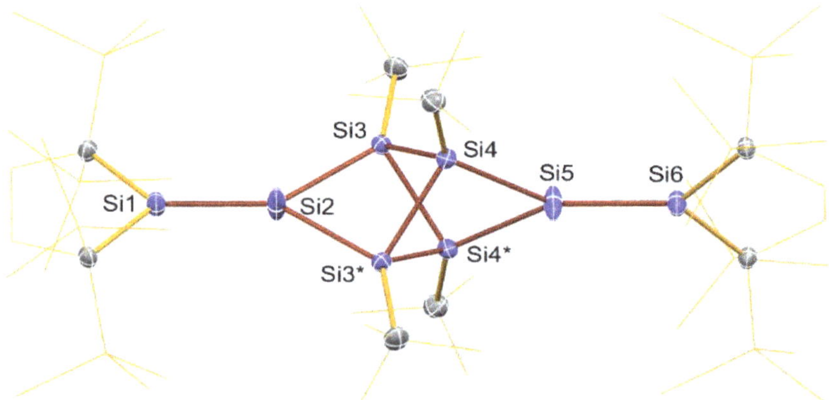

Figure 4. ORTEP drawing of **8** (atomic displacement parameters set at 50% probability; hydrogen atoms and thermal ellipsoids of selected carbon and silicon atoms omitted for clarity). Selected distances [Å] and angles [°]: Si1–Si2 2.1570(12), Si2–Si3 2.3570(9), Si3–Si4 2.3496(8), Si4–Si5 2.3641(9), Si5–Si6 2.1641(12), Si1–Si2–Si3 142.020(19), Si3–Si2–Si3* 75.96(4), Si2–Si3–Si4 90.74(3), Si3–Si4–Si5 91.17(3), Si4–Si5–Si6 142.41(2), Si1–Si2–Si3–Si4 142.134(19), Si2–Si3–Si4–Si5 128.51(2).

In the UV–Vis spectrum of **8** in hexane, a distinct absorption band (λ_{max} 451 nm) with a shoulder band (500 to 600 nm) was observed in the visible region (Figure 5a). This band is considerably redshifted compared to that observed for a structurally related bis(disilene) in which two Si=Si double bonds were connected by a Si–Si bond (**13**, λ_{max} = 403 nm, Figure 2). DFT calculation suggested that the band in the visible region would be overlapped by a few bands due to π(Si=Si) to π*(Si=Si) transitions. For compound **8** optimized at the B3PW91-D3/6-31G(d) level of theory (**8**$_{opt}$), HOMO (−4.47 eV) and HOMO−1 (−4.50 eV) are π(Si=Si) orbitals, while LUMO (−1.47 eV) and LUMO+1 (−1.17 eV) are π*(Si=Si) orbitals (Figure 5b). While HOMO and HOMO−1 are almost degenerated, the difference in energy between LUMO and LUMO+1 was relatively large (0.30 eV). This large energy difference would result from considerable interaction between two π*(Si=Si) orbitals through the central silicon cage. TD-DFT calculation of **8**$_{opt}$ at the B3LYP/6-311G(d) [hexane]//B3PW91-D3/6-31G(d) level of theory predicts that the major peak observed at 451 nm involves a combination of HOMO−1 to LUMO+1 and HOMO to LUMO+1 transitions, while the shoulder band involves HOMO to LUMO and HOMO−1 to LUMO transitions. The spectral feature of **8** is consistent with the substantial interactions between two Si=Si double bonds through the central silicon cages.

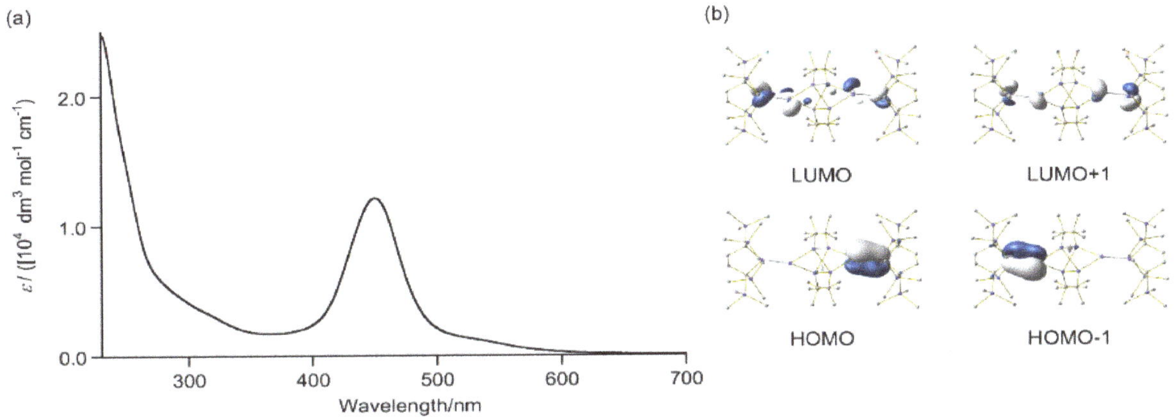

Figure 5. (**a**) UV–Vis spectrum of **8** in hexane at room temperature. (**b**) Frontier Kohn–Sham orbitals of **8** at the B3LYP-D3/6-311G(d)//B3PW91-D3/6-31G(d) level of theory (isosurface value = 0.05 e$^-$/Å^3).

2.4. Molecular Structure of 9

The XRD analysis indicates that **9** is also a silicon cluster classified as a siliconoid having unsubstituted vertices at Si2 and Si3 atoms with a hexasila[2.1.1]propellane skeleton in which two wings (Si4 and Si5–Si6) are bridged by Si7 atom (Figure 6). To the best of our knowledge, this is the first example of a silicon cluster having [2.1.1]propellane skeleton, although persilapropellane family including hexamesitylpentasila[1.1.1]propellane **11** (Figure 2) [18] and its bridged siliconoids [13,14] have been synthesized as isolable molecules. Cluster **9** can also be seen as a hexasilaprismane (Si2–Si7) in which two skeletal silicon atoms (Si2 and Si3) are bridged by one silicon atom (Si1). The Si2 and Si3 atoms adopt an umbrella-type inverted tetrahedral geometry as observed for bridgehead silicon atoms of pentasila[1.1.1]properanes [13,14,18]. The Si2–Si3 distance [2.6829(8) Å)] is comparable with the distances between silicon atom with an umbrella geometry in hexamesitylpentasila[1.1.1]propellane **11** [2.636(1) Å] [18] and siliconoids such as **12** [2.7076(8) Å] (Figure 2) [13,14].

The ^{29}Si NMR spectrum of **9** showed five signals due to skeletal silicon nuclei at −159.5 (Si2, Si3), −60.3 (Si7), 2.3 (Si5, Si6), 26.5 (Si1), 123.3 (Si4) ppm, which were assigned on the basis of a ^{1}H–^{29}Si HMBC spectrum. The highfield-shifted ^{29}Si NMR resonance due to the unsubstituted vertices of **9** (δSi = −159.5) is similar to those of pentasila[1.1.1]propellane **11** [δSi = −273.2 [18]] and its bridged siliconoid **12** [δSi = −274.2 [13]], while the considerably downfieldshift of Si4 nuclei in **9** resembles that of **12** [δSi = 174.6] [13]. The geometry around the hexasilaprismane framework in **9** except for the Si2 and Si3 atoms resembles to those of the reported hexasilaprismanes [10,36,37].

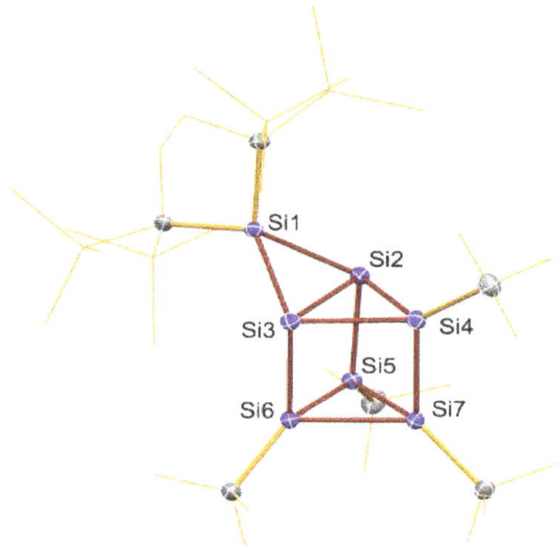

Figure 6. ORTEP drawing of **9** (atomic displacement parameters set at 50% probability; hydrogen atoms and thermal ellipsoids of selected carbon and silicon atoms omitted for clarity). Selected distances [Å] and angles [°]: Si1–Si2 2.3813(8), Si2–Si3 2.6829(8), Si2–Si4 2.3645(8), Si2–Si5 2.3738(8), Si3–Si4 2.3566(8), Si3–Si6 2.3804(8), Si4–Si7 2.3051(8), Si5–Si6 2.3292(8), Si5–Si7 2.3435(9), Si6–Si7 2.3511(8), Si2–Si1–Si3 68.86(3), Si1–Si2–Si3 55.26(2), Si1–Si2–Si4 102.08(3), Si1–Si2–Si5 117.64(3), Si4–Si2–Si5 88.85(3), Si1–Si3–Si2 55.88(2), Si1–Si3–Si4 102.85(3), Si1–Si3–Si6 115.41(3), Si4–Si3–Si6 88.32(3), Si1–Si2–Si3–Si4 142.33(4).

In the UV–Vis spectrum in hexane, **9** exhibits a distinct absorption band at 511 nm (ε 2700) (Figure 7), which is similar to the absorption bands observed for siliconoids **11** and **12** (Figure 2) but in contrast to those for hexasilaprismanes showing normally only weak absorption tailing to 500 nm [10,36,37]. TD-DFT calculation of **9** predicted that this band should be assigned to HOMO → LUMO and HOMO−1 → LUMO transitions. HOMO−1 and LUMO involve mainly σ and σ* orbitals of the interbridgehead Si2–Si3 bond with the umbrella geometry, while HOMO is σ orbitals of the Si–Si bond in the silicon framework. The characteristic absorption band observed in **9** would result from the presence of the unsubstituted silicon vertices with an inverted geometry.

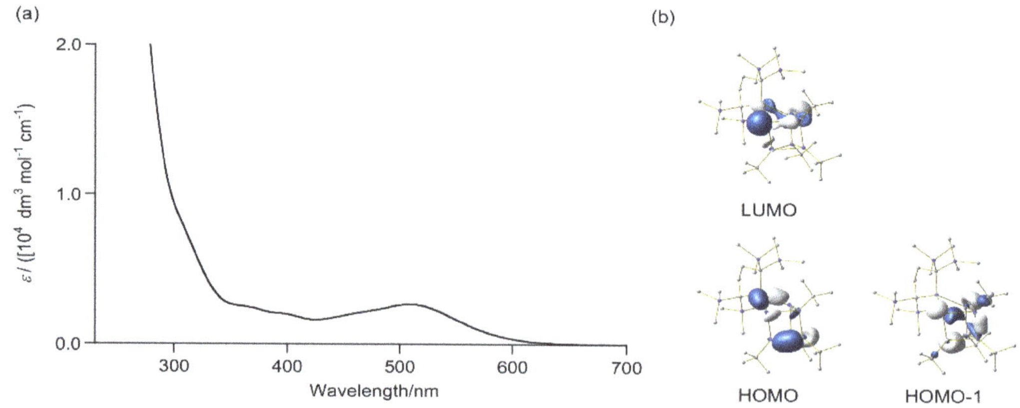

Figure 7. (**a**) UV–Vis spectrum of **9** in hexane at room temperature. (**b**) Frontier Kohn–Sham orbitals of **9** at the B3LYP-D3/6-311G(d)//B3PW91-D3/6-31G(d) level of theory (isosurface value = 0.05 $e^-/Å^3$).

2.5. Theoretical Study

Although the detailed mechanism for transformation from **4** to **7**–**9** remains unclear at this moment, it should be worthy to discuss the relative stability of Si_8R_8 isomers **4**, **6**, **7**, **8**. In Table 1, relative energies of the clusters calculated at the B3PW91-D3/6-31G(d) level are shown and all optimized structures (**4**$_{\mathbf{opt}}$,

6_{opt}, 7_{opt}, 8_{opt}, and 9_{opt}) were in agreement with the experimentally observed structures. Formations of 6_{opt} and 7_{opt} from 4_{opt} are predicted to be very slightly endoergonic, while generation of 8_{opt} and $9_{opt}+2_{opt}$ and from 4_{opt} are exoergonic at 348.14 K, which is consistent with the experimental results. In contrast to the predicted high thermal stability of 8_{opt}, the yield of **8** was considerably low. As silylene **2** decomposes quickly to **10** at 348 K [38], formation of **9** is expected to be irreversible under the thermal conditions. The irreversible formation of **9** and the higher activation barrier for formation of **8** compared to those for other reactions may be responsible for the low yield of **8**. Although the reason for the higher stability of 8_{opt} compared to other isomers remains unclear, the absence of the highly strained structure such as bicyclo[1.1.0]tetrasilane moiety and the umbrella geometry as found in 4_{opt}, 6_{opt}, and 7_{opt} may be mainly responsible for the relative stability of 8_{opt}.

Table 1. Relative free energies for Si_8R_8 isomers calculated at the B3PW91-D3/6-31G(d) level of theory.

Compound	ΔG(298.15 K) [kJ/mol]	ΔG(348.15 K) [kJ/mol]
4_{opt}	0.0	0.0
6_{opt}	+16.2	+14.6
7_{opt}	+14.2	+14.6
8_{opt}	−71.8	−77.5
$9_{opt} + 2_{opt}$	+14.9	−0.1

It is also noteworthy to compare with other possible Si_8R_8 isomers. Theoretical study of Si_8H_8 species demonstrated that a silicon cluster bearing a bicyclo[1.1.0]butane moiety with C_2 symmetry (**14^H^** in Figure 8) has been located theoretically as the most stable isomer which is 30.9 kJ/mol lower in energy than the second stable octasilacubane **15^H^** at the B3LYP/6-31G(d) level of theory [39]. While isolable Si_8R_8 species such as octasilacubane **15** [40–45] and octasilacuneane **16** [19] have been reported, the Si_8R_8 clusters in our study such as **4**, **6**, **7**, and **8** do not provide such stable isomers. Two silicon vertices of the Si_8R_8 clusters in this study have two alkyl groups, while two silicon vertices are unsubstituted, which would lead to a unique transformation reaction without involving the stable isomers such as octacubane.

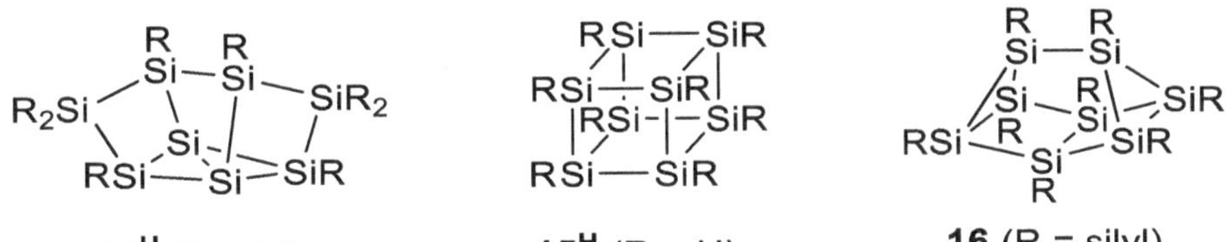

Figure 8. Examples of related Si_8R_8 isomers.

3. Materials and Methods

3.1. General Procedures

All reactions treating air-sensitive compounds were carried out under nitrogen atmosphere using a high-vacuum line and standard Schlenk techniques, or a glove box, as well as dry and oxygen-free solvents. NMR spectra were recorded on a Bruker Avance 500 FT NMR spectrometer (Bruker Japan, Yokohama, Japan). The ^1H and ^{13}C NMR chemical shifts were referenced to residual ^1H and ^{13}C signals of the solvents, benzene-d_6 (^1H δ 7.16; ^{13}C δ 128.0). The ^{29}Si NMR chemical shifts were relative to Me_4Si in ppm (δ 0.00). Sampling of air-sensitive compounds was carried out using a VAC NEXUS 100027 type glove box (Vacuum Atmospheres Co., Hawthorne, CA, USA). Reactions at low temperatures were performed by using an EYELA PSL-1400 cryobath (Tokyo Rikakikai Co, Ltd, Tokyo, Japan). UV–Vis spectra were recorded on a JASCO V-660 spectrometer (JASCO, Tokyo, Japan).

3.2. Materials

Benzene, benzene-d_6, diethyl ether, hexane, tetrahydrofuran (THF), and toluene were dried over $LiAlH_4$, and then distilled prior to use by using the vacuum line. Compound **4** was prepared according to the procedure reported in the literature [22].

3.3. Thermolysis of 4

In a J. Young NMR tube, **4** (5.0 mg, 0.0044 mmol) was dissolved in benzene-d_6 (0.5 mL). The mixture was heated for 22 h at 75 °C, monitored by ^{1}H NMR spectroscopy (Figure S16). After heating the mixture for 3 h, compounds **7** (26%), **8** (3%), and **9** (28%) were formed (Figure S1). After heating for a total of 22 h, only **8** (6%) and **9** (59%) were observed (Figure S2). The yields of **7–9** were determined by ^{1}H NMR spectroscopy using ferrocene as an internal standard.

3.4. Isolation of 7

In a Schlenk tube (30 mL) equipped with a magnetic stir bar, compound **4** (150 mg, 0.131 mmol) and benzene (15 mL) were placed. The reaction mixture was heated at 75 °C for 3 h. Benzene was removed in vacuo. Recrystallization from hexane at −35 °C gave **4**. Recrystallization of the mother liquor from toluene at −35 °C gave cluster **7** as reddish orange crystals (11.4 mg, 0.0100 mmol) in 8.0% yield.

7: reddish orange crystals; mp. 138 °C (decomp.); ^{1}H NMR (C_6D_6, 500 MHz, 298 K) δ 0.40 (s, 18H, $SiMe_3$), 0.45 (s, 18H, $SiMe_3$), 0.46 (s, 18H, $SiMe_3$), 0.52 (s, 18H, $SiMe_3$), 1.52 (s, 18H, *t*-Bu), 1.53 (s, 9H, *t*-Bu), 1.57 (s, 9H, *t*-Bu), 1.94–2.00 (m, 4H, CH_2), 2.03 (s, 4H, CH_2); ^{13}C NMR (C_6D_6, 125 MHz, 298 K) δ 4.8 ($SiMe_3$), 5.8 ($SiMe_3$), 6.2 ($SiMe_3$), 6.6 ($SiMe_3$), 17.5 (C), 19.5 (C), 20.6 (C), 26.8 (C), 32.0 ($C(CH_3)_3$), 32.3 (C), 33.4 ($C(CH_3)_3$), 34.2 (C), 35.1 (CH_2), 35.6 (CH_2), 35.9 ($C(CH_3)_3$), 36.3 (CH_2); ^{29}Si NMR (C_6D_6, 99 MHz, 298 K) δ −191.9 (Si4), −66.3 (Si8), −20.0 (Si2), −12.0 (Si5,Si7), 2.60 ($SiMe_3$), 3.01 ($SiMe_3$), 3.66 (2 × $SiMe_3$), 7.53 (Si3), 42.2 (Si1), 64.7 (Si6); UV/vis (in hexane): λ_{max}/nm (ε) 492 (1400), 405 (sh, 3100), 351 (sh, 6500); HRMS (APCI) *m/z* ([M]$^+$ was not observed but [M + H + O_2]$^+$ was found probably because **7** is so reactive that it was partially oxidized and/or hydrolyzed before injecting the sample into the instrument) calcd for $C_{48}H_{117}O_2Si_{16}$: 1173.5356; found: 1173.5353; anal. calcd for $C_{48}H_{116}Si_{16}$: C, 50.45; H, 10.23; found: C, 50.83; H, 9.90.

3.5. Isolation of 8 and 9

In a Schlenk tube (30 mL) equipped with a magnetic stir bar, compound **4** (104.0 mg, 0.0910 mmol) and benzene (15 mL) were placed. The reaction mixture was heated at 75 °C for 20 h. The volatiles were removed in vacuo. Recrystallization from hexane at −35 °C gave **8** as an orange powder (2.6 mg, 0.0023 mmol) in 3% yield. Recrystallization of the mother liquor from diethyl ether at −35 °C gave **9** as red crystals (20.4 mg, 0.026 mmol) in 29% yield.

8: orange crystals; ^{1}H NMR (C_6D_6, 500 MHz, 299 K) δ 0.39 (s, 72H, $SiMe_3$), 1.71 (s, 36H, *t*-Bu), 1.98 (s, 8H, CH_2); UV/vis (in hexane): λ_{max}/nm (ε) 540 (sh, 1100), 451 (12000).

9: red crystals; mp. 164 °C (decomp); ^{1}H NMR (C_6D_6, 500 MHz, 299 K) δ 0.31 (s, 18H, $SiMe_3$), 0.54 (s, 18H, $SiMe_3$) 1.30 (s, 18H, *t*-Bu), 1.35 (s, 9H, *t*-Bu), 1.39 (s, 9H, *t*-Bu), 1.90–2.06 (m, 8H, CH_2); ^{13}C NMR (C_6D_6, 125 MHz, 300 K) δ 5.1 ($SiMe_3$), 5.6 ($SiMe_3$), 15.5 (C), 17.7 (C), 24.9 (C), 25.8 (C), 30.4 (C), 31.3 ($C(CH_3)_3$), 31.5 ($C(CH_3)_3$), 32.0 ($C(CH_3)_3$), 35.5 (CH_2), 36.9 (CH_2); ^{29}Si NMR (C_6D_6, 99 MHz, 299 K) δ −159.5 (Si2, Si3), −60.3 (Si7), 2.3 (Si5, Si6), 3.6 ($SiMe_3$), 5.3 ($SiMe_3$), 26.5 (Si1), 123.3 (Si4); UV/vis (in hexane): λ_{max}/nm (ε) 511 (2700), 260 (28000); HRMS (APCI) *m/z* ([M]$^+$ was missing but [M + H_3 + O_4]$^+$ was found probably because **9** is highly reactive and it was partially oxidized and/or hydrolyzed before injecting the sample into the instrument) calcd for $C_{32}H_{79}O_4Si_{11}$: 835.3435; found: 835.3437; anal. calcd for $C_{32}H_{76}Si_{11}$: C, 49.92; H, 9.95; found: C, 49.90; H, 9.81.

3.6. X-ray Analysis

Recrystallization from toluene (**7**), hexane (**8**), and diethyl ether (**9**) at −35 °C gave single crystals suitable for data collection. The single crystals coated by Apiezon® grease were mounted on the glass fiber and transferred to the cold gas stream of the diffractometer. X-ray data were collected on a BrukerAXS APEXII diffractometer (Bruker Japan, Yokohama, Japan) with graphite monochromated Mo Kα radiation (λ = 0.71073 Å). The data were corrected for Lorentz and polarization effects. An empirical absorption correction based on the multiple measurement of equivalent reflections was applied using the program *SADABS* [46]. The structures were solved by direct methods and refined by full-matrix least squares against F^2 using all data (*SHELXL*-2014/7) [47]. CCDC-1863288–1863290 contain the supplementary crystallographic data for this paper. These data can be obtained free of charge via http://www.ccdc.cam.ac.uk/conts/retrieving.html (or from the CCDC, 12 Union Road, Cambridge CB2 1EZ, UK.

7: CCDC-1863288; [code si302a]; 100 K; $C_{48}H_{116}Si_{16}\cdot(C_7H_8)_{2.5}$; Fw 1373.17; triclinic; space group $P{-}1$ (#2), a = 11.5327(18) Å, b = 17.883(3) Å, c = 21.560(3) Å, α = 69.416(2)°, β = 85.458(2)°, γ = 76.379(2)°, V = 4045.5(11) Å^3, Z = 2, D_{calcd} = 1.127 mg/m^3, $R1$ = 0.0357 ($I > 2\sigma(I)$), $wR2$ = 0.0941 (all data), GOF = 1.062.

8: CCDC-1863289: [code si303a]: 100 K; $C_{48}H_{116}Si_{16}$; Fw 1142.84; monoclinic; space group $C2/c$ (#15), a = 21.538(2) Å, b = 17.9167(19) Å, c = 20.792(3) Å, β = 118.8750(10)°, V = 7026.1(15) Å^3, Z = 4, D_{calcd} = 1.080 mg/m^3, $R1$ = 0.0363 ($I > 2\sigma(I)$), $wR2$ = 0.0910 (all data), GOF = 1.077.

9: CCDC-1863290: [code si301a]; 100 K; $C_{32}H_{76}Si_{11}$; Fw 769.92; orthorhombic; space group $Pbca$ (#61), a = 11.3287(13) Å, b = 17.976(2) Å, c = 46.325(5) Å, V = 9433.6(19) Å^3, Z = 8, D_{calcd} = 1.084 mg/m^3, $R1$ = 0.0377 ($I > 2\sigma(I)$), $wR2$ = 0.0839 (all data), GOF = 1.111.

3.7. Theoretical Study

All theoretical calculations were performed using a Gaussian 09 [48] and GRRM14 programs [49,50]. Geometry optimization of **2**, **4**, **6**, **7**, **8**, and **9** ($\mathbf{2_{opt}}$, $\mathbf{4_{opt}}$, $\mathbf{6_{opt}}$, $\mathbf{7_{opt}}$, $\mathbf{8_{opt}}$, and $\mathbf{9_{opt}}$) were carried out at the B3PW91-D3/6-31G(d) level of theory. The atomic coordinates of all optimized structures were summarized in a .xyz file (optimized structures.xyz).

4. Conclusions

Si_8R_8 cluster **4** transforms into novel silicon clusters **7**, **8**, and **9**, bearing unprecedented silicon frameworks accompanied by elimination of the R_2Si (silylene) unit under a mild thermal condition. Reversible isomerization between **4** and **7** was observed. XRD analyses exhibit that both **7** and **9** have two unsubstituted skeletal silicon atoms classified as siliconoids, while **8** has two Si=Si double bonds. A large dispersion of ^{29}Si chemical shifts was observed for **7** and **9** as found in the reported siliconoids. The observed transformation of molecular silicon clusters involving substantially different molecular structures and electronic properties suggests that unsaturated molecular silicon clusters have a great potential to provide various molecular silicon clusters bearing unprecedented structures and properties.

Author Contributions: Investigation, N.A. and S.I.; Project administration, T.I.; Writing—original draft, N.A.; Writing—review & editing, N.A., S.I., and T.I.

References

1. Hengge, E.; Janoschek, R. Homocyclic silanes. *Chem. Rev.* **1995**, *95*, 1495–1526. [CrossRef]
2. Sekiguchi, A.; Sakurai, H. Cage and cluster compounds of silicon, germanium, and tin. *Adv. Organomet. Chem.* **1995**, *37*, 1–38. [CrossRef]
3. Sekiguchi, A.; Nagase, S. Polyhedral silicon compounds. In *The Chemistry of Organic Silicon Compounds*; Rappoport, Z., Apeloig, Y., Eds.; Wiley: Chichester, UK, 1998; Volume 2, pp. 119–152.

4. Hengge, E.; Stüger, H. Recent advances in the chemistry of cyclopolysilanes. In *The Chemistry of Organic Silicon Compounds*; Rappoport, Z., Apeloig, Y., Eds.; Wiley: Chichester, UK, 1998; Volume 2, pp. 2177–2216.
5. Marschner, C. Oligosilanes. In *Structure and Bonding 156, Functional Molecular Silicon Compounds I. Regular Oxidation States*; Scheschkewitz, D., Ed.; Springer: New York, NY, USA, 2014; Volume 156, pp. 163–228.
6. Lyon, J.T.; Gruene, P.; Fielicke, A.; Meijer, G.; Janssens, E.; Claes, P.; Lievens, P. Structures of silicon cluster cations in the gas phase. *J. Am. Chem. Soc.* **2009**, *131*, 1115–1121. [CrossRef] [PubMed]
7. Peppernick, S.J.; Gunaratne, K.D.D.; Castleman, A.W. The relative abundances of silicon hydride clusters, $Si_nH_x^-$ (n = 8–12 and $0 \leq x \leq 25$), investigated with high-resolution time-of-flight mass spectrometry. *Int. J. Mass. Spectr.* **2010**, *290*, 65–71. [CrossRef]
8. Haertelt, M.; Lyon, J.T.; Claes, P.; de Haeck, J.; Lievens, P.; Fielicke, A. Gas-phase structures of neutral silicon clusters. *J. Chem. Phys.* **2012**, *136*, 064301. [CrossRef] [PubMed]
9. Berger, R.J.; Rzepa, H.S.; Scheschkewitz, D. Ring currents in the dismutational aromatic Si_6R_6. *Angew. Chem. Int. Ed.* **2010**, *49*, 10006–10009. [CrossRef] [PubMed]
10. Abersfelder, K.; Russell, A.; Rzepa, H.S.; White, A.J.P.; Haycock, P.R.; Scheschkewitz, D. Contraction and expansion of the silicon scaffold of stable Si_6R_6 isomers. *J. Am. Chem. Soc.* **2012**, *134*, 16008–16016. [CrossRef] [PubMed]
11. Scheschkewitz, D. A molecular silicon cluster with a "naked" vertex atom. *Angew. Chem. Int. Ed.* **2005**, *44*, 2954–2956. [CrossRef] [PubMed]
12. Abersfelder, K.; White, A.J.P.; Rzepa, H.S.; Scheschkewitz, D. A tricyclic aromatic isomer of hexasilabenzene. *Science* **2010**, *327*, 564–566. [CrossRef] [PubMed]
13. Abersfelder, K.; White, A.J.P.; Berger, R.J.F.; Rzepa, H.S.; Scheschkewitz, D. A stable derivative of the global minimum on the Si_6H_6 potential energy surface. *Angew. Chem., Int. Ed.* **2011**, *50*, 7936–7939. [CrossRef] [PubMed]
14. Willmes, P.; Leszczynska, K.; Heider, Y.; Abersfelder, K.; Zimmer, M.; Huch, V.; Scheschkewitz, D. Isolation and versatile derivatization of an unsaturated anionic silicon cluster (siliconoid). *Angew. Chem. Int. Ed.* **2016**, *55*, 2907–2910. [CrossRef] [PubMed]
15. Fischer, G.; Huch, V.; Mayer, P.; Vasisht, S.K.; Veith, M.; Wiberg, N. $Si_8(SitBu_3)_6$: A hitherto unknown cluster structure in silicon chemistry. *Angew. Chem. Int. Ed.* **2005**, *44*, 7884–7887. [CrossRef] [PubMed]
16. Klapötke, T.M.; Vasisht, S.K.; Fischer, G.; Mayer, P. A reactive Si_4 cage: $K(SitBu_3)_3Si_4$. *J. Organomet. Chem.* **2010**, *695*, 667–672. [CrossRef]
17. Klapötke, T.M.; Vasisht, S.K.; Mayer, P. Spirocycle $(SitBu_3)_6Si_9Cl_2$: The first of its kind among group 14 elements. *Eur. J. Inorg. Chem.* **2010**, *2010*, 3256–3260. [CrossRef]
18. Nied, D.; Koppe, R.; Klopper, W.; Schnockel, H.; Breher, F. Synthesis of a pentasilapropellane. Exploring the nature of a stretched silicon–silicon bond in a nonclassical molecule. *J. Am. Chem. Soc.* **2010**, *132*, 10264–10265. [CrossRef] [PubMed]
19. Ishida, S.; Otsuka, K.; Toma, Y.; Kyushin, S. An organosilicon cluster with an octasilacuneane core: A missing silicon cage motif. *Angew. Chem. Int. Ed.* **2013**, *52*, 2507–2510. [CrossRef] [PubMed]
20. Tsurusaki, A.; Iizuka, C.; Otsuka, K.; Kyushin, S. Cyclopentasilane-fused hexasilabenzvalene. *J. Am. Chem. Soc.* **2013**, *135*, 16340–16343. [CrossRef] [PubMed]
21. Tsurusaki, A.; Kamiyama, J.; Kyushin, S. Tetrasilane-bridged bicyclo[4.1.0]heptasil-1-ene. *J. Am. Chem. Soc.* **2014**, *136*, 12896–12898. [CrossRef] [PubMed]
22. Iwamoto, T.; Akasaka, N.; Ishida, S. A heavy analogue of the smallest bridgehead alkene stabilized by a base. *Nat. Commun.* **2014**, *5*, 5353. [CrossRef] [PubMed]
23. Hamers, R.J.; Köhler, U.K.; Demuth, J.E. Nucleation and growth of epitaxial silicon on Si(001) and Si(111) surfaces by scanning tunneling microscopy. *Ultramicroscopy* **1989**, *31*, 10–19. [CrossRef]
24. Swihart, M.T.; Girshick, S.L. Thermochemistry and kinetics of silicon hydride cluster formation during thermal decomposition of silane. *J. Phys. Chem. B* **1999**, *103*, 64–76. [CrossRef]
25. Swihart, M.T.; Girshick, S.L. Ab initio structures and energetics of selected hydrogenated silicon clusters containing six to ten silicon atoms. *Chem. Phys. Lett.* **1999**, *307*, 527–532. [CrossRef]
26. Li, C.P.; Li, X.J.; Yang, J.C. Silicon hydride clusters Si_5H_n (n = 3–12) and their anions: Structures, thermochemistry, and electron affinities. *J. Phys. Chem. A* **2006**, *110*, 12026–12034. [CrossRef] [PubMed]

27. Singh, R. Effect of hydrogen on ground state properties of silicon clusters (Si_nH_m; n = 11–15, m = 0–4): A density functional based tight binding study. *J. Phys. Condens. Matt.* **2008**, *20*, 045226. [CrossRef]
28. Adamczyk, A.J.; Broadbelt, L.J. Thermochemical property estimation of hydrogenated silicon clusters. *J. Phys. Chem. A* **2011**, *115*, 8969–8982. [CrossRef] [PubMed]
29. Gapurenko, O.A.; Minyaev, R.M.; Minkin, V.I. Silicon analogues of pyramidane: A quantum-chemical study. *Mendeleev Commun.* **2012**, *22*, 8–10. [CrossRef]
30. Thingna, J.; Prasad, R.; Auluck, S. Photo-absorption spectra of small hydrogenated silicon clusters using the time-dependent density functional theory. *J. Phys. Chem. Solids* **2011**, *72*, 1096–1100. [CrossRef]
31. Kaftory, M.; Kapon, M.; Botoshansky, M. The structural chemistry of organosilicon compounds. In *The Chemistry of Organic Silicon Compounds*; Rappoport, Z., Apeloig, Y., Eds.; Wiley: Chichester, UK, 1998; Volume 2, pp. 181–266.
32. Okazaki, R.; West, R. Chemistry of stable disilenes. *Adv. Organomet. Chem.* **1996**, *39*, 231–273. [CrossRef]
33. Kira, M.; Iwamoto, T. Progress in the chemistry of stable disilenes. *Adv. Organomet. Chem.* **2006**, *54*, 73–148. [CrossRef]
34. Kabe, Y.; Kuroda, M.; Honda, Y.; Yamashita, O.; Kawase, T.; Masamune, S. Reductive oligomerization of 1,2-di-*tert*-butyl-1,1,2,2-tetrachlorodisilane: The tricyclo[$2.2.0.0^{2,5}$]hexasilane and tetracyclo[$3.3.0.0^{2,7}.0^{3,6}$]octasilane systems. *Angew. Chem. Int. Ed. Engl.* **1988**, *27*, 1725–1727. [CrossRef]
35. Iwamoto, T.; Uchiyama, K.; Kabuto, C.; Kira, M. Synthesis of tricyclo[$3.1.0.0^{2,4}$]hexasilane and its photochemical isomerization to tricyclo[$2.2.0.0^{2,5}$]hexasilane. *Chem. Lett.* **2007**, *36*, 368–369. [CrossRef]
36. Li, Y.; Li, J.; Zhang, J.; Song, H.; Cui, C. Isolation of R_6Si_6 dianion: A bridged tricyclic isomer of dianionic hexasilabenzene. *J. Am. Chem. Soc.* **2018**, *140*, 1219–1222. [CrossRef] [PubMed]
37. Sekiguchi, A.; Yatabe, T.; Kabuto, C.; Sakurai, H. The missing hexasilaprismane: Synthesis, x-ray analysis and photochemical reactions. *J. Am. Chem. Soc.* **1993**, *115*, 5853–5854. [CrossRef]
38. Kira, M.; Ishida, S.; Iwamoto, T.; Kabuto, C. The first isolable dialkylsilylene. *J. Am. Chem. Soc.* **1999**, *121*, 9722–9723. [CrossRef]
39. Tang, M.; Lu, W.; Wang, C.Z.; Ho, K.M. Search for most stable structure of Si_8H_8 cluster. *Chem. Phys. Lett.* **2003**, *377*, 413–418. [CrossRef]
40. Matsumoto, H.; Higuchi, K.; Hoshino, Y.; Koike, H.; Naoi, Y.; Nagai, Y. The first octasilacubane system: Synthesis of octakis-(*t*-butyldimethylsilyl)pentacyclo[$4.2.0.0^{2,5}.0^{3,8}.0^{4,7}$]octasilane. *J. Chem. Soc. Chem. Commun.* **1988**, 1083–1084. [CrossRef]
41. Furukawa, K.; Fujino, M.; Matsumoto, N. Cubic silicon cluster. *Appl. Phys. Lett.* **1992**, *60*, 2744–2745. [CrossRef]
42. Sekiguchi, A.; Yatabe, T.; Kamatani, H.; Kabuto, C.; Sakurai, H. Preparation, characterization, and crystal structures of octasilacubanes and octagermacubanes. *J. Am. Chem. Soc.* **1992**, *114*, 6260–6262. [CrossRef]
43. Matsumoto, H.; Higuchi, K.; Kyushin, S.; Goto, M. Octakis(1,1,2-trimethylpropyl)octasilacubane: Synthesis, molecular structure, and unusual properties. *Angew. Chem. Int. Ed. Engl.* **1992**, *31*, 1354–1356. [CrossRef]
44. Furukawa, K.; Fujino, M.; Matsumoto, N. Superlattice structure of octa-*tert*-butylpentacyclo[$4.2.0.0^{2,5}.0^{3,8}.0^{4,7}$]octasilane found by reinvestigation of x-ray structure analysis. *J. Organomet. Chem.* **1996**, *515*, 37–41. [CrossRef]
45. Unno, M.; Matsumoto, T.; Mochizuki, K.; Higuchi, K.; Goto, M.; Matsumoto, H. Structure and oxidation of octakis(*tert*-butyldimethylsilyl)octasilacubane. *J. Organomet. Chem.* **2003**, *685*, 156–161. [CrossRef]
46. Sheldrick, G.M. *SADABS*; Empirical Absorption Correction Program; Institute for Inorganic Chemistry: Göttingen, Germany, 1996.
47. Sheldrick, G.M. *SHELXL-2014/7*; Program for the Refinement of Crystal Structures; University of Göttingen: Göttingen, Germany, 2014.
48. Frisch, M.J.; Trucks, G.W.; Schlegel, H.B.; Scuseria, G.E.; Robb, M.A.; Cheeseman, J.R.; Scalmani, G.; Barone, V.; Mennucci, B.; Petersson, G.A.; et al. *Gaussian 09, Revision D.01*; Gaussian, Inc.: Wallingford, CT, USA, 2009.
49. Maeda, S.; Harabuchi, Y.; Osada, Y.; Taketsugu, T.; Morokuma, K.; Ohno, K. Available online: http://grrm.chem.tohoku.ac.jp/GRRM/ (accessed on 1 October 2018).
50. Maeda, S.; Ohno, K.; Morokuma, K. Systematic exploration of the mechanism of chemical reactions: The global reaction route mapping (GRRM) strategy using the ADDF and AFIR methods. *Phys. Chem. Chem. Phys.* **2013**, *15*, 3683–3701. [CrossRef] [PubMed]

19

Synthesis and Characterization of *N*-Heterocyclic Carbene-Coordinated Silicon Compounds Bearing a Fused-Ring Bulky Eind Group

Naoki Hayakawa [1], Kazuya Sadamori [1], Shinsuke Mizutani [1], Tomohiro Agou [2,3], Tomohiro Sugahara [2], Takahiro Sasamori [2,4], Norihiro Tokitoh [2], Daisuke Hashizume [5] and Tsukasa Matsuo [1,*]

1. Department of Applied Chemistry, Faculty of Science and Engineering, Kindai University 3-4-1 Kowakae, Higashi-Osaka, Osaka 577-8502, Japan; 1544320202x@kindai.ac.jp (N.H.); 1733320213x@kindai.ac.jp (K.S.); ihsnusek555@yahoo.co.jp (S.M.)
2. Institute for Chemical Research, Kyoto University, Gokasho, Uji, Kyoto 611-0011, Japan; tomohiro.agou.mountain@vc.ibaraki.ac.jp (T.A.); sugahara@boc.kuicr.kyoto-u.ac.jp (T.Su.); sasamori@nsc.nagoya-cu.ac.jp (T.Sa.); tokitoh@boc.kuicr.kyoto-u.ac.jp (N.T.)
3. Department of Biomolecular Functional Engineering, College of Engineering, Ibaraki University, 4-12-1 Nakanarusawa, Hitachi, Ibaraki 316-8511, Japan
4. Graduate School of Natural Sciences, Nagoya City University Yamanohata 1, Mizuho-cho, Mizuho-ku, Nagoya, Aichi 467-8501, Japan
5. RIKEN Center for Emergent Matter Science (CEMS), 2-1 Hirosawa, Wako, Saitama 351-0198, Japan; hashi@riken.jp

* Correspondence: t-matsuo@apch.kindai.ac.jp

Abstract: The reactions of the fused-ring bulky Eind-substituted 1,2-dibromodisilene, (Eind)BrSi=SiBr(Eind) (**1a**) (Eind = 1,1,3,3,5,5,7,7-octaethyl-*s*-hydrindacen-4-yl (**a**)), with *N*-heterocyclic carbenes (NHCs) (Im-Me_4 = 1,3,4,5-tetramethylimidazol-2-ylidene and Im-$^iPr_2Me_2$ = 1,3-diisopropyl-4,5-dimethylimidazol-2-ylidene) are reported. While the reaction of **1a** with the sterically more demanding Im-$^iPr_2Me_2$ led to the formation of the mono-NHC adduct of arylbromosilylene, (Im-$^iPr_2Me_2$)$\rightarrow$SiBr(Eind) (**2a′**), a similar reaction using the less bulky Im-Me_4 affords the bis-NHC adduct of formal arylsilyliumylidene cation, $[(\text{Im-Me}_4)_2\rightarrow\text{Si(Eind)}]^+[\text{Br}^-]$ (**3a**). The NHC adducts **2a′** and **3a** can also be prepared by the dehydrobromination of Eind-substituted dibromohydrosilane, (Eind)SiHBr_2 (**4a**), with NHCs. The NHC-coordinated silicon compounds have been characterized by spectroscopic methods. The molecular structures of bis-NHC adduct, $[(\text{Im-}^i\text{Pr}_2\text{Me}_2)_2\rightarrow\text{Si(Eind)}]^+[\text{Br}^-]$ (**3a′**), and **4a** have been determined by X-ray crystallography.

Keywords: silicon; *N*-heterocyclic carbenes; bromosilylenes; silyliumylidenes; dehydrobromination

---

1. Introduction

Over many years, a number of unsaturated silicon compounds have been successfully obtained by virtue of the complexation of metal ions and/or coordination of ligands (mainly Lewis bases) in addition to steric protection with bulky substituents [1–11]. Among them, the coordination chemistry of highly-reactive halosilylenes, i.e., halogen-substituted divalent Si(II) species, have attracted a lot of attention as potentially useful precursors for the construction of a wide range of silicon-containing compounds [12–17]. Figure 1 shows recent examples of coordination-stabilized arylhalosilylenes and their derivatives [18–24]. In 2010, Filippou's group reported the first *N*-heterocyclic carbene (NHC) adducts of arylchlorosilylenes bearing sterically large *m*-terphenyl groups, (Im-Me_4)$\rightarrow$SiCl(Ar) (**Ic** and **Id**) (Ar = 2,6-$(Mes)_2C_6H_3$ (Mes = 2,4,6-$Me_3C_6H_2$) (**c**) and 2,6-$(Tip)_2C_6H_3$ (Tip = 2,4,6-$^iPr_3C_6H_2$) (**d**)), which were prepared by the dehydrochlorination of the aryldichlorohydrosilanes, (Ar)$SiHCl_2$, with NHC (Im-Me_4 = 1,3,4,5-tetramethylimidazol-2-ylidene) along with the formation of imidazolium chloride, $[(Im\text{-}Me_4)H]^+[Cl^-]$ [18]. The NHC adduct **Id** reacted with $[Li^+][CpMo(CO)_3]^-$ to afford the silylidene complex, $Cp(CO)_2Mo{=}Si(Ar)(Im\text{-}Me_4)$ (**IId**), and the subsequent treatment with $B(C_6H_4\text{-}4\text{-}Me)_3$ produced the first silylidyne complex, $Cp(CO)_2Mo{\equiv}Si(Ar)$ (**IIId**), featuring a metal-silicon triple bond [19,25].

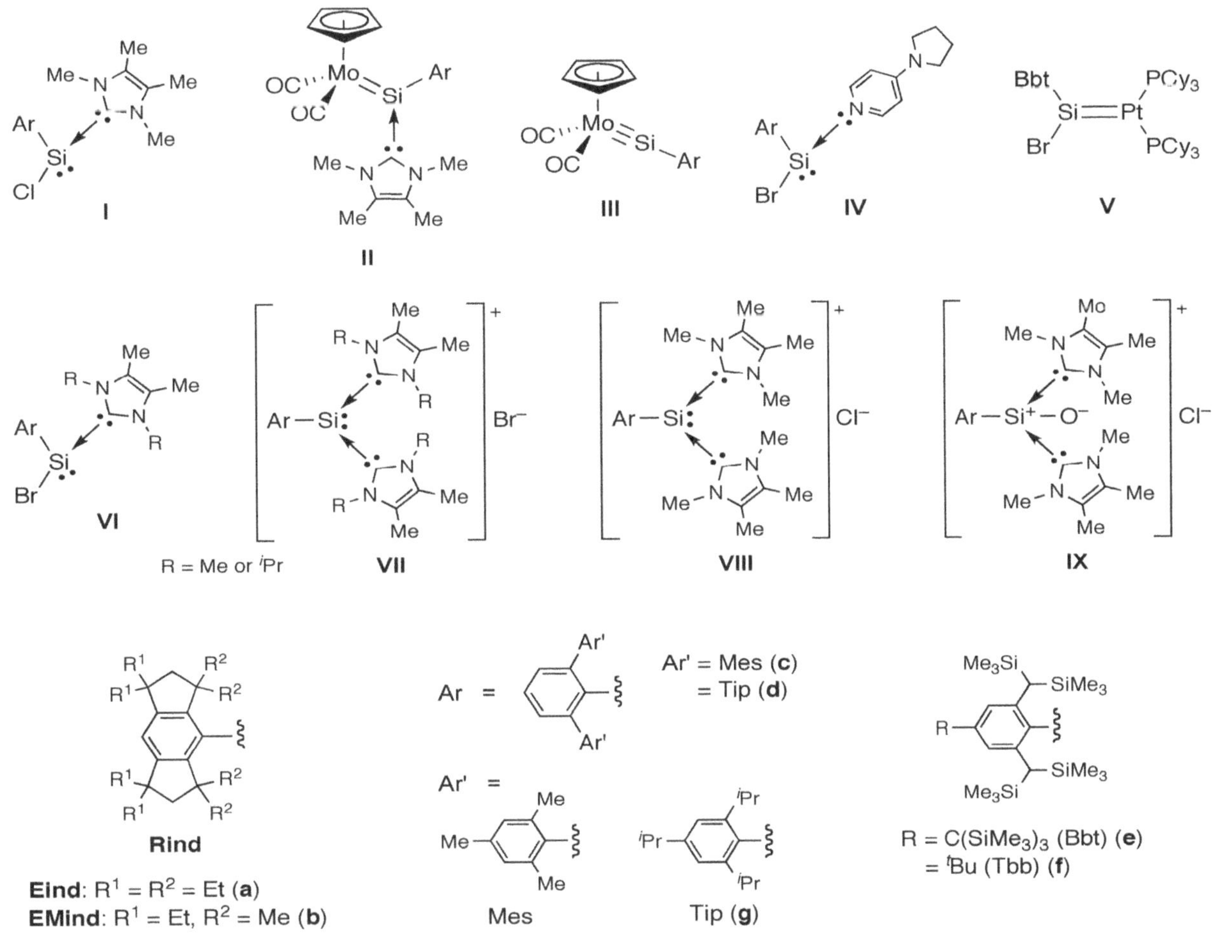

Figure 1. Examples of coordination-stabilized arylhalosilylenes and their derivatives. Each one of the possible canonical forms is depicted.

In 2011, we reported on the 4-pyrrolidinopyridine (PPy) adducts of arylbromosilylenes with fused-ring bulky Rind groups, PPy$\rightarrow$SiBr(Rind) (**IVa** and **IVb**) (Rind = 1,1,3,3,5,5,7,7-octa-R-substituted s-hydrindacen-4-yl; Eind (**a**: $R^1 = R^2$ = Et) and EMind (**b**: R^1 = Et, R^2 = Me)) [26,27], which were formed by the addition of PPy to 1,2-dibromodisilenes, (Rind)BrSi=SiBr(Rind) (**1a** and **1b**) [20]. Also in 2011, Cui reported on the related NHC-coordinated aminochlorosilylene [28], and recently Driess's group reported on the aminochlorosilylene-nickel complex [29]. In 2012, a platinum complex of arylbromosilylene, $(Bbt)BrSi{=}Pt(PCy_3)_2$ (**Ve**), was synthesized by the treatment of 1,2-dibromodisilene, (Bbt)BrSi=SiBr(Bbt) (**1e**), bearing the bulky Bbt groups (Bbt = 2,6-$\{CH(SiMe_3)_2\}_2$-4-$C(SiMe_3)_3$-C_6H_2 (**e**)) [30] with $Pt(PCy_3)_2$ [21,31]. In this context, some arylbromosilylidene and arylsilylidyne complexes of nickel and platinum were reported [32–34]. In 2013, Filippou's group reported the unprecedented dicationic NHC complexes of silicon(II) and NHC adducts of iodesilyliumylidene cation SiI^+ [35]. Subsequently, we reported the reaction of 1,2-dibromodisilenes (**1b**, **1e**, and **1f**) having EMind, Bbt, and Tbb groups (Tbb = 2,6-$\{CH(SiMe_3)_2\}_2$-4-^tBu-C_6H_2 (**f**)) with NHCs (Im-Me_4 and Im-$^iPr_2Me_2$) (Im-$^iPr_2Me_2$ =

1,3-diisopropyl-4,5-dimethylimidazol-2-ylidene) leading to the formation of mono-NHC adducts of arylbromosilylenes, NHC→SiBr(Ar) (**VIb′** and **VIe**), and the bromide salts of the bis-NHC adducts of formal arylsilyliumylidene cations, $[(NHC)_2 \rightarrow Si(Ar)]^+[Br^-]$ (**VIIb**, **VIIb′**, and **VIIf**) [22]. Inoue's group also reported the synthesis of chloride salts of bis-NHC adducts of formal arylsilyliumylidene cations, $[(Im\text{-}Me_4)_2 \rightarrow Si(Ar)]^+[Cl^-]$ (**VIIIc** and **VIIIg**) (Ar = Tip (**g**)) [23] and their unique conversion to silicon analogues of acylium ions, $[(Im\text{-}Me_4)_2 \rightarrow Si(O)(Ar)]^+[Cl^-]$ (**IXc** and **IXg**) [24]. Recently, Inoue's group reported on the chalcogen-atom transfer and exchange reactions of NHC-bound heavier silaacylium ions, $[(Im\text{-}Me_4)_2 \rightarrow Si(E)(Ar)]^+[Cl^-]$ (E = S, Se, and Te) [36].

In this article, we describe the preparation and characterization of NHC-coordinated silicon compounds bearing the bulky Eind group, which have been obtained by two different synthetic procedures, i.e., the NHC-induced fragmentation of the Eind-based 1,2-dibromo-disilene and the dehydrobromination of the Eind-based dibromohydrosilane with NHCs.

2. Results and Discussions

*2.1. Reactions of (Eind)BrSi=SiBr(Eind) (**1a**) with NHCs*

We first performed an NMR tube scale reaction of Eind-based 1,2-dibromodisilene, (Eind)BrSi=SiBr(Eind) (**1a**) [20], in C_6D_6 with two equivalents of the sterically more bulky NHC, Im-$^iPr_2Me_2$, relative to Im-Me_4. The progress of the reaction was monitored by ^1H NMR spectroscopy, indicating the selective formation of the mono-NHC adduct of the arylbromosilylene, (Im-$^iPr_2Me_2$)→SiBr(Eind) (**2a′**), after overnight heating at 70 °C. In the ^{29}Si NMR spectrum, only one signal was observed at δ = 18.0 ppm, which is comparable to those of (Im-$^iPr_2Me_2$)→SiBr(EMind) (**VIb′**) (δ = 13.1 ppm) and (Im-Me_4)→SiBr(Bbt) (**VIe**) (δ = 10.9 ppm) [22]. The ^{13}C signal at δ = 170.6 ppm for **2a′** is characteristic of a carbene carbon atom, similar to those for **VIb′** (δ = 169.7 ppm) and **VIe** (δ = 167.5 ppm) [22]. Based on the NMR tube experiment, the mono-NHC adduct **2a′** was synthesized as an orange solid in 88% crude yield (Scheme 1).

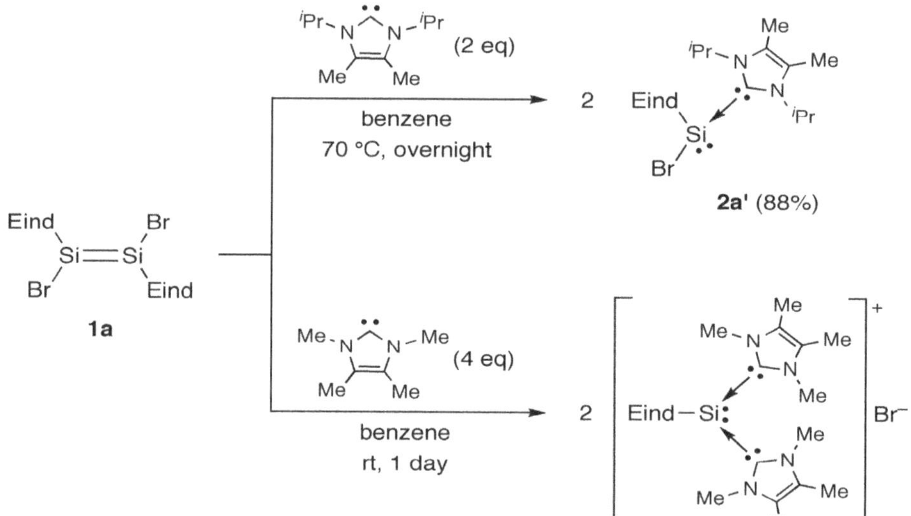

Scheme 1. Reactions of **1a** with *N*-heterocyclic carbenes (NHCs).

We also examined the reaction of **1a** with four equivalents of Im-$^iPr_2Me_2$. After 1-day heating at 70 °C in C_6D_6, the ^{29}Si NMR spectrum indicated the formation of a mixture containing the mono-NHC adduct **2a′** (δ = 18.0 ppm) as the major product and the bis-NHC adduct of the arylsilyliumylidene cation, $[(Im\text{-}^iPr_2Me_2)_2 \rightarrow Si(Eind)]^+[Br^-]$ (**3a′**) (δ = −59.6 ppm), as the minor product. The latter ^{29}Si signal was shifted upfield compared to the former, and was similar to those of the formal arylsilyliumylidene cations, $[(NHC)_2{\rightarrow}Si(EMind)]^+[Br^-]$ (**VIIb** and **VIIb′**) (δ = −60.8 and −75.9 ppm) and $[(Im\text{-}Me_4)_2{\rightarrow}Si(Tbb)]^+[Br^-]$ (**VIIf**) (δ = −70.9 ppm) [22]. However, we found that the reaction was not completed even after prolonged heating (longer than 1 week), probably due to the severe steric repulsion between the Eind group and Im-$^{i}Pr_2Me_2$ molecules. Thus, we were unable to isolate **3a′**. Nevertheless, single red crystals of **3a′** could be obtained from the reaction mixture, whose structure was determined by X-ray crystallography (Figure 2).

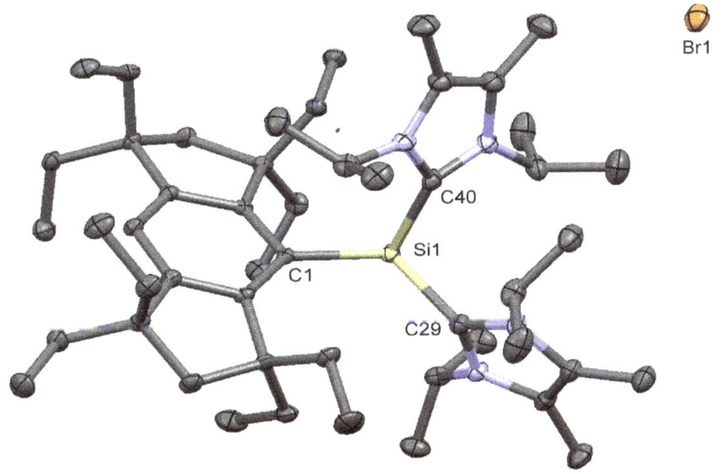

Figure 2. Molecular structure of **3a′**. The thermal ellipsoids are shown at the 50% probability level. All hydrogen atoms and benzene molecule are omitted for clarity.

Figure 2 shows a separated ion pair of **3a′** in the crystal. The closest Si···Br distance (7.6669(6) Å) is analogous to that of **VIIb′** (7.732(3) Å) [22], thus being much longer than the sum of the van der Waals radii of Si and Br (3.95 Å). The Si atom is three-coordinate adopting a distorted pyramidal geometry, which can be explained by the presence of a lone pair of electrons. The sum of the surrounding angles around the Si atom (ΣSi = 327.3°) is almost the same as that of **VIIb′** (ΣSi = 327.0°) [22]. The Si–C(Rind) bond length in **3a′** (Si1–C1 = 1.9482(19) Å) is similar to that in **VIIb′** [1.927(8) Å] [22] and longer than typical Si–C bonds (ca. 1.88 Å), suggesting the high s-character of the lone pair of electrons on the Si atom and the high p-character of the Si–C(Rind) bond. The Si←C(NHC) coordination distances in **3a′** (Si1–C29 = 1.953(2) and Si1–C40 = 1.942(2) Å) are comparable to those observed in **VIIb′** (1.955(9) and 1.979(8) Å) [22].

We next investigated the reaction of **1a** with two equivalents of the less bulky NHC, Im-Me_4, and C_6D_6. After 1 day at room temperature, two signals mainly appeared at δ = 73.3 and −63.3 ppm in the ^{29}Si NMR spectrum, corresponding to the unreacted **1a** and the bis-NHC adduct of the arylsilyliumylidene cation, $[(Im\text{-}Me_4)_2{\rightarrow}Si(Eind)]^+[Br^-]$ (**3a**). This indicated that the NHC-arylbromosilylene adduct, $(Im\text{-}Me_4){\rightarrow}SiBr(Eind)$ (**2a**), which serves as a potential intermediate, is more reactive toward Im-Me_4 compared to **1a**. When the dibromodisilene **1a** was treated with four equivalents of Im-Me_4 in benzene, the bis-NHC adduct **3a** was efficiently formed (Scheme 1). We obtained **3a** as an orange powder in 54% crude yield. The upfield-shifted ^{29}Si resonance for **3a** (δ = −63.3 ppm) suggests the contribution of the canonical form due to the bis(imidazolium) adduct of a silyl anion, whose electronic structure was previously supported by the theoretical calculations of **VIIb′** and **VIIf** [22]. In the ^{13}C NMR spectrum of **3a** in CD_3CN, one NHC carbene signal was observed at δ = 162.0 ppm, comparable to those for **VIIb** (δ = 160.5 ppm), **VIIb′** (δ = 162.4 ppm), and **VIIf** (δ = 160.4 ppm) [22].

2.2. *Reactions of (Eind)SiHBr$_2$* (**4a**) *with NHCs*

We also examined another synthetic route for the NHC-coordinated silicon compounds, i.e., the dehydrobromination of the Eind-substituted dibromohydrosilane, $(Eind)SiHBr_2$ (**4a**), with NHCs

(Scheme 2). The precursor (**4a**) was prepared as pale brown crystals by the dibromination of the Eind-based trihydrosilane, (Eind)SiH_3 [37,38], with allyl bromide in the presence of a catalytic amount of $PdCl_2$ [39]. We found that this reaction exclusively afforded **4a** even using an excess amount of allyl bromide with prolonged heating (longer than 1 week), most likely due to the steric bulkiness of the Eind group. In this context, Kunai, Ohshita, and their co-workers previously reported the selective dibromination of trihydrosilanes with $CuBr_2$ in the presence of CuI [40]. The formation of **4a** was deduced on the basis of the spectroscopic data (Figures S1–S4). In the ^{1}H NMR spectrum, the Si–H signal was found at δ = 6.89 ppm with satellite signals, due to the ^{29}Si nuclei [$^{1}J(^{29}Si–^{1}H)$ = 288 Hz]. The ^{29}Si NMR signal appeared at δ = −28.7 ppm, similar to that of (Bbt)$SiHBr_2$ (δ = −28.47 ppm) [41]. The infrared spectrum exhibited a Si–H stretching band at 2317 cm^{-1} in the KBr-pellet (Figure S4) and at 2298 cm^{-1} in THF [42,43]. The molecular structure of **4a** was determined by single-crystal X-ray diffraction analysis (Figure 3). The hydrogen atom on the silicon atom was located on difference Fourier maps and isotropically refined. In the crystal, the $SiHBr_2$ group is fixed in one conformation with respect to the rotamer around the Si–C bond. A similar conformation was also observed in the crystal of (Eind)PCl_2 [44]. The Si–C bond length for **4a** (1.8746(18) Å) is comparable to those of typical Si–C bonds (ca. 1.88 Å).

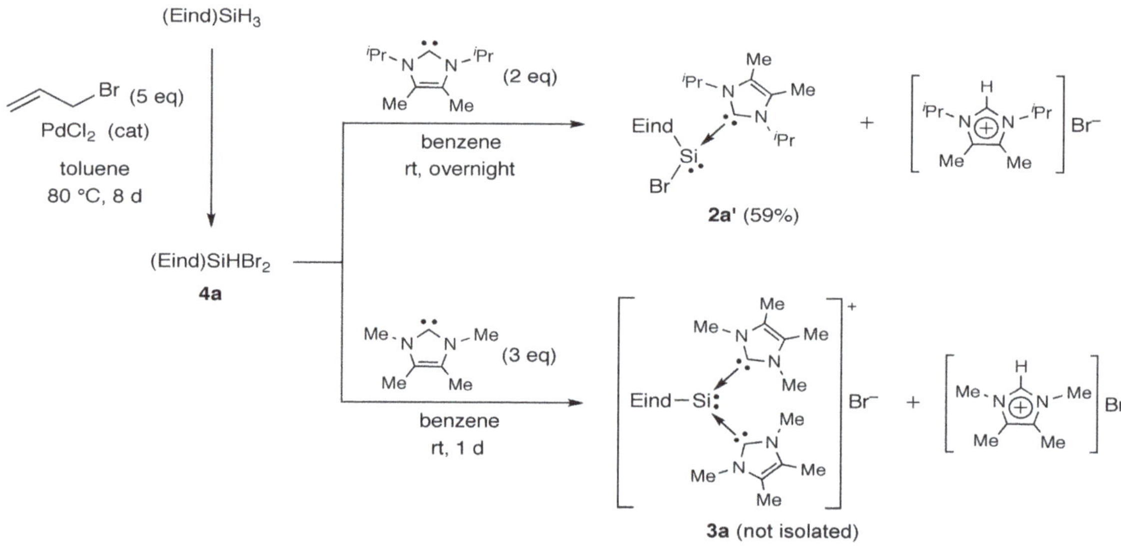

Scheme 2. Reactions of **4a** with NHCs.

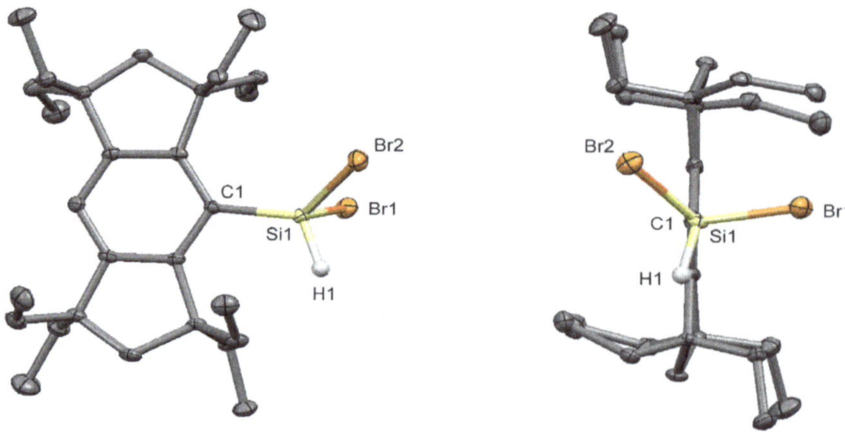

Figure 3. Molecular structure of **4a**: Side view (**left**), front view (**right**). The thermal ellipsoids are shown at the 50% probability level. The hydrogen atoms, except for the Si–H group, are omitted for clarity.

As shown in Scheme 2, the reaction of **4a** with two equivalents of Im-$^{i}Pr_2Me_2$ proceeded more smoothly at room temperature in comparison to the reaction of **1a** with Im-$^{i}Pr_2Me_2$ (Scheme 1), producing the mono-NHC adduct **2a′** in 59% crude yield. The reaction of **4a** with three equivalents of Im-Me_4 also afforded the bis-NHC adduct **3a** on the basis of the NMR data. In these reactions, it is essential to remove the byproducts, imidazolium bromides, $[(NHC)H]^+[Br^-]$, for the isolation procedure of the silicon products, which may be considered as a disadvantage when compared to the no-byproduct strategy of using **1a** as a precursor (vide supra). Actually, the separation of **3a** and $[(Im\text{-}Me_4)H]^+[Br^-]$ was found to be difficult in our experiments. However, dibromodisilene **1a** can only be obtained by a two-step synthesis from the trihydrosilane, (Eind)SiH_3; thus the bromination of (Eind)SiH_3 with *N*-bromosuccinimide (NBS) first affords the tribromosilane, (Eind)$SiBr_3$, then the reduction of (Eind)$SiBr_3$ with two equivalents of lithium naphthalenide (LiNaph) produces **1a** [20]. Therefore, the dehydrobromination of **4a** with NHCs can be considered as a convenient short-step synthesis for NHC-coordinated silylene derivatives.

3. Materials and Methods

3.1. General Procedures

All manipulations of the air- and/or moisture-sensitive compounds were performed either using standard Schlenk-line techniques or in a glove box under an inert atmosphere of argon. Anhydrous hexane, benzene, and toluene were dried by passage through columns of activated alumina and supported copper catalyst supplied by Nikko Hansen & Co., Ltd. (Osaka, Japan). Anhydrous pentane and acetonitrile were purchased from Wako Pure Chemical Industries, Ltd. (Osaka, Japan), and used without further purification. Deuterated benzene (C_6D_6, benzene-d_6) was dried and degassed over a potassium mirror in vacuo prior to use. Deuterated acetonitrile (CD_3CN, acetonitrile-d_3) was dried and distilled over calcium hydride (CaH_2) prior to use. (Eind)SiH_3 [37,38], (Eind)BrSi=SiBr(Eind) (**1a**) [20], 1,3,4,5-tetramethylimidazol-2-ylidene (Im-Me_4) [45] and 1,3-diisopropyl-4,5-dimethylimidazol-2-ylidene(Im-$^{i}Pr_2Me_2$) [45] were prepared by the literature procedures. All other chemicals and gases were used as received.

Nuclear magnetic resonance (NMR) measurements were carried out using a JEOL ECS-400 spectrometer (399.8 MHz for ^1H, 100.5 MHz for ^{13}C, and 79.4 MHz for ^{29}Si) or JEOL JNM AL-300 spectrometer (300 MHz for ^1H, 75 MHz for ^{13}C, and 59 MHz for ^{29}Si) (JEOL Ltd., Tokyo, Japan). Chemical shifts (δ) are given by definition as dimensionless numbers and relative to ^1H chemical shifts of the solvents for ^1H (residual C_6D_5H in C_6D_6, $^1H(\delta) = 7.15$, residual CD_2HCN in CD_3CN, $^1H(\delta) = 1.94$), and ^{13}C chemical shifts of the solvent for ^{13}C (C_6D_6: $^{13}C(\delta) = 128.06$ and CD_3CN: $^{13}C(\delta) = 118.26$). The signal of tetramethylsilane ($^{29}Si(\delta) = 0.0$) was used as an external standard in the ^{29}Si NMR spectra. The absolute values of the coupling constants are given in Hertz (Hz) regardless of their signs. Multiplicities are abbreviated as singlet (s), doublet (d), triplet (t), quartet (q), multiplet (m), and broad (br). The mass spectra were recorded by a JEOL JMS-T100LC AccuTOF LC-plus 4G mass spectrometer (ESI-MS) with a DART source. The elemental analyses were performed in the Microanalytical Laboratory at the Institute for Chemical Research (Kyoto University, Uji, Japan). Melting points (m.p.) were determined by a Stanford Research Systems OptiMelt instrument. We were unable to obtain a satisfactory elemental analysis for **2a′** and **3a**, probably due to their extremely high air- and moisture-sensitivity as well as a contamination of NHCs and unidentified compounds associated with some thermal decomposition (Figures S5–S10).

3.1.1. Synthesis of (Eind)$SiHBr_2$ (*4a*)

To a solution of (Eind)SiH_3 (4.09 g, 9.91 mmol) in toluene (30 mL) was added $PdCl_2$ (38.0 mg, 0.21 mmol) and allyl bromide (4.2 mL, 48.5 mmol). The reaction mixture was heated at 80 °C for 8 days. After the solvent was removed in vacuo, the residue was dissolved in hexane and the resulting mixture was centrifuged to remove the insoluble materials. The supernatant was concentrated to dryness and the resulting residue was recrystallized from pentane to afford **4a** as pale brown crystals in 81% yield (4.58 g, 8.02 mmol).

^{1}H NMR (399.8 MHz, C_6D_6, 30 °C): δ = 0.78 (t, *J* = 7.3 Hz, 12 H, CH_2CH_3), 0.80 (br. s, 12 H, CH_2CH_3), 1.48–1.65 (m, 8 H, CH_2CH_3), 1.76 (s, 4 H, CH_2), 2.11 (br. s, 8 H, CH_2CH_3), 6.89 (s, 1 H, satellite, $J_{Si\text{–}H}$ = 288 Hz, Si*H*), 7.01 (s, 1 H, Ar*H*). ^{13}C NMR (100.5 MHz, C_6D_6, 25 °C): δ = 9.3, 10.5 (br), 33.5, 34.2 (br), 42.6 (br ×1), 44.8 (br ×1), 47.9 (br ×2), 54.7 (×2), 123.6, 125.1, 150.8 (one aromatic peak is broadened at 155–158); ^{29}Si NMR (79.4 MHz, C_6D_6, 30 °C): δ = −28.7 (d, $J_{Si\text{–}H}$ = 288 Hz). IR (KBr, cm^{-1}): ν = 2317 (Si−H); IR (THF, cm^{-1}): ν = 2298 (Si−H). DART-HRMS (positive-mode) Calcd. for $C_{28}H_{46}Br_2Si$ + H: 569.1814. Found: 569.1820. Anal. Calcd. for $C_{28}H_{46}Br_2Si$: C, 58.94; H, 8.13. Found: C, 59.41; H, 8.19. Melting point (argon atmosphere in a sealed tube) 102–105 °C.

3.1.2. Synthesis of (Im-^{i}Pr$_2$Me$_2$)→SiBr(Eind) (**2a′**)

(Method A) Reaction of (Eind)BrSi=SiBr(Eind) (**1a**) with Im-^{i}Pr$_2$Me$_2$

A mixture of **1a** (158 mg, 0.16 mmol) and Im-^{i}Pr$_2$Me$_2$ (63.0 mg, 0.35 mmol) was dissolved in benzene (5 mL). The reaction mixture was heated overnight at 70 °C. After the solvent was removed in vacuo, the residue was washed with pentane to afford **2a′** as an orange solid in 88% crude yield (190 mg, 0.28 mmol). We were unable to isolate **2a′** in pure form, because **2a′** was not thermally stable in solution, gradually giving Im-^{i}Pr$_2$Me$_2$ and unidentified compounds (Figure S5).

^{1}H NMR (399.8 MHz, C_6D_6, 60 °C): δ = 0.81–1.00 (m, 24 H, CH_2CH_3), 1.12 (br. s, 6 H, CH(CH_3)$_2$–(Im-^{i}Pr$_2$Me$_2$)), 1.19 (d, *J* = 7.0 Hz, 6 H, CH(CH_3)$_2$–(Im-^{i}Pr$_2$Me$_2$)), 1.61 (s, 6 H, CH_3–(Im-^{i}Pr$_2$Me$_2$)), 1.62–1.95 (m, 20 H, CH_2 + CH_2CH$_3$), 5.00–5.27 (m, 2 H, *CH*(CH$_3$)$_2$–(Im-^{i}Pr$_2$Me$_2$)), 6.71 (s, 1 H, Ar*H*). ^{13}C NMR (100.5 MHz, C_6D_6, 70 °C): δ = 9.3, 9.4, 9.9, 10.0, 10.5, 20.7, 21.4, 24.6, 33.7 (br, overlapped, Im-^{i}Pr$_2$Me$_2$ and CH_2CH_3), 42.8, 48.4, 51.1 (Im-^{i}Pr$_2$Me$_2$), 54.5, 119.9, 125.9 (Im-^{i}Pr$_2$Me$_2$), 147.2, 148.5, 153.7, 170.6 (Im-^{i}Pr$_2$Me$_2$); ^{29}Si NMR (79.4 MHz, C_6D_6, 25 °C): δ = 18.0. HRMS (ESI, positive) Calcd. for $C_{39}H_{65}BrN_2Si$ + H: 669.4179. Found: 669.4211. Melting point (argon atmosphere in a sealed tube) 152–156 °C (dec.).

(Method B) Reaction of (Eind)SiHBr$_2$ (**4a**) with Im-^{i}Pr$_2$Me$_2$

A mixture of **4a** (476 mg, 0.97 mmol) and Im-^{i}Pr$_2$Me$_2$ (352 mg, 1.95 mmol) was dissolved in benzene (7 mL). After stirring overnight at room temperature, the resulting orange suspension was filtered through a polytetrafluoroethylene (PTFE) syringe filter to remove the insoluble materials. The filtrate was concentrated to dryness and the resulting residue was washed with pentane to afford **2a′** as an orange solid in 59% crude yield (197 mg, 0.29 mmol).

3.1.3. Synthesis of [(Im-Me$_4$)$_2$→Si(Eind)]$^+$[Br]$^-$ (**3a**)

(Method A) Reaction of (Eind)BrSi=SiBr(Eind) (**1a**) with Im-Me$_4$

A mixture of **1a** (102 mg, 0.11 mmol) and Im-Me$_4$ (54 mg, 0.43 mmol) was dissolved in benzene (6 mL). After stirring for 1 day at room temperature, the resulting orange solid was separated and washed with a mixture of hexane and benzene to afford **3a** as an orange powder in 54% crude yield (85.2 mg, 0.12 mmol). We were unable to isolate **3a** in pure form, because **3a** was not thermally stable in solution leading to the contamination of Im-Me$_4$ and unidentified compounds (Figure S8).

^{1}H NMR (399.8 MHz, CD_3CN, 20 °C): δ = 0.61 (br. t, *J* = 6.3 Hz, 12 H, CH_2CH_3), 0.80 (t, *J* = 7.3 Hz, 12 H, CH_2CH_3), 1.50–1.70 (m, 8 H, CH_2CH$_3$), 1.75 (br. s, 4 H, CH_2), 1.93−1.97 (m, overlapped, 8 H, CH_2CH$_3$), 2.15 (s, 12 H, CH$_3$–(Im-Me$_4$)), 3.25 (br. s, 12 H, CH$_3$–(Im-Me$_4$)), 6.75 (s, 1 H, Ar*H*). ^{13}C NMR (100.5 MHz, CD_3CN, 19 °C): δ = 9.2, 9.4, 10.0 (br, Im-Me$_4$), 34.1 (br, overlapped, Im-Me$_4$ and CH_2CH_3), 42.8, 48.1, 54.2, 121.8, 128.5 (Im-Me$_4$), 136.5, 150.7, 162.0 (Im-Me$_4$) (one aromatic peak is overlapped). ^{29}Si NMR

(79.4 MHz, CD_3CN, 25 °C): $\delta = -63.3$. DART-HRMS (positive-mode) Calcd. for $C_{42}H_{69}BrN_4Si + H$: 737.4553. Found: 737.4562. Melting point (argon atmosphere in a sealed tube) 169–174 °C (dec.).

(Method B) Reaction of (Eind)$SiHBr_2$ (**4a**) with Im-Me_4

A mixture of **4a** (70.3 mg, 0.12 mmol) and Im-Me_4 (50.6 mg, 0.41 mmol) was dissolved in benzene (6 mL). After stirring for 1 day at room temperature, an orange suspension was formed. An insoluble orange solid was collected by filtration, whose ^{1}H NMR spectrum indicated the formation of a mixture of **4a** and [(Im-Me_4)H]$^+$[Br^-].

3.2. X-ray Crystallographic Studies of **3a′** *and* **4a**

Single crystals suitable for X-ray diffraction measurements were obtained from benzene for **3a′** and from hexane for **4a**. Intensity data were collected using a Rigaku XtaLAB P200 with a PILATUS 200K detector for **3a′** and a Rigaku AFC-8 with a Saturn 70 CCD detector for **4a** (Rigaku Corporation, Tokyo, Japan). All measurements were carried out using Mo $K\alpha$ radiation (λ = 0.71073 Å). The integration and scaling of the diffraction data were carried out using the programs CrysAlisPro [46] for **3a′** and CrystalClear [47] for **4a**. Lorentz, polarization, and absorption corrections were also performed. The structures were solved by an iterative method with the program of SHELXT [48], and refined by a full-matrix least-squares method on F^2 for all the reflections using the program SHELXL-2017/1 [49]. The non-hydrogen atoms were refined by applying anisotropic temperature factors. Positions of all the hydrogen atoms were geometrically calculated, and refined as riding models. The Si–H hydrogen atom was located on difference Fourier maps and isotropically refined. Full details of the crystallographic analysis and accompanying CIF files can be obtained free of charge from the Cambridge Crystallographic Data Centre (CCDC numbers 1811699 and 1811700) via http://www.ccdc.cam.ac.uk/conts/retrieving.html (or from the CCDC, 12 Union Road, Cambridge CB2 1EZ, UK.

3.2.1. [(Im-^{i}Pr$_2$Me$_2$)$_2$→Si(Eind)]$^+$[Br^-] (**3a′**)

$C_{50}H_{85}BrN_4Si{\cdot}C_6H_6$, M = 928.32, crystal size 0.36 × 0.15 × 0.12 mm, triclinic, space group $P-1$ (#2), a = 10.6548(3) Å, b = 12.0149(2) Å, c = 21.7366(5) Å, α = 80.5199(18)°, β = 82.343(2)°, γ = 72.816(2)°, V = 2611.54(11) Å^3, Z = 2, D_x = 1.181 g cm^{-3}, μ(Mo $K\alpha$) = 0.849 mm^{-1}, 65594 reflections collected, 13761 unique reflections, and 579 refined parameters. The final $R(F)$ value was 0.0491 [$I > 2\sigma(I)$]. The final $R_w(F^2)$ value was 0.1364 (all data). The goodness-of-fit on F^2 was 1.031.

3.2.2. (Eind)$SiHBr_2$ (**4a**)

$C_{28}H_{46}Br_2Si$, M = 570.56, crystal size 0.16 × 0.17 × 0.41 mm, triclinic, space group $P-1$ (#2), a = 7.972(3) Å, b = 11.070(4) Å, c = 16.621(5) Å, α = 89.972(4)°, β = 80.770(3)°, γ = 73.314(5)°, V = 1385.1(8) Å^3, Z = 2, D_x = 1.368 g cm^{-3}, μ(Mo $K\alpha$) = 2.992 mm^{-1}, 22599 reflections collected, 6332 unique reflections, and 292 refined parameters. The final $R(F)$ value was 0.0298 [$I > 2\sigma(I)$]. The final $R_w(F^2)$ value was 0.0803 (all data). The goodness-of-fit on F^2 was 1.016.

4. Conclusions

We have synthesized some new NHC-coordinated silicon species having the fused-ring bulky Eind group by two methods; one is via the reactions of the stable diaryldibromodisilene, (Eind)BrSi=SiBr(Eind) (**1a**), with NHCs, and the other is the dehydrobromination of the aryldibromohydrosilane, (Eind)$SiHBr_2$ (**4a**), with NHCs. In both synthetic pathways, we have mainly obtained the mono-NHC adduct of the arylbromosilylene, (Im-^{i}Pr$_2$Me$_2$)→SiBr(Eind) (**2a′**), and the bis-NHC adduct of the formal arylsilyliumylidenecation, [(Im-Me_4)$_2$→Si(Eind)]$^+$[Br^-] (**3a**), depending on the steric bulk of the NHCs (Im-^{i}Pr$_2$Me$_2$ vs. Im-Me_4). Further studies on the reactivities of the NHC-coordinated silicon compounds are now in progress.

Acknowledgments: This study was supported by Grants-in-Aid for Scientific Research on Innovative Areas "Stimuli-responsive Chemical Species for the Creation of Functional Molecules (No. 2408)" (JSPS KAKENHI Grant Nos. JP20109003 for Tsukasa Matsuo, JP15H00964 for Daisuke Hashizume, and JP20109013 for Norihiro Tokitoh), Scientific Research (B) (No. JP15H03788 for Tsukasa Matsuo and JP15H03777 for Takahiro Sasamori), Young Scientists (A) (No. 15H05477 for Tomohiro Agou) and Challenging Exploratory Research (No. 16K13953 for Tomohiro Agou). This study was partially supported by a MEXT-Supported Program for the Strategic Research Foundation at Private Universities 2014–2018 subsidy from MEXT and Kindai University. We thank the Collaborative Research Program of The Institute for Chemical Research, Kyoto University (grants #2016-94 and #2017-99). Naoki Hayakawa and Tomohiro Sugahara acknowledge the support by Grants-in-Aid for JSPS Fellows from JSPS (Nos. JP16J01036 and JP16J05501).

Author Contributions: Naoki Hayakawa, Kazuya Sadamori, Shinsuke Mizutani, Tomohiro Agou, and Tomohiro Sugahara performed the experiments. Naoki Hayakawa and Daisuke Hashizume carried out the X-ray crystallographic analysis. Tomohiro Sugahara, Norihiro Tokitoh, and Tsukasa Matsuo designed the experiments and co-directed the project. Naoki Hayakawa, Kazuya Sadamori, and Tsukasa Matsuo co-wrote the paper. Takahiro Sasamori, Norihiro Tokitoh, Daisuke Hashizume, and Tsukasa Matsuo reviewed and approved the final manuscript. All authors contributed to the discussions.

References

1. Kira, M.; Iwamoto, T. Progress in the chemistry of stable disilenes. *Adv. Organomet. Chem.* **2006**, *54*, 73–148.
2. Wang, Y.; Robinson, G.H. Unique homonuclear multiple bonding in main group compounds. *Chem. Commun.* **2009**, 5201–5213. [CrossRef] [PubMed]
3. Mizuhata, Y.; Sasamori, T.; Tokitoh, N. Stable heavier carbene analogues. *Chem. Rev.* **2009**, *109*, 3479–3511. [CrossRef] [PubMed]
4. Fischer, R.C.; Power, P.P. π-Bonding and lone pair effect in multiple bonds involving heavier main group elements: Developments in the new millennium. *Chem. Rev.* **2010**, *110*, 3877–3923. [CrossRef] [PubMed]
5. Lee, V.Y.; Sekiguchi, A. *Organometallic Compounds of Low-Coordinate Si, Ge, Sn and Pb: From Phantom Species to Stable Compounds*; John Wiley & Sons Ltd.: Chichester, UK, 2010; ISBN 978-0-470-72543-6.
6. Asay, M.; Jones, C.; Driess, M. *N*-Heterocyclic carbene analogues with low-valent group 13 and group 14 elements: Syntheses, structures, and reactivities of a new generation of multitalented ligands. *Chem. Rev.* **2011**, *111*, 354–396. [CrossRef] [PubMed]
7. Yao, S.; Xiong, Y.; Driess, M. Zwitterionic and donor-stabilized *N*-heterocyclic silylenes (NHSis) for metal-free activation of small molecules. *Organometallics* **2011**, *30*, 1748–1767. [CrossRef]
8. Kira, M. Bonding and structure of disilenes and related unsaturated Group-14 element compounds. *Proc. Jpn. Acad. Ser. B Phys. Biol. Sci.* **2012**, *88*, 167–191. [CrossRef] [PubMed]
9. Scheschkewitz, D. *Functional Molecular Silicon Compounds II: Low Oxidation States*; Springer: Basel, Switzerland, 2014; ISBN 978-3-319-03734-9.
10. Wang, Y.; Robinson, G.H. *N*-heterocyclic carbene—Main-group chemistry: A rapidly evolving field. *Inorg. Chem.* **2014**, *53*, 11815–11832. [CrossRef] [PubMed]
11. Ghadwal, R.S.; Azhakar, R.; Roesky, H.W. Dichlorosilylene: A high temperature transient species to an indispensable building block. *Acc. Chem. Res.* **2013**, *46*, 444–456. [CrossRef] [PubMed]
12. Ghadwal, R.S.; Roesky, H.W.; Merkel, S.; Henn, J.; Stalke, D. Lewis base stabilized dichlorosilylene. *Angew. Chem. Int. Ed.* **2009**, *48*, 5683–5686. [CrossRef] [PubMed]
13. Filippou, A.C.; Chernov, O.; Schnakenburg, G. $SiBr_2$(Idipp): A stable *N*-heterocyclic carbene adduct of dibromosilylene. *Angew. Chem. Int. Ed.* **2009**, *48*, 5687–5690. [CrossRef] [PubMed]
14. Roy, S.; Stollberg, P.; Herbst-Irmer, R.; Stalke, D.; Andrada, D.M.; Frenking, G.; Roesky, H.W. Carbene-dichlorosilylene stabilized phosphinidenes exhibiting strong intramolecular charge transfer transition. *J. Am. Chem. Soc.* **2015**, *137*, 150–153. [CrossRef] [PubMed]
15. Hickox, H.P.; Wang, Y.; Xie, Y.; Wei, P.; Schaefer, H.F., III; Robinson, G.H. Push-pull stabilization of parent monochlorosilylenes. *J. Am. Chem. Soc.* **2016**, *138*, 9799–9802. [CrossRef] [PubMed]
16. Geiβ, D.; Arz, M.I.; Straβmann, M.; Schnakenburg, G.; Filippou, A.C. Si=P double bonds: experimental and theoretical study of an NHC-stabilized phosphasilenylidene. *Angew. Chem. Int. Ed.* **2015**, *54*, 2739–2744.
17. Ghana, P.; Arz, M.I.; Das, U.; Schnakenburg, G.; Filippou, A.C. Si=Si double bonds: Synthesis of an NHC-stabilized disilavinylidene. *Angew. Chem. Int. Ed.* **2015**, *54*, 9980–9985. [CrossRef] [PubMed]

18. Filippou, A.C.; Chernov, O.C.; Blom, B.; Stumpf, K.W.; Schnakenburg, G. Stable *N*-heterocyclic carbene adducts of arylchlorosilylenes and their germanium homologues. *Chem. Eur. J.* **2010**, *16*, 2866–2872. [CrossRef] [PubMed]
19. Filippou, A.C.; Chernov, O.C.; Stumpf, K.W.; Schnakenburg, G. Metal-silicon triple bonds: The molybdrnumsilylidyne complex [Cp(CO)$_2$Mo≡Si-Ar]. *Angew. Chem. Int. Ed.* **2010**, *49*, 3296–3300. [CrossRef] [PubMed]
20. Suzuki, K.; Matsuo, T.; Hashizume, D.; Tamao, K. Room-temperature dissociation of 1,2-dibromodisilenes to bromosilylenes. *J. Am. Chem. Soc.* **2011**, *133*, 19710–19713. [CrossRef] [PubMed]
21. Agou, T.; Sasamori, T.; Tokitoh, N. Synthesis of an arylbromosilylene-platinum complex by using a 1,2-dibromodisilene as a silylene source. *Organometallics* **2012**, *31*, 1150–1154. [CrossRef]
22. Agou, T.; Hayakawa, N.; Sasamori, T.; Matsuo, T.; Hashizume, D.; Tokitoh, N. Reactions of diaryldibromodisilenes with *N*-heterocyclic carbenes: Formation of formal bis-NHC adducts of silyliumylidene cations. *Chem. Eur. J.* **2014**, *20*, 9246–9249. [CrossRef] [PubMed]
23. Ahmad, S.U.; Szilvási, T.; Inoue, S. A facile access to a novel NHC-stabilized silyliumylidene ion and C–H activation of phenylacetylene. *Chem. Commun.* **2014**, *50*, 12619–12622. [CrossRef] [PubMed]
24. Ahmad, S.U.; Szilvási, T.; Irran, E.; Inoue, S. An NHC-stabilized silicon analogue of acylium ion: Synthesis, structure, reactivity, and theoretical studies. *J. Am. Chem. Soc.* **2015**, *137*, 5828–5836. [CrossRef] [PubMed]
25. Mork, B.V.; Tilley, T.D. Multiple bonding between silicon and molybdenum: A transition-metal complex with considerable silylyne character. *Angew. Chem. Int. Ed.* **2003**, *42*, 357–360. [CrossRef] [PubMed]
26. Matsuo, T.; Suzuki, K.; Fukawa, T.; Li, B.L.; Ito, M.; Shoji, Y.; Otani, T.; Li, L.C.; Kobayashi, M.; Hachiya, M.; et al. Synthesis and structures of a series of bulky "Rind-Br" based on a rigid fused-ring *s*-hydrindacene skeleton. *Bull. Chem. Soc. Jpn.* **2011**, *84*, 1178–1191. [CrossRef]
27. Matsuo, T.; Tamao, K. Fused-ring bulky "Rind" groups producing new possibilities in elemento-organic chemistry. *Bull. Chem. Soc. Jpn.* **2015**, *88*, 1201–1220. [CrossRef]
28. Cui, H.; Cui, C. Silylation of *N*-heterocyclic carbene with aminochlorosilane and -disilane: Dehydrohalogenation vs. Si–Si bond cleavage. *Dalton Trans.* **2011**, *40*, 11937–11940. [CrossRef] [PubMed]
29. Hadlington, T.J.; Szilvási, T.; Driess, M. Silylene-nichkel promoted cleavage of B–O bonds: From catechol borane to the hydroborylene ligand. *Angew. Chem. Int. Ed.* **2017**, *56*, 7470–7474. [CrossRef] [PubMed]
30. Sasamoti, T.; Hironaka, K.; Sugiyama, Y.; Takagi, N.; Nagase, S.; Hosoi, Y.; Furukawa, Y.; Tokitoh, N. Synthesis and reactions of a stable 1,2-diaryl-1,2-dibromodisilene: A precursor for substituted disilenes and a 1,2-diaryldisilyne. *J. Am. Chem. Soc.* **2008**, *130*, 13856–13857. [CrossRef] [PubMed]
31. Liao, W.-H.; Ho, P.-Y.; Su, M.-D. Mechanisms for the reactions of Group 10 transition metal complexes with metal-group 14 element bonds, Bbt(Br)E=M(PCy$_3$)$_2$ (E = C, Si, Ge, Sn, Pb; M = Pd and Pt). *Inorg. Chem.* **2013**, *52*, 1338–1348. [CrossRef] [PubMed]
32. Filippou, A.C.; Hoffmann, D.; Schnakenburg, G. Triple bonds of niobium with silicon, germanium and tin: The tetrylidyne complexes [(κ^3-tmps)(CO)$_2$Nb≡E–R] (E = Si, Ge, Sn; tmps = MeSi(CH$_2$OMe$_2$)$_3$; R = aryl). *Chem. Sci.* **2017**, *8*, 6290–6299. [CrossRef]
33. Blom, B. Reactivity of Ylenes at Late Transition Metal Centers. Dissertation, University of Bonn, Göttingen, Germany, 2011.
34. Papazoglou, I. Unprecedented Tetrylidyne Complexes of Group 6 and Group 10 Metals. Dissertation, University of Bonn, Dr. Hut Verlag, München, Germany, 28 May 2017.
35. Filippou, A.C.; Lebedev, Y.N.; Chernov, O.; Straβmann, M.; Schnakenburg, G. Silicon(II) coordination chemistry: *N*-Heterocyclic carbene complexes of Si^{2+} and SiI^{+}. *Angew. Chem. Int. Ed.* **2013**, *52*, 6974–6978. [CrossRef] [PubMed]
36. Sarkar, D.; Wendel, D.; Ahmad, S.U.; Szilvási, T.; Pöthig, A.; Inoue, S. Chalcogen-atom transfer and exchange reactions of NHC-stabilized heavier silaacylium ions. *Dalton Trans.* **2017**, *46*, 16014–16018. [CrossRef] [PubMed]
37. Fukazawa, A.; Li, Y.; Yamaguchi, S.; Tsuji, H.; Tamao, K. Coplanar oligo(*p*-phenylenedisilenylene)s based on the octaethyl-substituted *s*-hydrindacenyl groups. *J. Am. Chem. Soc.* **2007**, *129*, 14164–14165. [CrossRef] [PubMed]
38. Kobayashi, M.; Hayakawa, N.; Nakabayashi, K.; Matsuo, T.; Hashizume, D.; Fueno, H.; Tanaka, K.; Tamao, K. Highly coplanar (*E*)-1,2-di(1-naphthyl)disilene involving a distinct CH–π interaction with the perpendicularly oriented protecting Eind group. *Chem. Lett.* **2014**, *43*, 432–434. [CrossRef]

39. Iwata, A.; Toyoshima, Y.; Hayashida, T.; Ochi, T.; Kunai, A.; Ohshita, J. $PdCl_2$ and $NiCl_2$-catalyzed hydrogen-halogen exchange for the convenient preparation of bromo- and iodosilanes and germanes. *J. Organomet. Chem.* **2003**, *667*, 90–95. [CrossRef]
40. Kunai, A.; Ochi, T.; Iwata, A.; Ohshita, J. Synthesis of bromohydrosilanes: Reactions of hydrosilanes with $CuBr_2$ in the presence of CuI. *Chem. Lett.* **2001**, 1228–1229. [CrossRef]
41. Agou, T.; Sugiyama, Y.; Sasamori, T.; Sakai, H.; Furukawa, Y.; Takagi, N.; Guo, J.-D.; Nagasa, S.; Hashizume, D.; Tokitoh, N. Synthesis of Kinetically Stabilized 1,2-Dihydrodisilenes. *J. Am. Chem. Soc.* **2012**, *134*, 4120–4123. [CrossRef] [PubMed]
42. Simons, R.S.; Haubrich, S.T.; Mork, B.V.; Niemeyer, M.; Power, P.P. The Syntheses and Characterization of the Bulky Terphenylsilanes and Chlorosilanes 2,6-$Mes_2C_6H_3SiCl_3$, 2,6-$Trip_2C_6H_3SiCl_3$, 2,6-$Mes_2C_6H_3SiHCl_2$, 2,6-$Trip_2C_6H_3SiHCl_2$, 2,6-$Mes_2C_6H_3SiH_3$, 2,6-$Trip_2C_6H_3SiH_3$ and 2,6-$Mes_2C_6H_3SiC1_2SiCl_3$. *Main Group Chem.* **1998**, *2*, 275–283. [CrossRef]
43. Weidemann, N.; Schnakenburg, G.; Filippou, A.C. Novel silanes with sterically demanding aryl substituents. *Z. Anorg. Allg. Chem.* **2009**, *635*, 253–259. [CrossRef]
44. Li, B.; Tsujimoto, S.; Li, Y.; Tsuji, H.; Tamao, K.; Hashizume, D.; Matsuo, T. Synthesis and characterization of diphosphenes bearing fused-ring bulky Rind groups. *Heteroat. Chem.* **2014**, *25*, 612–618. [CrossRef]
45. Kuhn, S.; Kratz, T. Synthesis of imidazol-2-ylidenes by reduction of imidazole-2(3*H*)-thiones. *Synthesis* **1993**, 561–562. [CrossRef]
46. *CrysAlisPro*; Agilent Technologies Ltd.: Yarnton, Oxfordshire, UK, 2014.
47. *CrystalClear*; Rigaku/MSC. Inc.: The Woodlands, TX, USA, 2005.
48. Sheldrick, G.M. *SHELXT*—Integrated space-group and crystal-structure determination. *Acta Crystallogr. Sect. A* **2015**, *A71*, 3–8. [CrossRef] [PubMed]
49. Sheldrick, G.M. Crystal structure refinement with *SHELXL*. *Acta Crystallogr. Sect. C* **2015**, *C71*, 3–8.

Molecular Structures of Enantiomerically-Pure (*S*)-2-(Triphenylsilyl)- and (*S*)-2-(Methyldiphenylsilyl)pyrrolidinium Salts

Jonathan O. Bauer † and Carsten Strohmann *

Anorganische Chemie, Fakultät für Chemie und Chemische Biologie, Technische Universität Dortmund, Otto-Hahn-Straße 6, D-44227 Dortmund, Germany; jonathan.bauer@ur.de

* Correspondence: carsten.strohmann@tu-dortmund.de

† Current address: Institut für Anorganische Chemie, Fakultät für Chemie und Pharmazie, Universität Regensburg, Universitätsstraße 31, D-93053 Regensburg, Germany

Abstract: Silyl-substituted pyrrolidines have gained increased interest for the design of new catalyst scaffolds. The molecular structures of four enantiomerically-pure 2-silylpyrrolidinium salts are reported. The perchlorate salts of (*S*)-2-(triphenylsilyl)pyrrolidine [(*S*)-**1**·$HClO_4$] and (*S*)-2-(methyldiphenylsilyl)pyrrolidine [(*S*)-**2**·$HClO_4$], the trifluoroacetate (*S*)-**2**·TFA, and the methanol-including hydrochloride (*S*)-**1**·HCl·MeOH were elucidated by X-ray crystallography and discussed in terms of hydrogen-bond interactions.

Keywords: hydrogen bonds; silicon; 2-silylpyrrolidines; stereochemistry; X-ray crystallography

---

1. Introduction

In 2010, we and others reported on the enantioselective synthesis of 2-silylpyrrolidines as organocatalysts for the asymmetric Michael addition of aldehydes to nitroolefines [1,2]. Since then, some impressive developments in the catalyst design have been achieved, overcoming synthetic challenges and introducing pyrrolidinylsilanols as bifunctional hydrogen bond-directing organocatalysts [3,4]. The stereochemical information of 2-substituted silylpyrrolidines was introduced by asymmetric deprotonation of *N*-(*tert*-butoxycarbonyl)pyrrolidine (*N*-Boc-pyrrolidine) with *sec*-butyllithium in the presence of (–)-sparteine [5,6], followed by a substitution reaction with a silyl halide or methoxide as the electrophile. Concerning the first successful preparation of enantiomerically-pure (*S*)-2-(triphenylsilyl)pyrrolidine [(*S*)-**1**], we established an indirect synthetic route via intermediate formation of 2-(methoxydiphenylsilyl)-*N*-Boc-pyrrolidine [1]. Recently, a detailed structural and kinetic investigation gave new insight into the structure-reactivity relation in enamines and iminium ions derived from 2-tritylpyrrolidine [7] and 2-(triphenylsilyl)pyrrolidine [8]. (*S*)-2-(Triphenylsilyl)pyrrolidine [(*S*)-**1**] and (*S*)-2-(methyldiphenylsilyl)pyrrolidine [(*S*)-**2**] have already been structurally characterized in the form of their hydrochloride [1,2] and their hydrobromide salts [3] (Figure 1).

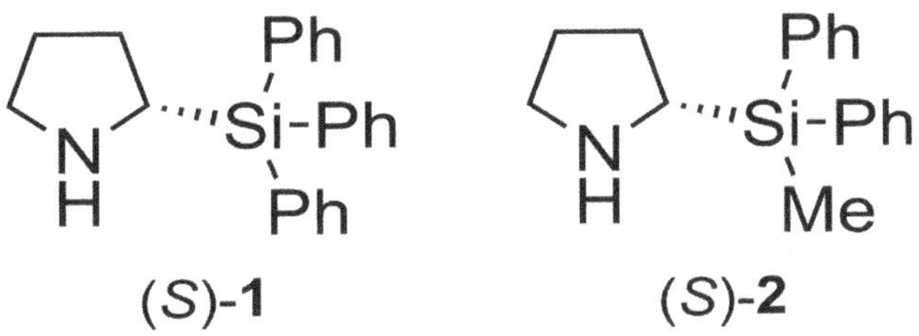

Figure 1. (*S*)-2-(triphenylsilyl)- [(*S*)-**1**] and (*S*)-2-(methyldiphenylsilyl)pyrrolidine [(*S*)-**2**].

Herein, we present the molecular structures of enantiomerically-pure (*S*)-2-(triphenylsilyl)-[(*S*)-**1**·$HClO_4$] and (*S*)-2-(methyldiphenylsilyl)pyrrolidinium perchlorate [(*S*)-**2**·$HClO_4$], which were obtained from the respective optically-pure chloride salts [1] by treatment with perchloric acid. In addition, we report on hydrogen-bonding motifs in the new enantiomerically-pure methanol inclusion compound (*S*)-**1**·HCl·MeOH and in the enantiomerically-pure trifluoroacetate (*S*)-**2**·TFA. Hydrogen bonding in enantiomerically-pure pyrrolidines is worth studying in order to explore new activation modes in organocatalytic transformations.

2. Results and Discussion

Compound (*S*)-**1**·$HClO_4$ crystallized in the monoclinic crystal system, space group $P2_1$, as colorless plates (Figure 2 and Table 1). The pyrrolidinyl nitrogen atom of (*S*)-**1**·$HClO_4$ has been protonated by perchloric acid and is involved in hydrogen bonds to two perchlorate anions via H(1N) and H(2N). The hydrogen bond N–H(1N)···O(1) is slightly stronger [N–H(1N) 0.911 Å, H(1N)···O(1) 1.966 Å, N···O(1) 2.845 Å, N–H(1N)···O(1) 161.79°] than the interaction between H(2N) and O(2) [N–H(2N) 0.773 Å, H(2N)···O(2) 2.230 Å, N···O(2) 2.940 Å, N–H(2N)···O(2) 152.82°]. The C–Si bond lengths of the triphenylsilyl moiety are comparable to those found in the hydrochloride species (*S*)-**1**·HCl [1] and in other triphenyl-substituted silanes [9,10]. The C(19)–Si bond between silicon and the heterocyclic carbon atom amounts to 1.9111(15) Å and is in the characteristic range for 2-(triphenylsilyl)pyrrolidines [1] (Figure 2). This significantly longer bond compared to the respective C–C bond in the carbon analogue 2-tritylpyrrolidine was considered a crucial parameter for the higher reactivity of (*S*)-**1** in enamine catalysis [1].

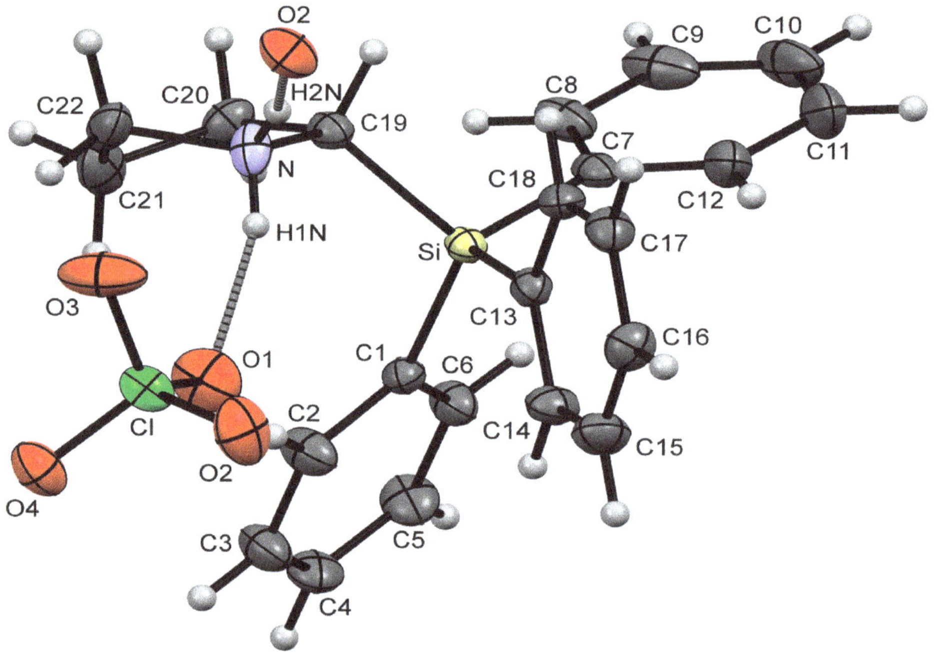

Figure 2. Part of the crystal structure (ORTEP plot) of compound (*S*)-**1**·$HClO_4$ in the crystal, with the displacement ellipsoids set at the 50% probability level. Selected bond lengths (Å) and angles (°): C(1)–Si 1.8727(15), C(7)–Si 1.8724(15), C(13)–Si 1.8595(15), C(19)–Si 1.9111(15), C(19)–N 1.5229(19), C(22)–N 1.4999(19), Cl–O(1) 1.4376(12), Cl–O(2) 1.4305(13), Cl–O(3) 1.4316(13), Cl–O(4) 1.4217(12), C(13)–Si–C(7) 111.31(7), C(13)–Si–C(1) 111.55(7), C(7)–Si–C(1) 108.77(7), C(13)–Si–C(19) 108.01(6), C(7)–Si–C(19) 105.67(7), C(1)–Si–C(19) 111.39(6). Hydrogen bond N–H(1N)···O(1): N–H(1N) 0.911, H(1N)···O(1) 1.966, N···O(1) 2.845, N–H(1N)···O(1) 161.79. Hydrogen bond N–H(2N)···O(2): N–H(2N) 0.773, H(2N)···O(2) 2.230, N···O(2) 2.940, N–H(2N)···O(2) 152.82. Symmetry transformations used to generate the equivalent atom O(2), hydrogen-bonded to H(2N): $1 - x, \frac{1}{2} + y, -z$.

Table 1. Crystal data and structure refinement of compounds (*S*)-**1**·$HClO_4$, (*S*)-**1**·HCl·MeOH, (*S*)-**2**·$HClO_4$, and (*S*)-**2**·TFA.

Compound	(S)-1·$HClO_4$	(S)-1·HCl·MeOH	(S)-2·$HClO_4$	(S)-2·TFA
Empirical formula	$C_{22}H_{24}ClNO_4Si$	$C_{23}H_{28}ClNOSi$	$C_{17}H_{22}ClNO_4Si$	$C_{19}H_{22}F_3NO_2Si$
Formula weight [g·mol^{-1}]	429.96	398.00	367.90	381.47
Crystal system	Monoclinic	Orthorhombic	Monoclinic	Orthorhombic
Space group	$P2_1$	$P2_1\ 2_1\ 2_1$	$P2_1$	$P2_1\ 2_1\ 2_1$
a [Å]	8.1535(2)	7.2981(3)	9.6006(6)	9.5760(14)
b [Å]	7.8737(2)	11.7818(5)	14.2054(8)	9.7100(17)
c [Å]	16.9410(4)	25.1513(11)	13.3890(8)	41.176(4)
β [°]	100.780(2)	90	91.537(6)	90
Volume [Å^3]	1068.39(5)	2162.63(16)	1825.34(19)	3828.7(9)
Z	2	4	4	8
Density (calculated) .9ρ [g·cm^{-3}]	1.337	1.222	1.339	1.324
Absorption coefficient μ [mm^{-1}]	0.263	0.245	0.295	0.163
F(000)	452	848	776	1600
Crystal size [mm^3]	0.20 × 0.20 × 0.10	0.40 × 0.30 × 0.20	0.20 × 0.20 × 0.10	0.40 × 0.20 × 0.10
Theta range for data collection θ [°]	2.45–25.99	2.37–26.00	2.56–26.00	2.18–25.00
Index ranges	$-10 \leq h \leq 10$	$-9 \leq h \leq 9$	$-11 \leq h \leq 11$	$-11 \leq h \leq 11$
	$-9 \leq k \leq 9$	$-14 \leq k \leq 14$	$-17 \leq k \leq 17$	$-11 \leq k \leq 10$
	$-20 \leq l \leq 20$	$-31 \leq l \leq 30$	$-16 \leq l \leq 16$	$-47 \leq l \leq 48$
Reflections collected	46,918	16,132	28,993	41,441
Independent reflections	4181 (R_{int} = 0.0369)	4244 (R_{int} = 0.0388)	7177 (R_{int} = 0.0540)	6661 (R_{int} = 0.0425)
Completeness to θ	100.0% (θ = 25.99°)	99.9% (θ = 26.00°)	99.9% (θ = 26.00°)	99.3% (θ = 25.00°)
Max. and min. transmission	0.9742 and 0.9493	0.9527 and 0.9085	0.9711 and 0.9433	0.9681 and 0.9376
Data/restraints/parameters	4181/1/270	4244/0/254	7177/1/435	6661/0/566
Goodness-of-fit on F^2	1.000	1.000	1.000	1.000
Final R indices [$I > 2\sigma(I)$]	$R1$ = 0.0237, w$R2$ = 0.0620	$R1$ = 0.0337, w$R2$ = 0.0682	$R1$ = 0.0596, w$R2$ = 0.1683	$R1$ = 0.0340, w$R2$ = 0.0572
R indices (all data)	$R1$ = 0.0257, w$R2$ = 0.0624	$R1$ = 0.0459, w$R2$ = 0.0700	$R1$ = 0.0809, w$R2$ = 0.1745	$R1$ = 0.0539, w$R2$ = 0.0592
Absolute structure parameter (Flack parameter)	0.01(4)	0.01(6)	0.08(9)	0.02(8)
Largest diff. peak and hole [e·Å^{-3}]	0.207 and −0.228	0.424 and −0.252	0.447 and −0.309	0.260 and −0.197

Previously-known single-crystal X-ray diffraction data of compound (*S*)-**1**·HCl correspond to the chloroform adduct (*S*)-**1**·HCl·$CHCl_3$ [1]. After recrystallization of compound (*S*)-**1**·HCl from methanol, single crystals of the methanol including compound (*S*)-**1**·HCl·MeOH were obtained (Figure 3 and Table 1). Compound (*S*)-**1**·HCl·MeOH crystallized in the orthorhombic crystal system, space group $P2_1\ 2_1\ 2_1$, as colorless needles. The chloride ion is involved in hydrogen bond interactions with H(1N), H(2N), and the methanol hydroxyl group, with the O–H(1O)···Cl hydrogen-bond [O–H(1O) 1.034 Å, H(1O)···Cl 2.071 Å, O···Cl 3.086 Å, O–H(1O)···Cl 166.42°] being essential for the formation of a defined inclusion compound. In addition, a short distance of 3.651 Å between chloride and the phenyl carbon atom C(14) gives a hint for a weak C(14)–H(14)···Cl interaction within the crystal structure of (*S*)-**1**·HCl·MeOH (Figure 3).

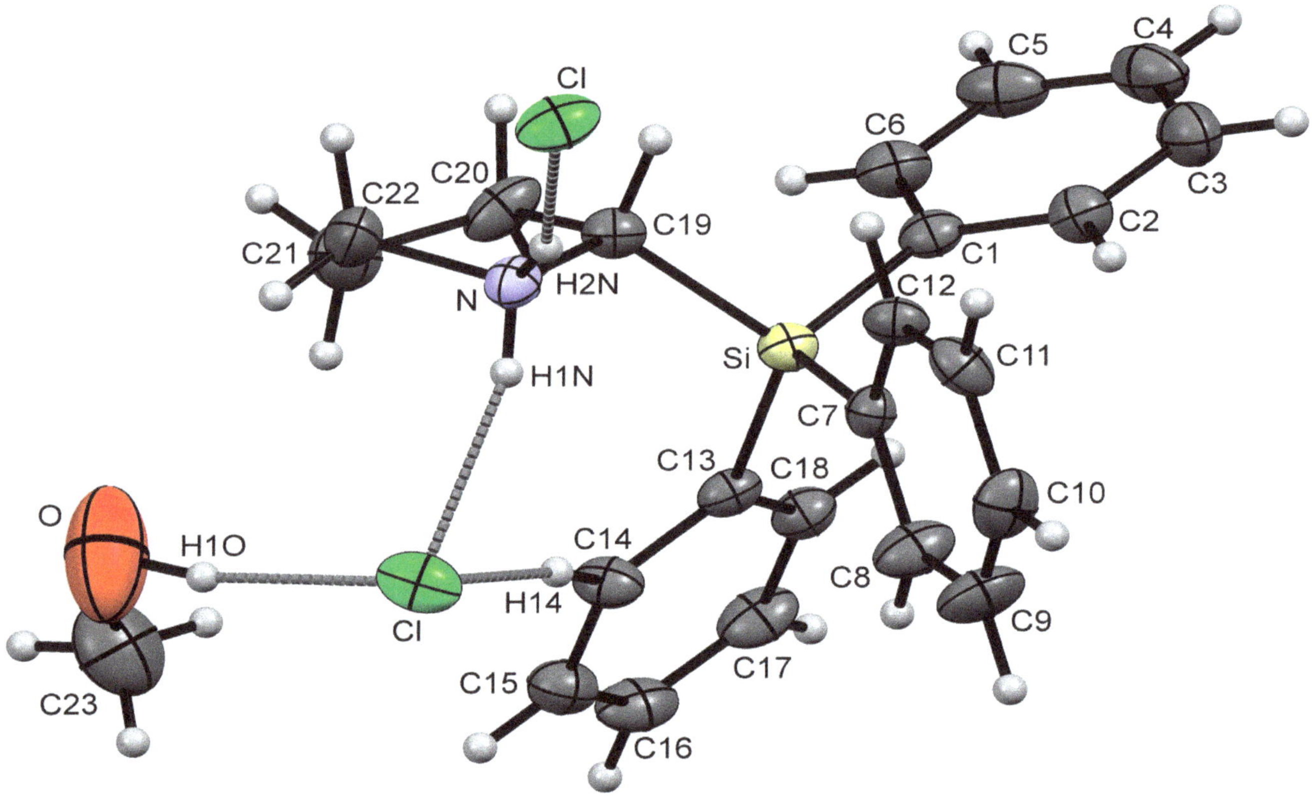

Figure 3. Part of the crystal structure (ORTEP plot) of compound (*S*)-**1**·HCl·MeOH in the crystal, with the displacement ellipsoids set at the 50% probability level. Selected bond lengths (Å) and angles (°): C(1)–Si 1.863(2), C(7)–Si 1.870(2), C(13)–Si 1.873(2), C(19)–Si 1.905(2), C(19)–N 1.498(3), C(22)–N 1.494(3), O–C(23) 1.385(3), C(13)–Si–C(7) 112.48(9), C(13)–Si–C(1) 108.46(9), C(7)–Si–C(1) 110.86(9), C(13)–Si–C(19) 110.34(9), C(7)–Si–C(19) 108.18(9), C(1)–Si–C(19) 106.34(9). Hydrogen bond N–H(1N)···Cl: N–H(1N) 0.870, H(1N)···Cl 2.310, N···Cl 3.125, N–H(1N)···Cl 155.93. Hydrogen bond N–H(2N)···Cl: N–H(2N) 0.913, H(2N)···Cl 2.184, N···Cl 3.062, N–H(2N)···Cl 160.99. Hydrogen bond O–H(1O)···Cl: O–H(1O) 1.034, H(1O)···Cl 2.071, O···Cl 3.086, O–H(1O)···Cl 166.42. C(14)–H(14)···Cl interaction: C(14)···Cl 3.651, C(14)–H(14)···Cl 171.50. Symmetry transformations used to generate the equivalent methanol molecule: x, $1 + y$, $1 + z$. Symmetry transformations used to generate the equivalent atom Cl, hydrogen-bonded to H(2N): $\frac{1}{2} + x$, $1.5 - y$, $2 - z$.

Compound (*S*)-**2**·$HClO_4$ crystallized in the monoclinic crystal system, space group $P2_1$, as colorless plates (Figure 4 and Table 1). The asymmetric unit of compound (*S*)-**2**·$HClO_4$ contains two independent molecules [(*S*)-**2**·$HClO_4$-**A** and (*S*)-**2**·$HClO_4$-**B**]. The bond lengths between the silyl group and the heterocycle with 1.878(5) Å [C(14)–Si(1), molecule **A**] and 1.889(5) Å [C(31)–Si(2), molecule **B**] differ slightly from each other, but are considerably shorter than the respective bond in the 2-triphenylsilyl derivative (*S*)-**1**·$HClO_4$ (compare Figures 2 and 4). The C–Si–C angles are in both independent molecules very close to the ideal tetrahedral angle (Figure 4).

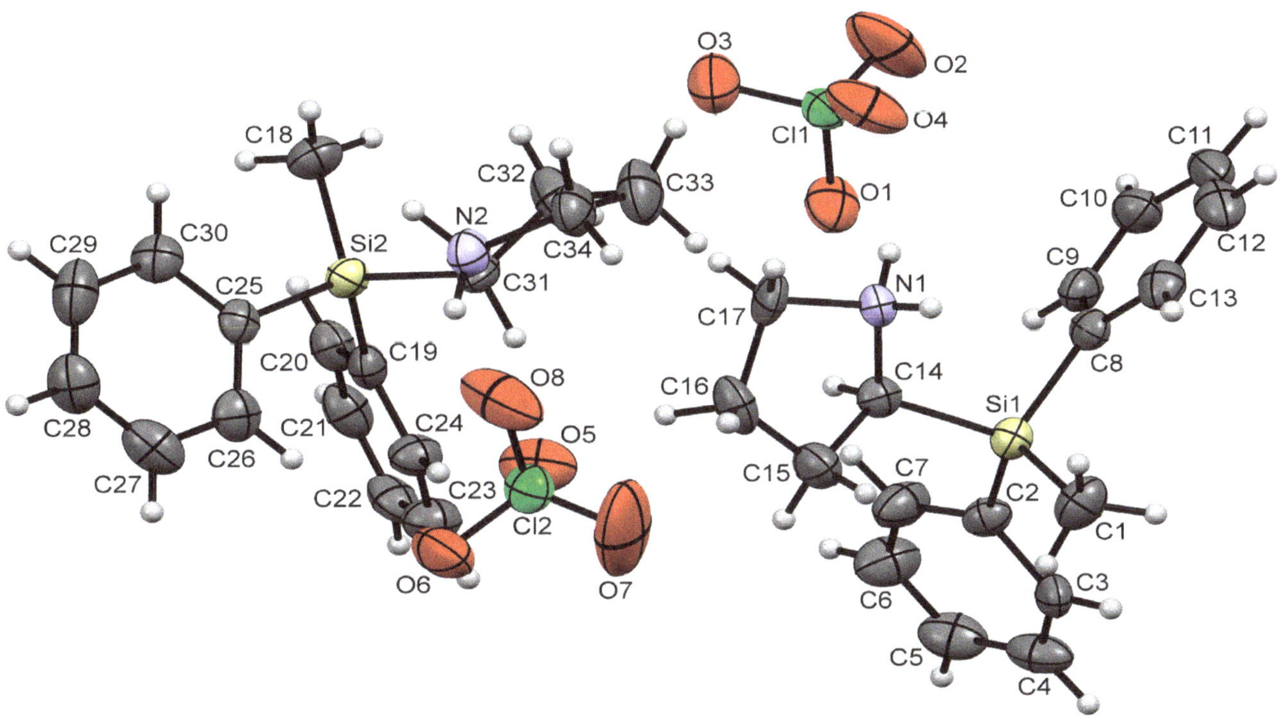

Figure 4. Molecular structure (ORTEP plot) of compound (*S*)-**2**·$HClO_4$ in the crystal, with the displacement ellipsoids set at the 50% probability level. Selected bond lengths (Å) and angles (°): Molecule (*S*)-**2**·$HClO_4$-**A**: C(1)–Si(1) 1.852(6), C(2)–Si(1) 1.855(6), C(8)–Si(1) 1.871(5), C(14)–Si(1) 1.878(5), C(14)–N(1) 1.508(7), C(17)–N(1) 1.508(6), Cl(1)–O(1) 1.416(4), Cl(1)–O(2) 1.413(5), Cl(1)–O(3) 1.426(5), Cl(1)–O(4) 1.413(5), C(1)–Si(1)–C(2) 112.5(3), C(1)–Si(1)–C(8) 109.7(3), C(2)–Si(1)–C(8) 110.6(2), C(1)–Si(1)–C(14) 109.4(3), C(2)–Si(1)–C(14) 106.4(2), C(8)–Si(1)–C(14) 108.1(2). Molecule (*S*)-**2**·$HClO_4$-**B**: C(18)–Si(2) 1.856(6), C(19)–Si(2) 1.875(6), C(25)–Si(2) 1.884(5), C(31)–Si(2) 1.889(5), C(31)–N(2) 1.509(7), C(34)–N(2) 1.510(7), Cl(2)–O(5) 1.400(5), Cl(2)–O(6) 1.405(5), Cl(2)–O(7) 1.365(6), Cl(2)–O(8) 1.431(5), C(18)–Si(2)–C(19) 109.9(3), C(18)–Si(2)–C(25) 111.5(3), C(19)–Si(2)–C(25) 110.1(2), C(18)–Si(2)–C(31) 110.4(3), C(19)–Si(2)–C(31) 106.0(2), C(25)–Si(2)–C(31) 108.8(2).

Treatment of enantiomerically-enriched (*S*)-2-(methyldiphenylsilyl)pyrrolidine [(*S*)-**2**] with trifluoroacetic acid in dichloromethane resulted in the formation of the trifluoroacetate (*S*)-**2**·TFA. Enantiomerically-pure single crystals of (*S*)-**2**·TFA were obtained after recrystallization from acetonitrile. Compound (*S*)-**2**·TFA crystallized in the orthorhombic crystal system, space group $P2_1\,2_1\,2_1$, as colorless needles (Figure 5 and Table 1). Like in the perchlorate (*S*)-**2**·$HClO_4$ (see Figure 4) and in the hydrochloride (*S*)-**2**·HCl [1], the asymmetric unit of compound (*S*)-**2**·TFA contains two independent molecules [(*S*)-**2**·TFA-**A** and (*S*)-**2**·TFA-**B**]. The fluorine atoms of the trifluoroacetate anions and the pyrrolidine carbon atom C(35) of molecule (*S*)-**2**·TFA-**B** are disordered. Each trifluoroacetate is hydrogen-bonded to two silylpyrrolidinium cations via an N–H···O interaction. The parameters of the found hydrogen-bonds differ from each other, with the N–H distances ranging from 0.847 Å [N(2)–H(4N)] to 1.082 Å [N(2)–H(3N)], and the N–H···O angles ranging from 159.48° [N(1)–H(2N)···O(2)] to 172.31° [N(2)–H(3N)···O(3)], although the N···O distances are quite similar (Figure 5).

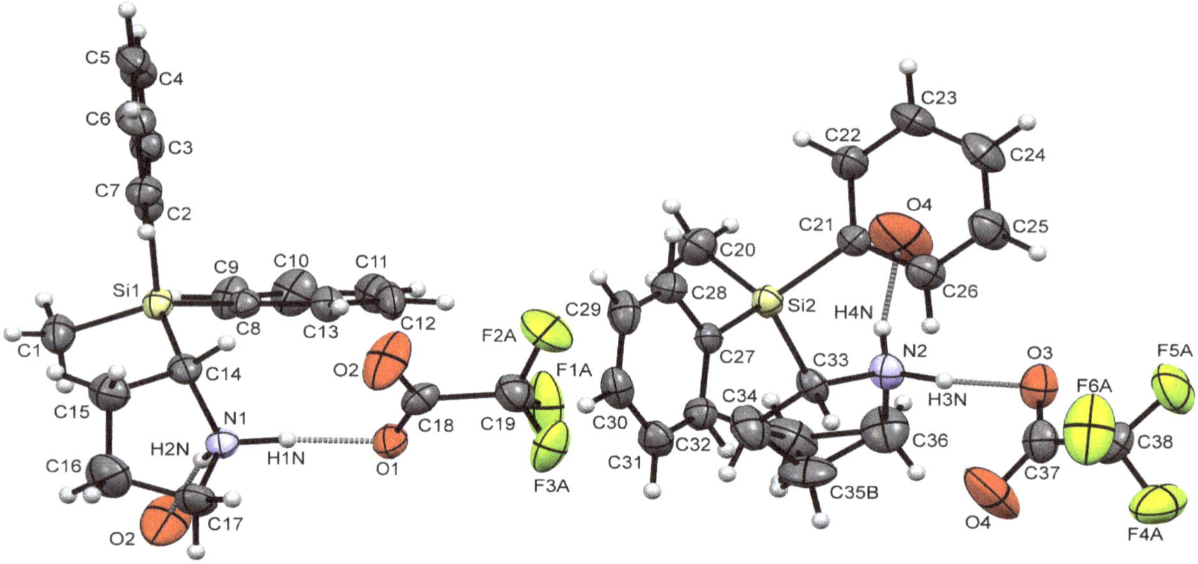

Figure 5. Part of the crystal structure (ORTEP plot) of compound (*S*)-**2**·TFA in the crystal, with the displacement ellipsoids set at the 50% probability level. The ellipsoids F(1B), F(2B), and F(3B) of molecule (*S*)-**2**·TFA-**A**, and the ellipsoids F(4B), F(5B), and F(6B) of (*S*)-**2**·TFA-**B** are omitted for clarity. Selected bond lengths (Å) and angles (°): Molecule (*S*)-**2**·TFA-**A**: C(1)–Si(1) 1.849(2), C(2)–Si(1) 1.863(2), C(8)–Si(1) 1.857(2), C(14)–Si(1) 1.889(2), C(14)–N(1) 1.505(2), C(17)–N(1) 1.491(3), C(18)–C(19) 1.520(3), C(18)–O(1) 1.225(3), C(18)–O(2) 1.206(3), C(1)–Si(1)–C(2) 111.77(10), C(1)–Si(1)–C(8) 110.07(10), C(2)–Si(1)–C(8) 108.11(9), C(1)–Si(1)–C(14) 109.69(10), C(2)–Si(1)–C(14) 107.16(10), C(8)–Si(1)–C(14) 109.98(9). Hydrogen bond N(1)–H(1N)···O(1): N(1)–H(1N) 1.050, H(1N)···O(1) 1.680, N(1)···O(1) 2.716, N(1)–H(1N)···O(1) 167.81. Hydrogen bond N(1)–H(2N)···O(2): N(1)–H(2N) 1.010, H(2N)···O(2) 1.769, N(1)···O(2) 2.738, N(1)–H(2N)···O(2) 159.48. Symmetry transformations used to generate the equivalent atom O(2), hydrogen-bonded to H(2N): $1 - x$, $-\frac{1}{2} + y$, $\frac{1}{2} - z$. Molecule (*S*)-**2**·TFA-**B**: C(20)–Si(2) 1.845(2), C(21)–Si(2) 1.850(2), C(27)–Si(2) 1.851(2), C(33)–Si(2) 1.884(2), C(33)–N(2) 1.507(3), C(36)–N(2) 1.491(3), C(37)–C(38) 1.522(3), C(37)–O(3) 1.220(3), C(37)–O(4) 1.205(3), C(20)–Si(2)–C(21) 110.03(11), C(20)–Si(2)–C(27) 110.46(10), C(21)–Si(2)–C(27) 108.38(9), C(20)–Si(2)–C(33) 111.26(11), C(21)–Si(2)–C(33) 110.30(9), C(27)–Si(2)–C(33) 106.30(10). Hydrogen bond N(2)–H(3N)···O(3): N(2)–H(3N) 1.082, H(3N)···O(3) 1.665, N(2)···O(3) 2.740, N(2)–H(3N)···O(3) 172.31. Hydrogen bond N(2)–H(4N)···O(4): N(2)–H(4N) 0.847, H(4N)···O(4) 1.903, N(2)···O(4) 2.728, N(2)–H(4N)···O(4) 164.58. Symmetry transformations used to generate the equivalent trifluoroacetate anion, hydrogen-bonded to H(3N): $\frac{1}{2} + x$, $1.5 - y$, $1 - z$. Symmetry transformations used to generate the equivalent atom O(4), hydrogen-bonded to H(4N): $1 + x$, y, z.

3. Experimental Details

3.1. Synthetic Methods

Synthesis and characterization data of compounds (*S*)-**1**·HCl and (*S*)-**2**·HCl were previously reported by our group [1]. The perchlorate salts (*S*)-**1**·$HClO_4$ and (*S*)-**2**·$HClO_4$ were prepared by treatment of (*S*)-**1**·HCl and (*S*)-**2**·HCl, respectively, with excess perchloric acid (60 wt %). Colorless single-crystals were formed overnight under normal atmosphere at room temperature within the remaining solvent. Single crystals of the inclusion compound (*S*)-**1**·HCl·MeOH were obtained after recrystallization of (*S*)-**1**·HCl from methanol. (*S*)-**2**·TFA was prepared by treatment of 800 mg (2.99 mmol) enantiomerically-enriched (*S*)-2-(methyldiphenylsilyl)pyrrolidine [(*S*)-**2**, e.r. = 89:11] with 341 mg (2.99 mmol) trifluoroacetic acid in 10 mL dichloromethane at room temperature. After removing of all volatiles in vacuo, the residue was dissolved in acetonitrile. Enantiomerically-pure single crystals of (*S*)-**2**·TFA were formed under normal atmosphere at room temperature within three months.

3.2. X-ray Crystallography

Single-crystal X-ray diffraction analyses were performed on an Oxford Diffraction Xcalibur S diffractometer (Oxford Diffraction Ltd. (Abingdon, UK)) at 173(2) K using graphite-monochromated Mo Kα radiation (λ = 0.71073 Å). The crystals were mounted at room temperature. The crystal structures were solved with direct methods (*SHELXS*-97 [11]) and refined against F^2 with the full-matrix least-squares method (*SHELXL*-97 [12,13]). A multi-scan absorption correction using the implemented CrysAlis RED program (Version 1.171.32.37, Oxford Diffraction Ltd.) was employed. The non-hydrogen atoms were refined anisotropically. The hydrogen atoms H(1N) and H(2N) in compound (*S*)-**1**·$HClO_4$, H(1N), H(2N), and H(1O) in compound (*S*)-**1**·HCl·MeOH, and H(1N), H(2N), H(3N), H(4N), H(34A), H(34B), H(36A), and H(36B) in compound (*S*)-**2**·TFA were located on the difference Fourier map and refined independently. All other hydrogen atoms were placed in geometrically-calculated positions and each was assigned a fixed isotropic displacement parameter based on a riding model. The fluorine atoms F(1)–F(6) and the carbon atom C(35) in (*S*)-**2**·TFA are disordered. The absolute configuration of (*S*)-**1**·$HClO_4$, (*S*)-**1**·HCl·MeOH, (*S*)-**2**·$HClO_4$, and (*S*)-**2**·TFA was determined by Flack's method based on resonant scattering [14]. Figures 2–5 were created using Mercury (Version 3.3). Crystal and refinement data are collected in Table 1. Crystallographic data of enantiomerically-pure 2-silylpyrrolidinium salts (*S*)-**1**·$HClO_4$, (*S*)-**2**·$HClO_4$, (*S*)-**1**·HCl·MeOH, and (*S*)-**2**·TFA have been deposited with The Cambridge Crystallographic Data Centre. CCDC 1582443 [(*S*)-**1**·HCl·MeOH], CCDC 1582444 [(*S*)-**2**·TFA], CCDC 1582445 [(*S*)-**1**·$HClO_4$], and CCDC 1582446 [(*S*)-**2**·$HClO_4$] contain the supplementary crystallographic data for this paper (see Supplementary Materials). These data can be obtained free of charge via http://www.ccdc.cam.ac.uk/conts/retrieving.html (or from the CCDC, 12 Union Road, Cambridge CB2 1EZ, UK.

4. Conclusions

In the context of studies concerning the effect of the counter anion in organocatalytic reactions with polar species involved, we were interested to synthesize salts of optically-pure 2-silylpyrrolidines with different counter anions for further reactivity studies, which might be of interest in terms of hydrogen-bond activation in the initial enamine formation step. By studying the hydrogen-bonding behavior of 2-silylpyrrolidines, interesting information about silicon-specific effects on the basicity of the pyrrolidine nitrogen center may be provided. Future studies will also address the respective salts of the carbon analogue 2-tritylpyrrolidine for comparison.

Acknowledgments: This research was supported by the Deutsche Forschungsgemeinschaft (DFG). Jonathan O. Bauer thanks the Alexander von Humboldt Foundation for the award of a Feodor Lynen Return Fellowship.

Author Contributions: Jonathan O. Bauer performed the experiments and wrote the manuscript. Carsten Strohmann was coordinating the project and performed the XRD analyses.

References

1. Bauer, J.O.; Stiller, J.; Marqués-López, E.; Strohfeldt, K.; Christmann, M.; Strohmann, C. Silyl-modified analogues of 2-tritylpyrrolidine: Synthesis and applications in asymmetric organocatalysis. *Chem. Eur. J.* **2010**, *16*, 12553–12558. [CrossRef] [PubMed]
2. Husmann, R.; Jörres, M.; Raabe, G.; Bolm, C. Silylated pyrrolidines as catalysts for asymmetric Michael additions of aldehydes to nitroolefins. *Chem. Eur. J.* **2010**, *16*, 12549–12552. [CrossRef] [PubMed]
3. Jentzsch, K.I.; Min, T.; Etcheson, J.I.; Fettinger, J.C.; Franz, A.K. Silyl fluoride electrophiles for the enantioselective synthesis of silylated pyrrolidine catalysts. *J. Org. Chem.* **2011**, *76*, 7065–7075. [CrossRef] [PubMed]
4. Min, T.; Fettinger, J.C.; Franz, A.K. Enantiocontrol with a hydrogen-bond directing pyrrolidinylsilanol catalyst. *ACS Catal.* **2012**, *2*, 1661–1666. [CrossRef]

5. Kerrick, S.T.; Beak, P. Asymmetric deprotonations: Enantioselective syntheses of 2-substituted (*tert*-butoxycarbonyl)pyrrolidines. *J. Am. Chem. Soc.* **1991**, *113*, 9708–9710. [CrossRef]
6. Strohfeldt, K.; Seibel, T.; Wich, P.; Strohmann, C. Synthesis and reactivity of an enantiomerically pure *N*-methyl-2-silyl-substituted pyrrolidine. In *Organosilicon Chemistry VI: From Molecules to Materials*; Auner, N., Weis, J., Eds.; Wiley-VCH: Weinheim, Germany, 2005; Volume 1, pp. 488–494.
7. Kano, T.; Mii, H.; Maruoka, K. Direct asymmetric benzoyloxylation of aldehydes catalyzed by 2-tritylpyrrolidine. *J. Am. Chem. Soc.* **2009**, *131*, 3450–3451. [CrossRef] [PubMed]
8. Erdmann, H.; An, F.; Mayer, P.; Ofial, A.R.; Lakhdar, S.; Mayr, H. Structures and reactivities of 2-trityl- and 2-(triphenylsilyl)pyrrolidine-derived enamines: Evidence for negative hyperconjugation with the trityl group. *J. Am. Chem. Soc.* **2014**, *136*, 14263–14269. [CrossRef] [PubMed]
9. Bauer, J.O.; Strohmann, C. *tert*-Butoxytriphenylsilane. *Acta Crystallogr.* **2010**, *E66*, o461–o462. [CrossRef] [PubMed]
10. Brendler, E.; Heine, T.; Seichter, W.; Wagler, J.; Witter, R. ^{29}Si NMR shielding tensors in triphenylsilanes—^{29}Si solid state NMR experiments and DFT-IGLO calculations. *Z. Anorg. Allg. Chem.* **2012**, *638*, 935–944. [CrossRef]
11. Sheldrick, G.M. *SHELXS-97, a Program for the Solution of Crystal Structures*; Universität Göttingen: Göttingen, Germany, 1997.
12. Sheldrick, G.M. *SHELXL-97, a Program for Crystal Structure Refinement*; Universität Göttingen: Göttingen, Germany, 1997.
13. Sheldrick, G.M. A short history of *SHELX*. *Acta Crystallogr.* **2008**, *A64*, 112–122. [CrossRef] [PubMed]
14. Flack, H.D.; Bernardinelli, G. The use of X-ray crystallography to determine absolute configuration. *Chirality* **2008**, *20*, 681–690. [CrossRef] [PubMed]

Permissions

All chapters in this book were first published in MDPI; hereby published with permission under the Creative Commons Attribution License or equivalent. Every chapter published in this book has been scrutinized by our experts. Their significance has been extensively debated. The topics covered herein carry significant findings which will fuel the growth of the discipline. They may even be implemented as practical applications or may be referred to as a beginning point for another development.

The contributors of this book come from diverse backgrounds, making this book a truly international effort. This book will bring forth new frontiers with its revolutionizing research information and detailed analysis of the nascent developments around the world.

We would like to thank all the contributing authors for lending their expertise to make the book truly unique. They have played a crucial role in the development of this book. Without their invaluable contributions this book wouldn't have been possible. They have made vital efforts to compile up to date information on the varied aspects of this subject to make this book a valuable addition to the collection of many professionals and students.

This book was conceptualized with the vision of imparting up-to-date information and advanced data in this field. To ensure the same, a matchless editorial board was set up. Every individual on the board went through rigorous rounds of assessment to prove their worth. After which they invested a large part of their time researching and compiling the most relevant data for our readers.

The editorial board has been involved in producing this book since its inception. They have spent rigorous hours researching and exploring the diverse topics which have resulted in the successful publishing of this book. They have passed on their knowledge of decades through this book. To expedite this challenging task, the publisher supported the team at every step. A small team of assistant editors was also appointed to further simplify the editing procedure and attain best results for the readers.

Apart from the editorial board, the designing team has also invested a significant amount of their time in understanding the subject and creating the most relevant covers. They scrutinized every image to scout for the most suitable representation of the subject and create an appropriate cover for the book.

The publishing team has been an ardent support to the editorial, designing and production team. Their endless efforts to recruit the best for this project, has resulted in the accomplishment of this book. They are a veteran in the field of academics and their pool of knowledge is as vast as their experience in printing. Their expertise and guidance has proved useful at every step. Their uncompromising quality standards have made this book an exceptional effort. Their encouragement from time to time has been an inspiration for everyone.

The publisher and the editorial board hope that this book will prove to be a valuable piece of knowledge for researchers, students, practitioners and scholars across the globe.

List of Contributors

Yang Pan and Shinobu Aoyagi
Graduate School of Science, Nagoya City University, Nagoya, Aichi 467-8501, Japan

Shogo Morisako
Division of Chemistry, Faculty of Pure and Applied Sciences, University of Tsukuba, Tsukuba, Ibaraki 305-8571, Japan

Takahiro Sasamori
Graduate School of Science, Nagoya City University, Nagoya, Aichi 467-8501, Japan
Division of Chemistry, Faculty of Pure and Applied Sciences, University of Tsukuba, Tsukuba, Ibaraki 305-8571, Japan

Rita C. C. Rangel, Nilson C. Cruz and Elidiane C. Rangel
Laboratory of Technological Plasmas (LaPTec), São Paulo State University (UNESP), Science and Technology Institute of Sorocaba (ICTS), Av. Três de Março, 511, 18087-180 Sorocaba, Brazil

Rabia Ayub, Kjell Jorner and Henrik Ottosson
Department of Chemistry – BMC, Uppsala University, SE-751 23 Uppsala, Sweden
Department of Chemistry-Ångström Laboratory Uppsala University, SE-751 20 Uppsala, Sweden

Fabian Dankert, Kirsten Reuter, Carsten Donsbach and Carsten von Hänisch
Fachbereich Chemie and Wissenschaftliches Zentrum für Materialwissenschaften (WZMW), Philipps-Universität Marburg, Hans-Meerwein Straße 4, D-35032 Marburg, Germany

Lisa Ehrlich, Robert Gericke and Jörg Wagler
Institut für Anorganische Chemie, TU Bergakademie Freiberg, D-09596 Freiberg, Germany

Erica Brendler
Institut für Analytische Chemie, TU Bergakademie Freiberg, D-09596 Freiberg, Germany

Yasunobu Egawa, Chihiro Fukumoto, Nobuhiro Takeda and Masafumi Unno
Department of Chemistry and Chemical Biology, Graduate School of Science and Technology, Gunma University, Kiryu, Gunma 376-8515, Japan

Koichiro Mikami
Functional Polymer Group, Sagami Chemical Research Institute, 2743-1 Hayakawa, Ayase, Kanagawa 252-1193, Japan

Norio Nakata, Nanami Kato, Noriko Sekizawa and Akihiko Ishii
Department of Chemistry, Graduate School of Science and Engineering, Saitama University, 255 Shimo-okubo, Sakura-ku, Saitama 338-8570, Japan

Debabrata Dhara and Anukul Jana
Tata Institute for Fundamental Research Hyderabad Gopanpally, Hyderabad-500107, Telangana, India

Volker Huch and David Scheschkewitz
Krupp-Chair of General and Inorganic Chemistry, Saarland University, 66123 Saarbrücken, Germany

Lisa Pecher and Ralf Tonner
Faculty of Chemistry and Material Sciences Center, Philipps-Universität Marburg, Hans-Meerwein Straße 4, 35032 Marburg, Germany

Laura-Alice Jantke and Thomas F. Fässler
Department of Chemistry, Technical University of Munich, Lichtenbergstr. 4, 85747 Garching, Germany

Tomohiro Sugahara and Norihiro Tokitoh
Institute for Chemical Research, Kyoto University, Gokasho, Uji, Kyoto 611-0011, Japan

Ravindra K. Raut, Sheikh Farhan Amin, Padmini Sahoo, Vikas Kumar and Moumita Majumdar
Indian Institute of Science Education and Research Pune, Dr. Homi Bhabha Road, Pashan, Pune-411008, India

Yohei Adachi, Daiki Tanaka, Yousuke Ooyama and Joji Ohshita
Department of Applied Chemistry, Graduate School of Engineering, Hiroshima University, Higashi-Hiroshima 739-8527, Japan

Ken-ichiro Kanno, Yumi Aikawa, Yuka Niwayama, Misaki Ino, Kento Kawamura and Soichiro Kyushin
Division of Molecular Science, Graduate School of Science and Technology, Gunma University, Kiryu, Gunma 376-8515, Japan

Yusuke Sunada
Institute of Industrial Science, The University of Tokyo, 4-6-1 Komaba Meguro-ku, Tokyo 153-8580, Japan
Department of Applied Chemistry, School of Engineering, The University of Tokyo, 4-6-1 Komaba Meguro-ku, Tokyo 153-8580, Japan

Nobuhiro Taniyama
Institute of Industrial Science, The University of Tokyo, 4-6-1 Komaba Meguro-ku, Tokyo 153-8580, Japan

Kento Shimamoto
Department of Applied Chemistry, School of Engineering, The University of Tokyo, 4-6-1 Komaba Meguro-ku, Tokyo 153-8580, Japan

Hideo Nagashima
Institute for Materials Chemistry and Engineering, Kyushu University, 6-1 Kasugakoen Kasuga, Fukuoka 816-8580, Japan

Naohiko Akasaka, Kaho Tanaka, Shintaro Ishida and Takeaki Iwamoto
Department of Chemistry, Graduate School of Science, Tohoku University, Aoba-ku, Sendai 980-8578, Japan

Jürgen Kahr and Ferdinand Belaj
Institut für Chemie, Karl-Franzens-Universität, NAWI Graz, Schubertstraße 1, 8010 Graz, Austria

Rudolf Pietschnig
Institut für Chemie, Karl-Franzens-Universität, NAWI Graz, Schubertstraße 1, 8010 Graz, Austria
Institut für Chemie und CINSaT, Universität Kassel, Heinrich-Plett-Straße 40, 34132 Kassel, Germany

Naoki Hayakawa, Kazuya Sadamori, Shinsuke Mizutani and Tsukasa Matsuo
Department of Applied Chemistry, Faculty of Science and Engineering, Kindai University 3-4-1 Kowakae, Higashi-Osaka, Osaka 577-8502, Japan

Tomohiro Agou
Institute for Chemical Research, Kyoto University, Gokasho, Uji, Kyoto 611-0011, Japan
Department of Biomolecular Functional Engineering, College of Engineering, Ibaraki University, 4-12-1 Nakanarusawa, Hitachi, Ibaraki 316-8511, Japan

Daisuke Hashizume
RIKEN Center for Emergent Matter Science (CEMS), 2-1 Hirosawa, Wako, Saitama 351-0198, Japan

Jonathan O. Bauer and Carsten Strohmann
Anorganische Chemie, Fakultät für Chemie und Chemische Biologie, Technische Universität Dortmund, Otto-Hahn-Straße 6, D-44227 Dortmund, Germany

Index

www.ingramcontent.com/pod-product-compliance
Lightning Source LLC
LaVergne TN
LVHW082216200926
848484LV00008B/2265